Cyclic Deterioration of High-Performance Concrete

Nadja Oneschkow · Ludger Lohaus
Editors

Cyclic Deterioration of High-Performance Concrete

Reports of the SPP 2020 Experimental-Virtual-Lab

SPP 2020
EXPERIMENTAL-VIRTUAL-LAB

Funded by
DFG Deutsche Forschungsgemeinschaft
German Research Foundation

Springer

Editors
Nadja Oneschkow
Institute of Building Materials Science
Leibniz University Hannover
Hannover, Germany

Ludger Lohaus
Institute of Building Materials Science
Leibniz University Hannover
Hannover, Germany

ISBN 978-3-032-19393-3 ISBN 978-3-032-19394-0 (eBook)
https://doi.org/10.1007/978-3-032-19394-0

This work was funded by the Deutsche Forschungsgemeinschaft (DFG, German Research Foundation)

This Springer imprint is published by the registered company Springer Nature Switzerland AG
The registered company address is: Gewerbestrasse 11, 6330 Cham, Switzerland

For our Students, Colleagues and all Researchers and Engineers interested in Concrete Fatigue.

Preface

Following earlier research projects on high-performance and ultra-high-performance concrete (i.e. Priority Programme SPP 1182), the need for broad-based coordinated research on the behaviour of such concretes under fatigue loading was identified and discussed in the early 2010s. Based on this, the initiators Ludger Lohaus (Leibniz University Hannover), Peter Wriggers (Leibniz University Hannover) and Steffen Anders (University of Wuppertal) applied for funding for a Priority Programme, which was finally granted in 2016 as SPP 2020 "Cyclic Deterioration of High-Performance Concrete in an Experimental-Virtual-Lab" by the German Research Foundation (DFG). The SPP 2020 was coordinated at the Institute of Building Materials Science, Leibniz University Hannover.

Twelve joint research projects with different scientific focuses and the Central Project were brought together in a joint research network involving 14 German universities and granted for a first funding period of three years in 2017. Nine research projects and the Central Project continued or, in one case, began their research work in the second funding period, which started in 2020. This second period extended unexpectedly till 2024 due to the hindered research possibilities during the corona pandemic. However, more than 90 researchers—doctoral and postdoctoral researchers and professors—worked enthusiastically together in those seven years.

The overall scope of the SPP 2020 was to capture, understand, describe, model and predict the material degradation of high-performance concretes. The damage processes in microstructure occur on a very small scale and, thus, can hardly be observed directly in experiments. The desired results were, therefore, developed in a close cooperation between materials and modelling science which represents the interconnection of experiment and computation in the Experimental-Virtual-Lab. The detection, characterisation and model-based description of the damage and crack development in the heterogeneous concrete microstructure at different scales over up to some millions of load cycles were the specific research challenges within this DFG Priority Programme.

This book contains the main results of the coordinated joint research projects in the SPP 2020 regarding the current state of knowledge on the subject of fatigue damage of high-performance concretes and its numerical modelling in a compact and, at the same time, in-depth form. All authors have put a huge amount of effort into writing the chapters of this book.

The research work in this Priority Programme was characterised in an exceptional way by mutual trust, open scientific exchange and goal-oriented cooperation, also across project boundaries. In addition, networks for their future careers or even friendships developed between the participants. In any case, bonds were forged between building materials and modelling scientists that will continue to exist beyond the SPP 2020 and certainly lead to many new and exciting research results in the future. Many thanks to everyone involved for their valuable cooperation and great engagement.

Furthermore, and on behalf of all scientists involved in the SPP 2020, we thank the Deutsche Forschungsgemeinschaft (DFG, German Research Foundation) for the financial support of this Priority Programme SPP 2020.

Hannover, Germany
July 2026

Ludger Lohaus
Coordinator of SPP 2020

Nadja Oneschkow
Manager of SPP 2020

SPP 2020
EXPERIMENTAL-VIRTUAL-LAB

The Team of SPP 2020:

Mohamed Abubakar Ali • Mario Aguilar • Fadi Aldakheel • Marreddy Ambati • Steffen Anders • Marco Basaldella • Gauravdatt Basutkar • Abedulgader Baktheer • Henrik Becks • Veit Birtel • Dominik Brands • Rolf Breitenbücher • Antonio Caggiano • Pietro Carrara • Rostislav Chudoba • Martin Classen • Laura De Lorenzis • Frank Dehn • Dieter Dinkler • Maksym Dosta • Karl-Heinz Drake • Dennis Dreger • Ante Džolan • Martin Empelmann • Michael Engelhard • Nikoleta Evangelidou-Spanos • Oliver Fischer • Harald Garrecht • Gregor Gebuhr • Lena Gietz • Christian Grosse • Vladislav Gudžulić • Alexander Gunkler • Michael Haist • Stefan Harenberg • Marvin Jentsch • Dominik Junger • Michael Kaliske • Wisam Karim • Josef Katzmann • Bianca Kern • Henrike Klodt • Eddie Koenders • Ursula Kowalsky • Jan-Paul Lanwer • Philipp Lauff • Thorsten Leusmann • Tanja Lochner • Stefan Löhnert • Ludger Lohaus • Dirk Lowke • Hamid Madadi • Markus Mahlbacher • Viktoria Malárics-Pfaff • Martin Markert • Svenja Matz • Viktor Mechtcherine • Günther Meschke • Melisa Mocevic • Steffen Müller • Roland Niedermeier • Nima Noii • Vincent Oettel • Nadja Oneschkow • Matthias Pahn • Rohan Patel • Malte Peter • Rodolfo Piña • Mangesh Pise • Anna-Lena Podhajecky • Polina Pugacheva • Manuel Raith • Majid Ranjbarian • Lisa Reischmann • Rydell Riko • Martin Ritter • Corinna Rozanski • Matthias Rutzen • Sebastian Rybczynski • Mohammad Sarhil • Gunnar Schaan • Niklas Schäfer • Silke Scheerer • Diego Schicchi • Frank Schmidt-Döhl • Jörg Schröder • Holger Steeb • Johannes Storm • Melchior Thieme • Tim Timmermann • Christoph Tomann • Ngoc Linh Tran • Dirk Volkmer • Ursula Weiss • Claudia Westphal • Kerstin Willers • Peter Wriggers • Sha Yang

Contents

Introduction

Nadja Oneschkow and Ludger Lohaus

1 Scientific-Historical Background

Concrete is one of the most widely used construction materials in the world, valued for its strength, durability, and versatility. Advances in concrete technology over the last 25 years have led to the development of increasingly efficient high-performance concretes. These high-performance concretes differ considerably from concretes used in traditional constructions in terms of their composition, properties and, of course, their spectrum of applications. Examples are self-compacting concretes with outstanding flow properties, high- and ultra-high-strength concretes with partly steel-similar properties or fibre- and textile-reinforced concretes with high-ductile load-bearing behaviour.

The usage of these innovative high-performance concretes allows for ever lighter, more filigree and resource-saving structures which, however, are more susceptible to cyclic loading due to their reduced dead weight. Furthermore, structures and components such as long-span bridges for high-speed trains, wind power plants or machine foundations are typically subjected to very large variable loads and high numbers of load cycles. Furthermore, fine-grained high- or ultra-high-strength concretes, partly with fibres or textile reinforcement, are increasingly used for the refurbishment and maintenance of infrastructure buildings utilising the high load-bearing capacity with only a little additional dead weight.

These materials experience fatigue during the lifespan of those buildings—a gradual deterioration caused by repeated loading with limited maximum stresses significantly below the ultimate strength. The fatigue behaviour of those high-performance concretes is, therefore, decisive for the design and realisation of such modern concrete structural applications. However, the backgrounds and mechanisms of the formation and propagation of fatigue damage have rarely been systematically

N. Oneschkow (✉) · L. Lohaus
Institute of Building Materials Science, Leibniz University Hannover, Hannover, Germany
e-mail: n.oneschkow@baustoff.uni-hannover.de

N. Oneschkow and L. Lohaus (eds.), *Cyclic Deterioration of High-Performance Concrete*, https://doi.org/10.1007/978-3-032-19394-0_1

investigated and characterised for either normal strength or those innovative high-performance concretes. Understanding the mechanisms behind concrete fatigue, its effects and ways to mitigate them is essential for engineers and researchers aiming to extend the lifespan of concrete structures and ensure their safety and reliability.

Earlier research work in the 1970s–1990s concentrated mainly on determining the numbers of cycles to failure, also in order to describe the fatigue behaviour of concrete in guidelines using so-called S/N-curves (e.g. [1–4]). These curves represent the numbers of cycles to failure in relation to the stress levels applied, which represent the stresses in relation to the concrete strength. The first S/N-curves were included in Model Code 90 [5] and also, a short time later, in Eurocode 2 [6, 7]. However, the continuous measurement of the permanently changing specimen deformations during the experiments presented the researchers with major challenges at that time. In order to accomplish these measurements, the fatigue tests had to be stopped and the measuring gauges attached to the test specimens had to be read and noted by the laboratory staff. Later, digital recording made it possible to continuously record the fatigue behaviour and, thus, enabled further insights.

The design rules considered at that time remained in place for decades: less research was ongoing because concrete fatigue had only rarely become relevant for design, i.e. for classical concrete construction with normal strength concrete. Much later, at the beginning of the new century, novel investigations, which were driven by the upcoming high- and ultra-high-strength concretes, showed that these design rules were very conservative, especially for concretes with higher strength [8]. It turned out that the use of innovative high-performance concretes in fatigue-loaded structures is uneconomical, severely restricted and sometimes even prevented in practice. However, more background knowledge was also required for the further development of standards and guidelines. Moreover, there was also a lack of numerical approaches and models to accurately predict the material degradation in concrete due to the lack of basic knowledge concerning fatigue damage development in concrete. In this respect, modelling the fatigue behaviour was scarcely possible, above all due to its very heterogeneous microstructure compared to metals, for example. The research idea for the DFG Priority Programme SPP 2020 arose against this overall background.

2 Influences on the Fatigue Behaviour of Concrete

It is known that the fatigue behaviour is influenced by many factors, which, in turn, influence each other. This fact makes it difficult to compare results from the literature, as the boundary conditions under which the results have been obtained almost always differ from one another. Various factors that significantly influence the numbers of cycles to failure and, thus, the fatigue resistance of concrete had already been identified in previous research projects before the start of the SPP 2020.

The numbers of cycles to failure depend significantly on the type of loading: the highest numbers of cycles and, thus, the longest lifetimes are reached under pure compressive cyclic loading. Significantly lower numbers of cycles to failure are determined under tensile or compressive-tensile cyclic loading. In the case of high compressive stresses greater than about $S_{max} \approx 0.75$, lower load frequencies lead to a lower fatigue resistance. By contrast, in the case of lower compressive stresses, lower load frequencies lead to a higher fatigue resistance. This effect might be related to the stress- and frequency-dependent increase of the concrete temperature due the cyclic mechanical loading. However, this effect is described in single papers but has not yet been understood. It is advisable to select test frequencies for experimental investigations so that the rise in temperature remains very limited. Furthermore, it was already known that a higher moisture content of concrete or water as a surrounding environment could lead to a significant reduction in the numbers of cycles to failure.

By contrast, other influences were less clear, and, in some cases, contradictory results were reported in the literature. Steel fibres as a type of evenly distributed, internal reinforcement have hardly any effect on the number of cycles to failure under pure high cycle compressive loading. However, positive effects are reported with respect to low compressive cyclic loading, for example, earthquake loading, and under bending tensile stresses. Contradictory results and statements can be found in the literature on a possible influence of the concrete strength on the fatigue resistance. However, the widespread opinion has been that concretes with higher strengths have an increased fatigue sensitivity and, thus, a poorer fatigue resistance has to be assumed compared to those concretes with lower strengths.

The compilation above does not claim to represent the current state of research. This is available in the individual chapters of this book. However, it shows how diverse the issues in the research field of concrete fatigue are. Therefore, the idea was to address these issues in a coordinated research network with a large number of scientists.

3 Damage Indicators

Due to the previous focus on determining the numbers of cycles to failure, there has been hardly any generalisable knowledge, either in Germany or internationally, specifically on the material degradation of concrete due to fatigue loading. This means that although effects on the numbers of load cycles to failure have been identified, the causes have not been systematically investigated. Therefore, the fundamental processes and mechanisms of the development and progression of fatigue damage have not become known. The possibilities for the collection of extensive data during experimental investigations are much more advanced today than in the 1990s and before. More damage indicators, such as deformations, stiffnesses, natural frequencies, temperature, acoustic emission or ultrasonic runtime, can be analysed. The advantage of these indicators is that they can be continuously recorded during the relatively long-lasting fatigue tests with limited effort.

It was not clear in former investigations to what extent the indicators mentioned are also influenced by the influences already known on the numbers of cycles to failure. Concrete fatigue was more or less like a black box: fatigue test—fatigue failure. It could be shown in [9] that fatigue strains and the concrete stiffness are also influenced by the maximum stress level, test frequency and waveform. Furthermore, indications were found that fatigue damage starts to propagate on extreme small scales. However, the macroscopically measured damage indicators can only provide indirect information about the development of damage in the microstructure and how it is influenced. Therefore, microstructural investigations are additionally required to visualise and quantify fatigue damage and, thus, to obtain direct information about the degradation of the microstructure. Thus, the idea arose to acquire new fundamental knowledge concerning fatigue damage mechanisms based on a combined usage of damage indicators and damage visualisation techniques.

4 Challenges in Experimental Concrete Fatigue Research

Performing fatigue tests is very time-consuming, also due to the high number of repetition tests required compared to static tests in order to capture scattering effects. In addition, the high loading rates and long test durations place high demands on the acquisition of measurement data. Furthermore, the fatigue tests must be interrupted at selected stages of the fatigue process and samples must be prepared for further microstructural investigations to visualise and quantify fatigue damage in the concrete's microstructure. Here too, the scattering of results must be captured by using a sufficient number of samples. However, each microstructural damage visualisation or quantification only represents a time-discrete insight of the fatigue damage process.

Light and raster electron microscopy, for example, enable high resolutions for the visualisation of damage. However, the higher the resolution is, the smaller the excerpt of the microstructure considered. Thus, numerous images have to be made, analysed in detail and damage development has to be quantified to enable reliable statements about the damage propagation. Methods such as micro- and nanoindentation, gas adsorption measurement, mercury pressure porosimetry or nuclear magnet resonance spectroscopy can be comparably applied to pristine specimens and specimens after fatigue loading in order to quantify the damage introduced by a selected fatigue loading under designated boundary conditions. However, it must always be critically considered to what extent the sample preparation could have influenced the results of the analyses.

Another imaging method is computer tomography. Although the experiment still has to be interrupted, no samples have to be destructively prepared from the specimen for further investigations. The specimen could even be subjected to further fatigue loading. The application of computer tomography in concrete research was associated with great hopes fifteen years ago regarding the visualisation of cracks and structural damage in concrete. Essential components of high-performance concretes, such as aggregates and fibres, and also pores and cracks, can, in fact, be displayed and measured separately in two and three dimensions. However, research in the SPP 2020 showed that commercially available computer tomographs can hardly achieve the resolution required to adequately analyse fatigue damage with the specimen diameter required, because the fatigue damage starts to develop on an extremely small scale. Hence, the determination of reliable and meaningful data in fatigue tests, on whatever scale, is, thus, very complex and expends great effort.

5 Computational Modelling of Concrete Fatigue

The modelling and simulation of fatigue processes in high-performance concrete is an additional challenging task. Experiments show that damage processes that take place on very fine scales may have a significant influence on the fatigue behaviour on coarser scales. The highly heterogeneous microstructure of high-performance concrete leads to inhomogeneous stress fields on these fine scales, necessitating multiscale techniques to resolve and understand these effects and consider them for the fatigue simulation of macroscopic concrete structures. Additionally, damage occurring under tensile stresses rapidly leads to localisation effects and discrete and oriented microcracks that may grow and form macroscopic fatigue cracks. By contrast, damage due to compression leads more to a distributed degradation field and a loosening of the material. Obviously, the generally anisotropic material behaviour of concrete is rather complex, especially under fatigue loading conditions. As a consequence, sophisticated material models are required to capture the relevant effects and eventually provide predictive computational models.

Fatigue models that have frequently been used previously, such as those based on linear damage accumulation, can only provide rough estimations for the fatigue behaviour. Their predictive capabilities are very limited and they are unable to robustly forecast the failure of concrete structures under fatigue loads with non-constant load levels or changing loading directions. This is especially relevant considering the future necessity of resource-efficient buildings providing high structural stiffness and comparatively low inertial mass, leading to a significant increase of eigenfrequencies and a much larger number of load cycles the structures must endure.

Modern fatigue material models considering the local, anisotropic and generally inelastic material history to capture the detailed state of fatigue damage are mandatory for the predictive character of a computational model usable until failure. The calibration of such complex material models and their parameters requires an intensive collaboration of modelling experts and experimentalists in an experimental-virtual lab. Polymer, carbon or steel fibres can be employed in different concentrations and with different fibre shapes to obtain significantly higher quasi-ductile material behaviour. The modelling of the interaction between the concrete matrix and fibres poses further difficulties, especially if fatigue loadings are to be considered. Fibre rupture, fibre debonding, and partial or complete fibre pull-out needs to be modelled adequately, posing an additional challenge also for the computational strategy.

Smart temporal discretisations are necessary for the efficient simulation of multiple thousands or even millions of load cycles. Time homogenisation and cycle jump techniques are classically used and perform quite well for constant fatigue load levels. However, the temporal discretisation constitutes additional challenges for the computational strategy if the load levels or loading directions change.

6 Scientific Approach

This book presents the results of the joint research in the DFG Priority Programme SPP 2020 "Cyclic Deterioration of High-Performance Concrete in an Experimental-Virtual-Lab", coordinated by Leibniz University Hannover. The fatigue damage mechanisms in high- and ultra-high-performance plain and fibre-reinforced concretes were investigated within this Priority Programme using the latest experimental and virtual numerical methods in tandem projects. The main research idea was to combine building material science with modelling science to meet the previously outlined challenges of concrete fatigue research and generate new fundamental knowledge. The scope of this programme was to capture, understand, describe, model, and predict the material degradation of high-performance concretes.

As is usual in Priority Programmes, each research project had its own thematic focus and research question to be answered. Various concrete strength grades under different types of loading (compression, tension, bending) or loading scenarios, plain concretes or concretes with steel, carbon or polyvinyl alcohol fibres with different geometries, were investigated within respective projects. These diversities were also reflected in the spectrum of numerical investigation methods applied. However, two reference concretes, one high- and one ultra-high-strength concrete, were provided by the Central Project to achieve a certain degree of comparability between the projects. Furthermore, round robin tests concerning the influence of the specimens' production and preparation, as well as the influence of the concrete strength on the fatigue behaviour were conducted with the participation of various laboratories guided by the Central Project of the SPP 2020.

Furthermore, the bending behaviour of fibre-reinforced ultra-high-strength concrete was investigated using different modelling approaches within the framework of a cross-project benchmark.

On the experimental side, modern measurement techniques, such as digital image correlation, acoustic emission analysis and microstructure analysis techniques, including micro- and nanoindentation, were used, in addition to light and raster electron microscopic examinations in both scanning and transmission electron microscopy as well as micro computer tomography devices. Furthermore, different microstructural analyses techniques were applied and comparably evaluated regarding their meaningfulness concerning damage capture [10]. On the modelling side, modern numerical approaches, such as phase field methods, the extended finite element method or bonded particle models, were used and further developed specifically for their application in cementitious systems.

The research results of the individual SPP 2020 projects are presented in the following chapters.

References

1. Aas-Jacobsen, K.: Fatigue of Concrete Beams and Columns. Bulletin No. 70-1, Institute for Betonkonstruksjoner, Trondheim, 148 (1970)
2. Holmen, J.O.: Fatigue of Concrete by Constant and Variable Amplitude Loading. Norwegian Institute of Technology, University of Trondheim (1979)
3. Petković, G., Rosseland, S., Stemland, H.: High Strength Concrete SP3—Fatigue. Report 3.2: Fatigue of High Strength Concrete. SINTEF Structural Engineering—FCB, Trondheim (1992)
4. Klausen, D.: Festigkeit und Schädigung von Beton bei häufig wiederholter Beanspruchung. [Strength and damage of concrete subjected to often repeated loading] (in German). Doctoral Thesis. Technische Hochschule Darmstadt (1978)
5. CEB—Comité Eurointernationale du Béton (ed.): CEBFIP Model Code 1990. Bulletin d'Information, No. 213/214, Thomas Telford Ltd., London (1993)
6. CEN—European Committee for Standardization (ed.): Eurocode 2: EN 1992-1-1:2004, Design of Concrete Structures—Part 1-1: General Rules and Rules for Buildings (2024)
7. CEN—European Committee for Standardization (ed.) Eurocode 2: EN 1992-2:2005, Design of Concrete Structures—Concrete Bridges—Design and Detailing Rules (2005)
8. Oneschkow, N., Lohaus, L.: Zum Nachweis von druckschwellbeanspruchtem Beton, Teil 2: Sicherheitsüberlegungen und Potenzial für Weiterentwicklungen. [About the fatigue design concept of concrete subjected to pure compressive fatigue loading. Part 2: Safety considerations and potential for further developments] (in German). Beton- und Stahlbetonbau **112**, 611–622 (2017). https://doi.org/10.1002/best.201700026
9. Oneschkow, N.: Analyse des Ermüdungsverhaltens von Beton anhand der Dehnungsentwicklung. [Analysis of the fatigue behaviour of concrete with respect to the development of strain] (in German). Doctoral Thesis, Reports of the Institute of Building Materials Science, 13, Leibniz Universität Hannover (2014). https://doi.org/10.15488/357
10. Engelhardt, M., Kalytta-Mewes, A., Volkmer, D., Lohmann, J., Ritter, M., Schaan, G., Schmidt-Döhl, F., Abubakar Ali, M., Basaldella, M., Haist, M., Kern, B., Lohaus, L., Oneschkow, N., Rozanski, C., Timmermann, T., Grosse, C.U., Birtel, V., Garrecht, H., Madadi, H., Markert, M., Steeb, H.: Micro- and nanostructural investigations of high and ultra-high performance concrete under fatigue. Int. J. Fatigue **199**, 109038 (2025). https://doi.org/10.1016/j.ijfatigue.2025.109038

Material Properties and Fatigue Resistance of the Reference High-Strength and Ultra-High-Strength Concrete of SPP 2020

Marco Basaldella, Bianca Kern, Ludger Lohaus, and Nadja Oneschkow

Abstract The compressive fatigue resistance of concrete is influenced by many effects from the composition, storage conditions, to boundary conditions of tests. According to the influence of the compressive strength, contradictory results can be found in the literature. A comparison of fatigue test results from literature is generally difficult to obtain due to the differences in the test boundary conditions and uncertainties due to laboratory-specific influences. A high-strength and an ultra-high-strength concrete were investigated comparatively within the central project of the Priority Programme SPP 2020 in cooperation with several laboratories. The objectives were to investigate the influences of concrete production and specimen preparation and the influence of the compressive strength on the fatigue resistance. It was found that the production technique has an influence on the compressive strength, especially in case of the high-strength concrete, but does not affect the mean number of cycles to failure in either case. Nevertheless, an accurate preparation of specimens is necessary as it has a positive effect on the scattering of results of monotonic and cyclic load tests. Furthermore, the results revealed that a concrete with a higher compressive strength can resist higher numbers of cycles to failure and shows an improved fatigue behaviour.

Keywords High-strength concrete · Ultra-high-strength concrete · Cyclic loading · Specimen preparation · Stiffness development · Strain development

1 Introduction

Advancements in concrete technology have paved the way for the use of concrete mixtures with increasingly higher compressive strengths. This has allowed the construction of more filigree and slender structures that have not been feasible previously. However, these structures are more susceptible to fatigue-induced loads

---

M. Basaldella · B. Kern · L. Lohaus · N. Oneschkow (✉)
Institute of Building Materials Science, Leibniz University Hannover, Hannover, Germany
e-mail: n.oneschkow@baustoff.uni-hannover.de

N. Oneschkow and L. Lohaus (eds.), *Cyclic Deterioration of High-Performance Concrete*, https://doi.org/10.1007/978-3-032-19394-0_2

compared to bulkier structures, as their ratio of dead-weight to non-static load is lower. As a result, the fatigue resistance of concrete has become a critical factor in the design of such structures. Additionally, there has been a growing demand for structures in recent years that are subjected to fatigue-related stresses, such as wind turbines or slender bridges. Consequently, research efforts [1–4] have been increasingly focused on understanding the fatigue resistance and fatigue behaviour of concretes in order to meet the needs of modern construction practices.

The results gained from fatigue tests generally show a significant scattering and, furthermore, a considerable variability between different investigations. This variability is attributed to various factors that impact the fatigue behaviour of concrete, including the estimation of the load level, the type of loading and the test conditions, such as load frequency, humidity and ambient conditions.

The influence of the compressive strength on the fatigue resistance of concrete has been discussed in the literature with conflicting findings. Some studies, such as those presented in [5, 6], have shown that high-strength concretes exhibit relatively lower fatigue resistance compared to concrete compositions with lower compressive strengths. Conversely, other investigations have suggested that compressive strength has no significant influence on the fatigue resistance as, for example, documented in [1, 7, 8]. Interestingly, even higher numbers of cycles to failure were observed for higher strength concrete specimens compared to lower strength ones in [9]. However, it should be noted that there are only a few comparative investigations on the fatigue resistance that have explored concretes with different compressive strengths, and there is especially a lack of studies that consider the fatigue behaviour described by damage indicators, such as strain and stiffness development.

The developments of strains at the maximum and minimum peak stresses in relation to the related number of cycles to failure N/N_f exhibit distinct S-shaped curves which can be divided into three phases (see Fig. 1a). The first phase is characterised by a disproportionate increase of deformation that turns linear in the second phase. In the third phase, the deformations increase again disproportionally up to failure of the specimen. In the case of normal-strength concretes, the transitions from phase I to II and from II to III occur at approximately 5–20% and 80–95%, respectively, of the number of cycles to failure N_f, as reported in previous studies [1, 3, 4, 10–14]. However, phase I and III for high-strength and ultra-high-strength concretes, tend to be significantly shorter in duration and less prominent compared to concretes with lower concrete strength [5, 8, 15, 16].

The stiffness of concrete during the fatigue process is typically characterised by the secant modulus, which is determined from the decreasing branch of the hysteresis loop [11, 14, 17, 18]. Similar to the strain developments, the development of stiffness per load cycle also follows an S-shaped curve (see Fig. 1b). Existing literature suggests that the reduction in stiffness until fatigue failure is comparatively lower in concretes with higher compressive strengths [11, 19].

The gradients of stiffness in phase II in relation to the number of cycles to failure documented in the literature [15, 17, 20] were compared with own investigations on a high-strength concrete in [16]. Despite the partial determination of differences, the influence of concrete strength could not be clearly attributed, as potential influences of

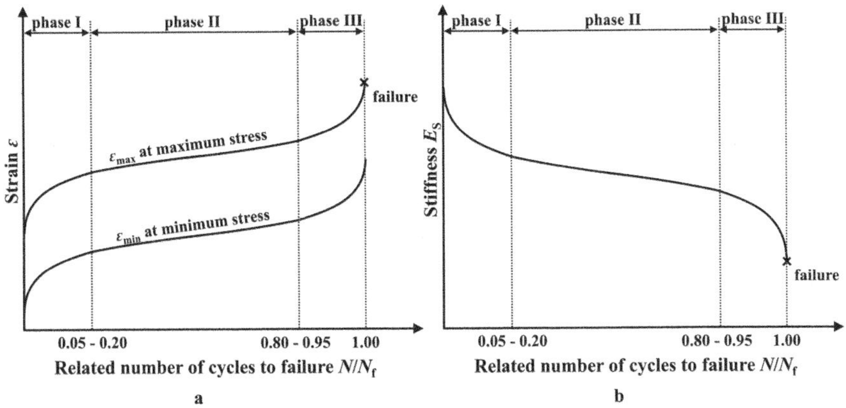

Fig. 1 Schematic diagram of S-shaped development of **a** strain at the maximum and minimum peak stresses and **b** stiffness due to fatigue loading

different testing laboratories could not be ruled out. A cautious approach is essential when conducting comparisons of test results obtained from different laboratories. It is important to consider not only the differences in the boundary conditions of the tests but also the potential impact of individual laboratory procedures on the scattering of results. The variability in laboratory-specific procedures can introduce additional sources of variability in the results obtained. It was observed in a previous study [21] that a significant scattering of results occurred due to the influence of factors related to the individual specimens used, such as the quality of load surfaces and centring in the test machine. In addition, the inhomogeneous nature of concrete itself, which can be partially influenced by production techniques [22], is also considered to contribute to the scattering of results. Although the scattering of empirical fatigue results is examined from a statistical point of view, there is no research in the literature that investigates the influence of concrete production and specimen preparation on the numbers of cycles to failure.

The field of civil engineering still generally lacks a comprehensive understanding of the potential differences in fatigue resistance and fatigue behaviour between high-strength and ultra-high-strength concrete. This knowledge gap gives rise to uncertainties in their implementation within the fatigue design rules. Consequently, the current design rules for compressive fatigue loading in widely used standards and guidelines, such as Eurocode 2 [23] or Model Code 2020 [24], presume that concretes become more susceptible to fatigue loading with increasing compressive strength [25–27]. Consequently, a strength-dependent reduction factor is incorporated, resulting in a reduction of the fatigue resistance applicable. This leads to uneconomical utilisation of high-strength concrete structures and the realisation of innovative concrete structures is considerably hindered, sometimes even prevented [26, 27].

The framework of the Priority Programme SPP 2020, where a total of 21 projects from different research institutes were participating in addition to the central project, offered the possibility of analysing the influence of the quality of the specimen. Thus, the influence of the concrete production and specimen preparation techniques of the participating laboratories on the compressive strength and fatigue resistance of a high-strength and an ultra-high-strength concrete were studied in the central project. Furthermore, the fatigue behaviour of the two different concrete compositions were investigated comparatively with special regard to the influence of the compressive strength. The investigations were conducted on the two reference concretes in the SPP 2020.

2 Materials and Methods

2.1 Concrete Composition

A high-strength and an ultra-high-strength concrete mixture were developed within the framework of the Priority Programme SPP 2020 and used as reference by the participating projects in order to link them together. The ultra-high-strength concrete (UHPC) named "RU1" was based on the M3Q mixture previously used in the SPP 1182 [28]. The high-strength concrete (HPC) called "RH1" was developed at the Institute of Materials Science, Leibniz University of Hannover (IfB). The compositions of both reference concretes are given in Table 1.

Table 1 Compositions of concrete mixtures (from [29], licensed under CC-BY 4.0)

Component	Unit	HPC (RH1)	UHPC (RU1)
CEM I 52.5 R-HS/NA (Holcim Sulfo, Lägerdorf, Germany)	[kg/m^3]	500	795
Silica fume (Sika® Silicoll P)	[kg/m^3]	–	169
Quartz powder (Quarzwerke MILLSIL® W12, Frechen, Germany)	[kg/m^3]	–	198
Quartz sand (0/0.5 mm) (Quarzwerke H33, Haltern, Germany)	[kg/m^3]	75	971
Sand (0/2 mm) (Tuendern, Germany)	[kg/m^3]	850	–
Basalt (2/5 mm) (Oelberg, Germany)	[kg/m^3]	350	–
Basalt (5/8 mm) (Oelberg, Germany)	[kg/m^3]	570	–
Superplasticiser (BASF MasterGlenium® ACE 460, Germany)	[kg/m^3]	5	–
Superplasticiser (BASF MasterGlenium® ACE 394, Germany)	[kg/m^3]	–	24
Stabiliser (BASF MasterMatrix® SDC 100, Germany)	[kg/m^3]	2.85	–
Water	[kg/m^3]	176	188
w/c ratio; w/c_{eq} ratio	[–]	0.35	0.19

The physical properties and the grain size distribution of the cement and quartz powder are shown in Table 2. The properties of the silica fume could not be determined due to the lack of adequate measuring equipment. The chemical properties of the cement are given in Table 3. Both superplasticisers used are based on a polycarboxylate-based admixture, while the stabiliser used to modify the viscosity is based on a synthetic copolymer.

The grain size distributions of the quartz sand, the sand and the basalt aggregates are shown in Fig. 2.

Seven participating laboratories have tested the standard 28-day cubic compressive strength of each concrete batch produced since the start of the SPP 2020 in 2017. The tests were performed according to DIN EN 12390-3:2019-10 [30], and the results were collected in a central database to perform a statistical analysis. The storage conditions after demoulding were either "under water until testing", according to

Table 2 Physical properties and grain size distribution of cement and fine aggregates (from [29], licensed under CC-BY 4.0)

Materials	Density	Fineness	d_{10}	d_{50}	d_{90}
	[kg/dm³]	[cm²/g]	[μm]	[μm]	[μm]
CEM I 52.5 R-HS/NA	3.18	3969	1.60	11.25	33.18
Quartz powder	2.66	–	2.09	12.71	37.18

Table 3 Chemical properties of cement CEM I 52.5 R-HS/NA (from [29], licensed under CC-BY 4.0)

Materials	Chemical composition [%]					
	SiO_2	Al_2O_3	Fe_2O_3	CaO	MgO	SO_3
CEM I 52.5 R-HS/NA	21.41	3.97	4.79	65.41	0.85	2.98

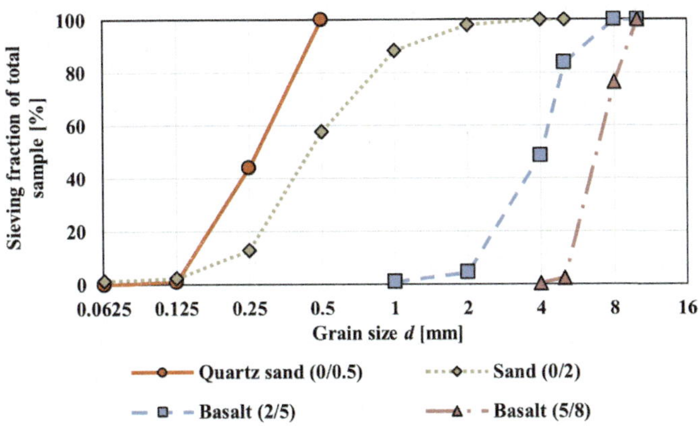

Fig. 2 Grain size distributions of quartz sand, sand and basalt aggregates

Table 4 Classification of concretes based on tests of seven SPP 2020 laboratories (from [29], licensed under CC-BY 4.0)

Concrete	Qty	f_{cm}	σ_x	f_{ck}	Classification
	[–]	[MPa]	[MPa]	[MPa]	
HPC (RH1)	75	111.8	7.7	99.2	C80/95[a]
UHPC (RU1)	72	154.8	8.1	141.5	C130/145[b]

[a] According to DIN EN 206:2021-06 [35]; [b] According to the draft DAfStb [36]

DIN EN 12390-2:2009-08 [31], or "in standard climate conditions" (20 °C/65% R.H.), due to the adaption of the last update of DIN EN 12390-2:2019-10 [32]. The edge length of the cubic specimens was either 100 or 150 mm, therefore, a conversion of compressive strength according to DIN 1045-2:2008-08 [33] was possible in principle. In view of the fact that a general applicability of the conversion factors given in standards is questionable and, furthermore, they are not necessarily applicable for UHPC, it was decided to consider the compressive strengths without conversion [29].

The characteristic concrete compressive strength f_{ck} was calculated (Eq. 1) as a 5%-fractile from the mean compressive strength f_{cm}, in accordance with DIN EN 1990:2021-10 [34].

$$f_{ck} = f_{cm} - k_n \cdot \sigma_x \approx f_{cm} - 1.64 \cdot \sigma_x \tag{1}$$

where k_n is the 5%-fractile factor and σ_x is the standard deviation. The characteristic compressive strengths calculated and the classifications are shown in Table 4 for both concrete types.

2.2 Experimental Programme

The experimental programme of the central project at IfB was divided into two work packages which were conducted in cooperation with the laboratories of the participating projects. In the first work package, the influences of the concrete production and specimen preparation on the compressive strength and fatigue resistance were investigated for both reference concretes RH1 and RU1. In the second work package, the influence of the compressive strength on the fatigue resistance of concrete was investigated in general.

Cylindrical specimens were tested with nominal dimensions of $d/h = 60/180$ mm for both work packages. The compressive strengths were tested force-controlled with a constant speed of 0.5 MPa/s directly before the respective fatigue tests were carried out. The mean compressive strengths were additionally used as reference strengths for the cyclic tests. The tests for the determination of the compressive strengths and the fatigue tests with sinusoidal cyclic loading were carried out using servo-hydraulic testing machines with 1 MN actuators, exemplarily shown in Fig. 3a. The axial deformations were measured continuously in all tests using three laser distance sensors positioned on the circumference of the specimen at 0°, 120° and 240° (Fig. 3b). The axial force and the displacement of the actuator were also recorded. In addition, the temperature on the specimen's surface and the ambient temperature were measured by means of thermocouples.

Work Package 1 A total of seven laboratories were involved in the investigation of the influence of the concrete production and specimen preparation on the compressive strength and the fatigue resistance: Laboratories numbered 1–4 produced HPC specimens (RH1), while those numbered 5–7 produced UHPC specimens (RU1) [37]. The raw materials and the moulds (PVC) were supplied from a single production batch. The specimens were stored under standard climate conditions (20 °C/65% R.H.) after concreting and demoulding, until testing.

In order to investigate the effects of the specimen preparation separated from those of the concrete production, half of the test specimens produced at each laboratory was prepared for testing by the laboratories themselves with their own technique and equipment. These specimens are denoted with the letter "O" for "own preparation technique". The other half of the specimens was each prepared for testing in the central project laboratory at IfB always using the same preparation technique and equipment. These specimen sets are denoted by the letter "R" for "reference preparation technique". In this way, the effects of the preparation technique used in each laboratory could be compared with the effects of one reference technique. The own preparation techniques of the laboratories generally included sawing a few

a b

Fig. 3 Test set-up at the IfB: **a** 1 MN servo-hydraulic testing machine, **b** measurement set-up (from [29], licensed under CC-BY 4.0) centimetres off the upper and lower ends of the specimens to remove areas that could have been disturbed by the production process. The test surfaces of the cylindrical specimens were then parallel ground and in case of Lab 2 additionally polished. The reference preparation included precise sawing and polishing to ensure parallel end surfaces using consistent equipment and setup.

The effects of the different techniques used in the laboratories for the concrete production were analysed by comparing only the results obtained from the specimens prepared with the reference preparation technique ("R"). The techniques for the concrete production differed in e.g. the mixing technique or compacting technique. The equipment and techniques used by the different laboratories for the production and preparation methods are documented in [37].

All experimental investigations were conducted on the two sets of specimens ("R" and "O") at IfB. They consisted of (a) tests with monotonically increasing loading to determinate the compressive strength and (b) cyclic compressive tests to determine the fatigue resistance. Six specimens from each laboratory and specimens' set were investigated per test type. All compressive cyclic tests were performed with a load frequency of $f_t = 1$ Hz and a compressive stress level of $S_{max} = 0.75$ and $S_{min} = 0.05$.

In order to analyse the effects of the preparation techniques on the specimens, selected geometrical parameters were measured before testing: the cylindrical specimen's height and diameter, the perpendicularity of the cylinder's lateral surface to the test surface, the straightness of the lateral surface and the surface flatness according to DIN EN 12390-1:2012-12 [38]. The latter two were verified by measuring the deviation from a straight edge using steel feeler gauges (min. thickness of 0.5 mm). The surface roughness of the test surface defined according to DIN EN ISO 25178:2016-12 [39] was measured using the three-dimensional laser scanning confocal microscope "VR-5000" from KEYENCE. In order to quantify possible edge breakages of the specimens, the effective surface area without missing portions of each specimen was measured. It should be mentioned here that the effective surface area was not used to recalculate the stresses applied in the cyclic tests. The density of the microstructure was additionally characterised by means of the sound velocity (pulse velocity) using the Impact Resonance Method [37]. According to this method, the specimens were each mechanically stimulated with the impulse of a small metal ball at one end of the specimen while a microphone at the other end of the specimen perceived the vibrations so that the sound velocity could be determined. The measurement was performed by using the non-destructive testing equipment "RA 100 Concrete" by Lang Sensorik (ASTM C215-19 [40], BS 1881-209:1990-04-30 [41]). A mean value out of three measurements was used. A detailed description of these complementary investigations can be found in [37].

Work Package 2 In the second work package, both reference concretes were investigated comparatively regarding the influence of the concrete strength on the fatigue resistance, and to generally characterise their fatigue behaviour [29]. The HPC specimens were produced and prepared at the IfB Hannover and the UHPC specimens at the Institute of Concrete Structures, TUD Dresden University of Technology, respectively. Two batches were produced for each concrete (HPC-a and -b and UHPC-a and -b, respectively). After concreting and demoulding, the specimens were stored in standard climate conditions (20 °C/65% R.H.) until testing. The specimens were prepared by sawing a few centimetres off the top and bottom surfaces to a length of 180 mm. The surfaces were then ground parallel and afterwards polished.

The fatigue tests were performed at two maximum stress levels for both concretes. The specimens of batches HPC-b and UHPC-b were tested at the stress levels $S_{max} = 0.85$, $S_{min} = 0.05$ and $S_{max} = 0.75$, $S_{min} = 0.05$ at the Materials Testing Institute, University Stuttgart (MPA). Specimens of concrete batches HPC-a and UHPC-a were additionally investigated at the IfB in Hannover at $S_{max} = 0.75$, $S_{min} = 0.05$. In this way, the influence of the stress level was investigated for HPC-b and UHPC-b without laboratory influence. Furthermore, the influence of the laboratory was included by testing at the lower stress level in both laboratories. The load frequency applied was $f_t = 1$ Hz. The number of tests at each stress level is summarised in Table 5.

The number of cycles to failure per specimen identifies its respective fatigue resistance. These numbers were statistical analysed regarding the influence of the batches of each concrete and the compressive strength of the two concrete mixtures. Additionally, the developments of the strains at maximum and minimum peak stress of the sinusoidal load cycles were analysed as damage indicators. This included the total growth of maximum and minimum strains up to failure and the strain gradients in phase II of the classical S-shaped strain development as parameters. Accordingly, phase II was considered between the fixed values of $N/N_f = 0.20$ and 0.80. Furthermore, the concrete stiffness, calculated as the secant modulus of the descending branch of the hysteresis loop of each load cycle, was analysed. Here, the gradient in phase II and the percentile reduction of stiffness up to failure were considered as parameters [29].

Table 5 Number of fatigue tests in work package 2

Concrete	HPC-a	HPC-b	UHPC-a	UHPC-b
S_{max}/S_{min}	IfB	MPA	IfB	MPA
0.75/0.05	7	3	7	4
0.85/0.05	–	6	–	6

3 Results and Discussion

3.1 Influence of the Concrete Production and Specimen Preparation

The influences of the concrete production and specimen preparation on the compressive strength and the fatigue resistance of the HPC and UHPC were investigated by testing a total of 48, respectively, 53 HPC specimens and a total of 37, respectively, 38 UHPC specimens (work package 1).

The influence of the concrete production process on the concrete microstructure was analysed using the sound velocity as a parameter describing the density of the concrete. The sound velocity is generally influenced by factors, such as the aggregate size, grading, type, content, concrete moisture and curing condition, and internal discontinuities, such as cracks and voids [42]. The specimens in this investigation were produced at the different participating laboratories following the given compositions of the concrete mixtures using raw materials from the same production batches and uniform curing conditions (cf. Sect. 2.2). Thus, the concrete microstructure density and heterogeneity could only be influenced by the different production techniques, which would result in differences in the sound velocity and concrete compressive strength. The correlation between the compressive strength and the sound velocity as a parameter is shown in Fig. 4 for both concrete types and specimen sets “R”.

It is visible in Fig. 4 that a linear correlation between the sound velocity and the compressive strength can be identified for the HPC. Three groups, i.e. Lab 1, Lab. 2 with Lab. 4 and Lab. 3, can be distinguished, thus, identifying three different concrete microstructure density ranges. Based on the production protocols from

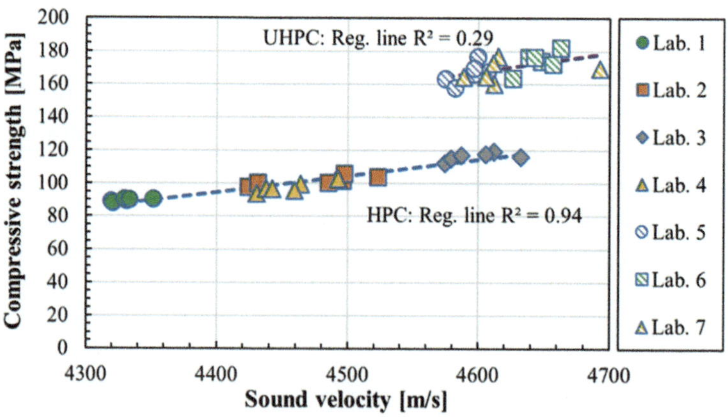

Fig. 4 Microstructure characterisation of HPC (Labs 1–4) and UHPC (Labs 5–7) (based on [37], licensed under CC-BY 4.0)

the laboratories, it is assumed that the main influence on the sound velocity and, thus, on the density of the microstructure and the compressive strength was due to the different compacting techniques (e.g. vibration frequency, vibration duration) used [37]. Figure 4 shows a weak correlation between the sound velocity and the compressive strength for the UHPC. The results of the specimens from Lab. 5 and 7 are not clearly distinguishable enough to be assigned to different groups, whereas they are barely distinguishable from those of Lab. 6. Thus, the influence of the concrete production on the compressive strength for the UHPC was not as marked as for the HPC.

Concerning the analysis of the geometrical properties of the specimens, no evidence of significant deviations was found concerning the height, diameter, surface flatness and straightness of the lateral surface. The analysis of the perpendicularity showed that Labs 2–7 achieved deviations comparable to those by using the reference preparation technique with a maximum mean deviation of less than 0.8 mm, while Lab. 1 only achieved an average deviation of about 1.2 mm. Furthermore, the polishing method of the reference technique provided the lowest surface roughness values. The usage of grinding by Labs 1, 3, 4, 6 and 7 led to a significantly higher roughness value. Although Lab. 5 indicated the usage of grinding, the roughness obtained was similar to that obtained by the polishing method of the reference technique. The analysis of the effective area showed that the specimen's edges after using the reference technique were not damaged or less damaged. Labs 3, 4, 5 and 7 caused slight breakages that reduced the contact surface by an average of about 1%, respectively, 2% for the specimens prepared by Lab. 2 and 6. The average reduction of the specimen's contact surface from Lab. 1 was about 8% and, thus, much higher than for all the other laboratories.

Overall, the analysis of the geometrical properties of the specimens showed that accurate sawing and polishing using techniques comparable to the reference technique "R" allowed for specimens with reduced irregularities and higher geometrical accuracy [37]. These irregularities of the specimens are implicitly included in the results analysed in the following.

Figure 5 shows the influence of the production and preparation techniques on the (reference) compressive strength and the numbers of cycles to failure of the HPC (Labs 1–4): The red and blue dots represent the mean values obtained from the specimen sets "O" and "R", respectively, while the whiskers indicate the standard deviations. The standard deviation is represented on the secondary coordinate system on the right vertical. Considering all the tested specimens, the overall mean compressive strength was $f_{cm,ref} = 98.7$ MPa with a standard deviation of 11.0 MPa.

In a first step, the results obtained from specimens prepared with the reference technique "R" are compared focussing the influence of the concrete production. Laboratories 2 and 4 produced specimens with (reference) compressive strengths very close to the overall mean value, while Labs 1 and 3 produced specimen sets with the lowest and highest compressive strengths, $f_{cm,ref} = 89.5$ MPa and $f_{cm,ref} = 116.2$ MPa, respectively. The differences in compressive strength were found to be statistically significant using ANOVA (Analysis of Variance) (p-value $\ll 0.05$). The comparison between the specimen sets "R" and "O"

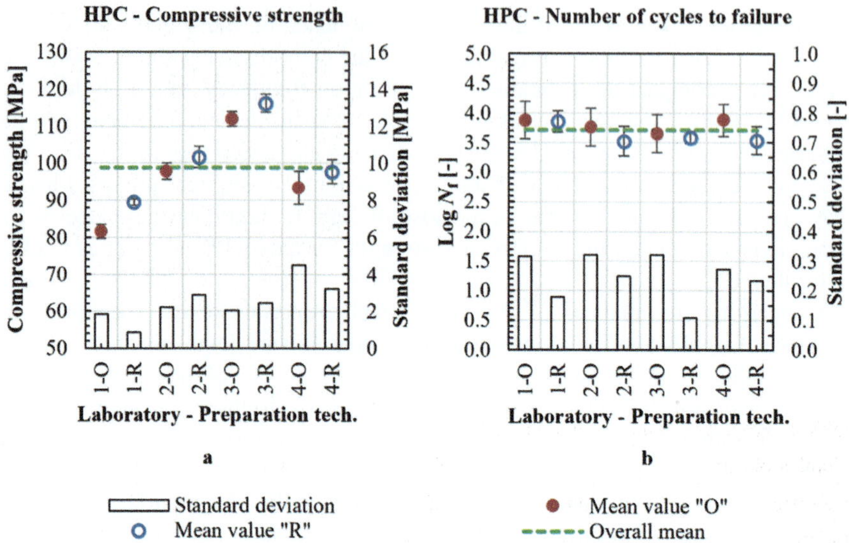

Fig. 5 HPC (RH1): **a** compressive strength, **b** number of cycles to failure (based on [37], licensed under CC-BY 4.0)

of each laboratory regarding the preparation technique shows that the mean compressive strengths from sets "R" were higher than those from the corresponding specimen sets "O". These differences were found using the classical t-test to be statistically significant (p-value $\ll 0.05$).

Regarding the fatigue tests, an overall mean log. number of cycles to failure of $\log N_f = 3.713$ with a standard deviation of 0.28 was determined (Fig. 5b). Furthermore, the mean values of each specimen set "O" and "R" do not deviate significantly from the overall mean. This points out that the clear influence of concrete production on the compressive strength (Fig. 5a) did not exist with respect to the numbers of cycles to failure. Furthermore, the preparation of the specimens did not significantly affect the mean value of the fatigue resistance. These findings match the explanation mentioned in [43]: The maximum and minimum stresses applied in the compressive fatigue tests were obtained using the mean compressive strength from specimens tested with the same characteristics as those specimens tested under fatigue loading. Therefore, the mean value of the numbers of cycles to failure obtained was nearly independent of the concrete production and specimen preparation technique used. Considering the standard deviation of each set in Fig. 5b, it appears that the reference preparation technique had a positive effect on the scattering of the results by reducing it in view of a higher geometrical uniformity and accuracy of the "R" specimens.

The results of the investigations on the UHPC specimens are shown in Fig. 6 using the same graphical presentation as for the HPC in Fig. 5. The overall mean (reference) compressive strength was $f_{cm,ref} = 170.1$ MPa with a standard deviation of 6.6 MPa. The mean values of the compressive strength of each set were close to the overall mean. The differences in the compressive strength were found to

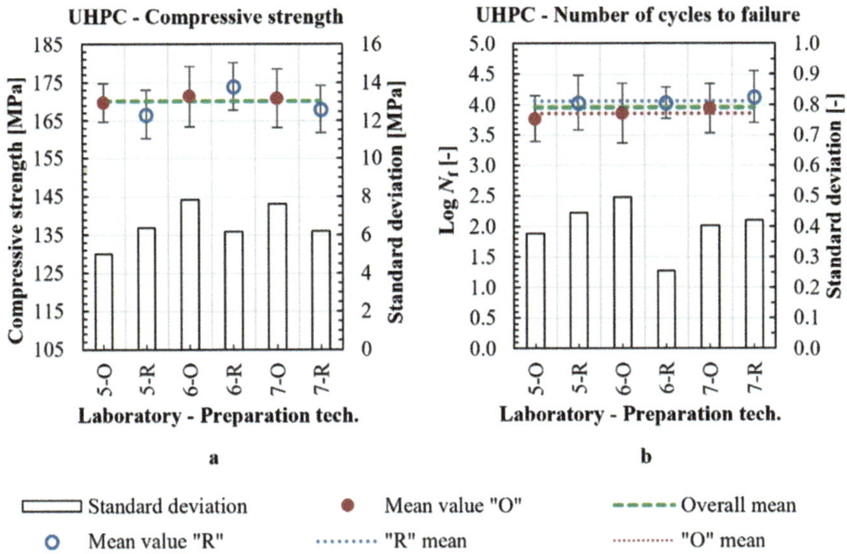

Fig. 6 UHPC (RU1): **a** compressive strength, **b** number of cycles to failure (based on [37], licensed under CC-BY 4.0)

be statistically not significant using ANOVA (p-value $\gg$ 0.05). This indicates that the production techniques used by Labs 5–7 led to very similar concrete batches. Furthermore, no statistically significant influence (p-value $\gg$ 0.05) of the specimen preparation techniques on the mean value of the compressive strength could be found by comparing the results from the couples of sets "O" and "R" of each laboratory and applying the statistical t-test. It can be seen from Fig. 6a that a slight reduction of scattering due to the reference technique "R" may have been present for specimens set of Labs 6 and 7.

The compressive fatigue tests gave an overall mean log. number of cycles to failure of $\log N_f = 3.960$ with an overall standard deviation of 0.40. Figure 6b shows that the mean values of the fatigue resistance of each specimen set were close to the global mean value. Thus, the concrete production techniques used by the three laboratories had no influence on the numbers of cycles to failure for the UHPC. This was confirmed by the statistical t-test where no statistical significance was found.

However, Fig. 6b shows that the mean values of the specimen sets prepared using the reference technique "R" are slightly higher than those obtained from the specimen sets "O" prepared by the different laboratories. This can be interpreted as a slight positive influence of the preparation technique "R" by careful sawing and polishing of the UHPC specimens. The analysis of the geometrical characteristics of the specimens of sets "O" showed that Labs 5 and 7 prepared accurate specimens, while Lab. 6 introduced small inaccuracies [37]. Those inaccuracies may have been the cause of the greater scattering of the results obtained for the specimens of set 6-O.

Overall, the comparison of the results of both concretes shows that the influence of the concrete production and specimen preparation techniques on the compressive strength was higher for the HPC compared to the UHPC. This could have been attributed to the higher uniformity of specimen production techniques among the laboratories which produced the UHPC specimens. Furthermore, rare influence existed on the numbers of cycles to failure due to the usage of the compressive strength as a reference for the determination of the stresses applied. However, careful sawing and polishing of UHPC specimens led to slightly higher numbers of cycles to failure. An additional observation was that the numbers of cycles to failure obtained for the UHPC were higher on average than those obtained for the HPC.

3.2 *Influence of the Compressive Strength on the Fatigue Behaviour*

The reference HPC and UHPC were investigated comparatively in order to generally characterise both concretes and, furthermore, examine the influence of the concrete strength on the fatigue resistance (work package 2). The reference compressive strengths were determined directly before the fatigue tests were carried out. The results are given in Table 6, where $f_{\mathrm{cm,ref}}$ is the mean reference compressive strength, σ_{x} is the respective standard deviation of $f_{\mathrm{cm,ref}}$, $\varepsilon_{\mathrm{cm}}$ is the mean strain at maximum stress, and $E_{0.15\text{–}0.80}$ is the mean stiffness determined as the secant modulus of elasticity between 15 and 80% of the maximum stress.

The difference in the compressive strength between the two HPC batches (HPC-a: $f_{\mathrm{cm,ref}} = 116.2$ MPa and HPC-b: $f_{\mathrm{cm,ref}} = 89.8$ MPa) was due to a batch influence despite the fact that they were produced in the same laboratory. The reason for the difference between the two UHPC batches (UHPC-a: $f_{\mathrm{cm,ref}} = 174$ MPa and UHPC-b: $f_{\mathrm{cm,ref}} = 200.6$ MPa) was not clear: batch influence and/or influence of the concrete age [29].

The numbers of cycles to failure of the specimens tested are presented in Fig. 7 as single and mean values. At the lower maximum stress level $S_{\mathrm{max}} = 0.75$, specimens of both batches were tested for each concrete mixture. Therefore, the logarithmic numbers of cycles to failure were statistically analysed to investigate a possible batch

Table 6 Mean values of the fatigue reference compressive strength, ultimate strain, and secant modulus of elasticity (from [29], licensed under CC-BY 4.0)

Concrete	Qty	Age	$f_{\mathrm{cm,ref}}$	σ_{x}	$\varepsilon_{\mathrm{cm}}$	$E_{0.15\text{–}0.80}$
	[–]	[d]	[MPa]	[MPa]	[‰]	[MPa]
HPC-a	6	79	116.2	2.2	– 3.67	36,700
HPC-b	4	213	89.8	3.0	– 3.31	33,200
UHPC-a	6	97	174.0	5.6	– 4.57	41,200
UHPC-b	5	249	200.6	1.9	– 5.24	41,700

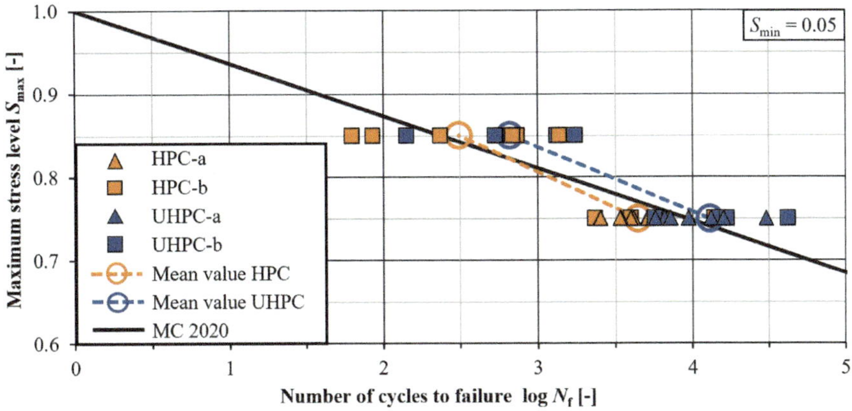

Fig. 7 Numbers of cycles to failure for the HPC and UHPC specimens (from [29], licensed under CC-BY 4.0)

influence on the fatigue resistance. No significant difference between the batches was identified by ANOVA (p-value $\gg 0.05$) for either couple of batches (HPC-a, HPC-b respectively UHPC-a, UHPC-b). Therefore, the mean number of cycles to failure of each concrete was calculated taking into account the single values of both batches at this stress level.

Figure 7 shows that the UHPC specimens reached higher mean numbers of cycles to failure compared to the HPC specimens at both stress levels tested. This difference of the mean values was again analysed by means of ANOVA. The difference was found to be statistically significant (p-value $\ll 0.05$) at the lower stress level $S_{max} = 0.75$. This means that the better fatigue resistance of the UHPC is significant and, therefore, confirmed. High scattering of the results is visible at the higher stress level $S_{max} = 0.85$ in Fig. 7. Therefore, it is not surprising that the difference of the mean values was found to be not statistically significant (p-value $\gg 0.05$) at this stress level.

In addition to the numbers of cycles to failure, the damage indicators strain and stiffness were investigated comparatively for both HPC and UHPC. The mean values of the total strain growth at maximum and minimum stress ($\Delta\varepsilon_{max}^{0.0-1.0}$, $\Delta\varepsilon_{min}^{0.0-1.0}$) as well as the gradients of the strain development in phase II ($\text{grad}\varepsilon_{max}^{0.2-0.8}$, $\text{grad}\varepsilon_{min}^{0.2-0.8}$) are displayed together with the gradient of the stiffness development in phase II ($\text{grad}E_s^{0.2-0.8}$) and the percentile reduction of stiffness within the complete fatigue process ($\Delta E_s^{0.0-1.0}$) in Table 7 as mean values. Hereby, the gradients were determined from a linear regression analysis [29].

It can be seen from Table 7 that the HPC specimens exhibited a higher total growth of the maximum and minimum strain than the UHPC specimens at both stress levels, although the HPC specimens reached lower numbers of cycles to failure. This correlates with the steeper gradients of strain of the HPC in phase II.

The single values of the gradients of the maximum strain development in phase II with respect to the numbers of cycles to failure are additionally shown in Fig. 8a.

Table 7 Mean values of parameters of fatigue damage indicators investigated (from [29], licensed under CC-BY 4.0)

Concrete			HPC-a	HPC-b	UHPC-a	UHPC-b
0.85/0.05	$\Delta\varepsilon_{max}^{0.0-1.0}$	[‰]	–	0.87	–	0.52
	$\Delta\varepsilon_{min}^{0.0-1.0}$	[‰]	–	0.50	–	0.22
	$\text{grad}\varepsilon_{max}^{0.2-0.8}$	[–]	–	2.57×10^{-3}	–	4.67×10^{-5}
	$\text{grad}\varepsilon_{min}^{0.2-0.8}$	[–]	–	1.59×10^{-3}	–	2.38×10^{-5}
	$\Delta E_{s}^{0.0-1.0}$	[%]	–	15.97	–	7.36
	$\text{grad}E_{s}^{0.2-0.8}$	[MPa]	–	– 16.38	–	– 2.61
0.75/0.05	$\Delta\varepsilon_{max}^{0.0-1.0}$	[‰]	1.21	1.17	0.99	0.78
	$\Delta\varepsilon_{min}^{0.0-1.0}$	[‰]	0.81	0.71	0.41	0.44
	$\text{grad}\varepsilon_{max}^{0.2-0.8}$	[–]	1.25×10^{-4}	1.16×10^{-4}	4.10×10^{-5}	2.92×10^{-5}
	$\text{grad}\varepsilon_{min}^{0.2-0.8}$	[–]	1.63×10^{-4}	7.66×10^{-5}	2.31×10^{-5}	1.82×10^{-5}
	$\Delta E_{s}^{0.0-1.0}$	[%]	16.62	21.80	16.08	9.52
	$\text{grad}E_{s}^{0.2-0.8}$	[MPa]	– 0.78	– 0.75	– 0.30	– 0.14

The gradients of strain decreased with decreasing stress levels, which was expected. Furthermore, the regression line of the UHPC specimens is located below the regression line of the HPC specimens, which corresponds to the findings in [15, 17, 19]. This means that smaller gradients of strain or, rather, smaller increases in strain per load cycle occurred for the UHPC compared to the HPC specimens for the same number of cycles to failure.

The mean gradients of stiffness in phase II were steeper for the HPC specimens than for UHPC specimens (cf. Table 7). The gradients of stiffness also decreased with the decreasing stress level as shown in Fig. 8b. The regression line of the UHPC is located below that of the HPC in the range of log $N_f < 4$. Thus, a lower stiffness reduction per load cycle was found for the UHPC specimens compared to the HPC ones for the same number of cycles to failure, which correlates to the findings in [19]. Without the high scattering of batch UHPC-a, this statement might have also been drawn for log $N_f > 4$. In addition, Table 7 shows that the HPC specimens exhibited a higher percentile reduction of stiffness compared to the UHPC ones, although the HPC reached a smaller mean number of cycles to failure.

Overall, the damage indicators revealed a deviating fatigue behaviour of the two concrete types investigated. The UHPC specimens with their higher compressive strengths showed a better fatigue resistance than the HPC ones. This was a surprising result because the approaches of the current standards and guidelines for fatigue design expects a lower fatigue resistance with increasing compressive strength. Further details of the analysis can be found in [29].

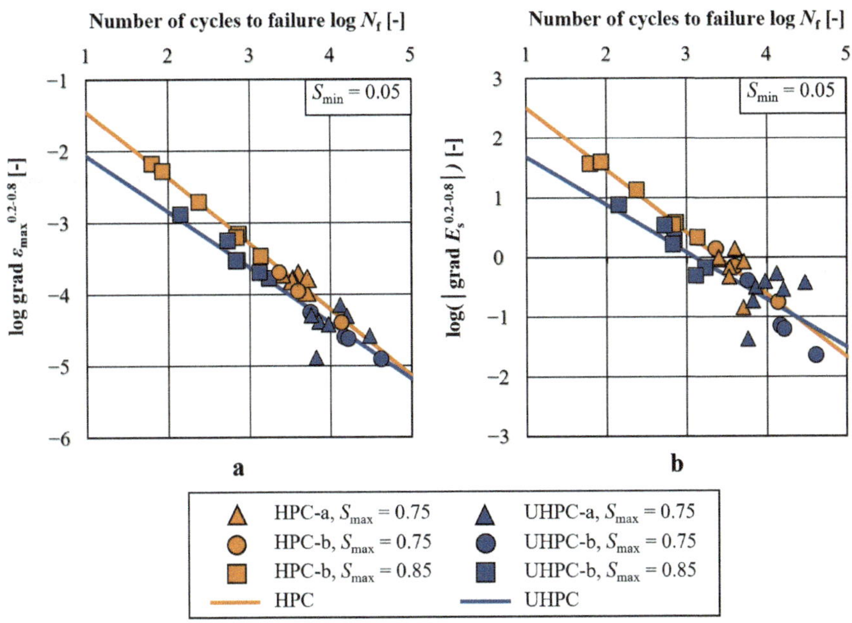

Fig. 8 HPC, UHPC: **a** log. gradients of maximum strain in phase II related to log. numbers of cycles to failure, **b** log. gradients of stiffness in phase II related to log. numbers of cycles to failure (based on [29], licensed under CC-BY 4.0)

4 Conclusions

The fatigue resistance of concrete is influenced by numerous factors, as shown in the literature. Within the central project of the Priority Programme SPP 2020, the influence of concrete production and specimen preparation on the compressive strength and fatigue resistance (work package 1) and the influence of the concrete strength on the fatigue resistance (work package 2) were thoroughly investigated involving the cooperation with several participating laboratories. The two reference concrete mixtures within the SPP 2020—one high-strength concrete ("RH1") and one ultra-high-strength concrete ("RU1")—were examined.

Regarding the investigation of the influence of the concrete production and specimen preparation, the laboratories produced and prepared specimens which were tested in the central laboratory in Hannover. The results of the tests under monotonically increasing load and compressive fatigue load at $S_{max} = 0.75$, $S_{min} = 0.05$ and $f_t = 1.0$ Hz and, additionally, the microstructural and geometrical characteristics of the concrete specimens were analysed. The main findings can be summarised as follows:

1. The concrete/specimen production techniques used to produce the HPC specimens had a significant influence on their compressive strength. A proper concrete production—indication for a positive influence of high compaction was found as

influence of the technique used—and specimen preparation can lead to a higher concrete strength. By contrast, the production techniques used to produce the UHPC specimens only rarely affected their compressive strengths. This could have been attributed to the higher uniformity of the specimen production techniques among the laboratories which produced the UHPC specimens. In the case of compressive fatigue loading, the production techniques did not significantly affect the mean numbers of cycles to failure reached for either concrete. This result can be traced back to the fact that the stress levels applied in the fatigue tests are generally determined using the reference compressive strength from specimens of the same batch. Thus, the effects of production are implicitly considered in the fatigue loading applied.

2. Inaccurate specimen preparation techniques can negatively influence the compressive strength and its scatter due to geometric deviations of the specimens. In particular, edge breakages caused by poor sawing and improper grinding/polishing had a significant negative influence on the mean value of the compressive strength of the specimens. However, an inaccurate preparation of the HPC specimens did not significantly affect the mean number of cycles to failure. By contrast, there might be a slight influence of the sawing and polishing technique on the mean number of cycles to failure of the UHPC specimens.

In addition, the influence of the compressive strength on the fatigue resistance was investigated comparatively considering the numbers of cycles to failure and the damage indicators strain and stiffness. Fatigue tests were performed for this purpose at two maximum stress levels of $S_{max} = 0.75$ and 0.85 with the same minimum stress level of $S_{min} = 0.05$ and a load frequency of $f_t = 1.0$ Hz.

3. The UHPC specimens reached higher mean numbers of cycles to failure in the comparative investigation than the HPC specimens at both stress levels tested. A negative influence of the higher compressive strength on the fatigue resistance, as assumed in current design rules, was not observed. This corresponds to results obtained in [9, 44].
4. The damage indicators revealed a different material-dependent fatigue behaviour in the comparison of the two reference concrete mixtures. The strain and stiffness indicators of the UHPC specimens showed a less pronounced damage evolution compared to the HPC.
5. A batch influence on the compressive strength was identified, especially for the UHPC. However, this influence was not present in the numbers of cycles to failure. The batch influence was equalised through the determination of the fatigue stress levels by means of the reference compressive strength.

In conclusion, the present study underlines the importance of an accurate determination of the reference compressive strength for the definition of the fatigue stresses based on the stress levels selected, which can equalise possible batch influences due to differences in the concrete production processes. In order to benefit from these positive effects, the reference compressive strength must be determined immediately prior to the start of fatigue testing. Furthermore, accurate prepared specimens allow the collection of results with a lower scattering. Finally, the higher fatigue sensitivity for higher strength concretes, i.e. the negative influence of the compressive strength on the fatigue resistance assumed and implemented in the actual design standards and guidelines, could not be confirmed. Further investigations of the influence of the compressive strength on the fatigue resistance of concrete, taking into account the comprehensive data available in the literature, confirm this result [45]. Based on this data, a basic concept for the fatigue design is presented.

Furthermore, results of a SPP 2020 round robin test with the objective to evaluate the significance of different microstructural analyses techniques with regard to fatigue damage on different scales were recently published in [46].

Associated data is published via the link: https://doi.org/10.25835/l9n63a7u.

References

1. Hohberg, R.: Zum Ermüdungsverhalten von Beton [On the fatigue behaviour of concrete] (in German). Doctoral Thesis, Technische Universität Berlin, Berlin, Germany (2004). https://doi.org/10.14279/depositonce-815 (in German)
2. Oneschkow, N.: Fatigue behaviour of high-strength concrete with respect to strain and stiffness. Int. J. Fatigue **87**, 38–49 (2016). https://doi.org/10.1016/j.ijfatigue.2016.01.008
3. Thiele, M.: Experimentelle Untersuchungen und Analyse der Schädigungsevolution in Beton unter Hochzyklischen Ermüdungsbeanspruchungen. [Experimental investigation and analysis of the damage development in concrete subjected to high-cycle fatigue] (in German). Doctoral Thesis, Technische Universität Berlin, Berlin, Germany (2016)
4. Viswanath, S., LaFave, J.M., Kuchma, D.A.: Concrete compressive strain behaviour and magnitudes under uniaxial fatigue loading. Constr. Build. Mater. **296**, 123718 (2021). https://doi.org/10.1016/j.conbuildmat.2021.123718
5. Kim, J.K., Kim, Y.Y.: Experimental study of the fatigue behavior of high strength concrete. Cem. Concr. Res. **26**, 1513–1523 (1996). https://doi.org/10.1016/0008-8846(96)00151-2
6. Zhao, G.Y., Wu, P.G., Bai, L.M.: Research on fatigue behavior of high-strength concrete under compressive cyclic loading. In: Proceeding of the 4th International Symposium on Utilization of High-Strength/High-Performance Concrete, Paris, France, 29–31 May 1996, vol. 2, pp. 757–764 (1996)
7. Bennett, E.W., Muir, S.E.S.J.: Some fatigue tests of high-strength concrete in axial compression. Mag. Concr. Res. **19**, 113–117 (1967)
8. Petković, G., Rosseland, S., Stemland, H.: High Strength Concrete SP3—Fatigue. In Report 3.2: Fatigue of High Strength Concrete; SINTEF Structural Engineering—FCB: Trondheim, Norway (1992)
9. Kono, S., Hasegawa, H., Mori, K., Ichioka, Y., Sakashita, M., Watanabe, F.: Low cycle fatigue characteristics of high strength concrete. In: Proceedings of the 8th International Symposium on Utilization of High-Strength and High-Performance Concrete, Tokyo, Japan, 27–29 October 2008, pp. 616–621
10. Klausen, D.: Festigkeit und Schädigung von Beton bei Häufig Wiederholter Beanspruchung. [Strength and damage of concrete subjected to often repeated loading] (in German). Doctoral Thesis, Technische Hochschule Darmstadt, Darmstadt, Germany (1978)
11. Holmen, J.O.: Fatigue of concrete by constant and variable amplitude loading. In: ACI Symposium Publication, vol. 75, pp. 71–110 (1982). https://doi.org/10.14359/6402
12. Dyduch, K., Szerszen, M., Destrebecq, J.-F.: Experimental investigation of the fatigue strength of plain concrete under high compressive loading. Mat. Struct. **27**, 505–509 (1994)

13. Hümme, J., von der Haar, C., Lohaus, L., Marx, S.: Fatigue behaviour of a normal-strength concrete—number of cycles to failure and strain development. Struct. Concr. **17**, 637–645 (2016). https://doi.org/10.1002/suco.201500139
14. Isojeh, B., El-Zeghayar, M., Vecchio, F.J.: Concrete damage under fatigue loading in uniaxial compression. ACI Mater. J. **114**, 225–235 (2017). https://doi.org/10.14359/51689477
15. Wefer, M.: Materialverhalten und Bemessungswerte von Ultrahochfestem Beton unter Einaxialer Ermüdungsbeanspruchung. [Material behaviour and design values of ultra-high-strength concrete subjected to uniaxial fatigue loading] (in German). Doctoral Thesis, Leibniz University Hannover, Hannover, Germany (2010)
16. Oneschkow, N.: Analyse des Ermüdungsverhaltens von Beton anhand der Dehnungsentwicklung. [Analysis of the fatigue behaviour of concrete with respect to the development of strain] (in German). Doctoral Thesis, Leibniz Universität Hannover, Hannover, Germany (2014). https://doi.org/10.15488/357
17. Petković, G.: Properties of concrete related to fatigue damage with emphasis on high strength concrete. Doctoral Thesis, University of Trondheim, Trondheim, Norway (1991)
18. Scheiden, T., Oneschkow, N.: Influence of coarse aggregate type on the damage mechanism in high-strength concrete under compressive fatigue loading. Struct. Concr. **20**, 1212–1219 (2019). https://doi.org/10.1002/suco.201900029
19. Do, M.T., Chaallal, O., Aïtcin, P.C.: Fatigue behavior of high-performance concrete. J. Mater. Civ. Eng. **5**, 96–111 (1993). https://doi.org/10.1061/(ASCE)0899-1561(1993)5:1(96)
20. Sparks, P.R.: The influence of rate of loading and material variability on the fatigue characteristics of concrete. Int. Concr. Abstr. Portal **75**, 331–342 (1982)
21. Urban, S., Wagner, R., Strauss, A., et al.: Monitoringbasierte Lebenszeitabschätzung von Betonstrukturen: Forschungsprojekt MOSES [Monitoring-based lifetime assessment of concrete structures—research project MOSES] (in German). Beton- Stahlbetonbau **108**, 630–640 (2013)
22. Ferraris, C.F.: Concrete mixing methods and concrete mixers: state of the art. J. Res. Natl. Inst. Stand. Technol. **106**, 391–399 (2001). https://doi.org/10.6028/jres.106.016
23. DIN EN 1992-2:2010-12. Eurocode 2: Design of Concrete Structures—Part 2: Concrete Bridges—Design and Detailing Rules; German version EN 1992-2:2005 + AC: 2008. German Institute for Standardisation, Berlin, Germany (2010). https://doi.org/10.31030/1723952
24. *fib*—Fédération Internationale du Béton (ed.): *fib* Model Code for Concrete Structures (2020), Lausanne (2024)
25. König, G., Danielewicz, I.: Ermüdungsfestigkeit von Stahlbeton- und Spannbetonbauteilen mit Erläuterungen zu den Nachweisen gemäß CEB-FIP Model Code 1990. [Fatigue strength of reinforced and prestressed concrete components with explanations of the verifications according to CEB-FIP Model Code 1990] (in German); Deutscher Ausschuss für Stahlbeton; German Committee for Reinforced Concrete Issue 439. Ernst & Sohn, Berlin, Germany (1994). https://doi.org/10.2366/3714933
26. Oneschkow, N., Lohaus, L.: Zum Ermüdungsnachweis von druckschwellbeanspruchtem Beton, Teil 1: Struktur des Ermüdungsnachweises [About the design concept for compressive fatigue loading of concrete, Part 1: Structure of the fatigue design concept] (in German). Beton-Stahlbetonbau **112**, 530–540 (2017). https://doi.org/10.1002/best.201700024
27. Oneschkow, N., Lohaus, L.: Zum Ermüdungsnachweis von druckschwellbeanspruchtem Beton, Teil 2: Sicherheitsüberlegungen und Potenzial für Weiterentwicklungen [About the design concept for compressive fatigue loading of concrete, Part 2: Safety considerations and potential of further developments] (in German). Beton- Stahlbetonbau **112**, 611–622 (2017). https://doi.org/10.1002/best.201700026
28. Schmidt, M., Fehling, E., Fröhlich, S., Thiemicke, J. (eds.): Nachhaltiges Bauen mit Ultrahochfestem Beton: Ergebnisse des Schwerpunktprogrammes 1182 gefördert durch die Deutsche Forschungsgemeinschaft (DFG) [Sustainable building with ultra-high-performance concrete: Results of the German Priority Programme 1182 funded by Deutsche Forschungsgemeinschaft (DFG)] (in German). In: Schriftenreihe Baustoffe und Massivbau—Structural Materials and Engineering Series, vol. 22. Kassel University Press, Kassel, Germany (2014)

29. Basaldella, M., Jentsch, M., Oneschkow, N., Markert, M., Lohaus, L.: Compressive fatigue investigation on high-strength and ultra-high-strength concrete within the SPP 2020. Materials **15**, 3793 (2022). https://doi.org/10.3390/ma15113793
30. DIN EN 12390-3:2019-10. Testing Hardened Concrete—Part 3: Compressive Strength of Test Specimens; German Version of EN 12390-3:2019. German Institute for Standardisation: Berlin, Germany (2019). https://doi.org/10.31030/3045735
31. DIN EN 12390-2:2009-08. Testing Hardened Concrete—Part 2: Making and Curing Specimens for Strength Tests; German Version EN 12390-2:2009. German Institute for Standardisation, Berlin, Germany (2009). https://doi.org/10.31030/1523171
32. DIN EN 12390-2:2019-10. Testing Hardened Concrete—Part 2: Making and Curing Specimens for Strength Tests; German version EN 12390-2:2019. German Institute for Standardisation, Berlin, Germany (2019). https://doi.org/10.31030/3045734
33. DIN 1045-2:2008-08. Concrete, Reinforced and Prestressed Concrete Structures—Part 2: Concrete—Specification, Properties, Production and Conformity—Application rules for DIN EN 206-1. German Institute for Standardisation, Berlin, Germany (2008). https://doi.org/10.31030/1453177
34. DIN EN 1990:2021-10. Eurocode 0: Basis of Structural Design; German version EN 1990-2002 + A1:2005 + A1:2005/AC:2010. German Institute for Standardisation, Berlin, Germany (2010). https://doi.org/10.31030/3291403
35. DIN EN 206:2021-06. Concrete—Specification, Performance, Production and Conformity. German version EN 206:2013+A2:2021. German Institute for Standardisation, Berlin, Germany (2021). https://doi.org/10.31030/3198971
36. Deutscher Ausschuss für Stahlbeton. German Committee for Reinforced Concrete. DAfStb-Richtlinie Ultrahochfester Beton, Teil 1 Bemessung und Konstruktion; DAfStb Guideline for Ultra-High-Strength Concrete, Part 1 Design and Construction—Draft, Version 14.08.2023; DAfStb, Berlin, Germany (2023) (in German)
37. Basaldella, M., Oneschkow, N., Lohaus, L.: Influence of the specimen production and preparation on the compressive strength and the fatigue resistance of HPC and UHPC. Mater. Struct. **54**, 99 (2021). https://doi.org/10.1617/s11527-021-01696-9
38. DIN EN 12390-1:2012-12. Testing hardened concrete—Part 1: Shape, dimensions and other requirements for specimens and moulds. German Institute for Standardisation, Berlin, Germany (2021). https://doi.org/10.31030/1917324
39. DIN EN ISO 25178:2016-12. Geometrical Product Specifications (GPS)—Surface Texture: Areal—Part 1: Indication of Surface Texture. German Institute for Standardisation, Berlin, Germany (2016). https://doi.org/10.31030/2324720
40. ASTM C215-19: Standard Test Method for Fundamental Transverse, Longitudinal, and Torsional Resonant Frequencies of Concrete Specimens. American Society for Testing and Materials (2019). https://doi.org/10.1520/C0215-19
41. BS 1881-209:1990-04-30: Testing Concrete—Part 209: Recommendations for the Measurement of Dynamic Modulus of Elasticity. British Standards Institution, London, United Kingdom (1990)
42. Popovic, S.: Ultrasonic testing of concrete structures. Mater. Eval. **63**, 50–55 (2005)
43. Hsu, T.T.: Fatigue of plain concrete. ACI J. Proc. **78**, 292–305 (1981)
44. Basaldella, M., Kern, B., Oneschkow, N., Lohaus, L.: Zum festigkeitsabhängigen Ermüdungswiderstand von Beton [On the Strength-Dependent Fatigue Resistance of Concrete] (in German), DAfStb-Heft 648. Beuth Verlag, Berlin, Germany (2024)
45. Basaldella, M.: Data-based design concept for compressive concrete fatigue. Doctoral Thesis, Leibniz University Hannover, Hannover, Germany (submitted, 2025)
46. Engelhardt, M., Kalytta-Mewes, A., Volkmer, D., Lohmann, J., Ritter, M., Schaan, G., Schmidt-Döhl, F., Abubakar Ali, M., Basaldella, M., Haist, M., Kern, B., Lohaus, L., Oneschkow, N., Rozanski, C., Timmermann, T., Grosse, C.U., Birtel, V., Garrecht, H., Madadi, H., Markert, M., Steeb, H.: Micro- and nanostructural investigations of high and ultra-high performance concrete under fatigue. Int. J. Fatigue **199**, 109038 (2025). https://doi.org/10.1016/j.ijfatigue.2025.109038

Composition Influenced Damage Development in High-Strength Concrete Due to Cyclic Loading

Tim Timmermann, Stefan Löhnert, and Nadja Oneschkow

Abstract In this chapter, results concerning the influence of the concrete composition on the compressive fatigue behaviour are presented and a mesoscopic modelling approach for fatigue damage is developed to capture these influences within numerical simulations. The influences of coarse aggregates generally, the substitution of basalt by granite coarse aggregate, the addition of silica fume and the variation of the water to cement (*w/c*) ratio were investigated systematically. The results show that the presence of coarse aggregates can influence the fatigue behaviour of concretes in a negative way at higher stress levels and in a positive way at lower stress levels compared to mortars. Silica fume improves the fatigue behaviour of concrete and mortar strongly. An increase of the *w/c* ratio and, thus, an increase in the porosity reduces the fatigue resistance of concrete and mortar significantly. The acoustic emission technique used additionally gives further valuable strain-independent information about the damage processes occurring on different scales. A modelling approach and simulations of compressive fatigue damage are additionally presented. They are based on a gradient-enhanced damage formulation with a reduction of the material strength dependent on the strain history.

Keywords Compressive fatigue behaviour · Damage process · Acoustic emission · Gradient-enhanced damage · Effective material behaviour

1 Introduction

High-strength concrete has a denser binder matrix, an increased compressive strength and improved interfacial transition zones compared to normal-strength concrete. However, contradictory findings are documented in the literature concerning the

T. Timmermann · N. Oneschkow (✉)
Institute of Building Materials Science, Leibniz University Hannover, Hannover, Germany
e-mail: n.oneschkow@baustoff.uni-hannover.de

S. Löhnert
Institute of Mechanics and Shell Structures, TUD Dresden University of Technology, Dresden, Germany

N. Oneschkow and L. Lohaus (eds.), *Cyclic Deterioration of High-Performance Concrete*, https://doi.org/10.1007/978-3-032-19394-0_3

influence of the concrete strength on the compressive fatigue resistance [1–4]. Nevertheless, only a few investigations have focused on the influence of the concrete composition or, rather, concrete microstructure on the fatigue resistance and, furthermore, the damage processes in the microstructure.

Mehmel and Kern [5] and Shah and Chandra [6] found that the interaction between the aggregates and the cement matrix is essential for the fatigue behaviour of normal-strength concrete. Breitenbücher et al. [7] observed for normal-strength concrete that increased differences between the modulus of elasticity of the coarse aggregate and the hardened cement paste lead to lower compressive fatigue resistance. Furthermore, the size and volume of the coarse aggregates influences the magnitude of peak stresses in the microstructure [5]. Mehmel and Kern assumed that these peak stress areas lead to damage and deterioration in the cement matrix [5]. A highly inhomogeneous three-dimensional stress distribution is also confirmed by results in [8]. To the best of author's knowledge, no results are documented concerning the influence of the water to cement (*w/c*) ratio and silica fume on the concrete compressive fatigue behaviour.

Concerning the fatigue damage process, indications were found for a diffuse, spread compressive damaging effect in the cement matrix and a localised, vertically orientated tensile damage [8]. It was previously hypothesised in [9] that damage also occurs on extremely small scales as precursors of damage on larger scales. Rybczynski et al. (see Sect. 3.3 "Characterization of Microstructural Changes Caused by Fatigue Damage" in the chapter by S. Rybczynski et al., this volume) found needle-shaped regions of lower density in the binder matrix of ultra-high-strength concrete using transmission electron microscopy and, thus, seem to confirm the findings and hypothesis of [8, 9]. The observation of the damage development on such small scales is generally challenging, especially due to fatigue loading, because fatigue experiments have to be stopped at a defined stage of damage, and samples prepared and analysed at great expense in order to get time-discrete insights. However, a promising continuously measurable and strain-independent damage indicator is the acoustic emission (AE). The analysis of AE hits has so far been applied to statically loaded reinforced concrete components (e.g. [10]), whereby, the cracks were often visible to the naked eye in these cases. Its applicability for the analyses of the damage development on a much smaller scale in the plain concrete microstructure due to fatigue loading has not yet been examined.

The modelling and simulation of the fatigue processes within the concrete structure on a mesoscale is a challenging task. Many models exist that are applicable only to certain cases, for example only a few load cycles or only loads with invariable directions. The goal, however, must be to develop a predictive model that is capable of reflecting all important observable effects, such as the influence of the change of the loading direction or sequence effects. A promising starting point is a microplane approach [11], which can capture the anisotropy due to damage within the concrete structure on a mesoscale. The combination with Drucker-Prager-Cap plasticity [12, 13] leads to a model which is able to predict the behaviour of concrete under monotonic loading or a few load cycles quite well. However, these models are not sufficient for the prediction of fatigue processes in concrete on a mesoscale because they do not contain material behaviour explicitly dependent on the strain history.

A microplane model has recently been proposed that captures the microscale and mesoscale behaviour of concrete effectively for the analysis of the fatigue behaviour of concrete structures on the macroscale [14]. Such models are difficult to calibrate and validate for arbitrary fatigue loads including multiple load levels and changes of the loading direction. Additionally, they do not contribute to the understanding of the mesostructural behaviour of concrete under fatigue loads.

In the following, results of conjoint research are presented. Systematically defined concrete compositions and their corresponding mortars were investigated regarding their fatigue behaviour and the damage development in the microstructure. Accordingly, continuously measurable strain-dependent damage indicators and the AEs, were analysed together with time discrete visualisations of damage in the microstructure to gain new knowledge about the damage development due to compressive fatigue loading and its influenceability. Furthermore, detailed analyses of the characteristics of AE hits were applied to concrete fatigue damage for the first time [15]. A model based on gradient-enhanced damage combined with a strain history-dependent reduction of the material strength is developed for the simulation of the fatigue damage development in concrete. This model serves as a starting point for the prediction of fatigue processes in concrete.

2 Investigation Programme

The high-strength reference concrete RH1 of the SPP 2020 (see Sect. 2.1 "Concrete Composition" in the chapter by M. Basaldella et al., this volume), here named RH1-B, and three additional high-strength concretes, whose compositions were based on the RH1-B but were systematically adjusted, were investigated. Concrete RH1-B-Si had the same composition as RH1-B but the cement was partially replaced by silica fume, following the approach of the same paste film thickness [16, Chapter by G. Basutkar et al., this volume]. In addition, a high-strength concrete RH1-B-w with an increased *w/c* ratio of 0.5 and, thus, a higher porosity as well as a lower compressive strength was considered. Regarding concrete RH1-G, the basalt aggregate was substituted by granite to gain different mechanical properties of the coarse aggregate. The compositions of the four high-strength concretes investigated are comparatively presented in Table 1.

Furthermore, corresponding mortars, which were produced by sieving off the concrete components with grain sizes $\geq$ 2 mm, were investigated. Additional information, also concerning the properties of the coarse aggregates used, can be found in [17]. Table 2 shows the mean values of the 28-day compressive strength $f_{cm,cube}$ determined according to DIN EN 12390-3 [18] and the modulus of elasticity E_{cm} according to DIN EN 12390-13 [19] of the four concretes and three corresponding mortars investigated. The mean reference compressive strengths $f_{cm,ref}$ of the concretes and mortars were determined force-controlled with a test velocity of 0.5 MPa/s immediately before conducting the fatigue tests on cylinders of *h*/*d* = 180/60 mm. The values of $f_{cm,ref}$ are additionally presented in Table 2.

Table 1 Concrete composition (from [17], licensed under CC-BY 4.0)

Component	RH1-B	RH1-B-Si	RH1-B-w	RH1-G
Portland Cement (CEM I 52.5 R HS/NA) [kg/m^3]	500	447	461	500
Microsilica [kg/m^3]	/	45	/	/
Quartz sand (0/0.5 mm) [kg/m^3]	75	75	69	75
Sand (0/2 mm) (Tuendern) [kg/m^3]	850	850	783	850
Basalt 2/5 mm (Oelberg) [kg/m^3]	350	350	322	/
Basalt 5/8 mm (Oelberg) [kg/m^3]	570	570	525	/
Granite 2/5 mm (Glensanda) [kg/m^3]	/	/	/	350
Granite 5/8 mm (Glensanda) [kg/m^3]	/	/	/	570
Superplasticiser [kg/m^3]	5.00	4.90	0.49	5.00
Stabiliser [kg/m^3]	2.85	2.80	/	2.85
w/c ratio; w/c_{eq} ratio [–]	0.35	0.35	0.50	0.35

Table 2 28-day compressive strength, modulus of elasticity and reference compressive strength (mean values)

Concrete	RH1-B	RH1-B-Si		RH1-B-w	RH1-G
	B1	B1	B2	B1	B1
$f_{cm,cube}$ [MPa]	113	133	129	71	109
$f_{cm,ref}$ [MPa]	96	123	120	58	110
E_{cm} [MPa]	40,000	44,400	–[a]	33,700	35,600
Mortar	M-B	M-B-Si		M-B-w	
	B1	B1		B1	
$f_{cm,cube}$ [MPa]	108	114		78	
$f_{cm,ref}$ [MPa]	97	116		75	
E_{cm} [MPa]	34,900	38,200		23,000	

[a] Not determined

The modulus of elasticity decreases due to the granite aggregate. The silica fume increases the modulus of elasticity and the compressive strength, whereby the effect on the compressive strength is more pronounced for the concrete compared to the mortar, revealing its positive effect on the interfacial transition zone. The higher w/c ratio leads to a decreased compressive strength and modulus of elasticity.

The fatigue investigations were conducted on cylindrical specimens with h/d = 180/60 mm [17]. The formworks of the concrete specimens were removed 7 days after concreting and the specimens were stored in standard climate (20 °C/65% R.H.) until testing. The mortar specimens were stored for 14 days in standard climate before the formworks were removed in order to avoid increased microcracking due to shrinkage. Afterwards, they were stored wrapped in plastic foil for a further 14 days and then for at least another 28 days without the plastic foil in standard climate (gentle drying) until testing. The test surfaces of all specimens were plane-parallel ground and polished to achieve a uniform stress distribution.

Two maximum stress levels of either $S_{max} = \sigma_{max}/f_{cm,ref} = 0.85$ or $S_{max} = 0.70$ with the same minimum stress level of $S_{min} = \sigma_{min}/f_{cm,ref} = 0.05$ were used in the compressive fatigue tests. The test frequency was $f_t = 1.0$ Hz in all tests. Two types of tests were conducted: full tests, where the specimens were fatigue loaded until failure, and partial tests where the fatigue loading was stopped in the middle of phase I or II or, instead, shortly before fatigue failure in phase III in order to prepare thin sections for subsequent microscopic analyses of damage [20]. The axial deformations were measured continuously using three laser distance sensors positioned on the circumference of the specimen at angles of 0°, 120° and 240°. In addition, the axial force, the axial displacement of the actuator and the temperature of the specimen's surface at mid-height were measured. Six acoustic sensors with a wide-band frequency response within the range of 250–1600 kHz were positioned at 60° from one another, alternating in the upper and lower third of the specimen.

The damage indicators maximum and minimum strain, stiffness, dissipated energy and AE hits were analysed. The stiffness was determined based on the secant modulus in the decreasing branch of the hysteresis loops. The dissipated energy describes the viscoelastic material behaviour within the fatigue process and, thus, the deformation capability of the material. Different parameters of each damage indicator such as the gradients of strain, stiffness and dissipated energy in phase II, the percentile reduction of stiffness or the average dissipated energy per load cycle in phase II, were analysed comparatively. More information concerning the analyses can be found in [17].

In addition, the AE hits were analysed in detail using parameter- and signal-based analyses. Here, the focus was on the development of the cumulated AE hits and the characteristics of the AE hits. Consequently, a clustering method using the Gaussian mixture modelling was successfully used based on the distribution of AE hits in the RA/AF space (RA = rise time/peak amplitude; AF = counts/duration), as shown in Fig. 1, to achieve a physically based classification of the damage mechanisms [20]. The AE hits in cluster I are characterised by low AE energy, and can be physically traced back to a tensile-like damage, whereas the AE hits in cluster II have a much higher AE energy, and result from a shear/friction-like damage [15, 21].

Furthermore, digitally transmitted light microscopy was applied to thin sections. Hence, the letter were prepared from unloaded specimens (reference) and specimens that had been previously subjected to compressive fatigue loading (partial tests). The thin sections were impregnated using blue epoxy resin for a better visualisation of the damage in the microstructure. Instead of microcracks expected, bluish impregnated areas were found, which represent regions with a higher porosity or more microdefects. These bluish areas change in quantity, size and area as a result of the compressive fatigue loading. Selected regions within these bluish areas were further evaluated using scanning electron microscopy [20].

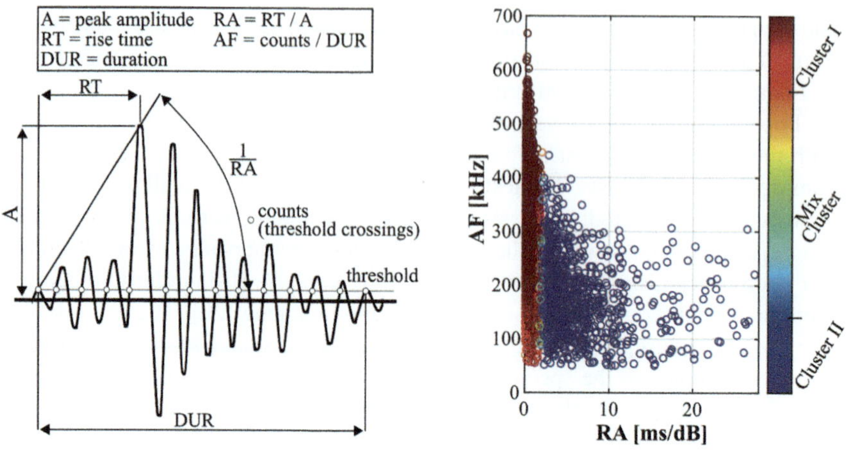

a b

Fig. 1 **a** Acoustic emission (AE) signal parameters following [10] (from [20], licensed under CC-BY 4.0) and **b** AE hits in the RA/AF space and assignment to clusters (exemplarily)

3 Experimental Results

The mean values of the numbers of cycles to failure are shown in Fig. 2. The concrete with silica fume RH1-B-Si reached the highest mean numbers of cycles to failure, followed by the reference concrete RH1-B and the concrete with the higher*w/c* ratio, RH1-B-w. The same order of numbers of cycles to failure was found for the mortars. The concretes reached lower numbers of cycles to failure at $S_{max} = 0.85$ compared to their corresponding mortars, except for RH1-B, respectively M-B, where the mean N_f was equal. By contrast, all concretes reached higher numbers of cycles to failure at $S_{max} = 0.70$ compared to the respective mortar. Furthermore, the increase in mean N_f due to the decreased stress level is higher for the concretes, thus, a more pronounced stress level dependency exists. The substitution of basalt aggregate by granite led to a higher mean N_f at $S_{max} = 0.85$ but a lower value at $S_{max} = 0.70$. The increased number of cycles to failure for the concrete and mortar with silica fume was accompanied by an increase in the scatter of results.

Figure 3a shows the averaged developments of stiffness and Fig. 3b, the developments of cumulated AE hits of the concretes exemplarily for $S_{max} = 0.70$. The averaged developments are only used for the purpose of graphical representation, knowing that the gradients of the curves are distorted in those graphs due to the averaging process (for further information, see [17]).

With respect to the developments of the damage indicators, the well-known relation of steeper gradients of strain, stiffness and dissipated energy in phase II for concrete or mortar that reached lower numbers of cycles to failure was found in these investigations. Furthermore, the analyses of the developments of strain, stiffness and dissipated energy clearly showed that the increase of *w/c* ratio degrades

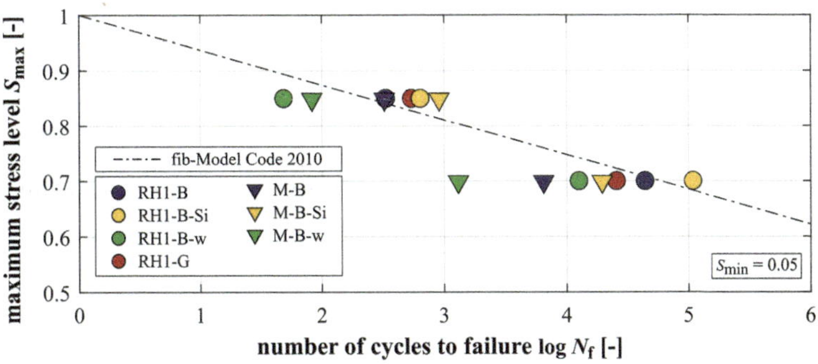

Fig. 2 Mean numbers of cycles to failure for the concretes and mortars (from [17], licensed under CC-BY 4.0)

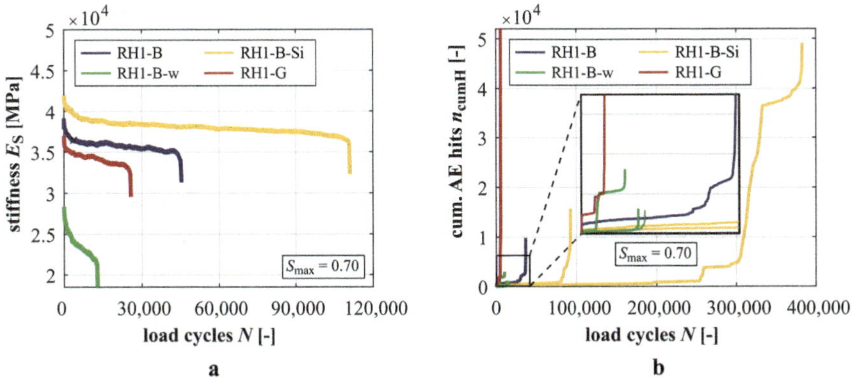

Fig. 3 **a** Averaged developments of stiffness and **b** developments of cumulated AE hits of the concretes investigated at $S_{max} = 0.70$ (based on [17], licensed under CC-BY 4.0)

the fatigue resistance of both mortar and concrete, while the addition of silica fume strengthens in both cases, leading to an improved fatigue resistance [17].

The gradients of strain and stiffness at $S_{max} = 0.85$ were higher for the concretes compared to their mortars, which is reversed at $S_{max} = 0.70$, revealing that the presence of basalt coarse aggregate impairs the fatigue behaviour at higher stress levels and improves it at lower stress levels. In comparison to the granite aggregate with its lower modulus of elasticity, the basalt aggregate led to steeper gradients of strain, stiffness and dissipated energy at the higher stress level, whereby the gradients were flatter at the lower stress level. Thus, the (comparatively stiffer) basalt coarse aggregate impairs the fatigue resistance at high stress levels due to the comparatively increased inhomogeneous stress distribution and improves it at low stress levels due to its fatigue resistance. Overall, the results indicate that the inhomogeneous stress distribution due to stiff coarse aggregates predominates the fatigue behaviour of concrete at high stress levels, whereby the attraction of stresses by a fatigue-resistant coarse aggregate predominates at low stress levels. All mortars showed higher magnitudes of strains at maximum peak stress compared to their corresponding concretes, which is also associated with the higher volume of mortar matrix in the respective specimen.

The averaged dissipated energy in phase II of the mortars was always higher compared to their corresponding concretes at both stress levels, thus, a higher deformation capability exists. The developments of cumulated AE hits often showed a three-phase development, whereby phase III started at lower numbers of cycles compared to phase III of the strain developments. Surprisingly, several developments of cumulated AE hits showed vertical courses, in which a huge number of AE hits arose, especially for the concrete with silica fume and at the lower stress level (cf. Figure 3b). Such vertical courses were not detectable in the developments of strain and stiffness [17]. The earlier onset of phase III and the vertical courses of AE hits are strong indications of damage occurring on extremely small scales as precursors of damage on larger scales. Furthermore, extremely high numbers of AE hits were detected for the concrete with granite coarse aggregate at the lower stress level, which can be attributed to a kind of “crackling” [22].

Overall, the number of AE hits arising due to fatigue loading is material- and stress level-dependent and, thus, the characteristics of the AE signals have to be additionally considered in order to describe the kind or quality of damage. Therefore, the AE hits were clustered based on the parameters RA and AF and subsequently analysed regarding their appearance in the fatigue process. Further information can be found in [20]. The AE hits in cluster I are generally characterised by much lower AE energy compared to the AE hits of cluster II. This indicates that hits in cluster I occur due to damage on a smaller scale compared to those in cluster II. The mixed cluster contains AE hits that cannot be clearly assigned to cluster I or II. Figure 4 shows the clustered numbers of AE hits for phases I–III exemplarily for the reference concrete RH1-B and both stress levels [20]. As the absolute number of AE hits depends on the numbers of load cycles and, thus, on the stress level, the corresponding percentages of AE hits per phase are added in parentheses.

At both stress levels, the numbers of AE hits in cluster I (absolute and percentage) are much higher (dominating) than those in cluster II and increase from phase to phase. The AE hits in cluster II mainly occur in phase I and especially in phase III. Furthermore, more hits in cluster II arise in phase I and II for the fatigue loading at the higher maximum stress level $S_{\max} = 0.85$ compared to $S_{\max} = 0.75$ (absolute and percentage). In phase III, a slightly higher percentage of AE hits in cluster II is observable at $S_{\max} = 0.75$ compared to $S_{\max} = 0.85$, which indicates a “simpler” propagation of cluster-II-damage. Similar correlations were found for all concretes and mortars analysed, whereby the number of AE hits in the respective cluster and the ratio between the two clusters is different [15]. Comparing the concretes with their corresponding mortars, it was found that the percentages of AE hits in cluster II is less for the mortars in all phases of the fatigue process, which must be related to the lack of coarse aggregate [15].

Further analyses of the characteristics of the AE signals showed that the AE hits of the vertical courses of their developments at the lower stress level (cf. Figure 3b) can

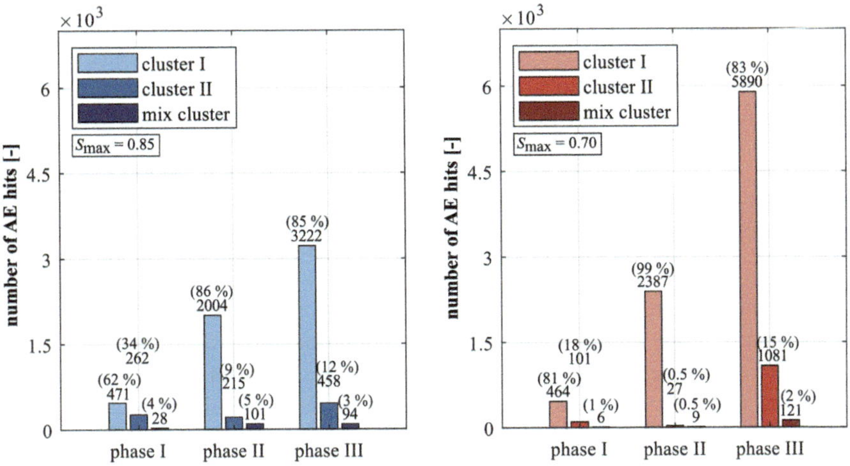

a **b**

Fig. 4 Cluster of AE hits within phases I–III at **a** $S_{max} = 0.85$ and **b** $S_{max} = 0.70$

be assigned to the dominating cluster I, which includes AE hits with low AE energy. Since these vertical courses cannot be seen in the strain and stiffness developments, this is an additional strong indication that the associated damage of hits in cluster I occurs on an extremely small scale.

However, the cumulated AE energy up to the transition from phase II to III is characteristic of concretes or mortars of a special composition and, thus, independent of the maximum stress level [15]. The transition from phase II to III, therefore, marks a state of damage in which the material is so severely damaged that larger-scale damage follows to a greater degree than before, resulting in the global failure after only a very few additional load cycles.

The microscopic analyses of unloaded specimens (reference) and preloaded/damaged specimens showed bluish impregnated areas (areas with increased porosity), which change in quantity, size and area instead of the microcracks expected [20]. Bluish areas already exist in the unloaded specimens. Due to the compressive fatigue loading, the number and size of those areas decrease up to phase I at $S_{max} = 0.85$ or, rather, phase II at $S_{max} = 0.70$, and then start to reincrease up to phase III. Using scanning electron microscopy for analysis on smaller scales, alterations of the microstructure in the form of the increase of voids in the mortar matrix within the fatigue process were found not only inside the bluish areas but also outside these areas with less concentration. This indicates that the mortar matrix suffers widespread degradation in general, with accumulation in highly stressed areas due to the heterogeneity of the microstructure [20].

Overall, the AE hits in the two clusters can be interpretated as belonging to different damage mechanisms which occur depending on the stress level and material due to compressive fatigue loading. Taking considerations and results from the literature additionally into account [5, 6, 8, 9], it is strongly assumed that AE hits in cluster I are due to diffusely distributed damage that begins to develop on a very small scale (sub-microscale). By contrast, AE hits in cluster II are associated with the development of more localised damage on the micro- to mesoscale, in the form of larger-scale damage or cracks, in the diffusely small-scale pre-damaged matrix.

4 Mechanical Model and Simulation

The goal of the mechanical model is to be able to simulate fatigue processes under compressive loads and various loading scenarios in order to eventually allow for simulative lifetime predictions of engineering parts made of concrete using finite elements. The starting point is the balance of momentum in its weak form and for quasistatic loadings

$$\int_{\Omega} \boldsymbol{\sigma} : \delta\boldsymbol{\varepsilon} d\Omega - \int_{\partial\Omega} \boldsymbol{t} \cdot \delta\boldsymbol{u} d\partial\Omega = 0 \tag{1}$$

where Ω is the domain of the body, $\partial\Omega$ is its boundary, $\boldsymbol{\sigma}$ is the stress tensor, $\boldsymbol{t}$ is the traction vector acting on the outer boundaries and $\delta\boldsymbol{\varepsilon}$ and $\delta\boldsymbol{u}$ are the variations of the strain and the displacement. Regarding the mesostructural components, firstly, a linear elastic material model including gradient-enhanced fatigue damage is chosen. The stresses for the gradient-enhanced fatigue damage model can be computed by

$$\boldsymbol{\sigma} = (1 - d)(2\mu\boldsymbol{\varepsilon} + \Lambda tr(\boldsymbol{\varepsilon})\mathbf{1}) \tag{2}$$

where μ and Λ are the Lamé constants, $\boldsymbol{\varepsilon}$ is the strain tensor and d is the scalar-valued damage variable, which is obtained from the classical damage model by Mazars and Pijaudier-Cabot [23]

$$d(\kappa) = 1 - \frac{\kappa_0}{\kappa}(1 - \alpha) - \alpha e^{-\beta(\kappa - \kappa_0)} \tag{3}$$

which is frequently used for concrete. The parameters α and β are purely phenomenological parameters adjusted according to experimental observations. The gradient-enhanced equivalent strain dependent history variable κ has the initial value κ_0. Once it exceeds this value, damage initialises and grows. To avoid mesh dependence, the history variable and, thus, also the amount of damage is a monotonic function of the gradient-enhanced equivalent strain quantity $\overline{\varepsilon}$, which is the solution of the weak form of the inhomogeneous Helmholtz equation

$$\int_{\Omega} (\overline{\varepsilon} - \tilde{\varepsilon}(\boldsymbol{\varepsilon})) \cdot \eta d\Omega + c \int_{\Omega} grad(\overline{\varepsilon}) \cdot grad(\eta) d\Omega = 0 \tag{4}$$

See [24] for its derivation. Here, c is the characteristic length, responsible for the width of a damage band that might form due to localisation. From the numerical point of view, it is important that the size of the finite elements is significantly smaller than c to achieve convergence. The function $\tilde{\varepsilon}$ is a scalar-valued function of the strain tensor $\boldsymbol{\varepsilon}$. It is a local equivalent strain measure to consider different damage characteristics, such as different damage behaviour under compression and tension or under purely volumetric deformations. The most widely used equivalent strain function for concrete is the so-called modified von Mises criterion [25].

$$\tilde{\varepsilon}(\boldsymbol{\varepsilon}) = \frac{(k-1)I_1}{2k(1-2\nu)} + \sqrt{\left(\frac{(k-1)I_1}{2k(1-2\nu)}\right)^2 + \frac{3J_2}{k(1+\nu)^2}} \tag{5}$$

Here, ν is Poisson's ratio, $I_1 = tr(\boldsymbol{\varepsilon})$ is the first invariant of the strain tensor and $J_2 = \frac{1}{2}\boldsymbol{\varepsilon} : \boldsymbol{\varepsilon} - \frac{1}{6}(tr(\boldsymbol{\varepsilon}))^2$ is the second invariant of the strain deviator. The dimensionless parameter k is used to emphasise the volumetric or the isochoric part of the deformation. Regarding concrete, choosing $k = 10$ leads to reasonable results. The actual damage criterion can be written as an inequality

$$f(\overline{\varepsilon}, \kappa, \gamma) = \overline{\varepsilon} - \phi(\gamma)\kappa \leq 0 \tag{6}$$

that needs to be fulfilled pointwise and is used to calculate the updated history variable κ in case the inequality is violated. The function $\phi(\gamma)$ can be regarded as a deformation history-dependent reduction factor for the strength of the material which allows for the simulation of fatigue processes. There are many options for the choice of $\phi(\gamma)$. According to [26] it is set:

$$\phi(\gamma) = \begin{cases} 1 & : \gamma < \gamma_t \\ (1 - a\log(\gamma/\gamma_t))^2 & : \gamma_t \leq \gamma \leq 10^{1/a}\gamma_t \\ 0 & : \gamma > 10^{1/a}\gamma_t \end{cases} \tag{7}$$

The parameters a and γ_t are responsible for the strain history-dependent rate of the reduction of the strength and for the threshold for γ beyond which a reduction of the material strength occurs. The monotonically growing scalar quantity γ describes the accumulated gradient-enhanced strain history and is here defined as

$$\gamma(\overline{\varepsilon}) = \int_0^t H\left(\overline{\varepsilon}\,\dot{\overline{\varepsilon}}\right)\left|\dot{\overline{\varepsilon}}\right| d\tau \tag{8}$$

Here, H is the Heaviside step function and $\dot{\overline{\varepsilon}}$ is the rate of the gradient-enhanced equivalent strain. This model is capable of simulating all three phases of fatigue damage growth, as will be demonstrated in the following section.

5 Numerical Results

In this section, two computational problems are investigated. Firstly, a comparison of the computational and experimental results is performed for the reference concrete RH1-B. Here, the macroscopic strain developments at the maximum stress level of $S_{\text{max}} = 0.85$ are compared. The minimum stress level is set to $S_{\text{min}} = 0.05$ for both the experiments and the simulation. Since this is a low cycle fatigue problem, and the number of cycles to failure is still rather small, it is not necessary to employ special temporal discretisation, such as cycle jump techniques. A cube-shaped material sample with an edge length of 23 mm was modelled for the simulation of the concrete. The mesostructure of the material sample was generated artificially. The mesostructural model generally follows the experimental approach of the coarse aggregate and surrounding mortar. On this scale, the mortar is assumed to be homogeneous. Basalt grains with a diameter $d \geq 2$ mm were modelled explicitly as spherical inclusions. The volume fraction and diameter of the inclusions follows the experimentally determined grain size distribution curve shown in Table 3.

The discretised mesostructural model is depicted in Fig. 5a. Standard Lagrangian tetrahedral finite elements with quadratic shape functions are used to discretise the mesostructure. Symmetry boundary conditions are applied in all directions, and a force-controlled external load is applied to a rigid diaphragm on the top surface. The basalt aggregate inclusions are assumed to behave isotropically and linear elastically without exhibiting damage. This assumption agrees with our own fatigue tests with comparable loads performed on the basalt stone, where no indications of occurring damage were found.

The Lamé constants for basalt and for mortar were obtained from macroscopic experiments performed on basalt and on mortar only, respectively. The damage and fatigue parameters for mortar were obtained by a fitting procedure for the concrete mesostructure to match the results obtained experimentally. Since a conventional least squares fitting procedure led to severe convergence problems, the parameters were found by a series of simulations with varying numbers for each parameter. The set of parameters eventually used for the simulations for both the basalt and the mortar, is provided in Table 4. The damage distribution after 306 load cycles which corresponds to $N/N_{\text{f}} = 0.99$, is shown in Fig. 5b.

The fatigue damage accumulates according to the size of the basalt grains and their distance to each other and, of course, also according to the loading direction.

Table 3 Grain size distribution for the concrete

Diameter of basalt grains (mm)	Number of basalt grains in specimen
2	165
4	16
5	29
8	2

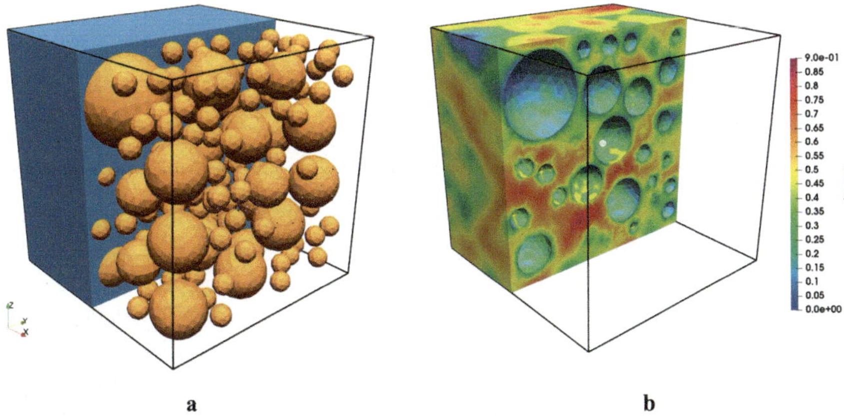

Fig. 5 **a** Mesostructure of the concrete sample (from [27], licensed under CC-BY 4.0) and **b** damage after 306 load cycles

Table 4 Material parameters for the basalt and the mortar

Material parameter	Basalt	Mortar
Shear modulus μ [MPa]	38,682	14,536
Lamé constant Λ [MPa]	53,418	9691
Length scale parameter c [mm^2]		0.3
Damage parameter κ_0 [–]		0.000155
Damage parameter α [–]		0.7
Damage parameter β [–]		2300
Fatigue parameter a [–]		0.022
Fatigue parameter γ_t [–]		0.00015

The gradient-enhanced fatigue damage model leads to a smeared damage distribution between the coarse aggregates, which was also observable in the experimental investigations [20]. Localisation bands develop in a 45° angle to the loading direction applied. Figure 6a depicts the experimentally and computationally determined development of the maximum strain in the loading direction, depending on the number of load cycles. The strain developments obtained experimentally are depicted for single tests performed (blue curves). Of course, the slope of the strain curves and the number of cycles to failure reached vary due to the "natural" scatter of fatigue tests. However, it can be seen that the computational results (red curve) are in good agreement with the experiments in the sense of mean results.

Especially the slope of the maximum strain development as a function of load cycles during phase II of the fatigue process but also the rapid initial damage development during phase I as well as the strains during phase III can be captured nicely. The number of cycles to failure calculated also shows good agreement with the experimental results in terms of a mean value.

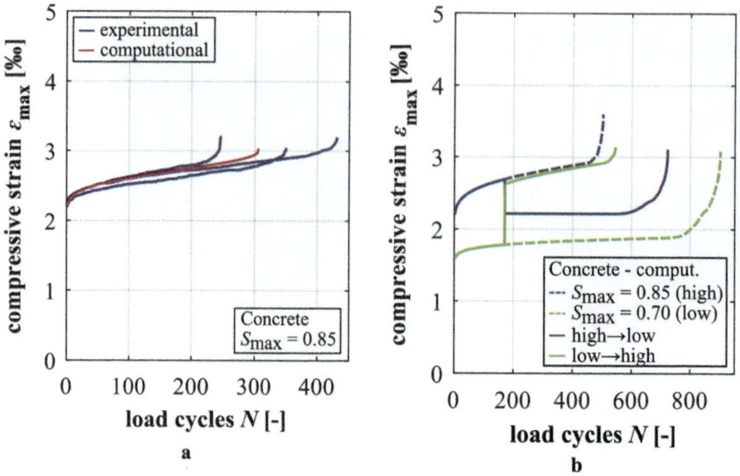

Fig. 6 **a** Experimentally and computationally determined maximum strain developments of the concrete due to constant fatigue loading (from [27], licensed under CC-BY 4.0) **b** Computationally determined maximum strain developments for sequences "high–low" and "low–high" as well as for constant maximum stress levels $S_{max} = 0.85$ and $S_{max} = 0.70$ for comparison

The second example demonstrates the ability of the model to capture sequence effects if load cycles of different amplitudes are applied. For this simulation, the same mesostructure and discretisation as those shown in Fig. 5 a as well as the same material parameters as those given in Table 4 were used. Four different simulations were performed, whereby the minimum stress level was kept constant with $S_{min} = 0.05$. Firstly, 170 load cycles with $S_{max} = 0.85$ are applied and, after that, the maximum stress level is decreased to $S_{max} = 0.70$ ("high–low") for the remaining load cycles until failure. For the second simulation, the first 170 load cycles with $S_{max} = 0.70$ are applied and, subsequently, the maximum stress level is increased to $S_{max} = 0.85$ ("low–high") until failure. These two simulations are complemented with the simulations with $S_{max} = 0.85$ and $S_{max} = 0.70$ until failure. The corresponding maximum strain developments up to failure are depicted in Fig. 6b. Failure occurs after 721 cycles for the "high–low" simulation. The material sample fails after 544 cycles for the "low–high" simulation. The reduction of numbers of cycles calculated for sequences "low–high" compared to "low" as well as the increase of numbers of cycles to failure for sequence "high–low" compared to "high" corresponds qualitatively to the experimental results and results from the literature. If the different sequence of applying the two different amplitudes had no effect, the state of damage should be the same after 340 cycles. This is clearly not the case, as can be seen in Fig. 7, where the damage state is depicted for both sequences and after 340 cycles. Thus, it can be concluded that the method is generally able to reflect the sequence effect that exists for the compressive fatigue behaviour of concrete.

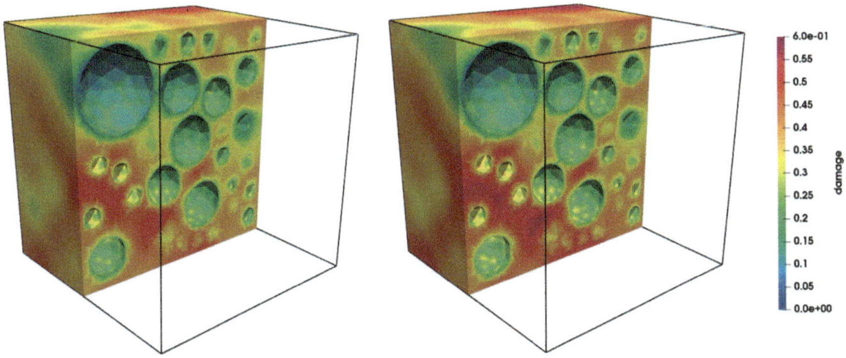

a b

Fig. 7 **a** Damage distribution within the mesostructure after 340 load cycles for the "high–low" sequence and **b** the "low–high" sequence

6 Conclusions

Comparative compressive fatigue investigations were conducted on systematically adjusted concrete (compositions) and their corresponding mortars. The objective was to investigate the influence of the concrete microstructure on the fatigue resistance and, furthermore, to gain basic knowledge about the damage mechanism due to compressive fatigue loading. The influence of coarse aggregates generally, the substitution of basalt by granite coarse aggregate, the addition of silica fume and the variation of the *w/c* ratio were considered. It was clearly shown that the fatigue resistance of concrete and mortar can be influenced by their composition. A mortar matrix that is as undisturbed as possible, a very good bond to the aggregates and a fatigue-resistant aggregate is essential for the fatigue resistance of concrete. The addition or, rather, part substitution of cement by silica fume improves the fatigue resistance of concretes and mortars. By contrast, the increase of the *w/c* ratio impairs the fatigue resistance in both cases.

A stiff and high-strength coarse aggregate generally leads to a more inhomogeneous stress distribution, on the one hand, but to an attraction of stresses and, thus, a relief of stresses in the mortar matrix, on the other hand. The results indicate that the first effect predominates the fatigue behaviour of concrete at high stress levels and the second effect at low stress levels. Thus, the fatigue-resistant coarse aggregate basalt improves the fatigue resistance at lower stress levels and impairs it at higher stress levels compared to granite aggregate. Regarding mortars, the positive influence of the coarse aggregate at a lower stress level outweighs the negative influence of the more inhomogeneous stress distribution created by it. Therefore, the lower fatigue resistance of the mortars at the lower stress level must be traced back to the missing basalt coarse aggregate. The extent of the influence of the coarse aggregate additionally depends on the properties of the mortar matrix, meaning a higher effect for a weaker mortar matrix and a lower effect for a stronger mortar matrix.

Detailed analyses of AE signals led to the differentiation of the AE hits into two clusters, whereby the AE hits of cluster I are characterised by low AE energies, while the AE hits in cluster II have higher AE energies. This indicates that hits in cluster I occur due to damage on a smaller scale compared to those in cluster II. Furthermore, the number of hits in cluster I is much higher than those in cluster II, and increases within the fatigue damage process. By contrast, the number of AE hits in cluster II is much smaller and these hits occur mainly in phase I and, especially, in phase III. The sum of the AE energy emitted up to the transition of phase II to phase III is characteristic of concretes and mortars of a special composition and independent of the maximum stress level [15].

Altogether, taking into account the developments of the macroscopic damage indicators, the detailed analyses of the AE hits and the results of the microscopic analyses, strong indications were found that two damage mechanisms in concrete with regular moisture content exist due to compressive fatigue loading: (1) diffusely distributed damage in the mortar matrix that begins to develop on an extremely small scale (sub-microscale) and accumulates in areas of stress concentration. (2) Within this pre-damaged matrix, more localised damage, in the form of cracks or larger areas of linked localised damage, develops on a micro- to mesoscale, especially in phase III of the fatigue damage process. The diffusely distributed, small-scale damage (1) predominates in the damage process due to compressive fatigue loading. However, the higher the maximum stress level or the greater the inhomogeneity of the stress distribution, the more localised, large-scale damage (2) occurs already in phase II.

It is remarkable that the combined analyses of macroscopic damage indicators, AE hits development and microscopic analyses enabled these new insights into the damage processes due to compressive fatigue loading. However, these damage processes can be influenced by the concrete or, rather, mortar composition.

According to the experimental investigations and findings, a computational model for the simulation of the fatigue behaviour of concrete at mesoscale was set up. On that scale, basalt grains larger than 2 mm in diameter were explicitly modelled as isotropic linear elastic spherical inclusions not exhibiting damage. The mortar including basalt grains smaller than 2 mm in diameter was modelled as an isotropic linear elastic material including a gradient-enhanced fatigue damage model. Unfortunately, this model is not (yet) able to capture any remaining (viscoplastic) deformations that can be observed experimentally. As it is known from other investigations [28], fatigue damage is detectable in the decrease of the elastic modulus, which is called "loosening of microstructure", and in the development of the remaining "plastic" deformation. However, the primary fatigue damage model employed here for the concrete can generally capture the portion of damage associated with the decrease of stiffness, which, indeed, will also be affected by remaining deformations. Furthermore, the three phases of fatigue and the sequence effects are quite well described. In the future, it is planned to extend this gradient-enhanced fatigue damage model with a Drucker-Prager-Cap microplane model [12, 13] to capture the remaining deformations after unloading as well as anisotropic damage. It is expected that this enhanced material model will be able to quantitatively reflect the sequence effects observable experimentally and that it will allow for predictive fatigue simulations for different concrete compositions.

Associated data is published via the link: https://doi.org/10.25835/r7l2jp9p.

References

1. Kono, S., Hasegawa, H., Mori, K., Ichioka, Y., Sakashita, M., Watanabe, F.: Low cycle fatigue characteristics of high strength concrete. In: Proceedings of the 8th International Symposium on Utilization of High-Strength and High-Performance Concrete, Tokyo, pp. 616–622 (2008)
2. Petković, G., Stemland, H., Rosseland, S.: High Strength Concrete SP 3—Fatigue, Report 3.2: Fatigue of High Strength Concrete. SINTEF Structural Engineering—FCB, Trondheim (1992)
3. Oneschkow, N., Lohaus, L.: Zum Nachweis von druckschwellbeanspruchtem Beton, Teil 2: Sicherheitsüberlegungen und Potenzial für Weiterentwicklungen. [About the fatigue design concept of concrete subjected to pure compressive fatigue loading. Part 2: Safety considerations and potential for further developments] (in German). Beton- und Stahlbetonbau **112**, 611–622 (2017). https://doi.org/10.1002/best.201700026
4. Basaldella, M., Oneschkow, N., Lohaus, L.: Influence of the specimen production and preparation on the compressive strength and the fatigue resistance of HPC and UHPC. Mat. Struct. **54**, 99 (2021). https://doi.org/10.1617/s11527-021-01696-9
5. Mehmel, A., Kern, E.: Elastische und plastische Stauchungen von Beton infolge Druckschwell- und Standbelastung [Elastic and plastic strains of concrete due to compressive fatigue and sustained loading] (in German). Deutscher Ausschuss für Stahlbeton, Issue 153, Wilhelm Ernst & Sohn, Berlin (1962)
6. Shah, S.P., Chandra, S.: Fracture of concrete subjected to cyclic and sustained loading. ACI J. **67**, 816–825 (1970)
7. Breitenbuecher, R., Ibuk, H., Yuceoglu, S.: Beeinflusst die Kornsteifigkeit der Gesteinskörnung im Beton den Degradationsprozess infolge zyklischer Druckbeanspruchung? [Does the aggregate stiffness in concrete affect the degradation process due to cyclic compressive loading?] (in German). Beton- und Stahlbetonbau **103**, 318–323 (2008). https://doi.org/10.1002/best.200800615
8. Thiele, M.: Experimentelle Untersuchungen und Analyse der Schädigungsevolution in Beton unter hochzyklischen Ermüdungsbeanspruchungen. [Experimental investigation and analysis of the damage development in concrete subjected to high-cycle fatigue] (in German). Doctoral Thesis, BAM-Dissertationsreihe Band 140, Technische Universität Berlin (2016). https://doi.org/10.14279/depositonce-4886
9. Oneschkow, N.: Analyse des Ermüdungsverhaltens von Beton anhand der Dehnungsentwicklung. [Analysis of the fatigue behaviour of concrete with respect to the development of strain] (in German). Doctoral Thesis, Reports of the Institute of Building Materials Science, 13, Leibniz Universität Hannover (2014). https://doi.org/10.15488/357.
10. Aggelis, D.G.: Classification of cracking mode in concrete by acoustic emission parameters. Mech. Res. Commun. **38**, 153–157 (2011)
11. Bažant, Z.P., Oh, B.H.: Microplane model for progressive fracture of concrete and rock. J. Eng. Mech. **111**, 559–582 (1985). https://doi.org/10.1061/(ASCE)0733-9399(1985)111:4(559)
12. Zreid, I., Kaliske, M.: A gradient enhanced plasticity-damage microplane model for concrete. Comput. Mech. **62**, 1239–1257 (2018). https://doi.org/10.1007/s00466-018-1561-1
13. Betz, P., Curoşu, V., Löhnert, S., Marx, S., Curbach, M.: Classification of multiaxial behaviour of fine-grained concrete for the calibration of a microplane plasticity model. Buildings **13**, 2704 (2023). https://doi.org/10.3390/buildings13112704

14. Baktheer, A., Aguilar, M., Chudoba, R.: Microplane fatigue model MS1 for plain concrete under compression with damage evolution driven by cumulative inelastic shear strain. Int. J. Plast. **143**, 102950 (2021). https://doi.org/10.1016/j.ijplas.2021.102950
15. Timmermann, T.: Charakterisierung der Ermüdungsschädigung in Beton mithilfe der Schallemissionsanalyse. [Characterisation of fatigue damage in concrete using acoustic emission analysis] (in German). Doctoral Thesis, Reports of the Institute of Building Materials Science, Leibniz Universität Hannover (submitted, 2025)
16. Leusmann, T., Basutkar, G., Lunardelli, M., Lowke, D.: Characterizing the 3D mesostructured of high performance concrete by computed tomography. In: Proceedings of RILEM Spring Convention and Sustainable Materials, Systems and Structures (SMSS), Rovinj, Croatia, 20–22 March 2019, pp. 176–184
17. Oneschkow, N., Timmermann, T.: Influence of the composition of high-strength concrete and mortar on the compressive fatigue behaviour. Mater. Struct. **55**, 83 (2022). https://doi.org/10.1617/s11527-021-01868-7
18. DIN EN 12390-3:2019-10. Testing Hardened Concrete—Part 3: Compressive Strength of Test Specimens, German version of EN 12390-3:2019
19. DIN EN 12390-13:2014-06. Testing Hardened Concrete—Part 13: Determination of Modulus of Elasticity in Compression, German version of EN 12390-13:2013
20. Oneschkow, N., Scheiden, T., Hüpgen, M., Rozanski, C., Haist, M.: Fatigue-induced damage in high-strength concrete microstructure. Materials **14**, 5650 (2021). https://doi.org/10.3390/ma14195650
21. Zhang, F., Yang, Y., Fennis, S.A.A.M., Hendriks, M.A.N.: Developing a new acoustic emission source classification criterion for concrete structures based on signal parameters. Constr. Build. Mater. **318**, 126163 (2022). https://doi.org/10.1016/j.conbuildmat.2021.126163
22. Scheiden, T., Oneschkow, N.: Influence of coarse aggregate type on the damage mechanism in high-strength concrete under compressive fatigue loading. Struct. Concr. **20**, 1212–1219 (2019). https://doi.org/10.1002/suco.201900029
23. Mazars, J., Pijaudier-Cabot, G.: From damage to fracture mechanics and conversely: a combined approach. Int. J. Solids Struct. **33**, 3327–3342 (1996). https://doi.org/10.1016/0020-7683(96)00015-7
24. Peerlings, R.H.J., de Borst, R., Brekelmans, W.A.M., de Vree, J.H.P.: Gradient enhanced damage for quasi-brittle materials. Int. J. Numer. Methods Eng. **39**, 3391–3403 (1996). https://doi.org/10.1002/(SICI)1097-0207(19961015)39:19%3c3391::AID-NME7%3e3.0.CO;2-D
25. de Vree, J.H.P., Brekelmans, W.A.M., van Gils, M.A.J.: Comparison of nonlocal approaches in continuum damage mechanics. Comput. Struct. **55**, 581–588 (1995). https://doi.org/10.1016/0045-7949(94)00501-S
26. Alessi, R., Vidoli, S., de Lorenzis, L.: A phenomenological approach to fatigue with a variational phase-field model: the one-dimensional case. Eng. Fract. Mech. **190**, 53–73 (2018). https://doi.org/10.1016/j.engfracmech.2017.11.036
27. Oneschkow, N., Timmermann, T., Löhnert, S.: Compressive fatigue behaviour of high strength concrete and mortar: experimental investigations and computational modelling. Materials **15**, 319 (2022). https://doi.org/10.3390/ma15010319
28. Kern, B., Podhajecky, A.-L., Lohaus, L., Haist, M., Oneschkow, N.: Deformation behaviour of concrete with different moisture contents subjected to compressive creep and cyclic loading. Mater. Struct. **57**, 6 (2024). https://doi.org/10.1617/s11527-024-02399-7

Multiscale Investigation of Fatigue Crack Growth in High-Performance Concrete Based on Computer Tomography and Phase-Field Fracture Modelling

Gauravdatt Basutkar, Marreddy Ambati, Pietro Carrara, Thorsten Leusmann, Laura De Lorenzis, and Dirk Lowke

Abstract High-performance concrete (HPC) is increasingly adopted in the construction of concrete structures for bridges, towers, offshore rigs, high-rise buildings and wind turbines. HPC enables slender cross-sections, lighter structural components and longer spans for structures. However, due to their inherently reduced self-weight, these structures are often subjected to cyclic loads throughout their lifetime. Despite being an advanced material, the potential of using HPC under fatigue loading cannot be fully exploited due to conservative design standards. The fracture mechanical behaviour of HPC, containing basalt aggregate (AG) and coarse mortar (CM), was investigated under static and cyclic tensile loading. The matrix material CM was developed as a concrete equivalent mortar using the excess paste theory. The test results show clear differences between the HPC and its components, basalt AG and CM, in terms of mechanical properties and fracture mechanical behaviour under static and cyclic loading. A novel and efficient hybrid formulation for phase-field modelling in brittle fracture is introduced. Both helping to understand the influence of the internal mesostructure on the fracture mechanical properties of the HPC on the macroscale.

Keywords High-performance concrete · Computed tomography · Fracture behaviour · Fatigue

G. Basutkar · T. Leusmann (✉)
Institute of Building Materials, Concrete Construction and Fire Safety (iBMB), Technische Universität Braunschweig, Braunschweig, Germany
e-mail: t.leusmann@ibmb.tu-bs.de

M. Ambati
Material Mechanics and Durability, GE Global Research, Cincinnati, OH, USA

P. Carrara · L. De Lorenzis
Department of Mechanical and Process Engineering, ETH Zürich, Zürich, Switzerland

D. Lowke
School of Engineering and Design, Technische Universität München, Munich, Germany

N. Oneschkow and L. Lohaus (eds.), *Cyclic Deterioration of High-Performance Concrete*, https://doi.org/10.1007/978-3-032-19394-0_4

1 Objectives and Methods

The objective of the study was to experimentally analyse the fatigue crack growth in HPC with a focus on the mesoscale. The mesoscale was defined as the scale at which the concrete consists of aggregates (AG) above a size threshold of 2 mm, pores (air voids) above the same size threshold, cement mortar (CM) and the interfacial transition zone (ITZ) i.e., the area of the material at the interface between aggregates and cement mortar. The scope of the study included a detailed experimental description of the phenomena of fatigue crack initiation and propagation. The schematic representation of the mesostructural composition of HPC is illustrated in Fig. 1.

The research comprised a structured methodology for achieving the research objectives. The initial phase focused on the selection of materials and the optimisation of contrasts between the two main components cement mortar and aggregates. To understand the interaction of aggregates and cement mortar in high-performance concrete, the material properties of the main components CM and AG and the properties of the high-performance concrete (HPC) were analysed. Further, fracture mechanical tests were performed under static and cyclic loading conditions, thus yielding insights into the mechanical properties, fracture patterns, and response of HPC under varying loading scenarios.

A novel framework to model fatigue in brittle materials based on the variational phase-field approach to fracture was also proposed. The new approach allows the simulation of crack nucleation and propagation under cyclic loading, while the monotonic case is included as a special case. The proposed approach can reproduce the main feature of the fatigue failure as obtained by classical theories while retaining the flexibility of the finite element method. Therefore, it can be used to study complex loading or geometric conditions and/or to optimize the design of structural components or of the concrete mixture.

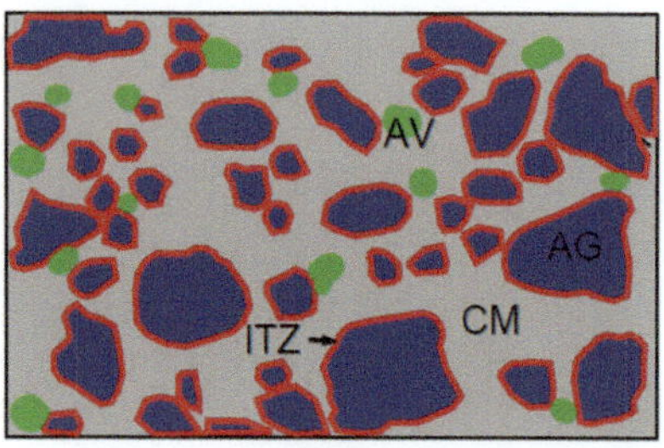

CM - Coarse Mortar
AG - Aggregate
ITZ - Interfacial Transition Zone
AV - Air Void

Fig. 1 Schematic representation of the idealised mesostructure of HPC (from [1], licensed under CC-BY 4.0)

2 Characterization of Materials

For the study, the reference HPC of the Priority Programme 2020 was used, see Table 1 in Sect. 2.1 "Concrete Composition" in the chapter by M. Basaldella et al., this volume and [2]. This consisted of basalt aggregate (AG) and a high-strength mortar (CM) at the meso level. The composition of the high-strength coarse mortar (CM) was not trivial and was carried out according to the excess paste theory, which was used to produce a concrete-equivalent mortar with the same excess paste layer thickness as the HPC. Subsequently, the mechanical properties of AG, CM and HPC were determined. The reference mix for HPC was based on the composition of the Priority Programme 2020. It was produced from the following constituents: crushed basalt aggregates from 2 to 8 mm; sand from 0 to 2 mm, quartz sand from 0 to 0.5 mm, Portland cement CEM I 52.5 R with a content of 500 kg/m^3. A water-cement ratio of 0.35 was used and no additives such as silica fume, fly ash or ground granulated blast furnace slag were included, see Table 1 in Sect. 2.1 "Concrete Composition" in the chapter by M. Basaldella et al., this volume and [2].

While conventional aggregates such as basalt are commonly employed in high-performance concrete production, the strength of the coarse aggregates themselves can be decisive for concretes with very high strength. Consequently, the strength of the aggregate is important, but the bond strength of the aggregate particles can also be a limiting factor. The mineralogical properties of the coarse aggregate have been found to influence the strength of the resulting concrete, but there is no simple recipe for aggregate selection. The requirement for aggregate strength applies when a high long-term strength of the concrete is required.

2.1 Composition of Concrete Equivalent Coarse Mortar (CM)

In order to determine the mechanical parameters of bulk material, a concrete equivalent mortar representative for the bulk material was developed with the same excess paste thickness as the reference concrete following the procedure described in [3–6]. The procedure comprised in substituting the coarse aggregate fraction by an amount of sand having an equivalent surface area. As a result, the total aggregate surface area to be coated with paste was the same for the mortar as for the coarse aggregates.

For the development of CM, concrete and mortar were regarded as a two-phase structure, in which the aggregates were distinguished as solid phase and the paste as fluid phase. The workability of concrete and mortar was determined by the excess paste thickness $t_{\mathrm{Paste,ex}}$, which is equal to half the distance between two adjacent

Table 1 Packing density of aggregate for HPC and CM

Aggregate composition	HPC	CM
Maximum packing density, $\varphi_{\max}$ [–]	0.755	0.685

aggregates. This distance is independent of the aggregate size. The excess paste thickness $t_{\text{Paste,ex}}$ was determined by enclosing the surface area of the aggregates A_{A} with the excess paste volume $V_{\text{Paste,ex}}$, which remains after filling the voids between the aggregates V_{void} with paste. The surface area A_{A} was calculated with Eq. 3.

$$t_{Paste,ex} = \frac{V_{Paste,ex}}{A_A} = \frac{V_{Paste} - V_{Void}}{A_A} \tag{1}$$

The volume V_{Void} was calculated from the packing density φ_{max}, determined with the compaction test by De Larrard [7], and the surface area of the aggregates A_{A} was determined with the help of CT-Scans.

A concrete mixture has been considered as a dry packing of aggregate particles, filled up with cement paste and its performance is significantly affected by the type and degree of packing of its constituents. In the current study, De Larrard's compaction packing model has been incorporated [7]. The maximum packing density is the maximum density attainable of a specified aggregate mixture under maximum compaction. During the De Larrard test, the maximum packing density has been achieved at the highest compaction level, by applying external energy such as vibration and top-weight. The maximum packing density φ_{max} was calculated as the ratio of the aggregate particle volume V_{A} to the volume of aggregates including voids in compacted state V_{C}. The maximum packing density of the CM and HPC aggregates was therefore evaluated and the results are summarised in Table 1.

The surface of the aggregate fractions was measured and evaluated three-dimensionally with computed tomography. For this purpose, the aggregate of the fractions basalt 5/8, basalt 2/5, sand 0/2 and quartz sand 0/0.5 was mixed with a low-density material to keep the aggregate particles at a distance and placed in a specimen container. For each fraction, four specimens were produced. For the reconstruction, 1000 X-ray images with a resolution of 1000×1000 pixels were generated. The reconstruction software reproduced the volume as an image stack with a voxel edge length between 10 and 40 μm using automatic scan optimization for correction of drift effects and beam hardening correction.

The greyscale difference between aggregate particles and the surrounding material was sufficient enough to identify the particles with the threshold value method. The CT image stack was segmented with a grayscale threshold. Voxels with a grey level above this threshold were assigned a value of 1 and the background material was assigned a value of 0. The watershed algorithm divided connected particles in the binarized image and the regions were subsequently analysed with the MATLAB function "regionprops3" to calculate the surface and the volume of individual particles.

The concrete equivalent mortar CM was developed from the HPC listed in Table 3. The water/cement ratio w/c is 0.35 for the HPC and the mortar CM [3]. The excess paste thickness $t_{\text{paste,ex}}$ of the reference HPC was calculated to 22 μm using Eqs. 1 and 2:

$$V_{Paste} = \frac{m_c}{\rho_c} + \frac{w}{c} \cdot m_c + \frac{0.01 \cdot m_c}{\rho_{sp}} + \frac{0.0057 \cdot m_c}{\rho_{st}} \tag{2}$$

Therefore, the air void volume VAV of 5 dm^3/m^3 was considered and the packing density $\varphi_{\max}$ from Table 1 was used. The superplasticizer and the stabilizer depended on the cement mass. A proportion of 1% of the cement mass was used for the superplasticizer and 0.57% for the stabilizer. The densities ρ_i can be found in Table 3. The aggregate surface area A_A was calculated from the ratio R_i listed in Table 2 and the aggregate fraction volume V_i for the HPC with Eq. 3:

$$A_A = \sum_i R_i \cdot V_i = \sum_i R_i \cdot \frac{m_i}{\rho_i} \tag{3}$$

The parameters necessary to solve Eq. 3 are listed in Table 3, which illustrates the mass of the fractions m_i and the corresponding particle density ρ_i. The cement content m_c was calculated conversely using equation and. It was varied until the paste thickness of 22 μm is attained.

Table 2 Ratio $R = A_A/V_A$ of the analysed aggregate fractions

Basalt 5/8	$R = 1.30\ \text{mm}^{-1}$
Basalt 2/5	$R = 2.43\ \text{mm}^{-1}$
Sand 0/2	$R = 14.7\ \text{mm}^{-1}$
Quartz 0/0.5	$R = 28.7\ \text{mm}^{-1}$

Table 3 Concrete composition of HPC and corresponding equivalent CM

Material	Density ρ (kg/dm³)	HPC (kg/m³)	CM (kg/m³)
Cement c	3.09	500	646
Water w	1.0	176	227
Superplasticizer sp	1.05	5.00	6.46
Stabilizer st	1.1	2.85	3.68
Quartz sand 0–0.5	2.7	75	118
Sand 0–2 mm	2.64	850	1336
Basalt 2–5 mm	3.06	350	–
Basalt 5–8 mm	3.06	570	–

Table 4 Mechanical properties and standard deviation of HPC, CM and basalt

Material	HPC	CM	Basalt
Density [kg/dm^3]	2.52	2.26	3.06
Compressive strength [N/mm^2]	116.8 ± 3.2	95.2 ± 1.1	312.8 ± 4.4
Tensile strength [N/mm^2]	5.7 ± 0.8	4.97 ± 0.7	19.20[a] ± 3.7
Young's modulus [kN/mm^2]	50.3 ± 1.8	39.5 ± 0.9	92.7 ± 7.6

[a] Split tensile strength according to EN 12390-6 [10]

2.2 Material Properties

The compressive strength was determined in accordance with EN 12390-3 [8] and the modulus of elasticity in accordance with EN 12390-13 [9] for cylinders with a diameter of 150 mm and a height of 300 mm. The tensile strength was determined in a direct tensile test on plain cylinders with a diameter of 80 mm and a height of 240 mm. Furthermore, indirect split tensile tests according to EN 12390-6 [10] were carried out to determine the tensile strength of basalt boulder. Table 4 summarizes the test results.

Table 4 shows that HPC reached a 16% higher compressive strength compared to CM. Moreover, HPC exhibits a modulus of elasticity and tensile strength that were approximately 15% and 12% higher than those for CM, respectively. Despite maintaining a consistent water-cement ratio of 0.35 across all concrete compositions, the enhanced mechanical properties of HPC were attributed to the presence of crushed basalt aggregates and an improved interfacial transition zone (ITZ). Conversely, in the context of the excess paste theory, the substitution of coarse aggregate (ranging from 2 to 8 mm) with an equivalent surface amount of sand in CM contributed to the observed strength decrease, primarily owing to the finer and naturally rounded sand particles. Additionally, the reduced strength of CM underscored the pivotal role of aggregate nature and morphology in dictating the mechanical performance of high-performance concrete compositions.

3 Fracture Mechanical Tests

The study followed an experimental approach, beginning with the setup and culminating in results and discussions. The experimental setup included the assembly of modified disk-shaped compact tension (MDCT) specimens and the implementation of appropriate testing protocols. The specimens were subjected to controlled loading conditions in order to assess their mechanical responses, fracture and fatigue behaviour. The obtained results were analysed and interpreted in the context of the research objectives. These findings were then discussed in detail, considering their implications, significance, and potential insights they offer.

3.1 Experimental Setup

The specifications and dimensions of the newly proposed MDCT specimen were based on ASTM standards [11], and their details were summarized in Fig. 2 and Table 5 correspondingly. For the analysis of crack opening displacements (COD) and deformations during both static and cyclic tensile tests under external loads, the digital image correlation (DIC) tool ARAMIS from GOM was employed. The measurement of crack opening displacement was performed across a 20 mm length along the loading axis formed by the GFRP threaded rods. Before conducting ARAMIS measurements, the specimen's surface was coated white and partly covered with a random black spray pattern, crucial for the system's detection of "facets." This approach enabled the system to identify and track facets based on distinct grey value levels, thereby facilitating analytical evaluation of the intricate behaviour of the test specimens. The free ends of the GFRP threaded rods were securely clamped into the electromechanical testing machine, approximately 80 mm away from the initial notch axis. The testing setup employed two 2448×2050-pixel cameras, directly linked with the ARAMIS software, to record displacements and strains. This was executed at a working distance of 320 mm during successive loading stages. The configuration of the test setup was depicted in Fig. 2 and Table 5 for static and cyclic tests, respectively.

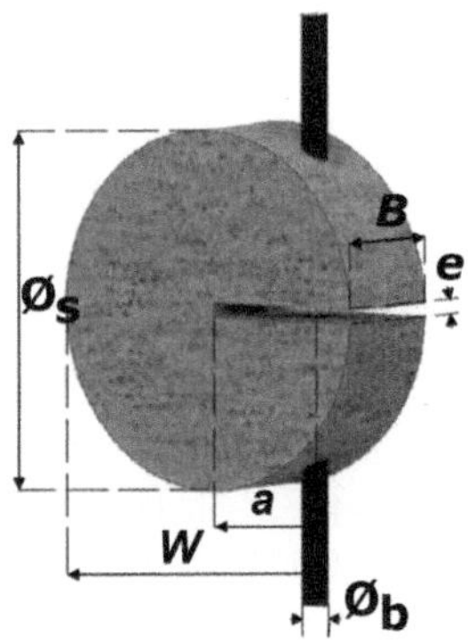

Fig. 2 Geometrical specifications of specimen (from [1], licensed under CC-BY 4.0)

Table 5 Characteristic dimensions of specimen

Specification	Dimensions (mm)
$Ø_s$	100
$Ø_b$	8
W	74
B	37
a	22
e	2

3.2 Fracture Mechanical Tests Under Static Loading

The static compact tensile tests were carried out under displacement control in an electro-mechanical testing machine at a rate of 0.01 mm/min. These static tests were performed on specimens with different material compositions such as HPC, CM and basalt. The digital image correlation (DIC) tool ARAMIS from GOM was implemented as discussed in Sect. 2.1 to analyse the crack opening displacements (COD) and deformations of the specimens under the external load of the static tensile tests. To evaluate the fracture characteristics of MDCT specimens, the initial step involves determining the relative notch length ratio α based on geometric parameters such as the initial notch length a and width W of the specimen. This ratio α was calculated using Eq. 4 [12].

Subsequently, the fracture toughness K_Q was calculated using Eq. 5 [12], where, P_Q represents the peak load, B is the thickness, and the geometric correction factor $f(a/W)$ is defined in Eq. 6 [12]. The calculation of $f(a/W)$ accounts for the relative notch length ratio and is a crucial component of the fracture toughness determination.

$$\alpha = \frac{a}{w} \tag{4}$$

$$K_Q = \frac{P_Q}{B\sqrt{W}} f\left(\frac{a}{W}\right) \tag{5}$$

$$f\left(\frac{a}{W}\right) = \frac{\left(2 + \frac{a}{W}\right)\left[0.76 + 4.8\frac{a}{W} - 11.58\left(\frac{a}{W}\right)^2 + 11.43\left(\frac{a}{W}\right)^3 - 4.08\left(\frac{a}{W}\right)^4\right]}{\left(1 - \frac{a}{W}\right)^{3/2}} \tag{6}$$

The adoption of nonlinear elastic fracture mechanics (NLFM) instead of linear elastic fracture mechanics (LEFM) is necessary for most concrete specimens, including MDCT geometry, as LEFM doesn't yield size-independent fracture properties [13, 14]. The total energy consumed to produce a unit area of fracture surface during the fracture process is defined as work of fracture W_f [15] and can be obtained from the area under the load-COD curve as shown in Fig. 3. Accordingly, the total fracture energy G_f was determined based on the area under the load-COD curve normalised to the uncracked ligament area A_{lig} and was expressed in Eq. 7. Fracture parameters were determined following the ASTM guidelines [11]. Table 6 summarizes the calculated fracture properties of concrete according to the work-of-fracture method applied to MDCT specimens.

In static tensile tests, performed under displacement control in an electromechanical testing machine at a rate of 0.01 mm/min, the specimens were subjected to tensile loading to failure. Fracture mechanical parameters were examined for each specimen. Notably, while HPC exhibited some degree of notch sensitivity, CM exhibited the most pronounced notch sensitivity, suggesting that cracks propagating through CM were less hindered due to the absence of larger aggregates. An increase in the size of the largest aggregate particles diminished this notch sensitivity [16], as supported

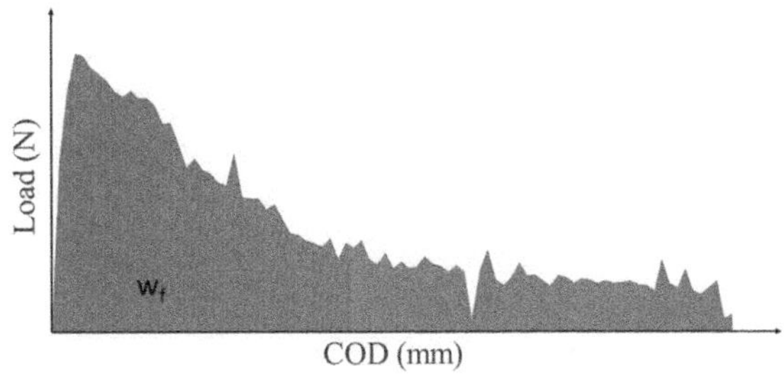

Fig. 3 P-COD curve for MDCT test illustrating the measured work-of-fracture W_f

Table 6 Experimental results of various material composition from MDCT tests

Material composition	Ultimate load P_Q (N)	Work of fracture W_f (N-mm)	Fracture toughness K_Q (MPa-m$^{1/2}$)	Fracture energy G_f (N/m)
HPC	1470 ± 223	319 ± 33	1.00 ± 0.17	165 ± 17
CM	1230 ± 170	248 ± 17	0.68 ± 0.11	141 ± 17
Basalt	2540 ± 131	819 ± 225	1.40 ± 0.07	425 ± 117

by the fracture toughness K_Q presented in Table 6.

$$G_f = \frac{W_f}{A_{lig}} \tag{7}$$

The experimental results of various material compositions from MDCT tests, including ultimate load P_Q, work of fracture W_f, fracture toughness K_Q, and fracture energy G_f, are listed in Table 6.

The static tensile tests on various material compositions revealed that basalt specimens have the highest average ultimate load of approximately 2540 N, while HPC and CM demonstrated average ultimate loads of about 1470 N and 1230 N, respectively. MDCT tests further emphasized the significant influence of material composition on fracture properties, with HPC exhibiting the highest fracture toughness K_Q value of around 1.00 MPa$\sqrt{m}$, indicating superior resistance to crack propagation, while CM displayed a lower K_Q value of about 0.68 MPa$\sqrt{m}$, signifying reduced crack resistance compared to HPC. Basalt boulder exhibited remarkable crack resistance with the highest K_Q value of approximately 1.40 MPa$\sqrt{m}$. Additionally, HPC demonstrated a fracture energy G_f of approximately 165 N/m, while CM exhibited lower G_f values of around 141 N/m. Notch sensitivity is found to be most pronounced in CM, implying less impeded crack propagation. This notch sensitivity was evident from the significant reduction in fracture toughness when notches were introduced, leading to easier crack initiation and propagation compared to unnotched specimens. In contrast, HPC and basalt specimens exhibited higher fracture toughness and resistance to crack propagation due to their improved aggregate interlock and stronger interfacial transition zone (ITZ). These findings underscore the importance of aggregate type and composition in governing the mechanical behaviours of concrete materials, with HPC superior properties attributed to basalt aggregates and an improved ITZ. Conversely, CM's reduced strength linked to the finer and naturally rounded aggregates, and Basalt boulder's exceptional properties stemming from its distinct aggregate nature and morphology.

3.3 *Fracture Mechanical Testing Under Cyclic Loading*

In high cycle fatigue testing, MDCT specimens were subjected to sinusoidal loading using a servo-hydraulic testing machine within predetermined minimum and maximum load levels. The fatigue testing process included a study of specimens composed of HPC and CM. Prior to conducting the fatigue tests, the ultimate tensile load of MDCT specimens was determined. At the onset of the test, the axial force was gradually increased in a continuous and consistent manner until it reached the mean load level after a duration of 80 s. The maximum load $P_{\max}$ applied was subject to variation, ranging from 50 to 90% of the mean ultimate load P_{Q}, as determined by the static tests previously elaborated. Furthermore, a minimum load level denoted as $P_{\min} = 0.05$ was upheld to ensure consistency and reliability. The testing process commenced with an initial set of four specimens subjected to cyclic loading, encompassing up to 100 cycles. A further set of four samples were then tested at a higher load cycle rate of up to 10,000 cycles. After that, a third set of four samples was examined, involving a broader range of load cycles varying from 1,000,000 to 2,000,000 cycles. Each individual sample within the testing process was subjected to cyclic loading at a constant frequency of 4 Hz. This frequency ensured uniformity and repeatability. Figure 4 provides a visual representation of different test cases in terms of the number of load cycles and the corresponding maximum load ratios displayed within the $P_{\max}/P_{\mathrm{Q}}$-N curves, which is comparable to the S–N curve/Wöhler curve. The determination of these maximum loads was guided by the average ultimate load values P_{Q} obtained from the static tests conducted on the respective MDCT samples, as outlined in Table 6.

4 Phase Field Modelling of Fatigue Fracture

The development of computational approaches able to predict the fatigue crack propagation is a hot topic in civil and mechanical engineering because it allows to reduce the costs related to extensive testing campaigns. In the following, the phase-field approach is extended to fracture so as to allow reproducing the major fatigue characteristics of brittle materials including, e.g., the Wöhler curve [17] and the Paris law [18].

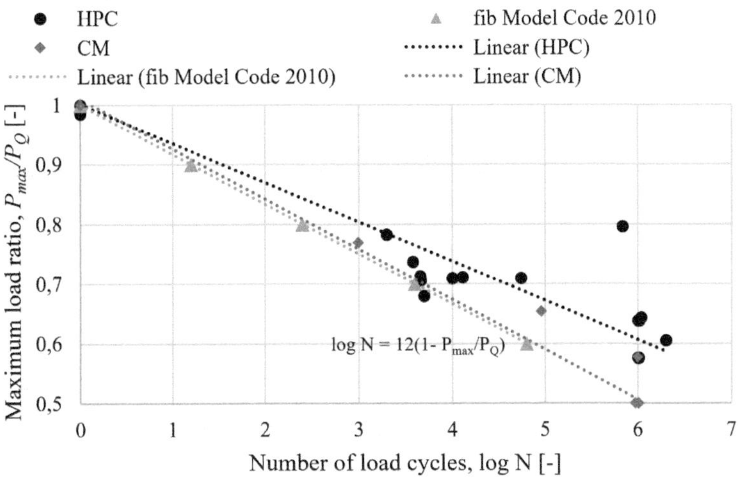

Fig. 4 Comparison of single experimental results with fib Model Code 2020 [16], considering the number of load cycles and the maximum load ratio

4.1 Formulation and Governing Equations

The starting point to develop the framework was the phase-field model for rate-independent brittle fracture under monotonic loading conditions, which describes the crack with a steep but smooth transition from intact to fully cracked material states through a phase-field variable d [19]. The free energy density of a linear elastic body Ω susceptible of brittle fracture can thus be written as [19]

$$E_\ell(\boldsymbol{u}, d) = \int_\Omega \psi_{el}(\boldsymbol{\varepsilon}(\boldsymbol{u}), d)d\boldsymbol{x} + \frac{G_c}{4c_w}\int_\Omega \frac{w(d)}{\ell} + \ell|\nabla d|^2 d\boldsymbol{x} \tag{8}$$

where x is the spatial coordinate, $\varepsilon(u)$ is the strain tensor related to the displacement field u by $\varepsilon = \nabla^s(u)$, where $\nabla^s\,(\cdot)$ is the symmetric gradient of $(\cdot)$. The phase-field parameter d ranges between 0 (sound material) and 1 (broken material). Also, the strain energy density $\psi_{el}(\varepsilon, d) = \psi_{el}^+(\varepsilon, d) + \psi_{el}^-(\varepsilon)$ decomposes into an active part $\psi_{el}^+(\varepsilon, d)$ contributing to the crack evolution and an inactive part $\psi_{el}^-(\varepsilon)$ linked to energy density stored upon crack closure and not participating to the evolution of d. Also, the monotonically decreasing degradation function $g(d)$ governs the transition of the material behaviour from the sound to the cracked state, while the last term in Eq. 8 is the dissipated energy density due to fracture, featuring the regularization length ℓ, the fracture energy G_c and the monotonically increasing dissipation function $w(d)$.

Minimizing the energy in Eq. 8 under the irreversibility constraint $\dot{d} \geq 0$, leads to the governing equations of the problem in terms of momentum balance and phase-field evolution along with the respective boundary conditions. The energy in Eq. 8

allows the crack to evolve only under monotonic loading conditions, making its application unfeasible in case of fatigue loadings [20].

To address this limitation, the total energy density is modified as follows [20]

$$E_\ell(\boldsymbol{\varepsilon}, d|\bar{\alpha}) = \int\limits_{\Omega} \psi_{el}(\boldsymbol{\varepsilon}, d)d\boldsymbol{x} + \int\limits_{\Omega}\int\limits_{0}^{t} f(\bar{\alpha}(\tau))\frac{G_c}{4c_w}\left(\frac{w(d)}{\ell} + 2\ell\nabla d \cdot \nabla\dot{d}\right)d\tau d\boldsymbol{x} \tag{9}$$

where t is the pseudo-time, $\bar{\alpha}(t)$ is an accumulated fatigue history variable, while $f(\bar{\alpha}(t))$ is a monotonically decreasing fatigue degradation function which forces the fracture toughness to decrease as $\bar{\alpha}(t)$ increases. The symbol | in Eq. 9 is used to highlight that the cumulated history variable acts as a parameter. Here, the history variable and the fatigue degradation function are assumed respectively as [20]

$$\bar{\alpha}(\boldsymbol{x}, t) = \int\limits_{0}^{t} H(\alpha\dot{\alpha})|\dot{\alpha}|d\tau \quad \text{with} \quad \alpha = \psi_{el}^{+}(\varepsilon, d) \tag{10}$$

$$f(\bar{\alpha}(\tau)) = \begin{cases} 1 & \text{if } \bar{\alpha}(t) < \alpha_T \\ \left(\frac{2\alpha_T}{\bar{\alpha}(t)+\alpha_T}\right)^2 & \text{if } \bar{\alpha}(t) \geq \alpha_T \end{cases} \tag{11}$$

where $H(\cdot)$ is the Heaviside function, defined ad $H(\cdot) = 1$ if $(\cdot) \geq 0$ and vanishing otherwise, while α_T is a threshold controlling when the fatigue effect is triggered. Also, the condition $\alpha\dot{\alpha} \geq 0$ indicate the loading condition, while $\alpha\dot{\alpha} < 0$ stands for an unloading branch.

The introduction of a time integral in Eq. 9 makes the energy time dependent. Therefore, the rate of free energy is minimised while considering $\bar{\alpha}(t)$ as a parameter. In fact, its value is frozen within a given time step. Note that, although this approach is able to qualitatively reproduce the main fatigue-related phenomena, this choice violates the variational nature of the approach.

The governing equations in terms of mechanical equilibrium result thus

$$\begin{cases} \text{div}(\boldsymbol{\sigma}) = \boldsymbol{0} \text{ in } \Omega \\ \boldsymbol{u} = \bar{\boldsymbol{u}} \quad \text{on } \partial\Omega_{\text{D}} \\ \boldsymbol{\sigma} \cdot \boldsymbol{n} = \boldsymbol{t}_n \quad \text{on } \partial\Omega_N \end{cases} \tag{12}$$

where $\sigma = \partial\psi_{el}(\varepsilon, d)\backslash\partial\varepsilon$ is the Cauchy stress, $\partial\Omega = \partial\Omega_D \cup \partial\Omega_N$, $\partial\Omega_D \cap \partial\Omega_N = \emptyset$ is the boundary of Ω including a Dirichlet $\partial\Omega_{\text{D}}$ and a Neumann part $\partial\Omega_{\text{N}}$, n is the associated unit outward normal, t_n are the applied external forces and $\bar{u}$ are the applied displacements. Also, the equations governing the evolution of the phase-field variable are

$$\begin{cases} \dot{d} \geq 0 & \text{in } \Omega \\ \frac{\partial \psi_{el}}{\partial d} - \frac{G_c}{2c_w}\left[f(\bar{\alpha})\left(\Delta d - \frac{w'(d)}{2\ell^2}\right) + \nabla f(\overline{\alpha}) \cdot \nabla d\right] \geq 0 & \text{in } \Omega \\ \left\{\frac{\partial \psi_{el}}{\partial d} - \frac{G_c}{2c_w}\left[f(\bar{\alpha})\left(\Delta d - \frac{w'(d)}{2\ell^2}\right) + \nabla f(\overline{\alpha}) \cdot \nabla d\right]\right\}\dot{d} = 0 & \text{in } \Omega \\ \nabla d \cdot \boldsymbol{n} = 0 & \text{on } \partial\Omega_N \end{cases} \tag{13}$$

Unlike the phase-field model in Eq. 8, the model described by Eqs. 12–13 is able to account for fatigue effects through the history variable $\bar{\alpha}$, which becomes thus an indicator of the fatigue "mileage" of the material.

4.2 Main Results

The capabilities of the proposed model were tested by predicting the behaviour of both three-point-bending (TPB) and compact tension (CT) specimens made of a material with Young's modulus $E = 6$ GPa, Poisson's ratio $\nu = 0.22$, critical fracture energy $G_c = 2.28$ N/mm, fracture toughness $K_{Ic} = 3.69$ MPa m$^{0.5}$, threshold variable $\alpha_T = 9.5 \times 10^{-1}$ N/mm^2 and regularization length $\ell = 0.2$ mm. Also, plane stress conditions were assumed while a force-controlled cyclic tensile load characterized by a vanishing minimum value $P_{\min} = 0$ and a range ΔP was applied. The tests were performed following the guidelines ASTM E 647 [21] for the CT specimens and ASTM E 1820 [22] for the TPB specimens. The crack growth rate curve relating the rate of crack advancement *da/dN* with the range of stress intensity factor range applied to the crack tip ΔK was obtained numerically, approximating the crack growth rate for sufficiently small crack length increment Δa as $da/dN \approx \Delta a/\Delta N$ where a constant value of $\Delta a \simeq 0.25$ mm was adopted as suggested in [21, 22]. The stress intensity factor range ΔK was obtained using the general expressions given in [21, 22]. The geometries and boundary conditions of the CT and TPB tests are illustrated in Fig. 5.

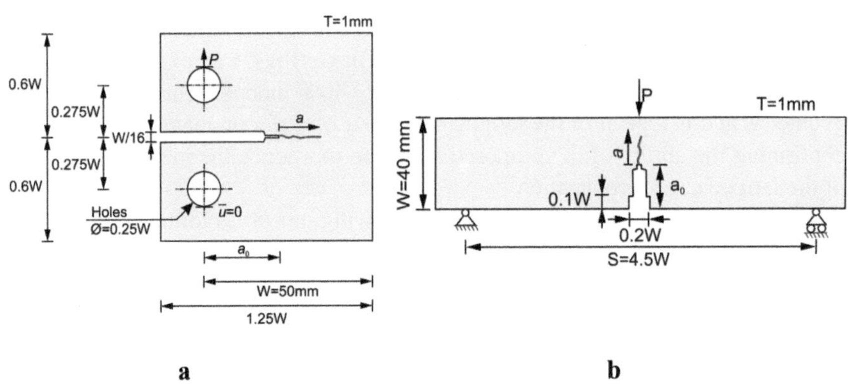

Fig. 5 Geometry and boundary conditions for **a** CT and **b** TPB specimens

According to [18], there is a regime, the so-called Paris regime, in which fatigue effects dominate and the crack grows stably following the so-called Paris law, which reads

$$\frac{da}{dN} = C\Delta K^m \quad (14)$$

where C and m are meant to be material parameters. In the customary double logarithmic plot of the crack growth rate curve, Eq. 14 leads to a linear trend depending only on the range of stress intensity factor applied at the crack tip during a load cycle ΔK, while it results independent on both load range ΔP and geometry, including the initial notch length a_0.

Figure 6 presents the results of different simulations of a cyclic test on the CT specimen varying either ΔP or a_0/W while keeping the other parameters fixed. Here it can be observed that the proposed framework is able to recover all the three branches of the experimental crack growth rate curve, namely an initial non-linear nucleation phase and the Paris regime followed by an unstable crack propagation (Fig. 6). Also, most of the linear Paris branches of the curves cluster together (curves marked with $\star$ in Fig. 6) as predicted by the Paris theory. In the few cases where the maximum load applied $P_{\max} = \Delta P$ is high and close to the monotonic peak load P_Q; this does not happen (curves marked with $\diamond$ in Fig. 6). For these tests the fatigue life N_u was limited (i.e., $N_u < 500$ cycles), leading to a failure mechanism where the damage process dominates over the fatigue effects and the nucleation and unstable crack propagation phases were so close to interfere with each other making thus the Paris branch vanishing.

The crack growth rate curves for the TPB are presented and compared to those obtained for the CT in Fig. 7. The results are similar both within the TPB series and between the latter and the CT ones. In particular, the Paris branches cluster together demonstrating that the proposed model agrees well with the prediction of the Paris law [18].

Next, the modified Wöhler or S–N curve [17] is analysed relating the ratio $P_{\max}/P_Q$ to N_u in Fig. 8. For both CT and TPB a transition from a non-linear damage-dominated trend to a linear fatigue-dominated one in correspondence with the switch from curves having a Paris regime to those lacking it (Figs. 6 and 7) can be observed. This change happens close to a ratio $\Delta P/P_Q \simeq 40\%$ independently of the adopted geometry, and is typical of the fatigue behaviour of different materials [17], further confirming the ability of the proposed model to reproduce the main characteristics of the fatigue crack propagation.

Consequently, Fig. 8 presents the crack growth rate curves for the TPB specimen and compares them with the CT results. The curves are close to each other, indicating that the crack growth behaviour is similar across different specimen geometries (CT-a/W = 0.22 and TPB a/W = 0.45) despite the different notch dimensions. This demonstrates the robustness of the proposed model in predicting fatigue crack growth. The Paris branches cluster together, showing that the model agrees well with the prediction of the Paris law. This consistency confirms the ability of the

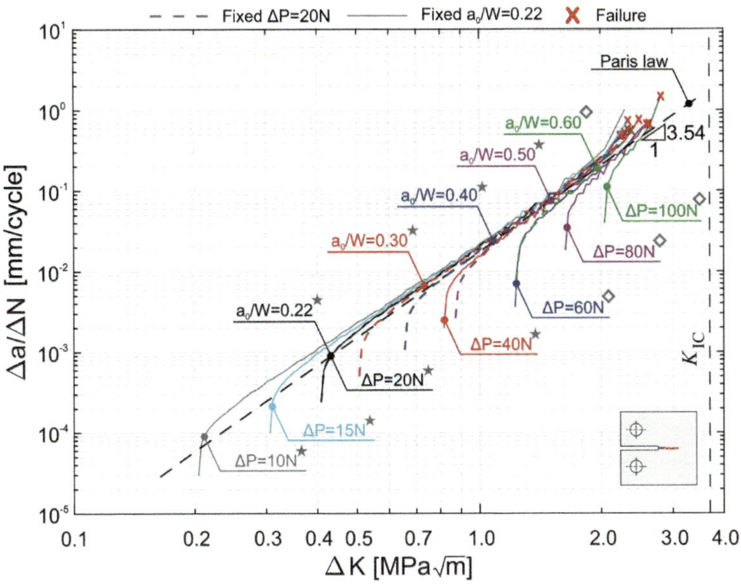

Fig. 6 Calculated crack growth rate curves for the CT specimen (⋆ curves with a fully developed Paris regime; ◇ curves lacking a Paris regime)

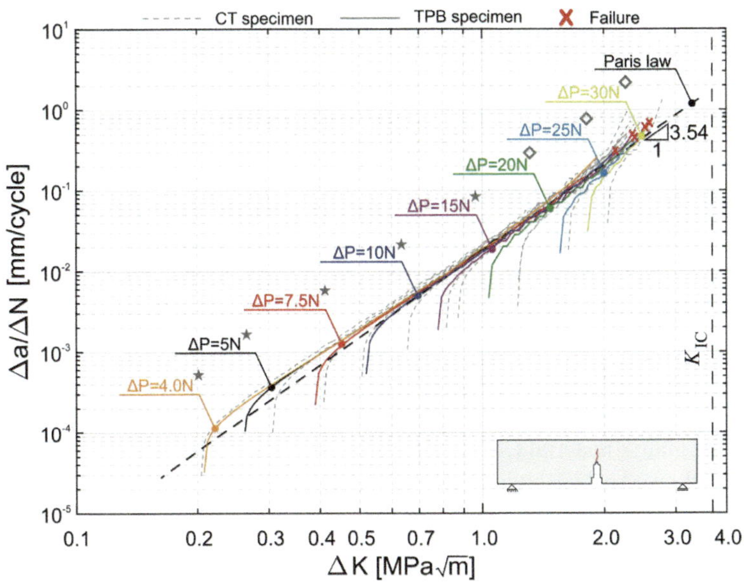

Fig. 7 Crack growth rate curves for the TPB specimen and comparison with the CT results (⋆ curves with a fully developed Paris regime; ◇ curves lacking a Paris regime)

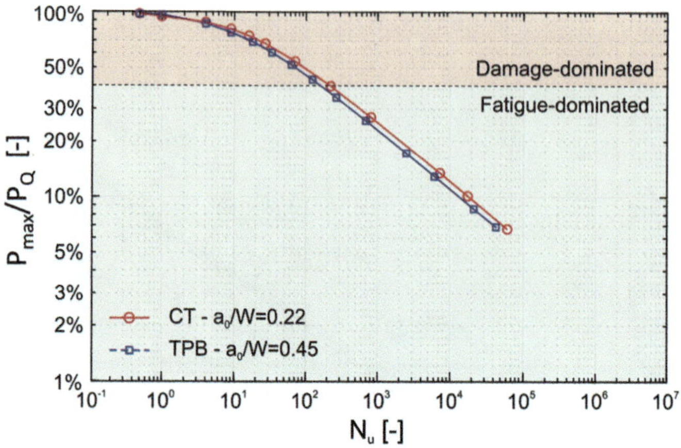

Fig. 8 Modified Wöhler or *S–N* curve for the CT and TPB specimens

model to reproduce the main characteristics of fatigue crack propagation in various configurations.

5 Summary and Conclusions

The experimental investigation aimed to comprehensively analyse the fracture and fatigue behaviour of high-performance Concrete (HPC) specimens subjected to various loading conditions. The study focused on two distinct material compositions: HPC with basalt aggregates (HPC) and Coarse Mortar (CM). Through a series of static, low-cycle fatigue, and high-cycle fatigue tests, valuable insights into the mechanical responses of these materials were obtained. The key conclusions drawn from the collected data, experimental results and numerical simulations are as follows:

- HPC showed 16% higher compressive strength, approximately 15% higher modulus of elasticity, and around 12% higher tensile strength than the corresponding values for CM. This proved that the basalt grain has a reinforcing effect on the matrix material CM.
- HPC showed a 47% higher fracture toughness compared to CM. This indicates a higher notch sensitivity of CM.
- High-cycle fatigue tests subjected specimens to sinusoidal tensile loading patterns. HPC achieved higher numbers of cycles than CM, as can be seen from the S–N curves. This was also evident for sinusoidal compressive loading, particularly at a stress level of $S_{min} = 0.70$, see Sect. 3 “Experimental Results” in the chapter by T. Timmermann et al., this volume. Thus, the reinforcing effect of the basalt

grains also has a positive effect on the fatigue resistance of the high-performance concrete.

- In addition to the experimental results, a novel framework to model fatigue in brittle materials based on the variational phase-field approach to brittle fracture is proposed. This framework allows for crack propagation and nucleation under cyclic loading and is demonstrated to reproduce the major features of fatigue failure, including both the Wöhler curve and the Paris law.

References

1. Basutkar, G., Leusmann, T., Lowke, D.: Emphasis of cyclic loading on the fracture mechanism and residual fracture toughness of high-performance concrete considering the morphological properties of aggregate. Constr. Mater. **4**(1), 292–314 (2024). https://doi.org/10.3390/constrmater4010016
2. Scheiden, T., Oneschkow, N.: Influence of two coarse aggregates on the damage mechanism in high-strength concrete under pure compressive fatigue loading (2019)
3. Leusmann, T., Basutkar, G., Lunardelli, M.: Characterizing the 3D mesostructure of high performance concrete with the help of computed tomography. In: Gabrijel, I., Grosse, C., Skazlic, M. (eds.) International Conference on Sustainable Materials, Systems and Structures (SMSS 2019)—Novel Methods for Characterization of Materials and Structures, Rovinj, Croatia, 20–22 Mar 2019, pp. 176–184 (2019)
4. Carrara, P., Kruse, R., Bentz, D., Lunardelli, M., Leusmann, T., Varady, P.A., De Lorenzis, L.: Improved mesoscale segmentation of concrete from 3D X-ray images using contrast enhancers. In: Cement and Concrete Composites, vol. 93, pp. 30–42 (2018)
5. Wüstholz, T.: Experimentelle und theoretische Untersuchungen der Frischbetoneigenschaften von Selbstverdichtendem Beton. Doctoral thesis in German, Universität Stuttgart, Institut für Werkstoffe im Bauwesen (2005)
6. Huss, A.: Mischungsentwurf und Fließeigenschaften von Selbstverdichtendem Beton (SVB) vom Mehlkorntyp unter Berücksichtigung der granulometrischen Eigenschaften der Gesteinskörnung. Universität Stuttgart, Materialprüfungsanstalt Otto-Graf-Institut (FMPA) (2010)
7. De Larrard, F.: Concrete Mixture Proportioning—A Scientific Approach. E & FN Spon, London (1999)
8. EN 12390-3: Testing hardened concrete—Part 3: Compressive strength of test specimens (2019)
9. EN 12390-13: Testing hardened concrete—Part 13: Determination of secant modulus of elasticity in compression (2013)
10. EN 12390-6: Testing hardened concrete—Part 6: Tensile splitting strength of test specimens (2009)
11. ASTM D7313-13: Standard Test Method for Determining Fracture Energy of Asphalt-Aggregate Mixtures Using the Disk-Shaped Compact Tension Geometry. ASTM International (2013)
12. Bažant, Z.P., Planas, J.: Fracture and Size Effect in Concrete and Other Quasibrittle Materials (2019)
13. Bažant, Z.P., Oh, B.H.: Crack band theory for fracture of concrete. Matériaux Constr. **16**(3), 155–177 (1983)
14. Jenq, Y., Shah, S.P.: Two parameter fracture model for concrete. J. Eng. Mech. **111**(10), 1227–1241 (1985)
15. Sakai, M., Ichikawa, H.: Work-of-fracture of brittle materials with microcracking and crack bridging. Int. J. Fract. **55**, 65–79 (1992)

16. Tada, H., Paris, P.C., Irwin, G.R.: The Stress Analysis of Cracks Handbook, 3rd edn. ASME Digital Collection, Lawrence, KS, USA (2000)
17. Wöhler, A.: Über die Festigkeitsversuche mit Eisen und Stahl. Zeitschrift für Bauwesen (XX), 73–106 (1870).
18. Paris, P., Erdogan, F.: A critical analysis of crack propagation laws. J. Basic Eng. **85**(4), 528–533 (1963)
19. Bourdin, B., Francfort, G., Marigo, J.J.: Numerical experiments in revisited brittle fracture. J. Mech. Phys. Solids **48**(4), 797–826 (2000)
20. Carrara, P., Ambati, M., Alessi, R., De Lorenzis, L.: A framework to model the fatigue behavior of brittle materials based on a variational phase-field approach. Comput. Methods Appl. Mech. Eng. **361**, 112731 (2020)
21. ASTM International: E647-13. Standard Test Method for Measurement of Fatigue Crack Growth Rates. ASTM International, West Conshohocken, PA, USA (2014)
22. ASTM International: E1820-01. Standard Test Method for Measurement of Fatigue Crack Growth Rates. ASTM International, West Conshohocken, PA, USA (2008)

Experimental and Numerical Framework for Characterization of High-Strength Concrete Fatigue Accounting for Local Dissipative Mechanisms at Subcritical Load Levels

Mario Aguilar, Henrik Becks, Abedulgader Baktheer, Martin Classen, and Rostislav Chudoba

Abstract The investigation presented in this chapter aim to contribute to the understanding of the fatigue phenomenology of a material zone under well-defined combination of compression and shear loading, a configuration with significant practical relevance. A comprehensive numerical-experimental analysis using the punch-through shear test (PTST) is presented, for prescribed levels of shear and compression in a tested ligament. The microplane model MS1 accurately simulates the degradation process in this multi-axial setup, capturing the dissipative behaviour under subcritical loading at the inter-aggregate level and the fatigue-induced triaxial stress redistribution. The combined numerical-experimental studies cover the PTST response for monotonic, stepwise increasing, and fatigue loading with constant and variable amplitude. In addition, the model provides an energetic interpretation of the effect of fatigue loading sequence which is made possible due to its thermodynamic formulation. The quantified energy dissipation attributed to the individual dissipative mechanisms over the fatigue lifetime shows an almost constant damage-induced dissipation across different loading histories.

Keywords Concrete · Fatigue · Experimental characterization · Constitutive modelling

---

M. Aguilar (✉) · H. Becks · M. Classen · R. Chudoba
Institute of Structural Concrete, RWTH Aachen University, Aachen, Germany
e-mail: maguilar@imb.rwth-aachen.de

A. Baktheer
Mechanics and Computational Mechanics, Leibniz Universität Hannover, Hannover, Germany

N. Oneschkow and L. Lohaus (eds.), *Cyclic Deterioration of High-Performance Concrete*, https://doi.org/10.1007/978-3-032-19394-0_5

1 Introduction

Accurate assessment of concrete fatigue life represents a critical challenge for meeting sustainability goals in the decades ahead. Minimizing material consumption while extending structural service life has become an urgent priority. However, our current knowledge regarding the fatigue response of concrete under multi-axial stress states remains incomplete, necessitating more targeted experimental and numerical investigations that would allow for an objective characterization of the fatigue behaviour in a material volume under well-defined stress conditions.

Recognizing the urgency for a deeper understanding of the fatigue behaviour of concrete, significant progress has been made in understanding concrete fatigue behaviour, with research focusing on factors such as loading frequency and temperature [1–3], moisture content [4–6], type of loading [7–9], and loading sequence and variable amplitude [10–13]. Recent developments in computational fatigue analysis have addressed the simulation of crack growth mechanisms in concrete under tensile and mixed-mode cyclic loading using cohesive interface formulations [14–16]. While these techniques perform well for Mode I dominant fracture, capturing fatigue behavior under multiaxial stress loading requires directionally-resolved constitutive frameworks such as the microplane approach [17]. For compression-dominated fatigue scenarios, researchers have employed continuum damage-plasticity formulations [18, 19]. These models offer computational advantages and can successfully replicate key experimental observations, including S–N relationships and fatigue-creep curves [20]. However, they are unable to reproduce local stress redistribution and anisotropic damage evolution under general triaxial stress states [17, 21].

The primary hypothesis of the authors regarding the modelling of fatigue in concrete and its introduction within the microplane framework are presented. This is followed by a presentation of the PTST setup, experimental investigations, and numerical modelling results. The presented simulations align well with the experimental data and provide an energetically based model interpretation of the loading sequence effect.

2 Cumulative Slip Hypothesis for Concrete Fatigue Modelling

The author's main hypothesis states that the fatigue damage evolution at subcritical load levels is governed by a cumulative measure of local strain, as originally postulated by Kirane and Bažant [17] and further refined by the authors [21–23]. This proposition finds support in the experimental results reported in [24] which documented that initiation and propagation of cracks during compressive fatigue loading occurred mainly along the interfaces between the aggregates and the cement paste.

This hypothesis was selected as the fundamental theoretical basis to be further developed and validated for high-strength concrete. The microplane framework has been chosen for formulating the constitutive model for fatigue in concrete.

It may be regarded as a semi-multiscale representation of the material behaviour. Indeed, it introduces an additional level of state representation compared to tensorial models. It can describe the directional evolution of damage occurring along weak interfaces within a heterogeneous material structure consisting of the cement paste and aggregates. While this additional level of resolution is not intended to explicitly model the skeleton of the material mesostructure, it can represent the evolution of smeared patterns of oriented damage distributed within the material structure. The authors introduced the hypothesis attributing concrete fatigue primarily to the cumulative effect of interaggregate sliding in the microplane framework, developing the microplane material model MS1 [21].

2.1 *Microplane Fatigue Model MS1*

The microplane fatigue model MS1 aims to capture the basic inelastic mechanisms that govern the fatigue behaviour of concrete and lead to triaxial stress redistribution and anisotropic damage evolution. The model is developed within the microplane framework, employing the kinematic projection to map the macroscopic strain onto directionally oriented microplanes. Stresses, damage variables, and plastic strains at the microplane level are integrated back to the macroscopic scale through the virtual work principle and energy equivalence. The detailed formulation of the model, which includes the relevant threshold functions and evolution laws, is available in [21]. In this publication, a systematic calibration and validation procedure of the model can be found, based on a test campaign with cylinders of different strength grades subjected to compressive fatigue loading, as summarized in Fig. 1. The numerical and experimental characterization of concrete fatigue included in the paper and performed by most of the researchers is done with cylinders and cubes. However, it is important to note, that on the one hand, these tests do not introduce a uniform macroscopic stress state within the specimen but induce the development of confined concrete cones at both ends of the specimen due to the contact with rigid loading plates. On the other hand, load configurations occurring in highly prestressed concrete structures include multiaxial stress configurations that can exhibit significant deviations from the fatigue life determined in the cylinder tests.

In fact, even in cylinder specimens a nonuniform stress profile and, thus, multiaxial stress state can occur at the later stages of fatigue loading. The global failure of the cylinder specimens is induced through the formation of inclined sliding planes along the boundary surfaces of confined and unconfined concrete regions. Therefore, the fatigue damage development in concrete cylinders and cubes is primarily driven by the shear sliding along the surface of the confined cones near the load introduction plates. Therefore, idealization of this test setup using a uniaxial stress state cannot deliver a fully objective description of the fatigue behaviour as it does

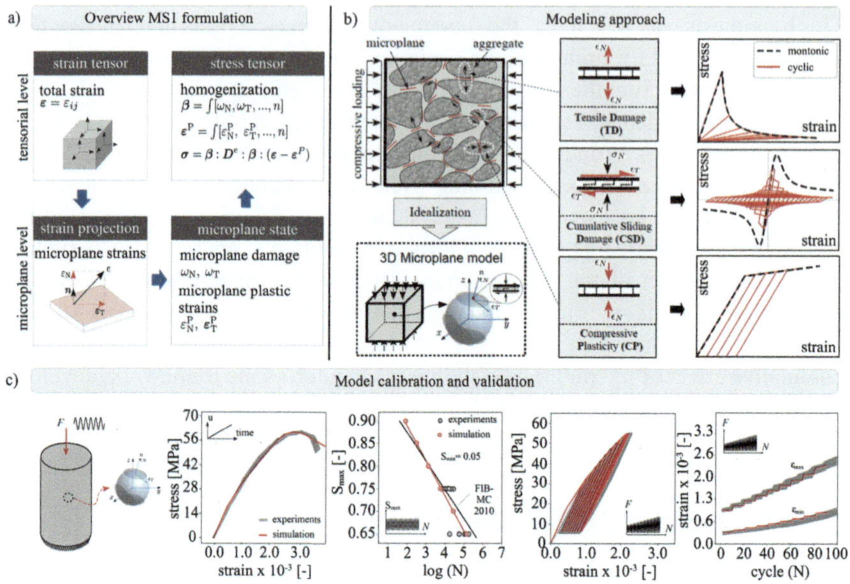

Fig. 1 Summary of the microplane fatigue MS1 model: **a** overview of the working principle of microplane model MS1, **b** dissipative mechanisms included in the microplane fatigue model MS1 and illustration of the microstructure containing a system of dissipative microplanes that are integrated within a 3D idealization [21, 25], **c** material model calibration and validation for monotonic, cyclic and fatigue behaviour (from [26], licensed under CC-BY 4.0)

not capture the key mechanism of the fatigue damage development. Although it is possible to reproduce the measured fatigue development of the entire specimen in terms of an anisotropic inelastic material model with distinguished volumetric and deviatoric strain components, it is impossible to successfully predict the response of a specimen with different sizes and slenderness values. This deficit clearly indicates that these methods are not able to deliver an objective characterization of the fatigue behaviour of a representative zone of material. The assumption that the fatigue damage development in cylindrical compressive specimens is driven by oscillating shear strain along the sliding planes, naturally leads to the idea to introduce shear and compressive loading in a controllable and uniform way.

3 Development of a Refined Test Setup

For a comprehensive experimental characterization of high-strength concrete under monotonic, cyclic, and fatigue shear loading, it is essential to recognize that shear (mode II) fracture can be classified into two distinct categories: pure mode II fracture under pure shear loading and compressive shear fracture under compressive shear loading [27, 28]. The main challenge in all experiments is to avoid mode I fracture and

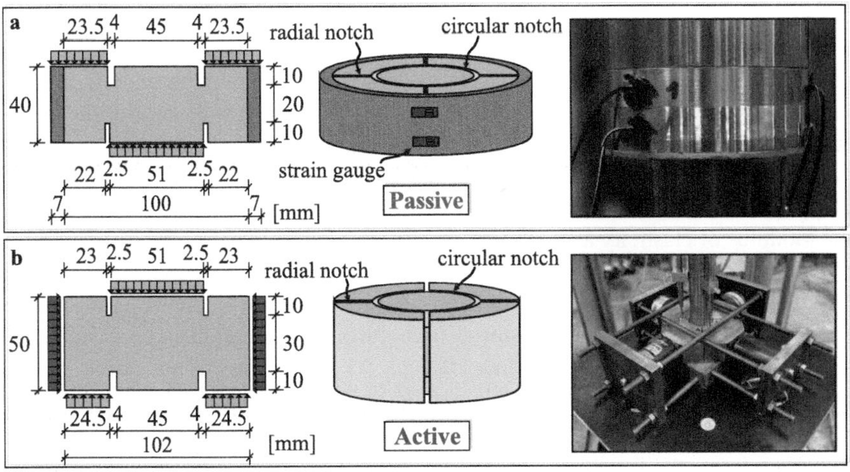

Fig. 2 Specimen geometry and test setup of both PTST types: shear loading with **a** passive (partly from [36], licensed under CC-BY 4.0), **b** active lateral compressive loading (partly from [39], licensed under CC-BY 4.0)

provoke macroscopic cracking along a distinguished material zone. According to Lin et al. [27], the punch-through shear test (PTST) is particularly effective in inducing *compressive shear fracture* and the cylindrical version of this test, as developed originally by Luong [29, 30] has been demonstrated in several studies to be well-suited for an in-depth investigation of shear behaviour [27, 31–35]. Consequently, the cylindrical PTST was adopted as the primary test setup in the current research.

When a body is subjected to shear, the volume increases, a phenomenon known as dilation. To prevent mode I failure during this process, it is necessary to suppress the tensile stresses developing within the specimen. Two distinct approaches were pursued for this purpose: firstly, concrete was cast within a steel ring to confine the dilation and subject the specimen to passive lateral compressive loading, measured using strain gauges (Fig. 2a) [36, 37]. Given that lateral compressive loading exerts a significant influence on material behaviour under shear loading, a novel test setup was subsequently developed that applies controlled lateral compressive loading to the specimen using four steel jaws (Fig. 2b) [38–40]. The latter was also used to study mortar and 3D-printed concrete.

4 Experimental Campaign

Within the scope of the presented research, multiple experimental campaigns were conducted. Due to space constraints, the following section focuses on presenting the main investigation campaign for characterizing material behaviour under monotonic and fatigue shear loading. Detailed results of these experiments and all accompanying evaluations are extensively documented in [39] and [40].

The tests were conducted using the SPP 2020 high-strength concrete mixture with a maximum grain size of 8 mm, see Sect. 2.1 "Concrete Composition" in the chapter by M. Basaldella et al., this volume. To evaluate the material properties, various specimens (material samples) were prepared and tested after 28 days. The compressive strength was tested on cube specimens (a = 150 mm) as 96 MPa, while the modulus of elasticity and splitting tensile strength were determined on cylinders (h/d = 300 mm/150 mm) as 39,226 MPa and 4.3 MPa, respectively.

The specimens were cast in steel molds, cured together with the material samples for one day, and subsequently stored until further preparation. Approximately a week before testing, the top surface of each specimen was ground to be parallel with the bottom surface. Subsequently, radial notches were cut into the specimens using a masonry saw and circular notches were created using a core drill. The entire preparation process is depicted in [40].

To obtain accurate information on fracture behaviour, damage evolution, and nonlinear shear-compression interaction, various loading scenarios were adopted. In the following, displacement-controlled monotonic tests, single stage fatigue tests with constant amplitude, and multistage fatigue tests with varying load increments will be presented. At the onset of each test, the specimen was first subjected to the respective lateral compressive loading, followed by the application of the shear loading. The loading rate was controlled by the displacement of the main hydraulic cylinder and stopped after reaching 6 mm.

5 Experimental Results and Discussion

The precise application of the lateral compressive loading σ_c was crucial for the systematic execution of the experiments. To maintain a consistent σ_c load level, the hydraulic cylinders were programmed to continuously readjust the applied force. This proactive measure prevents any increase in σ_c that would otherwise occur due to concrete dilatation [36]. To assess the accuracy and temporal variation of σ_c, the force exerted by the hydraulic cylinders was determined using two load cells (Fig. 2b). Further details on calculating σ_c (simplistically assumed to be uniformly distributed) can be found in [39, 40].

In all the following diagrams, the measured testing machine force F is depicted in kN—without postprocessing. Minor variations in the height of the shear surface (due to drilling inaccuracies), as well as the gradual reduction of the shear surface (due to the continuously increasing displacement), were not taken into account. An evaluation of the approximated shear stress τ—where both these influences were considered—can be found in publications [39] and [40].

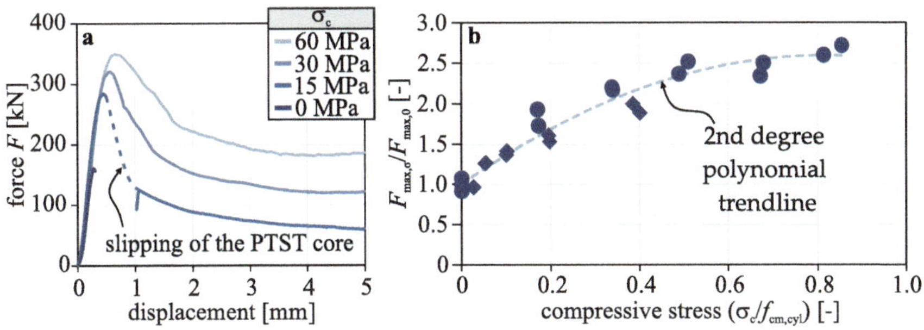

Fig. 3 Monotonic shear tests with various levels of lateral compressive loading: **a** force versus displacement and **b** influence of lateral compressive stress on maximum shear stress (adapted from [39], licensed under CC-BY 4.0)

5.1 *Monotonic Behaviour*

Monotonic shear failure is characterized by fracture occurring between the interior of the upper circular notch and the exterior of the lower notch, leading to the formation of an almost straight shear surface. Both shear strength and post-peak strength—largely influenced by friction between the inner concrete core and outer concrete ring—increase with higher lateral compressive stress (Fig. 3a). While the pre-peak loading behaviour appears unaffected by lateral compressive stress, there are distinct differences in post-peak behaviour. Higher levels of lateral compressive stress lead to a uniform decrease in load after reaching maximum shear strength, whereas specimens subjected to lower lateral compressive stress typically exhibit a brief post-peak slippage of the inner PTST core (a phenomenon already documented in [36] which does not affect the shape of the stress-displacement curve). Specimens without lateral compressive loading fail immediately after reaching maximum shear capacity, showing no post-peak behaviour. As shown in Fig. 3b, the shear strength increases sub-proportionally with increasing lateral compressive stress, with no further increase in shear strength beyond 45 MPa. The low scatter of individual results suggests effective homogenization of loading, absence of side effects, and activation of the entire shear surface.

5.2 *Fatigue Behaviour Under Constant Amplitude Loading*

To quantify the influence of a lateral compressive loading on the fatigue shear behaviour, multiple tests were conducted with varied levels of σ_c. The compressive loading was kept constant throughout the entire test, while the shear loading oscillated sinusoidally around a mean load ($S_{min} = 0.05$ and $S_{max} = 0.85$). It should be noted that both S_{max} and S_{min} always refer to the maximum shear stress under the respective lateral compressive loading i.e., the increase in shear strength due to σ_c (cf. Figure 3) is taken into account.

The force–displacement curve depicted in Fig. 4a illustrates a uniform and continuous loading and unloading behaviour. The displacement at shear fatigue failure approximately corresponds to the displacement observed in the softening branch of the reference test at an equivalent load level. Notably, the loading stiffness in all fatigue tests exceeds that of the monotonic reference tests, primarily due to the significantly higher loading rate. The displacement progression throughout the fatigue life exhibits the characteristic three-phased S-curve, with an almost linear increase during the second phase (Fig. 4b), while the reduction in reloading stiffness—calculated as described in [39]—is relatively moderate and reflects the phase-dependent behaviour (Fig. 4c). Over the entire fatigue life, there is a decrease of about 30% across all studied lateral compressive load levels, which correlates well with other investigations [36].

Figure 4d depicts the experimentally determined number of cycles with respect to the lateral compressive load level. Evidently, lateral compressive stress does not affect the fatigue life, indicating that fatigue life scales proportionally with monotonic shear strength. However, whether this behaviour applies to any S_{max} and S_{min} level—and not just to the tested combination of $S_{max} = 0.85$ and $S_{min} = 0.05$—remains to be verified. Additionally, it appears that the scatter of results diminishes with increasing level lateral compressive loading. This reduction in scatter could be attributed to the increasing homogenization of stresses along the shear surface with higher lateral compressive loading. A comprehensive overview of applied loads, achieved fatigue lives, and fracture behaviour can be found in [39].

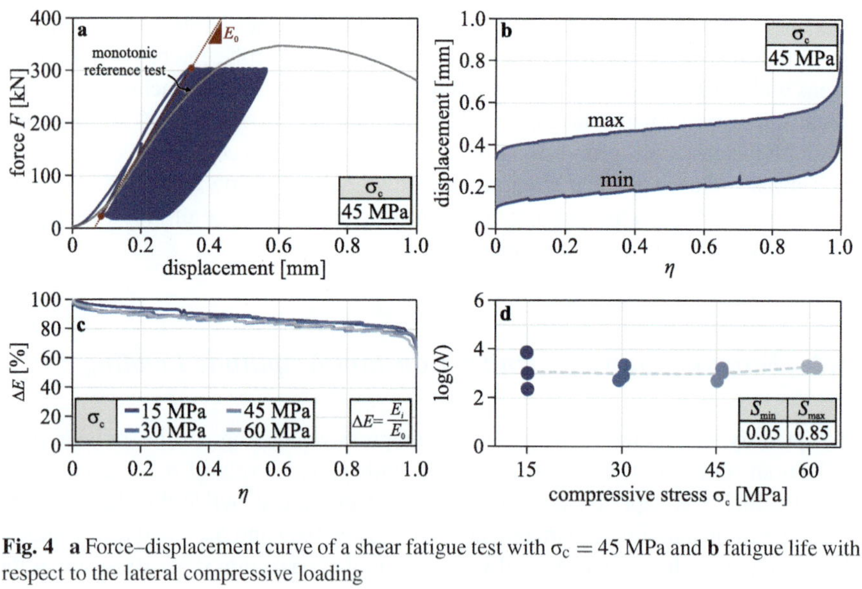

Fig. 4 **a** Force–displacement curve of a shear fatigue test with $\sigma_c = 45$ MPa and **b** fatigue life with respect to the lateral compressive loading

5.3 *Fatigue Behaviour Under Variable Amplitude Loading*

To quantify the influence of a varying load amplitude on the fatigue life of concrete and capture precise insights into the accompanying fracture behaviour and damage progression, more than 30 multistage fatigue tests were conducted in the scope of an extensive experimental campaign [40]. As previous tests have shown that the scatter of results decreases with increasing lateral compressive loading (Fig. 4a), σ_c was set to 60 MPa for all multistage fatigue experiments. Eight different load sequences with four distinct transition points (jump from one upper load level to another) were investigated. An exemplary case is depicted in Fig. 5a, showing a stress-displacement curve of a test with an L-H (low to high) transition at 52%. Further tests examined three distinct load sequences, with a maximum of 3 and 19 load jumps, respectively. Figure 5c illustrates a representative experiment with three load jumps. Each jump occurred after 22% of the reference lifespan, starting at the low upper load level (L). The individual load jumps are clearly visible, and the behaviour resembles that under constant amplitude loading.

Corresponding to the tests depicted in Fig. 5a, c, the results of all three repetitions of the respective load sequences are plotted as fatigue-displacement curves in Fig. 5b, d. The lifetime depicted on the horizontal axis of these fatigue-displacement curves is normalized relative to the predicted lifespan estimated by the P-M rule [41, 42]. The curves indicate that the existing scatter—though exceptionally low—can be explained by the varying displacement levels. While a high displacement level leads to accelerated failure, a low displacement level extends fatigue life. The difference in displacement level can be attributed to natural material scattering as well as notch accuracy.

According to the P-M rule, each PTST is expected to fail once it reaches $\eta = 1$. Therefore, in Fig. 5e, f, the fatigue life of each specimen is depicted using η_{H} and η_{L} (the number of cycles at the respective high/low load level relative to the reference fatigue life) and is compared with the P-M rule. The results exhibit notable deviations from the dashed line. Hence, in alignment with findings from other researchers, it appears that the P-M rule is not applicable to concrete subjected to variable amplitude loading. Specifically focusing on $\eta \approx 0.5$ (Fig. 5e), they seem to validate existing literature, suggesting a sequence effect, i.e., an L-H loading sequence prolongs lifespan, whereas an H-L loading sequence reduces it. This trend persists for H-L loading sequences even with $\eta < 0.5$, but it's reversed for L-H loading sequences, leading to progressively shorter lifespans with earlier load jumps (Fig. 5e).

Looking at Fig. 5e, f, it's evident that the majority of experiments are overestimated by the P-M rule. Additionally, Fig. 5f indicates that a significant increase in the number of load jumps (L-H mult. 22% = up to 3 jumps, L-H mult. 4% = up to 19 jumps) does not result in a noteworthy reduction of fatigue life.

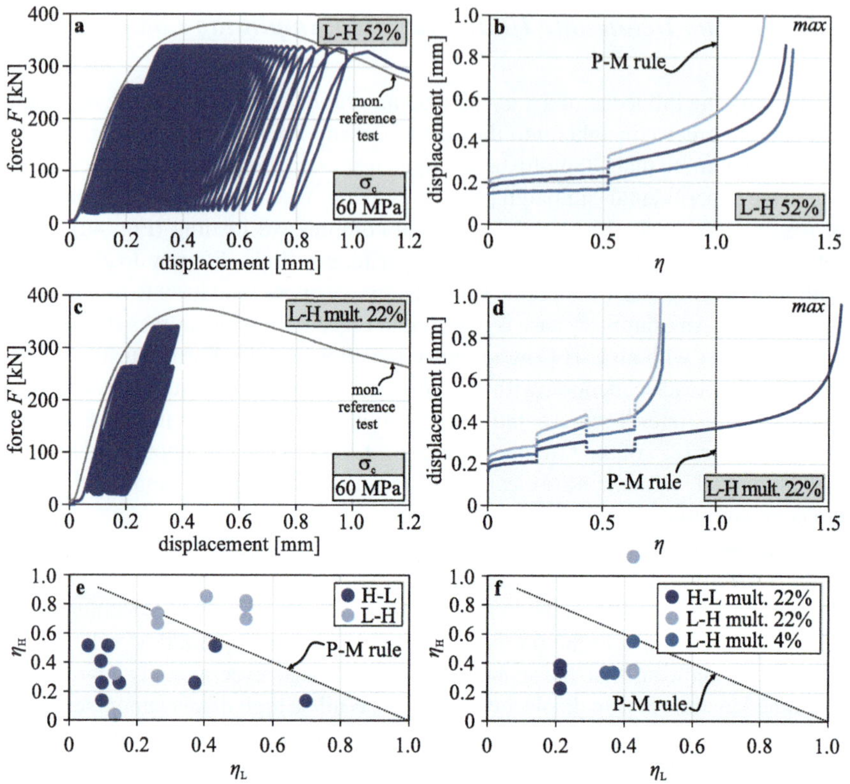

Fig. 5 Variable amplitude fatigue tests: **a,c** stress–displacement curves; **b, d** fatigue-displacement curves, and **e, f** fatigue life versus P-M rule

6 Numerical Modelling of Multiaxial Fatigue

The computational analysis employed the microplane constitutive model MS1, which integrates a thermodynamically-based fatigue model of cementitious interfaces [22, 23]. This model framework is formulated at the mesoscale level and has demonstrated robust predictive capabilities across diverse loading scenarios including uniaxial and multiaxial configurations, encompassing both monotonic and subcritical cyclic loading conditions [21, 26, 43–46]. The results presented in this section represent a condensed summary of the comprehensive numerical investigations detailed in [26], where in-depth analysis and interpretation can be found.

In order to achieve an efficient cycle-by-cycle simulation of the fatigue behavior of the PTST, a reduced-complexity model was created that exploits the symmetries of the specimen [43], as seen in Fig. 6a. Using boundary conditions that allow for an isolation of the significant effects, the finite element model includes one-eighth of the specimen. Specifically, on the side of the radial notch, corresponding to surfaces

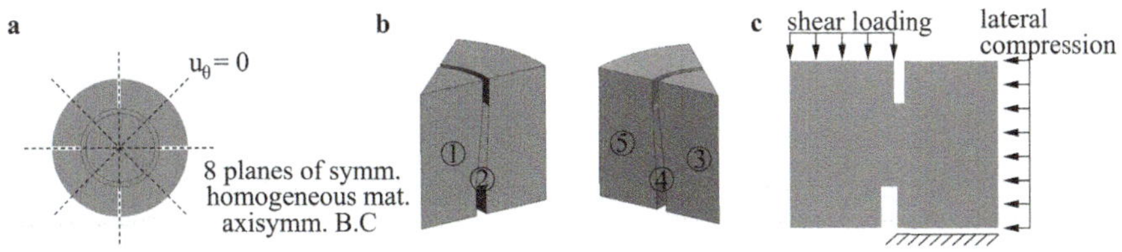

Fig. 6 Idealization of the PTST: **a** planes of symmetry, **b** surfaces in which the displacements perpendicular to them are constrained, **c** schematic representation of the loading and support

1 and 2 of Fig. 6b, the deformations perpendicular to the notch were constrained in the region corresponding to the inner cylinder and the ligament. On the opposite side, corresponding to surfaces 3–5 of Fig. 6b, all deformations perpendicular to the plane were constrained. In the outer regions, where the material exhibits elastic behavior, the mesh size was left coarse, while in the ligament it was refined. A total of 1732 nodes were used as linear tetrahedral elements. The loading applied in the model was performed analogously to the experiment, as schematically depicted in Fig. 6c. In the first step, the confinement pressure was applied to the outer area of the specimen, and in the second step, the shear loading was applied to the top of the inner cylinder, while keeping the confinement level constant. The computing time required for the simulation of 411 cycles was around 44 h on a computer with 8 cores and 16 GB of memory.

Hysteretic Behaviour and Fatigue Life. The presented computational analyses examined PTST specimens under radial confinements of 15 and 30 MPa with maximum load levels S_{max} of 0.95 and 0.90, comparing numerical calibration against experimental measurements [39]. Figure 7 presents the comparative numerical-experimental results. Force–displacement curves for 15 MPa confinement at S_{max} equal to 0.9 are shown in Fig. 7a. The corresponding creep-fatigue curves are shown in Fig. 7b. Figure 7c compares predicted and measured fatigue lives for both confinement levels (15 and 30 MPa) for various shear loading amplitudes. Extended results and detailed analysis can be found in [26].

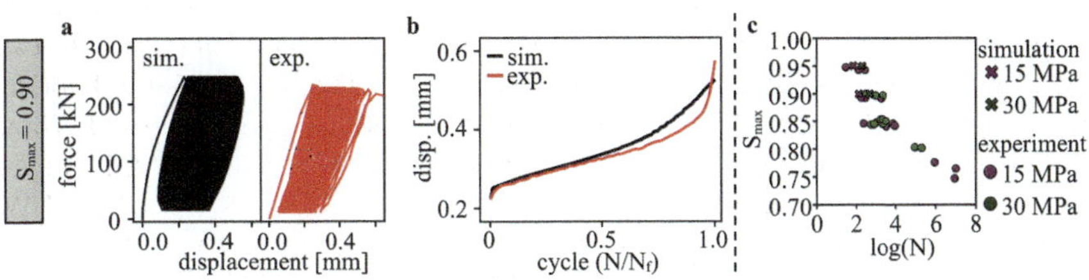

Fig. 7 Numerical-experimental characterization of PTST subjected to constant loading amplitude: **a** corresponds to the simulated and experimental force–displacement curves for $S_{max} = 0.90$, **b** presents the corresponding creep fatigue curve, **c** presents the experimental and simulated fatigue life for different amplitudes and confinement levels (partly adapted from [26], licensed under CC-BY 4.0)

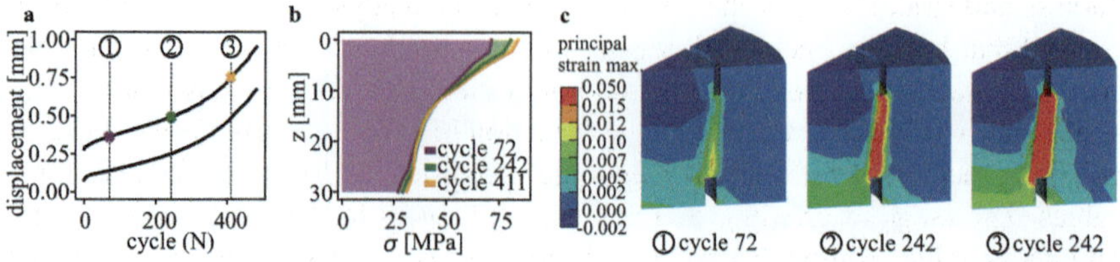

Fig. 8 Numerically obtained shear stress profiles in PTST specimens subjected to different constant loading amplitudes. **a** Correspond to the creep fatigue curve for loading scenarios with S_{max} equal to 0.95 and 0.90, while S_{min} is held constant at 0.05 for the specified confinement level, **b** shows the evaluated shear stress profile along the ligament of the specimen at the three specified cycles, **c** shows the evolution of ε_I during the fatigue life of the specimen (adapted from [26], licensed under CC-BY 4.0)

Stress Redistribution and Damage Localization. Figure 8b shows stable shear stress profiles throughout fatigue life for 30 MPa confinement at S_{max} equal to 0.9, indicating uniform damage evolution within the ligament. Strain contours (Fig. 8c) reveal strain concentrations in the lower ligament region from early cycles (cycle 72). Extended results and detailed analysis can be found in [26].

Effect of Fatigue Loading Sequence and Analysis of Energy Dissipation. Structures experience variable-amplitude fatigue loading during service. In order to account for this, the Fib Model Code [47] prescribes the Palmgren-Miner (P-M) rule, which assumes linear damage accumulation to estimate fatigue life, regardless of the sequence in which different amplitudes are applied. However, experimental data show that the P-M rule can provide unconservative predictions [10, 12, 40], requiring further investigation of variable amplitude effects.

The effect of loading sequence was investigated by studying two load levels: high (H, $S_{max} = 0.95$) and low (L, $S_{max} = 0.90$) with $S_{min} = 0.05$. The P-M rule predicts that after consuming 50% of H life (0.50 N_H^f), the remaining capacity equals 50% of L life (0.50 N_L^f). Both H-L and L-H sequences were simulated using FE models with MS1 (Fig. 9). Figure 9 compares the H-L (left) and L-H (right) sequences. Figure 9a, b present creep-fatigue curves for sequence loading (solid line) versus constant amplitude loading (dashed line, $S_{max} = 0.95/0.90$), showing transitions at 50% fatigue life.

Simulations reveal that H-L sequences fail earlier than P-M predictions (reduced life shown in red), while L-H sequences extend life beyond P-M estimates (shown in green). The sequence effect can be explained through cumulative energy damage dissipation analysis shown in Fig. 9c, d. For H-L loading, when the material reaches 50% of H cycles, over 50% damage dissipation occurs, causing premature failure when switching to L loading. By comparing energy states between sequence and constant amplitude loading—by shifting the transition point (point 1) to the constant-L energy curve (point 2), it becomes clear that over 50% of damage has occurred, leading to premature failure. Conversely, for L-H loading, at 50% of L cycles only 35% of the total damage dissipation occured. When switching to H loading, more capacity remains than P-M predicts, extending the fatigue life. Energy contour

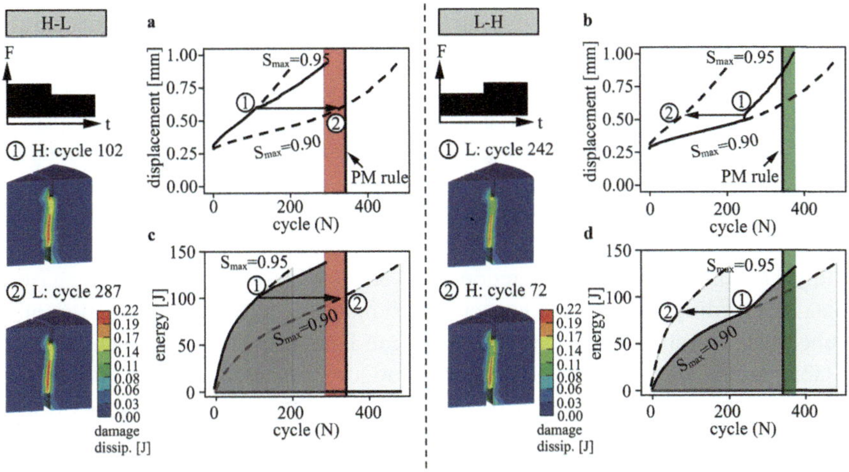

Fig. 9 Numerical prediction of a PTST subjected to a varying amplitude fatigue loading and energy dissipation analysis. **a**, **b** Show the creep fatigue curve for the considered H-L and L-H scenarios, accompanied by the creep fatigue curve for constant amplitudes $S_{\max}$ 0.95 and 0.90. Similarly, **c**, **d** show the energy dissipation profiles for the considered scenarios. In addition, contour plots are shown for each loading scenario displaying the damage dissipation for the specified cycles at constant amplitudes (from [26], licensed under CC-BY 4.0)

comparisons at specific cycles validate this interpretation, showing similar damage patterns between different loading histories at equivalent energy dissipation levels. Notably, the energy dissipation analysis reveals nearly constant damage dissipation across all loading cases studied, a finding consistent with previous investigations [45, 48]. Extended results and detailed analysis can be found in [26].

These computational findings validate the authors' proposed hypothesis. Given spatially uniform damage distribution within a dissipative zone, residual lifetime under sequential loading can be estimated by translating along energy dissipation profiles obtained from uniform-amplitude loading.

7 Conclusions

In order to induce inter-aggregate relative displacement, which the authors hypothesize to be the main mechanical variable that induces fatigue in concrete, a novel punch-through shear test setup was developed. This test allowed the systematic study of concrete fatigue in a material zone subjected to controllable levels of constant compression and oscillating shear. This is made possible because the setup allows independent application and control of the loading in the vertical and radial directions of a notched disk, providing better control of the stress state at the tested ligament, which improves the reproducibility of the test. The experimental campaign was designed and further interpreted using the microplane model MS1, developed by the authors, which incorporates the cumulative shear hypothesis in its microplane thermodynamical formulation. MS1 features anisotropic damage evolution and stress redistribution at the single material point level, allowing to study certain aspects of the concrete fatigue phenomenology with idealizations of maximum simplicity and to realistically reproduce stress redistribution due to fatigue at the structural level.

The computational-experimental investigations encompassed monotonic response, stepwise cyclic loading, and both constant and variable amplitude fatigue loading. Numerical predictions showed good correspondence with measured data, confirming the model's capacity to simulate progressive degradation under subcritical pulsating loads. Simulated fatigue lives matched experimental values. Stable shear stress profiles throughout fatigue life indicated uniform damage evolution without stress redistribution, confirming the suitability of the test for material characterization.

Investigations of loading sequence effects combined experimental observations with computational life predictions across variable amplitude histories. Findings aligned with published cylinder test data under comparable conditions. Predictions indicated unsafe lifetimes for high-to-low (H-L) sequences compared to Palmgren-Miner rule, while low-to-high (L-H) sequences demonstrated the opposite trend. These results establish a foundation for enhanced lifetime estimation in concrete infrastructure subjected to real variable amplitude loadings.

References

1. Schneider, S., Vöcker, D., Marx, S.: Zum Einfluss der Belastungsfrequenz und der Spannungsgeschwindigkeit auf die Ermüdungsfestigkeit von Beton. BuSt (2012). https://doi.org/10.1002/best.201200054
2. Elsmeier, K., Hümme, J., Oneschkow, N., Lohaus, L.: Prüftechnische Einflüsse auf das Ermüdungsverhalten hochfester feinkörniger Vergussbetone. Beton- und Stahlbetonbau (2016). https://doi.org/10.1002/best.201500065
3. Deutscher, M., Tran, N.L., Scheerer, S.: Experimental investigations on temperature generation and release of ultra-high performance concrete during fatigue tests. Appl. Sci. (2020). https://doi.org/10.3390/app10175845
4. Abubakar Ali, M., Tomann, C., Aldakheel, F., Mahlbacher, M., Noii, N., Oneschkow, N., Drake, K.-H., Lohaus, L., Wriggers, P., Haist, M.: Influence of moisture content and wet environment on the fatigue behaviour of high-strength concrete. Materials (2022). https://doi.org/10.3390/ma15031025
5. Markert, M., Katzmann, J., Birtel, V., Garrecht, H., Steeb, H.: Investigation of the influence of moisture content on fatigue behaviour of HPC by using DMA and XRCT. Materials (Basel, Switzerland) (2021). https://doi.org/10.3390/ma15010091
6. Tomann, C., Oneschkow, N.: Influence of moisture content in the microstructure on the fatigue deterioration of high-strength concrete. Struct. Concr. (2019). https://doi.org/10.1002/suco.201900023
7. Gallwoszus, J., Claßen, M.: Ermüdung von Verbunddübelleisten in UHPC unter zyklischer Pull-out-Beanspruchung. Bautechnik (2015). https://doi.org/10.1002/bate.201500020

8. Baktheer, A., Spartali, H., Chudoba, R., Hegger, J.: Concrete splitting and tip-bearing effect in the bond of anchored bars tested under fatigue loading in the push-in mode: an experimental investigation. Mater. Struct. (2022). https://doi.org/10.1617/s11527-022-01935-7
9. Subramaniam, K.V., Popovics, J.S., Shah, S.P.: Fatigue fracture of concrete subjected to biaxial stresses in the tensile C-T region. J. Eng. Mech. (2002). https://doi.org/10.1061/(ASCE)0733-9399(2002)128:6(668)
10. Holmen, J.O.: Fatigue of Concrete by Constant and Variable Amplitude Loading. Norwegian Institue of Technology, University of Trondheim, Norway (1979). Accessed 3 May 2020
11. Petkovic, G., Lenschow, R., Stemland, H., Rosseland, S.: Fatigue of high-strength concrete. In: Hester, W.T. (ed.) SP-121: High-Strength Concrete. Second International Symposium. Second International Symposium on the Utilization of High Strength Concrete, Berkeley, CA, USA, pp. 505–526. American Concrete Institute, Michigan, USA (1990). https://doi.org/10.14359/3740
12. Baktheer, A., Chudoba, R.: Experimental and theoretical evidence for the load sequence effect in the compressive fatigue behavior of concrete. Mater. Struct. (2021). https://doi.org/10.1617/s11527-021-01667-0
13. Hilsdorf, H., Kesler, C.: Fatigue Strength of Concrete Under Varying Flexural Stresses. JP (1966). https://doi.org/10.14359/7662
14. Alrayes, O., Könke, C., Ooi, E.T., Hamdia, K.M.: Modeling cyclic crack propagation in concrete using the scaled boundary finite element method coupled with the cumulative damage-plasticity constitutive law. Materials (2023). https://doi.org/10.3390/ma16020863
15. Baktheer, A., Martínez-Pañeda, E., Aldakheel, F.: Phase field cohesive zone modeling for fatigue crack propagation in quasi-brittle materials. Comput. Methods Appl. Mech. Eng. (2024). https://doi.org/10.1016/j.cma.2024.116834
16. Gudzulic, V., Daadouch, K., Meschke, G.: Modelling of damage process in concrete under monotonic and cyclic loading. In: 11th International Conference on Fracture Mechanics of Concrete
17. Kirane, K., Bažant, Z.P.: Microplane damage model for fatigue of quasibrittle materials: sub-critical crack growth, lifetime and residual strength. Int. J. Fatigue (2015). https://doi.org/10.1016/j.ijfatigue.2014.08.012
18. Alliche, A.: Damage model for fatigue loading of concrete. Int. J. Fatigue (2004). https://doi.org/10.1016/j.ijfatigue.2004.02.006
19. Desmorat, R., Ragueneau, F., Pham, H.: Continuum damage mechanics for hysteresis and fatigue of quasi-brittle materials and structures. Int. J. Numer. Anal. Meth. Geomech. (2007). https://doi.org/10.1002/nag.532
20. Baktheer, A., Chudoba, R.: Classification and evaluation of phenomenological numerical models for concrete fatigue behavior under compression. Constr. Build. Mater. (2019). https://doi.org/10.1016/j.conbuildmat.2019.06.022
21. Baktheer, A., Aguilar, M., Chudoba, R.: Microplane fatigue model MS1 for plain concrete under compression with damage evolution driven by cumulative inelastic shear strain. Int. J. Plast. (2021). https://doi.org/10.1016/j.ijplas.2021.102950
22. Baktheer, A., Chudoba, R.: Modeling of bond fatigue in reinforced concrete based on cumulative measure of slip. In: Meschke, G., Pichler, B., Rots, J.G. (eds.) Computational Modelling of Concrete and Concrete Structures. Conference on Computational Modelling of Concrete and Concrete Structures (EURO-C 2018), Bad Hofgastein, Austria, 26.02.–01.03., pp. 767–776. CRC Press, Taylor & Francis Group, Boca Raton (2018)
23. Chudoba, R., Vořechovský, M., Aguilar, M., Baktheer, A.: Coupled sliding–decohesion–compression model for a consistent description of monotonic and fatigue behavior of material interfaces. Comput. Methods Appl. Mech. Eng. (2022). https://doi.org/10.1016/j.cma.2022.115259
24. Skarżyński, Ł, Marzec, I., Tejchman, J.: Fracture evolution in concrete compressive fatigue experiments based on X-ray micro-CT images. Int. J. Fatigue (2019). https://doi.org/10.1016/j.ijfatigue.2019.02.002

25. Baktheer, A., Aguilar, M., Hegger, J., Chudoba, R.: Microplane damage plastic model for plain concrete subjected to compressive fatigue loading. In: Pijaudier-Cabot, G., Grassl. P., La Borderie, C. (eds.) Proceedings of the 10th International Conference on Fracture Mechanics of Concrete and Concrete Structures, FraMCoS-10, Bayonne, France, 23.06.–26.06. (2019). https://doi.org/10.21012/FC10.233196
26. Aguilar, M., Baktheer, A., Chudoba, R.: Multi-axial fatigue of high-strength concrete: model-enabled interpretation of punch-through shear test response. Eng. Fract. Mech. (2024). https://doi.org/10.1016/j.engfracmech.2024.110532
27. Lin, Q., Wang, S., Pan, P.-Z., Ji, W.-W., Lu, Y.: Fracture initiation under pure shear revisited: Remarks on the mode II fracture in quasi-brittle materials. Theoret. Appl. Fract. Mech. (2020). https://doi.org/10.1016/j.tafmec.2020.102700
28. Liu, H.-Z., Lin, J.-S., He, J.-D., Xie, H.-Q.: Dominant mode of planar fractures and the role of material properties. Eng. Fract. Mech. (2018). https://doi.org/10.1016/j.engfracmech.2018.03.024
29. Luong, M.P.: Direct tensile and direct shear strengths of fontainebleau sandstone. In: Cundall, P.A., Starfield, A.M., Sterling, R.L. (eds.) Key Questions in Rock Mechanics. 29th U.S. Symposium on Rock Mechanics (USRMS), Minneapolis, Minnesota, 13.06.–15.06., pp. 237–246. CRC Press, Rotterdam, The Netherlands (1988)
30. Luong, M.P.: Fracture testing of concrete and rock materials. Nucl. Eng. Des. (1992). https://doi.org/10.1016/0029-5493(92)90093-B
31. Backers, T., Stephansson, O., Rybacki, E.: Rock fracture toughness testing in mode II—punch-through shear test. Int. J. Rock Mech. Min. Sci. (2002). https://doi.org/10.1016/S1365-1609(02)00066-7
32. Montenegro, O.I., Sfer, D., Lopez, C.M., Carol, I.: Experimental tests and numerical modeling to identify the asymptotic shear-compression mode IIa of concrete fracture. In: Mier, J. van, Ruiz, G., Andrade, R.C., Yu, R.C., Zhang, X.X. (eds.) Proceedings of the 8th International Conference on Fracture Mechanics of Concrete and Concrete Structures, FraMCoS-8, Toledo, Spanien, 10.05.–14.05. (2013)
33. Lukić, B., Forquin, P.: Experimental characterization of the punch through shear strength of an ultra-high performance concrete. Int. J. Impact Eng. (2016). https://doi.org/10.1016/j.ijimpeng.2015.12.009
34. Backers, T., Stephansson, O.: ISRM suggested method for the determination of mode II fracture toughness. Rock Mech. Rock Eng. (2012). https://doi.org/10.1007/s00603-012-0271-9
35. Abdul Rahman, R., Forquin, P.: An experimental method to characterize the shear behaviour of concrete under high confinement. Exp. Mech. (2023). https://doi.org/10.1007/s11340-023-00948-9
36. Becks, H., Classen, M.: Mode II behavior of high-strength concrete under monotonic, cyclic and fatigue loading. Materials (2021). https://doi.org/10.3390/ma14247675
37. Becks, H., Aguilar, M., Baktheer, A., Chudoba, R., Classen, M.: Experimental and numerical investigations on the fatigue behavior of high-strength concrete under combined shear-compression loading. In: IABSE (ed.) Proceedings of IABSE Symposium: Challenges for Existing and Oncoming Structures. IABSE Symposium, Prague, Czech Republic, 25.05–27.05., pp. 532–540 (2022)
38. Becks, H., Aguilar, M., Hegger, J., Chudoba, R., Classen, M.: Experimental and numerical characterization of high-strength concrete under monotonic and fatigue mode II loading with active control of lateral compression. In: Fédération internationale du béton (fib) (ed.) Building for the future: Durable, Sustainable, Resilient. Proceedings of the fib Symposium 2023, Istanbul, Turkey, 05.06.–07.06., pp. 429–439 (2023). https://doi.org/10.1007/978-3-031-32519-9_41
39. Becks, H., Aguilar, M., Chudoba, R., Classen, M.: Characterization of high-strength concrete under monotonic and fatigue mode II loading with actively controlled level of lateral compression. Mater. Struct. (2022). https://doi.org/10.1617/s11527-022-02087-4
40. Becks, H., Classen, M.: New insights into the load sequence effect: Experimental characterization and incremental modeling of plain high-strength concrete under mode II fatigue loading with variable amplitude. Int. J. Fatigue (2024). https://doi.org/10.1016/j.ijfatigue.2024.108334

41. Palmgren, A.: Die Lebensdauer von Kugellagern. Zeitschrift des Vereines Deutscher Ingenieure (ZDVI) **68**, 339–341 (1924)
42. Miner, M.A.: Cumulative damage in fatigue. J. Appl. Mech. (1945). https://doi.org/10.1115/1.4009458
43. Aguilar, M., Baktheer, A., Becks, H., Classen, M., Chudoba, R.: Fatigue-induced concrete fracture under combined compression and shear studied using standard cylinder and refined punch-through shear test setup. In: 11th International Conference on Fracture Mechanics of Concrete and Concrete Structures (FraMCoS-11)
44. Aguilar, M., Becks, H., Baktheer, A., Claßen, M., Chudoba, R.: Fatigue of concrete studied using a confined punch- through shear test. In: Claßen, M., Hegger, J., Matschei, T., Raupach, M. (eds.) Beiträge zur 10. DAfStb-Jahrestagung mit 62. Forschungskolloquium. 62. DAfStb Forschungskolloquium, Aachen, 26.10.–27.10., pp. 217–224 (2023). https://doi.org/10.18154/RWTH-2023-06653
45. Aguilar, M., Baktheer, A., Chudoba, R.: On the energy dissipation in confined concrete subjected to shear cyclic loading. Proc. Appl. Math. Mech. (PAMM) (2023). https://doi.org/10.1002/pamm.202200301
46. Baktheer, A., Esfandiar, S., Aguilar, M., Becks, H., Claßen, M., Chudoba, R.: Fatigue-induced stress redistribution in prestressed concrete beams modeled using the constitutive hypothesis of inter-aggregate degradation. Fatigue Fract. Eng. Mater. Struct. (2024). https://doi.org/10.1111/ffe.14388
47. International Federation for Structural Concrete: fib Model Code for Concrete Structures. Ernst & Sohn, Berlin (Model Code 2010) (2013)
48. Aguilar, M., Baktheer, A., Chudoba, R.: Numerical investigation of load sequence effect and energy dissipation in concrete due to compressive fatigue loading using the new microplane fatigue model MS1. In: Oñate, E., Perić, D., Chiumenti, M., de Souza Neto, E.A. (eds.) Proceedings of the 16th International Conference on Computational Plasticity (COMPLAS), Barcelona, Spain, 06.09.–07.09., pp. 1–12 (2021). https://doi.org/10.23967/complas.2021.053

Fatigue Mechanisms in Cementitious Composites: Experimental and Numerical Analyses

Stefan Harenberg, Matthias Pahn, Kerstin Willers, Frank Dehn, Antonio Caggiano, Diego Schicchi, Sha Yang, and Eddie Koenders

Abstract This research reports an experimental and numerical study on the fatigue behaviour of high-performance concrete (UHPC) subjected to static and cyclic bending. The damage progress is monitored through digital image correlation in combination with clip gauges. Static tests are performed with the same loading rate as the cyclic tests. The latter were thus tested by adopting the similar maximum and minimum stress levels, but with different loading rates. A multiscale approach is proposed for analysing cyclic failures of concrete prisms under either low- or high-cycle fatigue actions. The combination of a micro-scale model, that describes the inner microstructural defects affecting the cyclic response, with an upscaled modelling approach is proposed to achieve the meso- and macro-scale failure patterns. For the micro-scale, projected representative volume elements, combined with a cracking model and equipped with continuous inelastic constitutive rules were employed to accumulate damages under cycle response. A plastic-damage-model was used for analysing specimens under cyclic loading. This has been formulated by merging fracture-energy theories with a stiffness damage degradation model, representing the key damage phenomenon in concrete under fatigue. The work demonstrates the potential of numerical techniques to explore the failure damage in UHPC due to low- and high cycle fatigue.

Keywords Fatigue · Cyclic · Experiments · Modelling · Fracture · DIC

S. Harenberg · M. Pahn
Fachgebiet Massivbau und Baukonstruktion, RPTU Kaiserslautern, Kaiserslautern, Germany

K. Willers · F. Dehn
Institut für Massivbau und Baustofftechnologie (IMB), KIT, Karlsruhe, Germany

A. Caggiano
Department of Civil, Chemical and Environmental Engineering, University of Genova, Genoa, Italy

D. Schicchi · S. Yang · E. Koenders (✉)
Institut für Werkstoffe im Bauwesen, TU Darmstadt, Darmstadt, Germany
e-mail: koenders@wib.tu-darmstadt.de

N. Oneschkow and L. Lohaus (eds.), *Cyclic Deterioration of High-Performance Concrete*, https://doi.org/10.1007/978-3-032-19394-0_6

1 Introduction

In recent years, the use of structural concrete has increased tremendously due to extensive infrastructure development worldwide [1]. Alongside plain and reinforced cement concrete, researchers and designers have developed structures made of ultra-high-performance concrete (UHPC) [2–4]. UHPCs are characterized by enhanced mechanical properties, particularly tensile resistance, ductility, resistance to cyclic and earthquake loading, and durability [5–7]. Therefore, it is extremely important to properly understand the behaviours of these materials under such loading conditions.

When subjected to repeated cyclic loading, structures suffer gradual stiffness reduction over time [8]. Due to this stiffness degradation, structures ultimately fail due to fatigue. Fatigue can thus be defined as the process of incremental damage in a structure resulting from cyclic loading, leading to structural failure without reaching ultimate strength [9]. Under repeated loading, cracks initiate in highly stressed regions, agglomerate to form macrocracks, and propagate through the structure with increasing irrecoverable strain, eventually resulting in complete failure. Fatigue damage in concrete and cementitious composites remains an area of ongoing scientific research. Due to fatigue, structures fail at stress levels lower than their strength before reaching their design life [10, 11]. Estimating the service life of structures under cyclic loading thus presents a significant challenge to engineers. For structures subjected to cyclic loading, such as bridge decks, concrete pavements, machine foundations, and offshore structures, fatigue is the primary cause of structural failure.

To explain the complex phenomenon of fatigue damage and predict service life, researchers have conducted numerous experimental [12–15] and theoretical investigations [16–29]. Guidelines and model codes addressing this topic are available at national and international levels [30–33]. However, understanding of fatigue in concrete and cementitious materials still lags behind that of metals, resulting in the development of several new models aimed at explaining this phenomenon more explicitly.

In this context, this work presents the results of experimental and numerical studies on UHPC materials and their degradation under cyclic tensile and bending tensile stress. First, an experimental programme is described that examines the fatigue behaviour of UHPC. The accuracy of static and cyclic strain measurements using photogrammetry as a function of measurement rate is evaluated. Then, a unified coupled damage-elastoplastic constitutive model for analysing specimens subjected to both Low Cycle Fatigue (LCF) and High Cycle Fatigue (HCF) is presented and discussed. The fracture-based damage model is combined with homogenized microscopic mechanics to capture damage accumulation in HCF.

2 Experimental Activities

2.1 Materials and Methods

The specimens were cast at the Institut für Massivbau und Baustofftechnologie (IMB)—KIT in Karlsruhe, Germany. For all test series, the standard UHPC mixture specified by the central project of the DFG Priority Programme SPP 2020 was utilized (see Sect. 2.1 "Concrete Composition" in the chapter by M. Basaldella et al., this volume). Table 1 of the named chapter presents the composition of the UHPC mixture, which consists primarily of Cement CEM I 52.5 R, microsilica, quartz powder, and sand. This mixture features a maximum aggregate size of 0.5 mm and a water-to-cement ratio of 0.24. The UHPC mixtures were prepared using a pan mixer following a specific sequence. Initially, cement and microsilica were thoroughly combined. Next, the aggregates were incorporated and mixed with the cement-microsilica blend. Subsequently, water and superplasticizer were added to complete the mixture. After moulding, the specimens were kept sealed in the formwork for 24 h, then immersed in water for 7 days. Finally, they were cured in a controlled laboratory environment maintained at 20 °C and 65% relative humidity until testing.

Bending tests were conducted on standard prisms measuring $40 \times 40 \times 160\ \text{mm}^3$ (in accordance with the EN 196-1 standard [34]). To ensure comparable measurement results, both Digital Image Correlation (DIC) measurements and clip gauges were employed at the bottom notch of the beams. The notch measured 6.7 mm in depth and 3 mm in width and was milled from the bottom surface of the specimens following EN 14651 [35] specifications (Fig. 1). At the time of testing, the specimens had reached between 50 and 54 days of hydration. Complementary compressive tests were performed on halves of the broken prisms, as prescribed by EN 196-1 [34], yielding a mean compressive strength of 107 N/mm^2 and a mean bending strength of 11.3 N/mm^2.

Table 1 Experimental programme

Test series	Specimen repetitions	Loading	S_{max} (%)	S_{min} (%)	Rate (Hz)	Speed (MPa/s)	DIC rate (Hz)	CG rate (Hz)
5	3 (A,B,C)	Static	–	–	–	0.06	15	1200
11	3 (A,B,C)	Static	–	–	–	0.06	15	1200
9	3 (A,B,C)	Cyclic	90	10	2.0	–	8	50
14	3 (A,B,C)	Static	–	–	–	0.06	15	1200
13	3 (A,B,C)	Cyclic	90	10	2.5	–	10	50
18	3 (A,B,C)	Static	–	–	–	0.06	15	1200
15	3 (A,B,C)	Cyclic	90	10	2.75	–	11	50
17	3 (A,B,C)	Cyclic	90	10	3.0	–	12	50

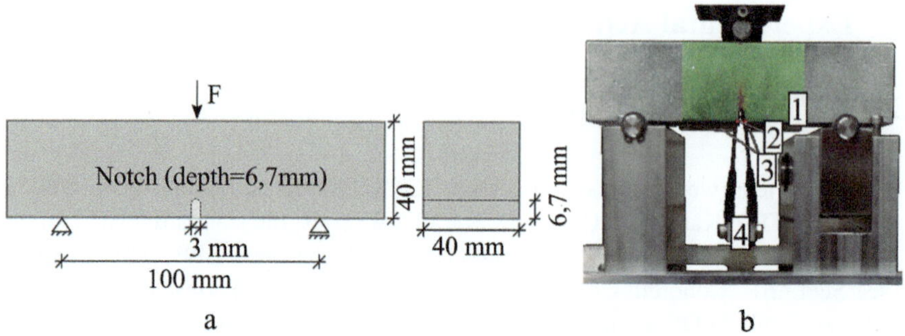

Fig. 1 **a** Specimen geometry, **b** Experimental setup (1 = DIC pattern, 2 = DIC-extensometer, 3 = glued metal plates and 4 = clip gauge) (from [15], licensed under CC-BY 4.0)

The objective of the experimental programme was to validate the strain measurements of UHPC under both static and cyclic bending, using a Digital Image Correlation (DIC) system (Fig. 2). A classical measurement technique, namely Clip Gauges (CGs), was employed for comparison with the DIC-based results. For this purpose, a CG UB-5A type (Tokyo Measuring Instruments Laboratory) was attached at the bottom of the notch to measure the Crack Mouth Opening Displacement (CMOD), which was then compared to the DIC results (Fig. 3). The clip gauge featured a measuring range of 3–8 mm and a measurement accuracy of 1000×10^6 strain/mm.

An ARAMIS 5 M industrial DIC system was employed, equipped with two cameras featuring monochrome CCD chips (2448×2050 pixels), electronic shutters, exposure times between 0.1 ms and 2 s, and adjustable frame rates of up to 15 frames per second. The DIC system provided a strain measurement range from 0.02% to > 100%, with a strain accuracy of 0.01%. In addition, the system enabled out-of-plane displacement measurements, requiring a calibration of the camera angles, distances, and the region of interest (ROI).

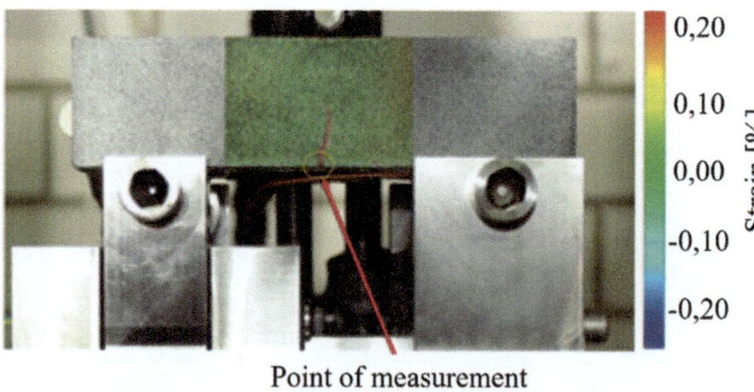

Fig. 2 DIC detail of measurements showing the area where the strain distribution during the bending test is monitored

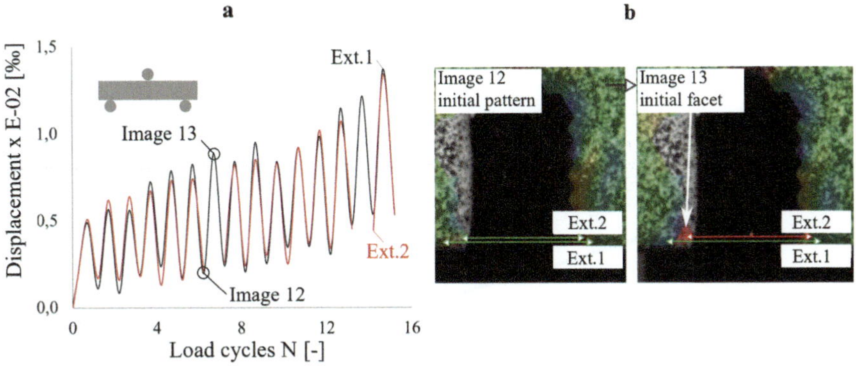

Fig. 3 **a** Detection of failure with DIC measurements in a cyclic test, **b** images of the measurement (from [15], licensed under CC-BY 4.0)

Prior to testing, the specimen surfaces were inspected for roughness, sanded if necessary, and subsequently coated with two or three layers of standard white acrylic paint. A black speckle pattern was then applied using graphite spray to ensure optimal DIC image quality.

Static and cyclic tests were conducted using a two-column frame from Instron, equipped with a servo-hydraulic actuator capable of applying static and cyclic forces up to 10 kN. Preliminary testing determined that reliable application of maximum and minimum stress levels was feasible only up to a loading frequency of 3 Hz.

The experiments were carried out at the Department of Concrete Structures and Structural Engineering of the RPTU Kaiserslautern. The experimental programme (Table 1) was subdivided into static and cyclic test series, with all specimens within a test series cast from the same formwork and UHPC batch.

Loading frequencies were varied between 0.5 Hz and 3.0 Hz, with a focus on high maximum stress levels (i.e., 90% of $S_{\max}$). In Table 1, the minimum and maximum stress levels are denoted as $S_{\min}$ (%) and $S_{\max}$ (%) respectively.

Static tests were performed under displacement control across four different series (i.e., Series 5, 11, 14, and 18). A loading rate of 0.06 MPa/s was applied, in accordance with EN 12390-5 [36]. During these tests, DIC measurements were taken at a rate of 15 frames per second, while the CMOD data from the clip gauge were recorded at 1.2 kHz. Displacement control was employed to precisely capture the peak strength and the post-cracking behaviour of the UHPC, allowing for accurate monitoring of the transition from elastic to brittle failure.

Conversely, cyclic tests were conducted under load control on five different series (i.e., Series 9, 13, 15, and 17), all featuring identical minimum and maximum stress levels but different loading rates. Given that the number of cycles to fatigue failure can reach several million, a minimum loading frequency of 0.5 Hz was selected, while the maximum frequency was determined by the machine's operational limits. Intermediate frequencies focused on achieving higher loading rates.

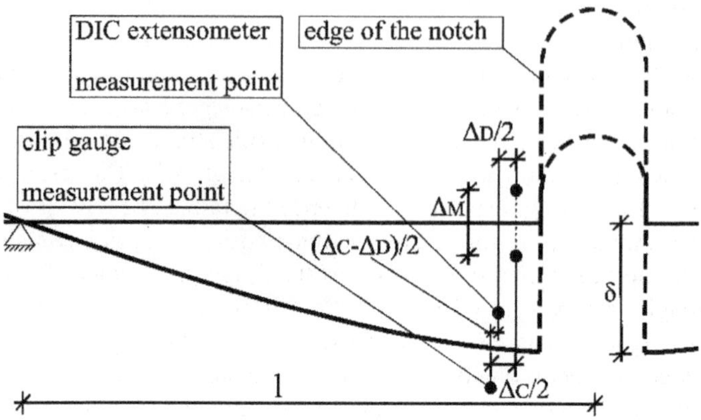

Fig. 4 Difference of measurements between DIC and the clip gauges (from [15], licensed under CC-BY 4.0)

In order to synchronize the DIC image acquisition with the minimum and maximum loads during cyclic testing, the camera's fade time (the time required to adjust exposure settings) was set to four times the loading rate. The fading time corresponds to the DIC measuring rate.

The displacement ΔC measured between two points by the clip gauge (Fig. 4) was converted into strain ε_C using an appropriate reference length L. To ensure direct comparison with the DIC results, two measurement points were defined within the DIC speckle pattern, close to the notch and specimen edges, where a virtual extensometer was placed. This extensometer provided a displacement output ΔD analogous to the clip gauge measurement. As with the clip gauge, the virtual extensometer could not be positioned exactly at the notch edge. Both strain values, ε_C (from CG) and ε_D (for DIC), were calculated using the notch width of 3 mm as the reference length.

Due to the different distances of the CG and DIC measurement points from the neutral axis of the specimen, discrepancies between the measured displacements at a given global deflection δ were expected (Fig. 4). These differences were estimated in a simplified manner, based on the deflection δ and the vertical distance between the clip gauge and DIC measurement points.

2.2 Experimental Results and Discussion

Figure 5 presents the results of the UHPC static tests. Each test series is shown separately, and the graphs are labelled according to the experimental programme. The label "CG" refers to the clip gauge, whilst "DIC" stands for Digital Image Correlation. It should be noted that no DIC results are available for specimen 5A, as the measurements could not be evaluated due to memory errors.

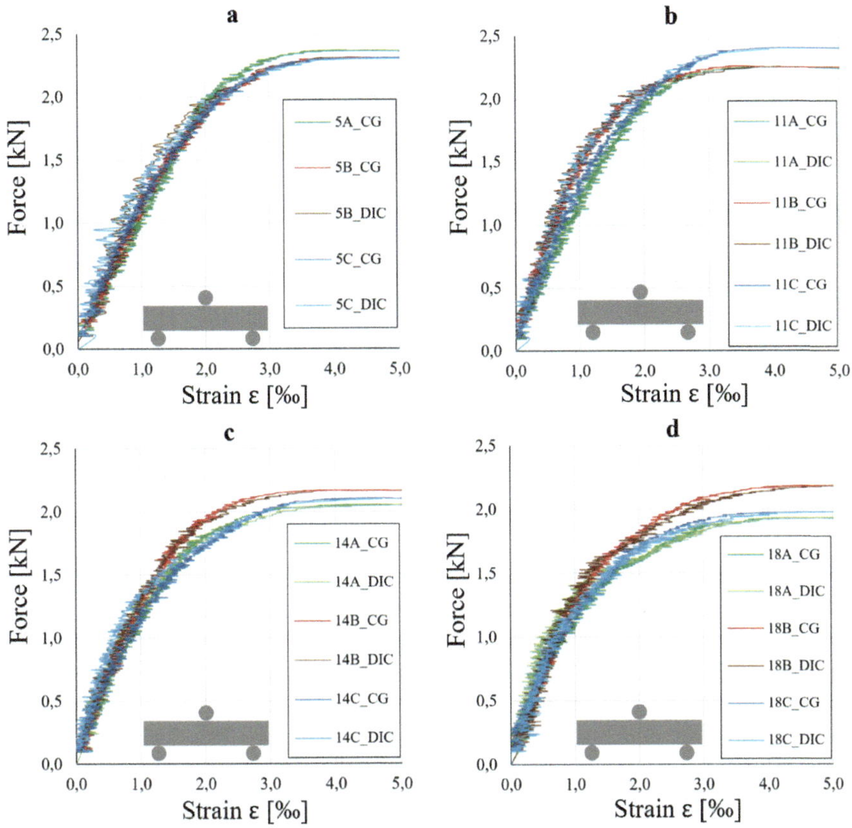

Fig. 5 F-strain curves of the static tests: **a** 5-type, **b** 11-type, **c** 14-type, **d** 18-type (from [15], licensed under CC-BY 4.0)

The strains measured by the CG and DIC are closely aligned for each specimen and in each test series. Furthermore, the strains of all specimens within the same group also show good agreement. This confirms that the experimental programme and the casting of the specimens were carried out with high precision. With regard to the linear behaviour observed up to forces of approximately 1.00–1.25 kN and fracture strains between 4‰ and 5‰, the results also show good consistency across the test series.

The accuracy of the DIC measurements is characterised by the standard deviation of the measured values. This was evaluated by calculating the deviation Δε of the measured strains from the mean strain curve (Fig. 6a), representing the measurement noise. Figure 6b presents the standard deviations of the clip gauge and DIC measurements for each specimen, along with a mean value curve.

Figure 7 displays the results of the fatigue tests, where strain under cyclic loading is plotted against the number of load cycles N. Selected results from each test series

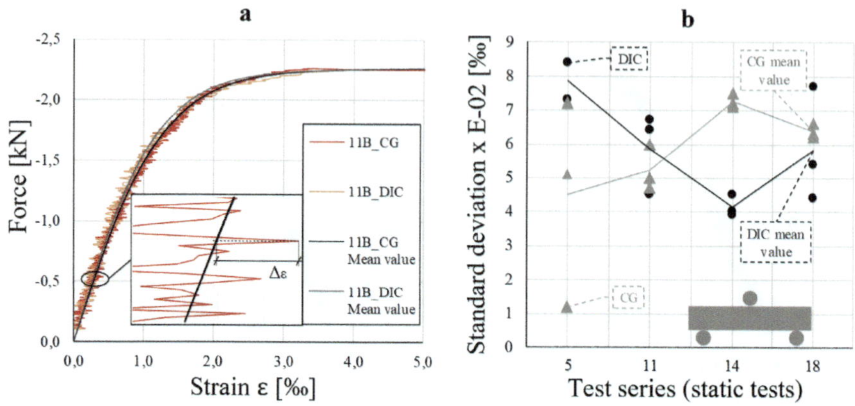

Fig. 6 **a** Deviation (Δε) of the measured values between CG and DIC of specimen type 11B, **b** standard deviation of the clip gauge and DIC for static tests (from [15], licensed under CC-BY 4.0)

are shown separately. It can be observed that, for test series 9, 15, and 17, the strains measured by CG and DIC were closely matched for each specimen.

In test series 13, where an additional image was captured per load cycle N, the strain amplitude measured by DIC is slightly lower than that measured by CG. It is assumed that the images taken between S_{min} and S_{max}, resulted in insufficient exposure times, leading to an underestimation of the measured deformations. Across all test series, it can be observed that strain predominantly increased with a decreasing number of load cycles N. Moreover, the curve showing increasing strain corresponds to the characteristic three-phase fatigue behaviour of UHPC after crack initiation. This indicates typical fatigue behaviour rather than a general measurement error.

The gradual increase in strain during the second phase after crack initiation also becomes flatter with increasing numbers of load cycles, with the fracture strain remaining within a comparable range (as schematically illustrated in Fig. 8).

The standard deviations of the CG and DIC measurements as a function of the DIC measuring rate are shown in Fig. 9. The CG measurements were taken consistently at a rate of 50 Hz across all test series and are plotted for comparison.

As with the static tests, the standard deviation was determined based on the deviation $\Delta\varepsilon$ of the measured values from the mean strain curves ε_{max} and ε_{min}. The results indicate that the standard deviation of the DIC measurements decreases as the measuring rate increases. In general, the standard deviation of the CG was 0.04 ‰, significantly lower than that of the DIC. For measuring rates of 10, 11, and 12 Hz, the variance of the CG's standard deviation was approximately 0.006 ‰, thus lower than the DIC values.

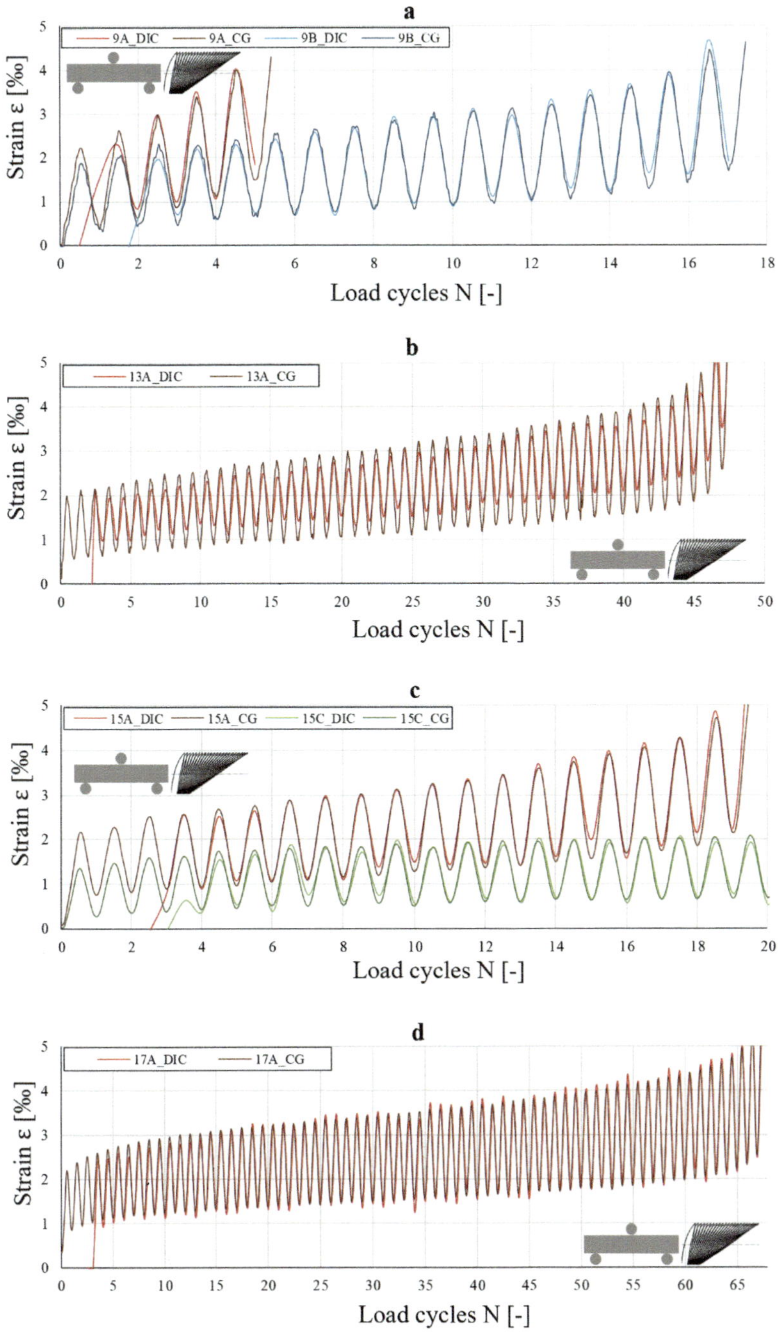

Fig. 7 Increasing strain due to cyclic loading: **a** 9-, **b** 13-, **c** 15-, **d** 17- series (from [15], licensed under CC-BY 4.0)

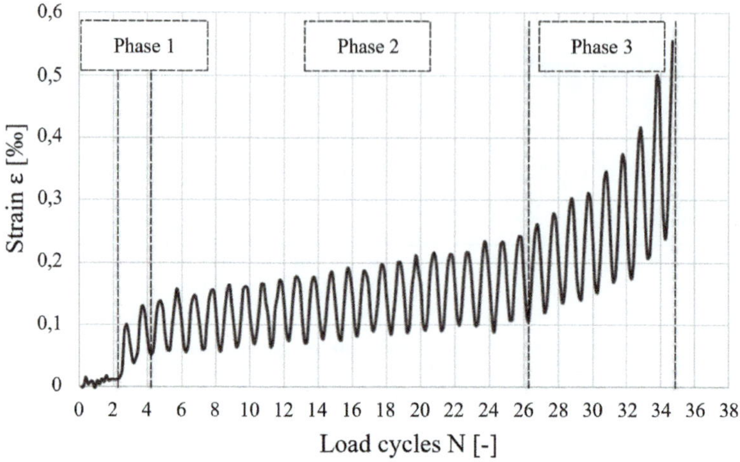

Fig. 8 Schematic measurements of the increasing strain due to cycling reversals

Fig. 9 Standard deviation of CG and DIC for cyclic tests (CG measurement rate = 50 Hz) (from [15], licensed under CC-BY 4.0)

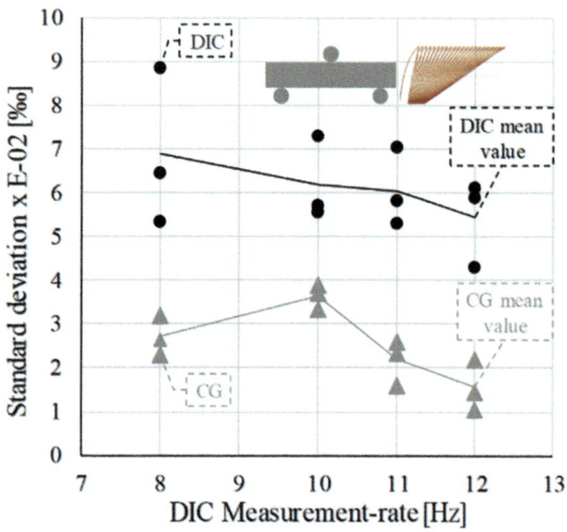

The close agreement between the CG and DIC results, alongside the slightly higher standard deviation observed in the DIC measurements, confirms the successful control of the DIC system via trigger lists during strain measurements on UHPC prisms under cyclic bending. Nevertheless, it is essential that DIC systems are controlled by trigger lists to achieve complete recording of the strain amplitude per load cycle.

3 Numerical Activities

3.1 Main Hypothesis

The numerical investigations were conducted at the Institut für Werkstoffe im Bauwesen (WiB), TU Darmstadt. The proposed approach for simulating the cyclic response of cement-based materials is based on the following assumptions:

- A unified coupled damage–plasticity model accounts for both low-cycle fatigue (LCF) and high-cycle fatigue (HCF)
- For LCF, stiffness degradation depends on the fracture work expended at each crack front
- For HCF, a micro-to-meso scale cracking mechanism describes fatigue phenomena
- Small displacements and quasi-static loading conditions are assumed, neglecting inertia forces
- Strain-rate effects are omitted.

3.2 Stress–Strain and Displacement Constitutive Rules

A discrete crack approach (DCA), based on the use of zero-thickness interface elements (0IEs) as illustrated in Fig. 10, is presented in a unified manner for the numerical analysis of concrete specimens subjected to either low-cycle fatigue (LCF) or high-cycle fatigue (HCF).

The coupled damage-elasto-plastic equations are proposed as follows (for further details, see Caggiano et al. [44, 45]):

$$\dot{\boldsymbol{t}} = \boldsymbol{C}_{\mathrm{d}} \cdot (\dot{\boldsymbol{u}} - \dot{\boldsymbol{u}}^{\mathrm{cr}}) \tag{1}$$

$$\dot{\boldsymbol{u}} = \dot{\boldsymbol{u}}^{\mathrm{el}} + \dot{\boldsymbol{u}}^{\mathrm{cr}} \tag{2}$$

where $\dot{\boldsymbol{t}} = (\dot{\sigma}_{\mathrm{N}}, \dot{\sigma}_{\mathrm{T}})^T$ is the rate of the interface stress vector, $\boldsymbol{C}_{\mathrm{d}}$ is the joint elastic-damage stiffness matrix, and $\dot{\boldsymbol{u}} = (\dot{u}, \dot{v})^T$ represents the rate of interface relative displacements, with $\dot{\boldsymbol{u}}^{\mathrm{el}}$ and $\dot{\boldsymbol{u}}^{\mathrm{cr}}$ being the elastic and cracking (inelastic) components, respectively. The effective interface stress, $\tilde{\boldsymbol{t}}$, is defined by [43]:

$$\dot{\boldsymbol{t}} = (\boldsymbol{I} - \boldsymbol{D}) \cdot \dot{\tilde{\boldsymbol{t}}} \tag{3}$$

where $\boldsymbol{I}$ is the second-order identity matrix, and $\boldsymbol{D}$ is the matrix representing the damage variables. Consequently, the damaged stiffness matrix is expressed as $\mathbf{C}_d = (\mathbf{I} - \mathbf{D}) \cdot \mathbf{C}$ where $\mathbf{C}$ is the undamaged elastic stiffness matrix.

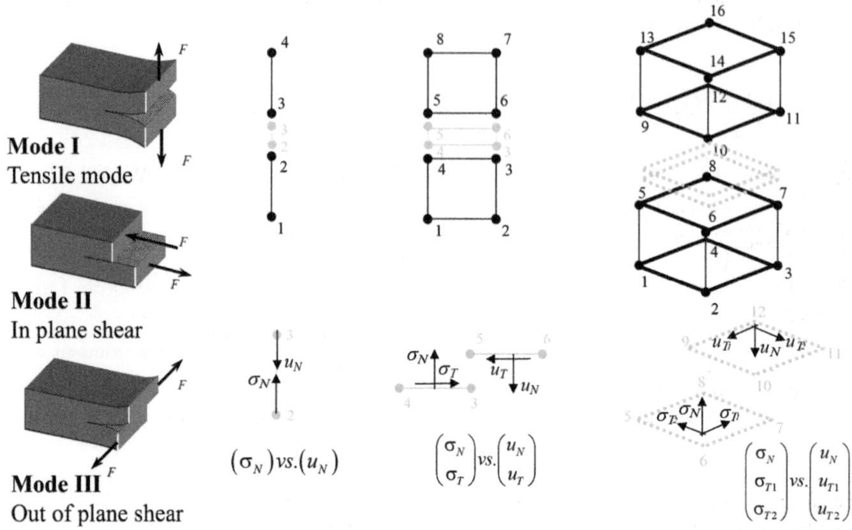

Fig. 10 Discontinuous-based approach based on the use of zero-thickness interface elements for 1D, 2D and 3D analyses [37–42]

Thus, the following unified relationship is proposed to capture the damage behaviour under general cyclic conditions (i.e., for both LCF and HCF):

$$\dot{\boldsymbol{D}} = \dot{\boldsymbol{D}}^{\mathbf{LCF}} + \dot{\boldsymbol{D}}^{\mathbf{HCF}} \tag{4}$$

where the time-accumulated damage $\boldsymbol{D}$ is evaluated as:

$$\boldsymbol{D} = \int_{t_0}^{t_f} \dot{\boldsymbol{D}}\, dt \in [\mathbf{0}, \mathbf{1}] \tag{5}$$

To drive LCF and HCF in uncoupled manner, it has been assumed that LCF is influenced by damage-elasto-plastic mechanisms, while HCF is accounted for by inner microplastic strains and micro-cracks modelled at the lover micro-level.

3.3 Modelling of Low Cycle Fatigue (LCF)

The fracture-based model by Caggiano et al. [44], combined with the continuous damage theory extension [45], is formulated within 0IE in order to account for the LCF effects. The failure surface shown in Fig. 11 defines the (effective) stress level at which the inelastic (cracking) displacements initiate:

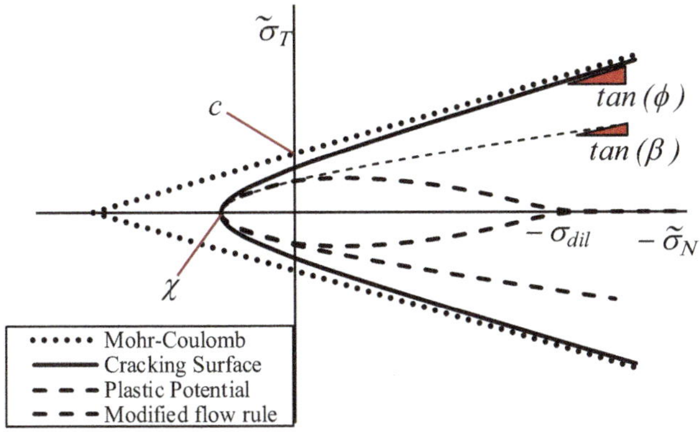

Fig. 11 Failure criterion, Mohr–Coulomb surface, plastic potential and modified flow rule

$$f = \tilde{\sigma}_{\mathrm{T}}^2 - (c - \tilde{\sigma}_{\mathrm{N}} \tan\phi)^2 + (c - \chi \tan\phi)^2 \tag{6}$$

where $\tilde{\sigma}_{\mathrm{T}}$ denotes the tensile strength, c the cohesion, and ϕ the frictional angle.

The crack displacement rates $\dot{\boldsymbol{u}}^{cr}$ are defined following a classical non-associated flow rule:

$$\dot{\boldsymbol{u}}^{cr} = \dot{\lambda} \cdot \boldsymbol{m} \tag{7a}$$

$$\boldsymbol{m} = \boldsymbol{A} \cdot \boldsymbol{n} \tag{7b}$$

where $\dot{\lambda}$ is the non-negative plastic multiplier, arising from the classical Kuhn–Tucker and consistency conditions ($\dot{\lambda} \geq 0, f \leq 0, \dot{\lambda} f = 0$ and $\dot{f} = 0$, respectively) ($\dot{\lambda} \geq 0,\ \ f \leq 0,\ \ \dot{\lambda} f = 0, \dot{f} = 0$, respectively). Here, $f = f[\tilde{\sigma}_{\mathrm{N}}, \tilde{\sigma}_{\mathrm{T}}]$ is the yield criterion, and $\boldsymbol{m}$ is the vector controlling the direction of the fracture displacements according to the non-associated flow rule.

The vector $\boldsymbol{n}$ is defined as the gradient of the yield function f with respect to the effective stress $\tilde{\sigma}_{\mathrm{T}}$, namely:

$$\boldsymbol{n} = \frac{\partial f}{\partial \tilde{t}} = \left(\frac{\partial f}{\partial \tilde{\sigma}_{\mathrm{N}}}, \frac{\partial f}{\partial \tilde{\sigma}_{\mathrm{T}}}\right)^{\mathrm{T}} \tag{8}$$

The post-cracking softening behaviour is governed by scaling functions (Fig. 12), which depend on the ratio between the plastic work W_{cr} and the available fracture energies $G_{\mathrm{f}}^{\mathrm{I}}$ or $G_{\mathrm{f}}^{\mathrm{IIa}}$:

$$\xi_{\mathrm{p_i}} = \begin{cases} \frac{1}{2}\left[1 - \cos\left(\pi \frac{W_{\mathrm{cr}}}{G_{\mathrm{f}}^{\#}}\right)\right], & \text{if } W_{\mathrm{cr}} \leq G_{\mathrm{f}}^{\#}, \# = \mathrm{I} \text{ or IIa} \\ 1, & \text{otherwise} \end{cases} \tag{9}$$

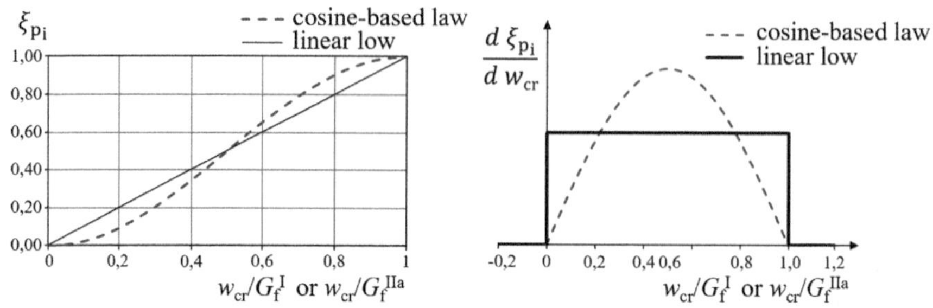

Fig. 12 Internal softening parameters and evolution rules

where p_i denotes χ, c and $\tan\phi$.

Finally, the following scalar equations describe the damage evolution in LCF:

$$d^{\#} = e^{-\beta_d}\left(\frac{W_{cr}}{G_f^{\#}}\right)^{\alpha_d}\left[1 + \left(e^{-\beta_d} - 1\right)\left(\frac{W_{cr}}{G_f^{\#}}\right)^{\alpha_d}\right]^{-1}, \quad \# = \mathrm{I, IIa} \tag{10a}$$

$$\boldsymbol{D}^{LCF} = \begin{pmatrix} d^{\mathrm{I}} & 0 \\ 0 & d^{\mathrm{IIa}} \end{pmatrix} \tag{10b}$$

where $\alpha_d \geq 0$ and β_d are parameters controlling the shape of the damage degradation induced by fracture.

3.4 *Modelling of High Cycle Fatigue (HCF)*

Inelastic strains and/or unrecoverable cracks are almost negligible in HCF failures when compared to LCF cases. Therefore, classical approaches based on fracture mechanics, plasticity theories, and continuum damage models cannot be directly applied. Instead, microscopic plastic strains and micro-cracking phenomena dominate damage accumulation under high- and very-high-cycle fatigue.

Following this reasoning, a two-scale (micro-to-meso) homogenisation approach (Fig. 13) is adopted. In this framework, no plastic strains occur at the meso- or macro-scales, while microplastic strains and micro-cracks are explicitly modelled at the micro-scale, primarily governing HCF failure. The following Hollomon hardening power law is assumed at the microscale:

$$\tilde{\sigma}_{eq} = K\delta_{micro}^{M} \tag{11a}$$

$$\tilde{\sigma}_{eq} = \begin{cases} \sqrt{(\tilde{\sigma}_N)^2 + (\tilde{\sigma}_T)^2} & \tilde{\sigma}_N \geq 0 \\ \tilde{\sigma}_T - |\tilde{\sigma}_N| \tan(\phi) & \tilde{\sigma}_N < 0 \end{cases}$$

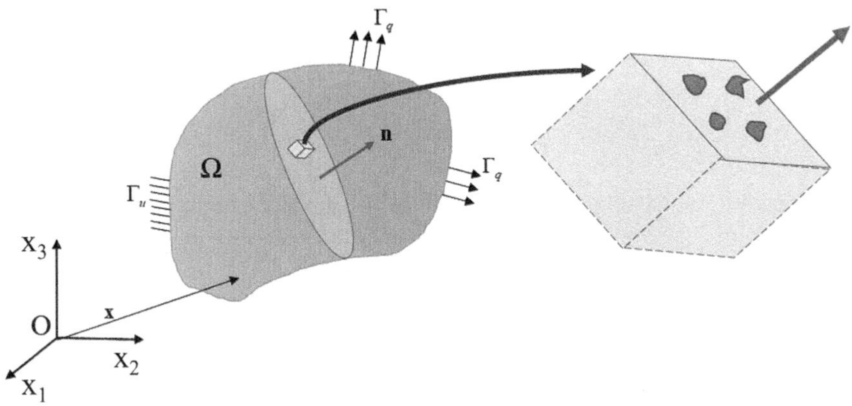

Fig. 13 Micro-to-meso scale homogenization approach

$$\wedge\delta_{\text{micro}} = \begin{cases} \sqrt{\left(u_{\text{micro}}^{\text{cr}}\right)^2 + \left(v_{\text{micro}}^{\text{cr}}\right)^2} & \tilde{\sigma}_{\text{N}} \geq 0 \\ v_{\text{micro}}^{\text{cr}} & \tilde{\sigma}_{\text{N}} < 0 \end{cases} \tag{11b}$$

where K is the strength index (also known as strength coefficient) and M the crack-hardening exponent, which, in general, is frequency-dependent $M = M[f]$. Here, u_{micro}^{cr} and v_{micro}^{cr} are the relative microscale displacements in the normal and tangential directions, respectively.

Following the isotropic damage approach, the following relation is obtained:

$$\frac{\sigma_{\text{eq}}}{1-D} = K\delta_{\text{micro}}^{\text{M}} \Rightarrow \delta_{\text{micro}} = \frac{\sigma_{\text{eq}}^{\frac{1}{\text{M}}}}{(1-D)^{\frac{1}{\text{M}}}K^{\frac{1}{\text{M}}}}$$
$$\Rightarrow \dot{\delta} = \frac{\sigma_{\text{eq}}^{1/\text{M}-1}}{M(1-D)^{1/\text{M}}K^{1/\text{M}}}\dot{\sigma}_{\text{eq}} \tag{12}$$

Thus, the microscopic damage can be evaluated as:

$$\dot{D} = \begin{cases} 0 & \text{if } \frac{\sigma_{\text{eq,M}} - \sigma_{\text{eq,m}}}{2} \leq \sigma_{\text{f}} \\ L_{\text{cl}^{-1}} \frac{Q[f]\sigma_{\text{eq}}^{q[f]}}{(1-D)^{q[f]+1}}\dot{\sigma}_{\text{eq}} & \text{if } \frac{\sigma_{\text{eq,M}} - \sigma_{\text{eq,m}}}{2} > \sigma_{\text{f}} \end{cases} \tag{13a}$$

$$\boldsymbol{D}^{\boldsymbol{HCF}} = \begin{pmatrix} \dot{D} & 0 \\ 0 & \dot{D} \end{pmatrix} \tag{13b}$$

where $q[f] = \frac{1}{M[f]} - 1$, $Q[f] = K^{-1/M[f]}/M[f]$ and $L_{cl^{-1}}$ ($\dot{D} = L_{cl^{-1}}\dot{\delta}_{micro}$) represents the characteristic length value. The latter three variables are the key drivers aiming at representing the link between the microscale and the macroscale. $\sigma_{eq,M}$

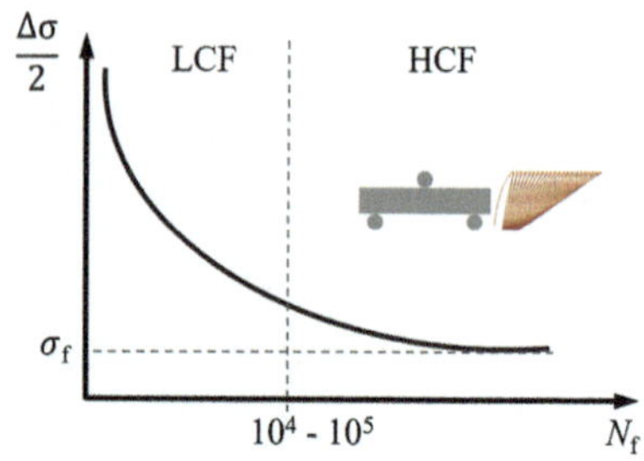

Fig. 14 Fatigue stress limit under cyclic loading

and $\sigma_{eq,m}$ in Eq. 12 are the "*sup*" and "*inf*" values of σ_{eq} under the cyclic protocols, while σ_{f} is the fatigue stress limit (Fig. 14).

3.5 *Test Data and Comparisons*

The numerical simulations focus on concrete flexural fatigue tests with a target compressive strength of approximately 40 MPa. Experimental data available in the literature [46] were considered for validation purposes. The reference experiments involve unnotched concrete beams (100 × 100 × 500 mm^3) subjected to four-point bending. The fatigue protocol, designed by Oh [46], specified that the stress in the bottom fibre varied from zero to a predetermined maximum stress. The maximum cyclic stress ratio, $f_R^{MAX}/f_{\mathrm{mon}}$ was analysed to determine the fatigue failure after a number of reversals N_{FAILURE} (denoted as N_{f}).

For numerical calibration, key geometric and material properties were adopted from experimental observations [46], while only two model parameters were fitted:

$$[f] = 24.30;\ \frac{q[f]+1}{2(q[f]+2)Q[f]L_{\mathrm{cl}^{-1}}} = 2.1\ \mathrm{MPa} \tag{14}$$

Additionally, the following assumptions were considered for model integration:

$$D = 0 \text{ (when the number of cycles is zero)} \tag{15a}$$

$$D = 1 \text{ reaching fatigue failure (i.e. } N = N_{\mathrm{f}}) \tag{15b}$$

The numerical S–N curves were predicted using the proposed approach and subsequently compared with the experimental data, as illustrated in Fig. 15. Specifically, the fatigue protocol, based on Oh's design [46], was implemented numerically, considering that the bottom fibre stress varies from zero to a predetermined maximum stress. To determine the maximum cyclic stress ratio, denoted as R, numerical analyses were conducted for each series. This methodology enabled the simulation of the loading conditions required to induce fatigue failure after a specified number of load

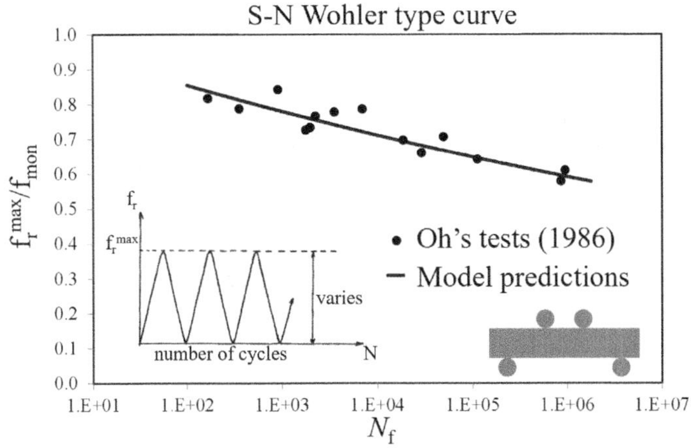

Fig. 15 Numerical examples in comparison to experimental data from [46]. Used with permission of Trans Tech Publications, Ltd., from Caggiano et al. [16]; permission conveyed through Copyright Clearance Center, Inc

reversals, referred to as N_f. By following this protocol within a numerical framework, the stress variations were reproduced and the material response under cyclic loading conditions was accurately modelled.

The numerical results and their comparison with experimental data are promising, demonstrating a good predictive capability. Furthermore, the outcomes follow the well-known Wöhler expression for describing the fatigue S–N curve:

$$f_R^{MAX}/f_{mon} = a + b \log_{10} N_f \quad (16)$$

The ability of the proposed formulation to capture the pronounced sensitivity of the mechanical behaviour of concrete to the stress levels influencing high-cycle fatigue (HCF) is thus demonstrated.

4 Concluding Remarks

This work presents the findings of a research study on the fatigue behaviour of ultra-high-performance concrete (UHPC). The static failure responses and fatigue degradation of UHPC were first investigated through cyclic bending-tensile tests. Various maximum and minimum stress levels (i.e., cycle reversals), characterised by different force and stress amplitudes, were considered and analysed using UHPC specimens. The influence of edge zones on stress cycles was examined by testing both notched and unnotched specimens to derive the so-called Wöhler curve for the material's fatigue behaviour. Damage progression during cyclic loading was monitored using a Digital Image Correlation (DIC) system to enhance measurement accuracy. In addition to the experimental investigations, macroscopic and mesoscale numerical simulations were performed. For this purpose, a unified numerical framework was proposed for the analysis of low-cycle fatigue (LCF) and high-cycle fatigue (HCF) behaviour in concrete. A discontinuous model, designed to describe the evolution of micro- to mesoscale crack growth under cyclic loading protocols, was developed. The model combines concepts from fictitious crack theory, damage-induced stiffness degradation, and microscopic inelasticity to predict failure mechanisms in concrete under cyclic loading conditions. The numerical studies were initially calibrated using the experimental data reported by Oh [46], which involved normal concrete (NC) specimens with varying dimensions, unnotched beams, and a different loading configuration (four-point bending rather than three-point bending). This calibration against existing literature data was intended to validate and refine the proposed modelling approach, thereby facilitating its application to a broader range of geometries, loading scenarios, and frequency domains. Future research will aim to achieve an even closer correspondence between the experimental results and numerical predictions based on the SPP 2020 experimental data.

References

1. Schneider, M., Romer, M., Tschudin, M., Bolio, H.: Sustainable cement production—present and future. Cem. Concr. Res. **41**, 642 (2011)
2. Chen, S., Huang, Y., Gu, P., Wang, J.: Experimental study on fatigue performance of UHPC-orthotropic steel composite deck. Thin-Walled Struct. **142**, 1 (2019)
3. Nodehi, M., Nodehi, S.E.: Ultra high performance concrete (UHPC): reactive powder concrete, slurry infiltrated fiber concrete and superabsorbent polymer concrete. Innov. Infrastruct. Solut. **7**(1), 1–22 (2022)
4. Classen, M., Gallwoszus, J., Stark, A.: Anchorage of composite dowels in UHPC under fatigue loading. Struct. Concr. **17**(2), 183–193 (2016)
5. Tian, H., Zhou, Z., Wei, Y., Zhang, L.: Experimental and numerical investigation on the seismic performance of concrete-filled UHPC tubular columns. J. Build. Eng. **43**, 103118 (2021)
6. Ren, L., Fang, B., Wang, K., Yuan, F.: Numerical investigation on plastic hinge length of ultra-high performance concrete column under cyclic load. J. Earthquake Eng. **26**(3), 1281–1299 (2022)
7. Aboukifa, M., Moustafa, M.A., Itani, A.M., Naeimi, N.: Durable UHPC Columns with High-Strength Steel (No. ABC-UTC-2013-C3-UNR02-Final). Accelerated Bridge Construction University Transportation Center (2019)
8. Lee, H.J., Daniel, J.S., Kim, Y.R.: Continuum damage mechanics-based fatigue model of asphalt concrete. J. Mater. Civ. Eng. **12**(2), 105–112 (2000)
9. Song, Z., Frühwirt, T., Konietzky, H.: Inhomogeneous mechanical behaviour of concrete subjected to monotonic and cyclic loading. Int. J. Fatigue **132**, 105383 (2020)
10. Fan, J., Jiang, D., Chen, J., Liu, W., Ngaha, W.T., Chen, J.: Fatigue performance of ordinary concrete under discontinuous cyclic loading. Constr. Build. Mater. **166**, 974–981 (2018)
11. Kachkouch, F.Z., Noberto, C.C., Babadopulos, L.F.D.A.L., Melo, A.R.S., Machado, A.M.L., Sebaibi, N., et al.: Fatigue behavior of concrete: a literature review on the main relevant parameters. Constr. Build. Mater. **338**, 127510 (2022)
12. Ortega, J.J., Ruiz, G., Rena, C.Y., Afanador-García, N., Tarifa, M., Poveda, E., et al.: Number of tests and corresponding error in concrete fatigue. Int. J. Fatig. **116**, 210–219 (2018)

13. Fan, Z., Sun, Y.: Detecting and evaluation of fatigue damage in concrete with industrial computed tomography technology. Constr. Build. Mater. **223**, 794–805 (2019)
14. Baktheer, A., Chudoba, R.: Experimental and theoretical evidence for the load sequence effect in the compressive fatigue behavior of concrete. Mater. Struct. **54**(2), 1–23 (2021)
15. Harenberg, S., Pahn, M., Malárics-Pfaff, V., Dehn, F., Caggiano, A., Schicchi, D.S., Yang, S., Koenders, E.: Digital image correlation strain measurement of ultra-high-performance concrete-prisms under static and cyclic bending-tensile stress. Struct. Concr. **20**(4), 1220–1230 (2019)
16. Caggiano, A., Schicchi, D.S., Yang, S., Harenberg, S., Malarics-Pfaff, V., Pahn, M., Dehn, F., Koenders, E.: A microscale approach for modelling concrete fatigue damage-mechanisms. Key Eng. Mater. **827**, 73–78 (2019)
17. Chen, A., Airey, G.D., Thom, N., Litherland, J., Nii-Adjei, R.A.: Modelling the stiffness development in asphalt concrete to obtain fatigue failure criteria. Constr. Build. Mater. **306**, 124837 (2021)
18. Oneschkow, N., Timmermann, T., Löhnert, S.: Compressive fatigue behaviour of high-strength concrete and mortar: experimental investigations and computational modelling. Materials **15**(1), 319 (2022)
19. Xu, L., Ma, M., Li, L., Xiong, Y., Liu, W.: Continuum-based approach for modelling the flexural behaviour of plain concrete beam under high-cycle fatigue loads. Eng. Struct. **241**, 112442 (2021)
20. Yadav, I.N., Thapa, K.B.: Fatigue damage model of concrete materials. Theor. Appl. Fract. Mech. **108**, 102578 (2020)
21. Chudoba, R., Vořechovský, M., Aguilar, M., Baktheer, A.: Coupled sliding–decohesion–compression model for a consistent description of monotonic and fatigue behavior of material interfaces. Comput. Methods Appl. Mech. Eng. **398**, 115259 (2022)
22. Song, Z., Konietzky, H., Herbst, M., Frühwirt, T., Wu, Y.: Fatigue and micro-seismic behaviors of concrete disks exposed to cyclic Brazilian testing: a numerical investigation based on a 3D particle-based model. Int. J. Fatigue **155**, 106629 (2022)
23. Li, D., Kaewunruen, S., You, R., Liu, P.: Fatigue life modelling of railway prestressed concrete sleepers. Structures **41**, 643–656 (2022)
24. Gebuhr, G., Pise, M., Anders, S., Brands, D., Schröder, J.: Damage evolution of steel fibre-reinforced high-performance concrete in low-cycle flexural fatigue: numerical modeling and experimental validation. Materials **15**(3), 1179 (2022)
25. Becks, H., Aguilar, M., Baktheer, A., Chudoba, R., Classen, M.: Experimental and numerical investigations on the fatigue behavior of high-strength concrete under combined shear-compression loading. In: Proceedings of IABSE Symposium: Prague (2022)
26. Shao, W., Wang, Y., Shi, D.: Corrosion-fatigue life prediction of reinforced concrete square piles in marine environments. Eng. Fail. Anal. **138**, 106324 (2022)
27. Baktheer, A., Aguilar, M., Chudoba, R.: Microplane fatigue model MS1 for plain concrete under compression with damage evolution driven by cumulative inelastic shear strain. Int. J. Plast **143**, 102950 (2021)
28. Carlesso, D.M., Cavalaro, S., de la Fuente, A.: Flexural fatigue of pre-cracked plastic fibre reinforced concrete: experimental study and numerical modeling. Cement Concr. Compos. **115**, 103850 (2021)
29. Renju, D.R., Keerthy, M.S.: A review on fatigue life prediction of plain concrete. IOP Conf. Ser. Mater. Sci. Eng. **936**(1), 012026 (2020)
30. CEN (European Committee for Standardization). Eurocode 2: design of concrete structures—Part 1-1: general rules and rules for buildings. EN 1992-1-1. European Committee for Standardization, Brussels (2004)
31. fib (Fédération Internationale du Béton). fib model code for concrete structures 2010. Ernst & Sohn, Berlin (2013)
32. AASHTO (American Association of State Highway and Transportation Officials). AASHTO LRFD bridge design specifications, 7th edn. AASHTO, Washington, DC (2014)

33. ACI (American Concrete Inst.). Considerations for the design of concrete structures subjected to fatigue loading. ACI PRC-215-92. American Concrete Institute, Farmington Hills, MI (1992)
34. CEN (European Committee for Standardization). EN 196-1:2016—Methods of testing cement—Part 1: determination of strength. European Committee for Standardization, Brussels (2016)
35. CEN (European Committee for Standardization). EN 14651:2005—test method for metallic fibred concrete—measuring the flexural tensile strength (limit of proportionality (LOP), residual). European Committee for Standardization, Brussels (2005)
36. CEN (European Committee for Standardization). EN 12390-5:2019—testing hardened concrete—Part 5: flexural strength of test specimens. European Committee for Standardization, Brussels (2019)
37. Carol, I., Prat, P.C., López, C.M.: Normal/shear cracking model: application to discrete crack analysis. J. Eng. Mech. **123**(8), 765–773 (1997)
38. Caballero, A., Carol, I., López, C.M.: A meso-level approach to the 3D numerical analysis of cracking and fracture of concrete materials. Fatigue Fract. Eng. Mater. Struct. **29**(12), 979–991 (2006)
39. López, C.M., Carol, I., Aguado, A.: Meso-structural study of concrete fracture using interface elements. I: Numerical model and tensile behavior. Mater. Struct. **41**(3), 583–599 (2008)
40. Carol, I., López, C.M., Roa, O.: Micromechanical analysis of quasi-brittle materials using fracture-based interface elements. Int. J. Numer. Meth. Eng. **52**(1–2), 193–215 (2001)
41. López, C.M., Carol, I., Aguado, A.: Meso-structural study of concrete fracture using interface elements. II: Compression, biaxial and Brazilian test. Mater. Struct. **41**(3), 601–620 (2008)
42. Caballero, A., Carol, I., López, C.M.: 3D meso-mechanical analysis of concrete specimens under biaxial loading. Fatigue Fract. Eng. Mater. Struct. **30**(9), 877–886 (2007)
43. Lemaitre, J.: How to use damage mechanics. Nucl. Eng. Des. **80**(2), 233–245 (1984)
44. Caggiano, A., Etse, G., Martinelli, E.: Zero-thickness interface model formulation for failure behavior of fiber-reinforced cementitious composites. Comput. Struct. **98**, 23–32 (2012)
45. Caggiano, A., Etse, G., Ferrara, L., Krelani, V.: Zero-thickness interface constitutive theory for concrete self-healing effects. Comput. Struct. **186**, 22–34 (2017)
46. Oh, B.H.: Fatigue analysis of plain concrete in flexure. J. Struct. Eng. **112**(2), 273–288 (1986)

Water-Induced Damage Mechanisms of Cyclically Loaded High-Performance Concrete

Mohamed Abubakar Ali, Christoph Tomann, Nadja Oneschkow, Markus Mahlbacher, Nima Noii, Karl-Heinz Drake, Ludger Lohaus, Peter Wriggers, Fadi Aldakheel, and Michael Haist

Abstract The role of fatigue in the design of concrete structures has increased significantly in recent years, particularly due to the expansion of offshore wind energy systems, where especially the moisture content plays a decisive role. Despite some contradictions in the literature, one trend is clear: Submerged concrete (wet) exhibits significantly lower fatigue resistance than air-stored concrete (dry). The underlying damage mechanisms responsible for this phenomenon yet remain unclear. In this chapter, the main results obtained from a joint project are reported. The results showed that the water in the microstructure, i.e. at a size of $\leq$ 10 nm, is primarily responsible for the water-induced fatigue damage. Additionally–and even more relevant for design–the effect of the water-induced damage is more significant at low load frequencies. The results provide clear indications that the threshold beyond which the water-induced damage mechanisms become decisive may be well in the range of storage at 85% relative humidity. Comprehensive microscopic investigations traced the water-induced damage mechanisms back to a water redistribution process at a scale of only a few nanometres. This leads to pore pressure, causing loosening effects (damages) in the dense structure of the calcium silicate hydrate phases.

Keywords Moisture content · Acoustic emission · Microstructure · NMR · Indentation techniques · Phase-field approach

---

M. Abubakar Ali (✉) · C. Tomann · N. Oneschkow · M. Mahlbacher · K.-H. Drake · L. Lohaus · M. Haist
Institute of Building Materials Science, Leibniz University Hannover, Hannover, Germany
e-mail: mohamed.Abubakar-Ali@mpa.uni-stuttgart.de

P. Wriggers
Institute of Continuum Mechanics, Leibniz University Hannover, Hannover, Germany

N. Noii · F. Aldakheel
Institute of Mechanics and Computational Mechanics, Leibniz University Hannover, Hannover, Germany

N. Oneschkow and L. Lohaus (eds.), *Cyclic Deterioration of High-Performance Concrete*, https://doi.org/10.1007/978-3-032-19394-0_7

1 Introduction

With the expansion of offshore wind energy systems, the number of fatigue-loaded structures submerged in water has significantly increased in the last few decades. High-strength fine-grained concrete (grout) is typically used as connection elements in the steel support structures of the offshore turbines. The offshore exposure of these structures leads to a significantly higher water content in the concrete's microstructure compared to structures in rather drier environments, such as normal outdoor exposure.

Comparatively few investigations of fatigue-tested concrete specimens immersed in water are documented in the literature. Despite considerable scatter in the results, a clear tendency can be observed: Specimens with a high moisture content or which are even immersed in water show a significantly lower fatigue resistance compared to those tested in dry conditions (e.g. [1, 2]). Yet the moisture content at which the transition from the "dry damage scenario" to the "wet scenario" occurs has not yet been investigated and remains unclear. Furthermore, concrete specimens in wet environments show a different fracture behaviour than dry specimens. This is indicated during testing by ascending air bubbles, the washout of fine particles into the surroundings and premature crack initiation [3]. Oneschkow et al. [4] found indications that the influence of the loading frequency is even higher for specimens stored and tested in a wet environment compared to those subjected to dry conditions.

In addition to the number of cycles to failure N, the specific shape of the well-known S-shaped strain and stiffness development curves with characteristic phases I–III is often used in the literature as an indicator for describing the fatigue damage of concrete. According to Hümme [5], a higher stiffness degradation per load cycle can be observed for wet specimens compared to dry specimens. Muguruma and Watanabe [1] additionally measured acoustic emission (AE) signals generated during the loading regime to quantify the fatigue damage. They stated that the number of AE hits increases with increasing moisture content, which they traced back to the accelerated development of microcracks.

Although water-induced damage due to cyclically loaded concrete has been recognised in the literature, there is still a pronounced lack of knowledge concerning the underlying damage mechanisms involved in the deterioration process. Various hypotheses regarding water-induced damage mechanisms have been documented in the literature (e.g. [5, 6]), however, without having been validated yet. According to Hümme [5] and Sørensen et al. [6], pore pressure in case of fully-saturated concrete can be build up within water-saturated pores due to fatigue loading. This pressure induces an additional local tensile stress, likely during the unloading phase, which leads to an accelerated fatigue damage progress.

Another explanation points to possible water transport within the pore system of the hardened cement paste (HCP) due to fatigue loading [7]. Such a water transport process has already been determined for creep loading using nuclear magnetic resonance (NMR) methods, as documented in [8]. However, it has not yet been experimentally proven whether a similar water redistribution occurs due to fatigue loading and at what scale this takes place. The numerical modelling and simulation of damage mechanisms influenced by water remain largely unexplored in the literature, primarily due to the complexity of material behaviour across multiple scales.

Well-instrumented fatigue investigations on high-strength concrete with varied storage and test conditions have been conducted in a joint project. The main aim of the project was to investigate and describe the water-induced damage mechanisms of fatigue-loaded high-strength concretes. The results revealed a significant reduction in the number of cycles to failure for specimens stored and tested underwater compared to those subjected to dry conditions. Additionally, it was shown that in the fully saturated case, the moisture content in the concrete's microstructure is mainly responsible for reducing the fatigue resistance of concrete. Moreover, there are clear indications for the existence of a moisture threshold beyond which the negative impact of moisture content increases significantly.

Several influencing parameters on the water-induced fatigue damage, such as the stress level and load frequency, were investigated on a macroscopic scale (roughly from a few millimeters to several centimeters). The key finding was that the negative effect of water on the fatigue resistance strongly intensifies as the stress level and load frequency decrease. This finding is contrary to the behaviours observed on dry specimens and makes the issue of water-induced fatigue damage particularly relevant for structures such as offshore wind energy systems.

Aside from the number of cycles to failure, the development of strain and stiffness as well as the AE signals were analysed as damage indicators to gain initial insights into the underlying damage mechanisms. The AE results showed that water-induced fatigue damage differs from dry fatigue by the timing of AE hits: in dry conditions, hits occur mainly during loading, specifically between the mean stress level (S_{mean}) and maximum stress level (S_{max}), while in wet conditions, they shift to the unloading phase, in the area between S_{mean} and minimum stress level (S_{min}). This trend clearly indicated that additional damage mechanisms are acting with increasing moisture content in the concrete.

In order to clarify the underlying microstructural mechanisms behind water-induced damage, a comprehensive test programme was conducted, combining NMR techniques, micro- and nanoindentation, as well as gas adsorption testing methods. The results clearly indicate that applying fatigue loading to concrete with a high moisture content leads to water transport on a scale smaller than approximately 10 nm. This process causes internal pore pressure, leading to an increase in porosity, which can be linked to a loosening effect (damage) in the dense structure (high density: HD) of the calcium silicate hydrate (C-S-H) phases (HD-C-S-H), as observed through nanoindentation. The deterioration of HD-C-S-H was identified as the key mechanism driving water-induced fatigue damage.

2 Materials and Methods

2.1 Macroscopic Investigations

All macroscopic investigations were carried out on the reference high-strength concrete (RH1) of the Priority Programme SPP 2020. The mixture is given in Sect. 2.1 "Concrete Composition" in the chapter by M. Basaldella et al., this volume.

The concrete was filled into polyvinyl chloride formworks (d/h = 60/250 mm) in two equal layers, which were mechanically compacted using a vibrating table. After two days, all specimens were demoulded. At the age of 14 days, the top and bottom surfaces of all specimens were cut to the height of 180 mm and prepared by grinding and polishing to guarantee a uniform stress distribution in the specimens. All specimens were sealed on the lateral surface using aluminium-coated butyl tape or epoxy resins to prevent drying during the fatigue tests. The different moisture contents of the concrete were adjusted directly after the concreting process (for details on, for example, sample preparation and storage condition, see [9, 10]). Table 1 contains an overview of the storage and test conditions.

The fatigue tests were carried out using a class 0.5 servo hydraulic testing machine, according to ISO 7500-1 [11], equipped with a 2.5 MN actuator (see Fig. 1a). The axial deformations were measured using three laser distance sensors and three linear variable differential transformers (LVDT) type SM20 from Schreiber Meßtechnik GmbH, which were placed on the circumference of the specimen at angles of 0°, 120° and 240°, respectively (see Fig. 1b). All strain measurements are given as mean values from the three sensors. Depending on the load frequency, the data sampling rate was set between 100 and 300 Hz, respectively. Six AE sensors from GMA Group with a frequency response range between 125 and 750 kHz were attached to the specimens at an angular distance of 60° from one another, with an alternating placement in the upper and lower third of the specimen height, to obtain insights into the damage mechanisms. A specific threshold of 40 dB was used to separate the useful signals from the background noise.

The moisture content was determined on parallel specimens using gravimetric analysis before the fatigue tests, following the standard DIN EN ISO 12571 [12]. In this process, three specimens for each storage condition were dried in a temperature chamber at 105 °C until mass constancy, which took several days. The reference compressive strength was tested on additional parallel specimens from the same

Table 1 Definition of the storage and test conditions

Sample designation	Storage conditions	Test conditions
D	Dried	Sealed
C	Climate chamber	Sealed
M	Sealed	Sealed
WS	Water storage	Sealed
WST	Water storage	Underwater

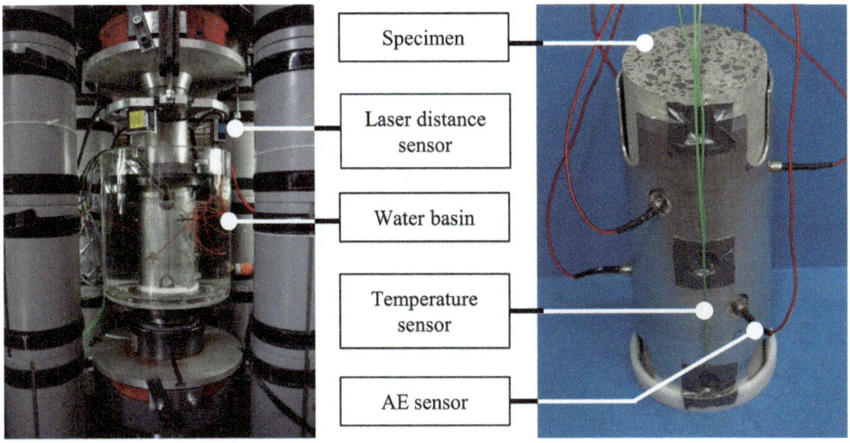

Fig. 1 **a** Experimental set-up and **b** measurement equipment (from [9, 14], licensed under CC-BY 4.0)

batch with the same moisture content as those used in the fatigue tests. For details, see [9, 13].

2.2 Microstructural Investigations

The microstructural investigations were carried out on HCP samples consisting of 1350 kg/m^3 of an Ordinary Portland Cement (CEM I 52.5 R-SR3 (Holcim) or CEM I 42.5 R (Dyckerhoff, white cement), 0.50 kg/m^3 of superplasticiser (Master Builders Solutions), and 472.5 kg/m^3 of water. The white cement was used for the NMR measurements, as it has a very low iron content in its composition, resulting in lower noise signals than grey cement. The samples were cylindrical, with a height of 30 mm and a diameter of 10 mm. The concreting process as well as the storage and test conditions were similar to the macroscopic samples (see Sect. 2.1). Four different microstructural investigation methods were primarily used in this project: ^{1}H-NMR, gas adsorption, and micro- and nanoindentation techniques.

The ^{1}H-NMR measurements were carried out on a BRUKER minispec mq10-rheo instrument with a magnetic field intensity of 0.235 Tesla (see Fig. 2a). The water in different pore spaces was measured using a Carr-Purcell-Meiboom-Gill pulse sequence with 1600 repetitions (for details, see [15]). An inverse Laplace transform analysis revealed different relaxation components attributed to different pore species, as described by Muller et al. [16]. Similar to the study of [8], the water redistribution process was determined using multiexponential fitting based on the T2 relaxation times obtained through the inverse Laplace transform analysis. In a prior experimental step, the water distribution was measured in a sample. Subsequently, ex situ fatigue loading was applied to the sample, with a maximum stress level of

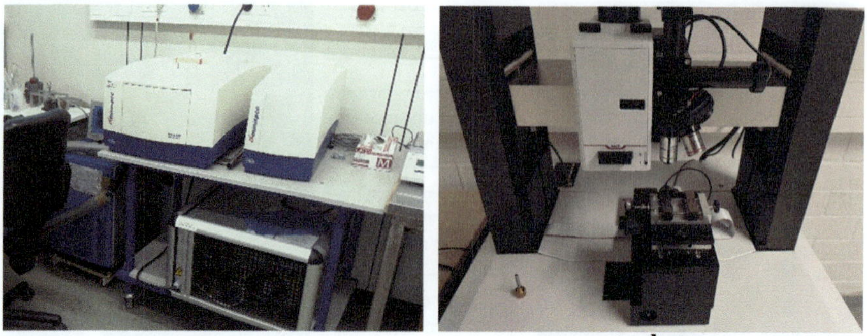

a b

Fig. 2 **a** Nuclear magnetic resonance measurement equipment and **b** microindentation set-up

$S_{max} = 0.70$ and a load frequency f_t of 1 Hz. The fatigue tests—conducted for all microstructural investigations—were stopped at different damage stages, either using a predefined number of cycles or automatically by evaluating the slope of the strain development. The latter method was exclusively used for interruption in phase III. Immediately after interrupting the fatigue test, the NMR measurement was conducted once again using the same settings as before loading. The change (i.e. difference) in the signal intensity for various frequency ranges before and after the fatigue loading represents the water redistribution or, rather, transport in the microstructure due to fatigue loading.

Regarding the indentation measurements, each sample was cut into subsamples using a Buehler IsoMet® 1000 Precision Saw (for details, see [13]). Directly before the indentation testing, the surface of the test sample was ground and polished using SiC papers [17]. The microindentation measurements were carried out using a micro combi tester machine (MCT3) with a three-sided Berkovich tip (see Fig. 2b). A total of 225 indentations were conducted on each microindentation sample within a 15 × 15 (spacing 400 μm) grid dispersed across the sample surface. The indentations were applied force-controlled with a maximum deformation of 25 μm and a 10 s holding phase at maximum load. The loading and unloading rates were kept constant at 10 N/min.

The nanoindentation measurements were conducted using an Anton Paar NHT3 nanoindentation tester machine with a three-sided Berkovich tip. A total of 625 indentations were applied to each sample within a 25 × 25 grid, spaced at 15 μm. The tests were performed force-controlled, consisting of a load ramp from 0 to 5 mN, at a loading and unloading rate of 30 mN/min.

For gas adsorption testing, samples were manually crushed into small particles and sieved. One gram of particles with a size of 2 mm was then treated using an alternating combination of vacuum and flowing gas techniques utilising VacPrep 061 degassing stations from Micromeritics. The nitrogen gas adsorption measurement was carried out using a TriStar II Plus analyser (Micromeritics). The surface area was determined using Brunauer–Emmett–Teller (BET) theory [18] and pore size distributions using the Barrer, Joiyner and Halenda (BJH) theory [19].

2.3 Computational Model

A boundary value problem for fluid-saturated porous media with fractures was formulated for material points $\boldsymbol{x} \in \Omega \in \mathcal{R}^{\delta}$ with dimension $\delta = \{2, 3\}$ and time $t \in \mathcal{T} = [0, T]$. The solution considers three coupled fields: displacement $\boldsymbol{u}(\boldsymbol{x}, t)$, fluid pressure $p(\boldsymbol{x}, t)$ and phase-field fracture $d(\boldsymbol{x}, t)$, where $d(\boldsymbol{x}, t) = 1$ represents complete fracture and $d(\boldsymbol{x}, t) = 0$ an intact state. The detailed mathematical formulation and governing equations can be found in [9, 20]. The total energy functional is decomposed into contributions from elasticity, fluid flow and fracture mechanics:

$$W = W_{\text{elas}}(\boldsymbol{\varepsilon}, d) + W_{\text{fluid}}(\boldsymbol{\varepsilon}, \theta) + \int_{0}^{t} f(\alpha) \frac{d}{dt} W_{\text{frac}}(d, \nabla d) \tag{1}$$

where the strain tensor under small deformations is given by:

$$\boldsymbol{\varepsilon} = \nabla_s \boldsymbol{u} = sym[\nabla \boldsymbol{u}] := \frac{1}{2}\left[\nabla \boldsymbol{u} + \nabla \boldsymbol{u}^T\right] \tag{2}$$

According to Terzaghi's theorem, the elastic contribution to the stress tensor is represented as the sum of an effective stress component and a fluid pressure term:

$$\boldsymbol{\sigma}(\boldsymbol{\varepsilon}, p, d) := \frac{\partial W}{\partial \boldsymbol{\varepsilon}} = \boldsymbol{\sigma}_{\mathit{eff}} - Bp\boldsymbol{I} \;\; with \;\; \boldsymbol{\sigma}_{\mathit{eff}} = g(d)[\lambda tr[\boldsymbol{\varepsilon}]\boldsymbol{I} + 2\mu\boldsymbol{\varepsilon}] \tag{3}$$

The fluid contribution is described by the fluid volume flux, which follows Darcy's law and relates it to the pressure gradient and permeability of the porous medium.

$$\boldsymbol{H} := -\boldsymbol{K}(\boldsymbol{\varepsilon}, d)\nabla p \tag{4}$$

where the permeability tensor $\boldsymbol{K}(\boldsymbol{\varepsilon}, d)$ [21] is additively decomposed into Darcy and Poiseuille-type flow components. The continuity equation for the fluid mass which reflects the fluid flow PDE [21] within a hydraulic fracturing setting is:

$$\dot{\theta} + \text{Div}[\boldsymbol{H}] = 0 \tag{5}$$

The regularised fracture energy density $W_{frac}(d, \nabla d)$ is scaled by a non-negative degradation function $f(\alpha)$. The second axiom of thermodynamics for the fatigue response is satisfied by the non-negativity of the dissipation rate. Hence, the following form was adopted according to [22, 23], where k is a constant to adapt the rate of decay of the fatigue degradation function.

$$f(\alpha) = \begin{cases} 1 & \text{if } \alpha(x,t) \le \alpha_c \\ \left[1 - k\log\left(\frac{\alpha(x,t)}{\alpha_c}\right)\right]^2 & \text{if } \alpha(x,t) \le \alpha_c 10^{1/k} \\ 0 & otherwise \end{cases} \tag{6}$$

Here, $\alpha_c \geq 0$ is a material threshold parameter below which no fatigue effects are triggered. The fatigue variable α is defined as an accumulation of strain energy density $W_{elas}(\boldsymbol{\varepsilon}, d)$ during loading stages, thus:

$$\alpha(\boldsymbol{x}, t) := \int_0^t \dot{\vartheta}(\boldsymbol{x}, s)\overline{H}\big(\dot{\vartheta}(\boldsymbol{x}, s)\big)ds \quad with \quad \vartheta(\boldsymbol{x}, t) := W_{\text{elas}}(\boldsymbol{\varepsilon}, d) \tag{7}$$

Here, $\overline{H}(x)$ refers to the Heaviside function. The third partial differential equation in the rate-dependent setting for the fatigue-induced damage yields:

$$\left[d - l^2\Delta d\right] + (d-1)\mathcal{H} = 0 \tag{8}$$

Here, $\mathcal{H}$ indicates the crack driving force which is derived through the maximum expression of the positive crack driving state function $\tilde{D}(\boldsymbol{x}, \boldsymbol{\varepsilon})$.

$$\mathcal{H}(\boldsymbol{\varepsilon}) = \max_{s\in[0,t]} \tilde{D}(\boldsymbol{x}, s) \geq 0 \tag{9}$$

For recent works on the phase-field modelling of fatigue failure behaviour, the interested reader is referred to [24, 25].

3 Results

3.1 Number of Cycles to Failure

The number of cycles to failure N_f is plotted in Fig. 3a against the moisture content, i.e., the amount of water expressed as a percentage of its total mass. The largest difference in the number of cycles to failure was found between the storage conditions D and WST (for abbreviation, see Table 1). It must, however, be mentioned that the real number of cycles to failure of series D is unknown, as the sample did not show a failure within the test duration. The number of cycles to failure obtained for the storage conditions C and M were between the conditions WST and D. Additionally, it was found that the number of cycles to failure of the series WS and WST were very close to each other. This indicates that, in the fully saturated case, the surrounding

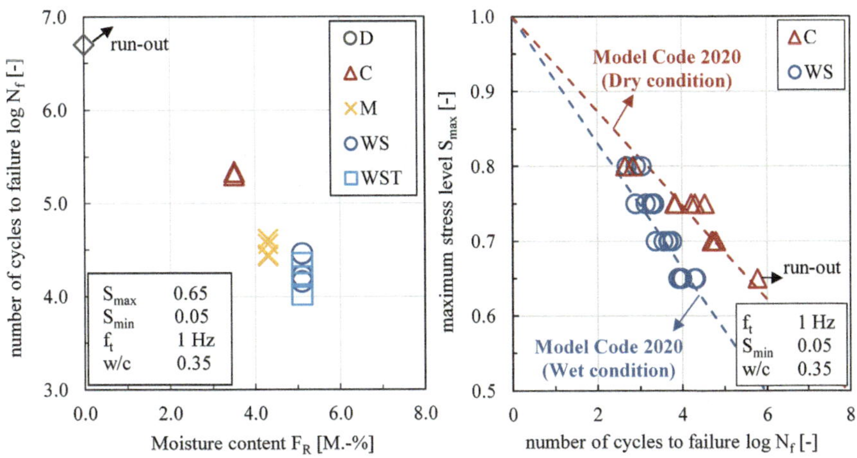

Fig. 3 **a** Influence of the moisture content on the fatigue resistance of concrete stored in different conditions (D: dried; C: stored in a climate chamber; M: sealed; WS: stored under water; WST: stored and tested under water) (from [14], licensed under CC-BY 4.0). **b** S/N curves of C and WS specimens, which were subjected to S_{max} between 0.80 and 0.65, a constant S_{min} of 0.05 and a constant f_t of 1 Hz

environment has no further significant influence on the fatigue resistance of concrete, unlike in samples previously stored in air.

Figure 3b shows the S/N curves of specimens stored in the climate chamber (C) and wet specimens (WS) subjected to different maximum stress levels S_{max} and a constant load frequency f_t of 1 Hz. The curves of the new *fib* Model Code 2020 [26] for dry and wet conditions are also included in the figure. It can be seen that the number of cycles to failure for both storage conditions corresponds to the values assumed by the *fib* Model Code 2020. At the highest maximum stress level (S_{max} = 0.80), specimens of both moisture contents exhibited nearly the same fatigue resistance. With decreasing S_{max} value, the specimens with higher moisture content showed increasingly lower fatigue resistance than the 'dry' ones.

The influence of load frequency f_t on the fatigue resistance of wet specimens with two different water to cement (w/c) ratios (0.35 and 0.50) subjected to two different maximum stress levels (S_{max} = 0.65 and 0.75) is presented in Fig. 4a. The fatigue resistance of specimens with both w/c ratios was similar across all load frequencies and stress levels. This may indicate that very small pores – rather than the larger capillary pores – are primarily responsible for water-induced fatigue damage in concrete. Additionally, it can be seen that the fatigue resistance of wet specimens strongly decreases with decreasing load frequency at both stress levels.

Figure 4b presents the S/N curve for wet specimens with a w/c ratio of 0.35 subjected to different load frequencies. It can be seen that the numbers of cycles to failure of the specimens loaded at frequencies of 1 Hz fit the *fib* curve quite well. The results obtained for load frequencies less than 1 Hz are lower than the wet *fib* curve at all stress levels. Specimens subjected to fatigue loading with S_{max} = 0.55 and a

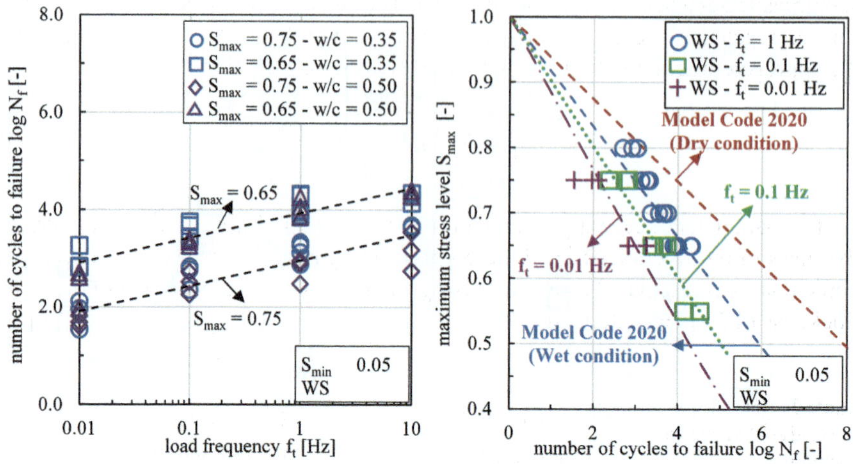

Fig. 4 **a** Influence of load frequencies on the number of cycles to failure of wet specimens (WS) with different w/c ratios (0.35 and 0.50) subjected to two different maximum stress levels. **b** S/N curves for wet specimens subjected to different load frequencies

frequency of $f_t = 0.1$ Hz, for instance, showed values that were 8 times lower than expected from the wet *fib* curve. The *fib* curves are applicable for load frequencies higher than 0.1 Hz. Furthermore, the tested concrete specimens with a permanent moisture content of 100% represent an extreme case with respect to real concrete structures. However, the new results presented here make it desirable to discuss the design curves, considering the effects of load frequency, particularly in interaction with moisture influence.

By contrast, dry specimens subjected to high load frequencies showed lower fatigue resistance, as published by the authors in [27]. This can be attributed to the excessive surface temperature increase during the fatigue tests, which may lead to a premature fracture, as postulated in [28]. The interaction between higher moisture contents and temperature increase is discussed in the chapter by S. Scheerer et al., this volume.

3.2 *Macroscopic Damage Indications*

The influence of the moisture content on the gradient of the strain at a maximum stress level, log grad ε_{max} (i.e. slope of maximum strain) in phase II (for $N/N_f = 0.20$–0.80) for specimens with a w/c ratio of 0.35 is shown in Fig. 5a. The results show a faster increase in the strain per load cycle when the moisture content in concrete increases. This can be explained by the presence of water between the gel particles, which acts as a lubricant and favours sliding processes between the particles, which may increase the viscosity of the cement paste.

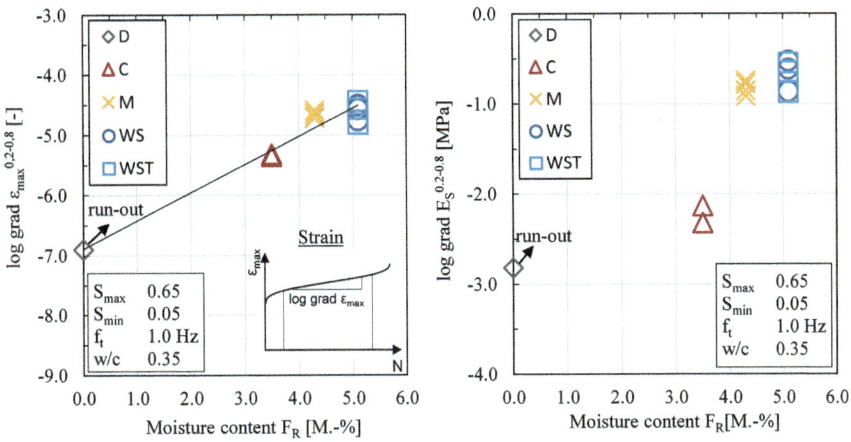

Fig. 5 **a** Gradient of strain log grad ε_{max} in phase II and **b** stiffness log grad $E_{s,}$ in phase II. Both plots refer to specimens stored in different conditions (D: dried; C: stored in a climate chamber at 65% relative humidity (RH); M: sealed; WS: stored under water; WST: stored and tested underwater) (from [14], licensed under CC-BY 4.0)

The stiffness development was analysed as a damage indicator and calculated for each load cycle as a secant modulus (E_s) in the decreasing branch of the sinusoidal strain curve using the strains at maximum and minimum (peak) stresses. The gradient of stiffness, log grad E_s, representing the stiffness degradation per load cycle, was analysed as a parameter.

The results are depicted in Fig. 5b. It can be seen that wet specimens (storage condition WS and WST) exhibited the highest stiffness degradation per load cycle. By contrast, the dried specimen (storage condition D) showed the lowest value of gradient of stiffness. The results from the storage conditions M and C were located in ascending order between the dried specimen (D) and wet specimens (WS and WST), with a sharp increase of the value of stiffness degradation from − 2.23 to − 0.82. This sharp increase might indicate that the water-induced damage mechanisms are strongly intensified when a certain, yet unknown, moisture content in the concrete is exceeded. A recent investigation by the authors [13, 29] indicates that the moisture threshold beyond which the water-induced damage mechanisms in the concrete are intensified may be well in the range of storage at 85% RH.

The AE activity was analysed as a further damage indicator to obtain insight into the moisture-dependent fatigue deterioration of concrete. Figure 6 illustrates the development of the AE activity throughout the test duration for one representative test specimen of the storage conditions C (Fig. 6a) and WS (Fig. 6b), respectively. The grey shapes schematically represent the well-known S-shaped strain development curve. The purple circles represent AE hits, which are located close to the maximum stress level, or rather, between the mean and maximum stress level. These hits are called Hit_{max} in the following. The green points represent hits situated near to the

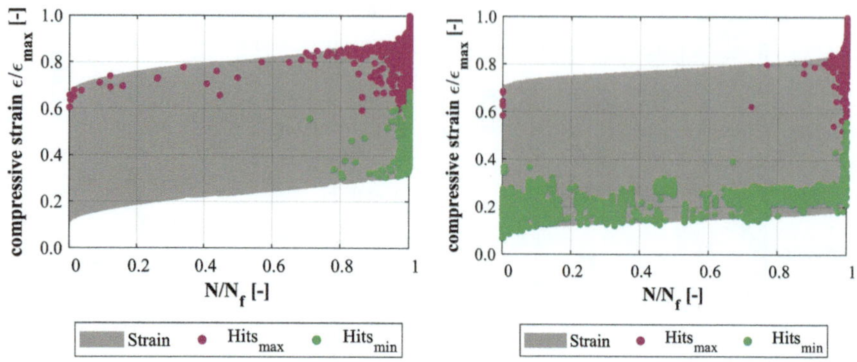

Fig. 6 Representative acoustic emission signals during fatigue loading: **a** Specimen stored in climate conditions at 65% RH (C) and **b** Specimen stored under water (WS) (from [14], licensed under CC-BY 4.0)

minimum stress level, or rather, between the mean and minimum stress level and are subsequently designated as Hit_{min}.

The results showed considerable differences in AE activity between the two storage conditions, C and WS. The number of AE hits in the dry specimen (see Fig. 6a) detected in phase I and approximately two-thirds of phase II was lower than in the wet specimen (see Fig. 6b). However, the number of AE hits in both storage conditions increased rapidly in the last third of phase II and in phase III. Additionally, the hits in the dry case occurred exclusively near the maximum stress level (Hit_{max}). By contrast, AE hits in the wet specimen were mainly observed near the minimum stress level (Hit_{min}). Hits near the maximum stress level in the wet case were only detected at the end of phase II and in phase III.

A further analysis of the AE activity regarding the point of occurrence of the hits in the sinusoidal load curve was carried out, as published by the authors in [27]. The results showed that recorded hits of the storage condition C are primarily located in the ascending branch of the sinusoidal load curve. Surprisingly, the majority of the hits shifted from the ascending to the descending branch of the load curve as the moisture content in the concrete was increased, which may indicate that the water-induced damage takes effect when the external load is removed. A plausible explanation for the occurrence of hits in the descending branch of the load curve in the wet case is an increase in pore pressure, as postulated in [6] and [5]. This may arise due to water transport within the cement paste [7] and finally leads to damage (hits) when the supporting external load is being removed.

3.3 Model for Water-Induced Damage

Based on the results presented above, a descriptive model was developed by Tomann [14] which strongly hints at pore water pressure within the structure of cement paste as the main driving mechanism for water-induced damage. Figure 7 displays the main elements of this model, which is described hereafter. For further details, see [14].

Water within the pore structure of HCP is located in different pore species with different sizes, labelled macro- and meso pores, which are interconnected via smaller micro pores. The pore classification employed in this model are according to the [30], which distinguishes between micro pores ($\leq$ 2 nm), meso pores (2 nm–50 nm), and macro pores ($\geq$ 50 nm).

The pressure in the pore space at a mean stress level S_{mean} is in a state of equilibrium and no pressure-related moisture relocation occurs (Fig. 7a). Water is moved from the macro pores towards the meso pores via the much smaller micro pores (Fig. 7b) due to the increased pressure in cyclic loading and inhomogeneous volumetric deformation of the different pore types. External pressure on the pore system resulting from the mechanical external loading prevents high tensile stresses in the matrix at this stage. Water relocates from the meso-towards the macro pores during unloading towards S_{min} to reach an equilibrium state once again, but backflow is hindered by the small size of the micro pores. A phase shift between the unloading of the external load and the remaining water pressure from the previously increased meso pore moisture content arises that may exceed the typical tensile strength of the pore walls in the meso pores (Fig. 7c). As a result, damages in the pore walls do not occur at the maximum external loading, i.e. when S_{max} is reached, but rather during the unloading phase. This leads to an accelerated fatigue damage process depending on the sample's moisture content, which explains the emergence of AE hits in the unloading phase of specimens with a high moisture content (cf. Figure 6).

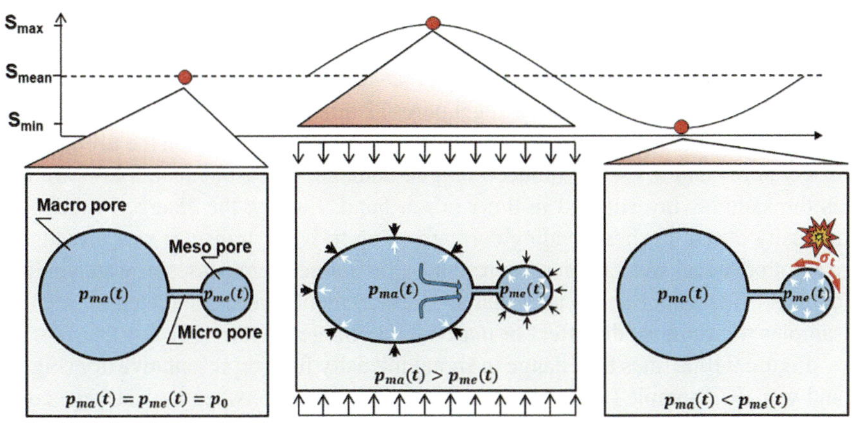

Fig. 7 Descriptive model for water-induced damage due to cyclic loading: **a** First status (S_{mean}), **b** Second status (S_{max}) and **c** Third status (S_{min}) (from [14], licensed under CC-BY 4.0)

3.4 Moisture Redistribution Measured by ^{1}H-NMR

Microstructural investigations using ^{1}H-NMR measurements were conducted on HCP samples to obtain more information on water redistribution due to cyclical loading and to verify the descriptive model presented above.

Firstly, the water content in non-loaded HCP samples stored under various conditions was evaluated. Figure 8a shows the NMR signal intensity, normalised to wet samples, as a function of the RH at which the samples were stored. Since the actual moisture content in the microstructure could not be measured experimentally, the storage RH serves as an indirect indicator, as a correlation exists between them (see [10, 31]). The results reveal a nearly linear increase in the total signal intensity with increasing moisture content in HCP samples, which was expected and consistent with the findings presented by Muller et al. [31]. Next, changes in the water distribution due to different storage conditions were analysed by considering four water species identified by evaluating the frequency spectrum of the relaxometric signal. The aim was to verify whether the degree of saturation of certain pores might be responsible for water-induced damage. The pore size definition given in [32, 33] was used here, which specifies the following sizes: < 2 nm for the pores between the C-S-H sheets (interlayer pores), about 2 nm–8 nm for gel pores, 8 nm–50 nm for interhydrate pores and ≥ 100 nm for the large capillary pores. The results presented in Fig. 8b show a very low to negligible water content in the range of interhydrate and capillary pores for samples of all moisture contents, which was expected due to the typically small number of large pores at a w/c ratio of 0.35. The highest water content was observed in the range of gel pores, followed by interlayer pores. Additionally, a clear influence of the storage condition on the small pores (< 10 nm) can be observed: Wet samples have the highest water content, followed by damp (90% RH) and moist (85% RH) samples. Dry samples showed the lowest water content in these pore species. These results suggest that storage in high RH increases the water content in concrete, particularly in gel pores (2 nm–8 nm), leading to water-induced damage. In other words, it seems that there is a correlation between the filling degree of gel pores and the water-induced fatigue damage, at least at such a low w/c ratio as the samples investigated in the work at hand. Finally, the change in the signal intensity due to cyclical loading corresponding to water transport was investigated for both dry and wet samples, representing the two extreme cases of water-induced damage. It is important to note that all measurements were conducted on identical samples to minimise the effect of material inhomogeneity.

Figure 9 illustrates the change in signal intensity for a representative dry (Fig. 9a) and wet HCP sample (Fig. 9b). Significant differences between both storage conditions can be observed. While the water distribution in the dry sample due to fatigue loading appears to be of minor importance, significant changes can be observed in the wet sample, with the gel pore water content continuously decreasing over the number of load cycles. The increase in the interlayer water appears to remain unchanged in phases I and II, despite the decrease in the gel water, whereas a sharp rise is observed in phase III, i.e. just before failure.

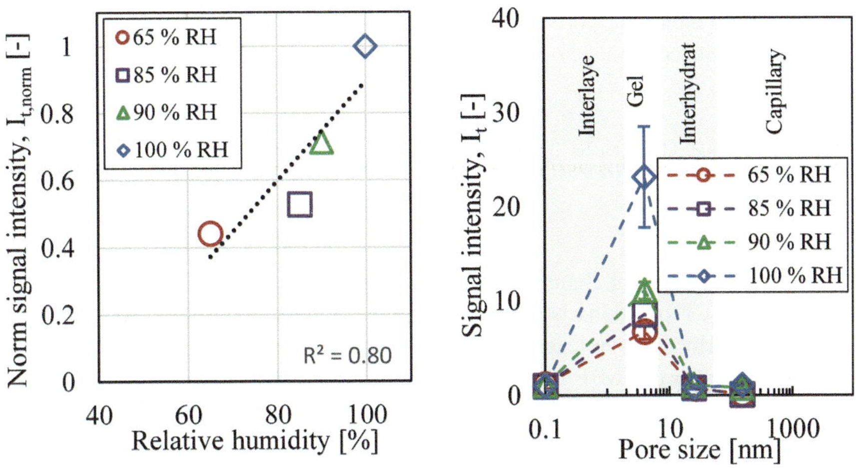

Fig. 8 **a** NMR signal intensity, normalised to wet samples (100% RH), as a function of the RH at which the HCP samples are stored. **b** The signal intensity of different water populations (interlayer pores between the C-S-H sheets with a size of < 2 nm, gel pores with a size of 2–8 nm, interhydrate with a size of 8–50 nm and a capillary with a size of ≥ 100 nm according to [32, 33]) of HCP samples stored in different conditions

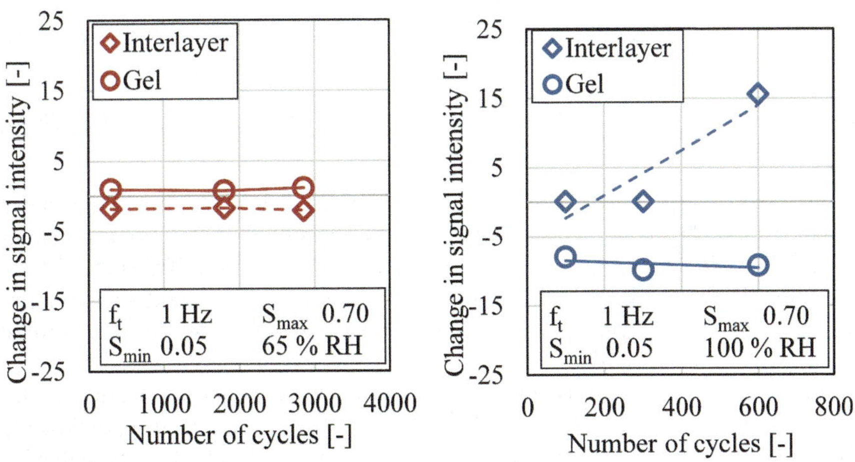

Fig. 9 Changes in signal intensity I_t due to fatigue loading ($I_{t,damaged} - I_{t,undamged}$) for **a** dry HCP sample (stored at 65% RH) and **b** Wet HCP sample (stored underwater)

Unfortunately, the method does not yet allow for a clear evaluation of the destination of the water transport. Yet, the findings give clear evidence that the water redistribution process due to fatigue loading, previously only speculated upon, takes place, which may lead to a more pronounced deterioration in the HCP [7]. Additionally, the reduction in the water population in the gel pores and the pronounced increase in the interlayer pores provide clear evidence that such water transport and, thus, water-induced damage occurs mainly at the nanoscale level (< approx. 10 nm).

3.5 Changes in Porosity Due to Water-Induced Fatigue Damage

Investigations regarding changes in the pore structure of HCP were conducted as a damage indicator to obtain a better and more detailed understanding of how the water transport described above leads to damages in cement paste. Accordingly, mercury intrusion porosimetry (MIP) and gas adsorption measurement were used. The focus was on the latter method, as it is better suited to cover changes in the porosity in the range between 2 and 50 nm, which is the range where the water distribution, and, thus, the water-induced damage, was expected to occur, as shown in Sect. 3.4.

Figure 10 shows the influence of the moisture content on the pore structure of HCP samples. It can be seen that HCP samples stored in an RH of 85% or higher exhibited almost the same pore volume, which, however, was much higher than samples stored at 65% RH (see Fig. 10a). The increase in the pore volume was clearly visible in pores smaller than about 10 nm. It can be seen in Fig. 10b that the surface area obtained by the BET method also increases with increasing moisture content. However, samples stored at a RH of 85% showed almost the same surface area as those stored at 90% RH.

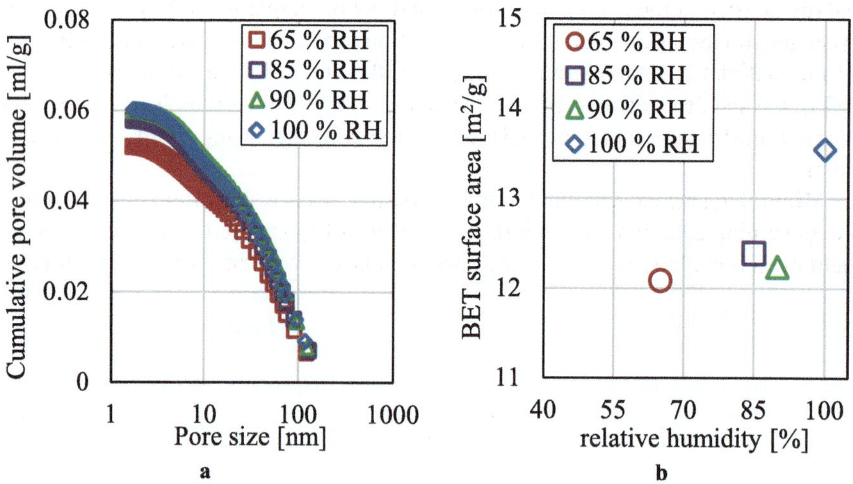

Fig. 10 **a** Cumulative pore volume obtained from N2 adsorption measurements analysed using the BJH method. **b** BET surface area for undamaged HCP samples stored in different conditions (65% RH, 85% RH, 90% RH and 100% RH)

The increase in pore volume, along with a simultaneous increase in the inner surface, suggests either an increase in the number or change in the size of pores, mainly in the scale range of < 10 nm. Both scenarios suggest, however, that storage in a humid environment creates more space in the microstructure of HCP, which allows greater water transport and, therefore, greater damage. An increase in the RH up to 85% seems to be sufficient to create such pore space and, thus, moisture-induced fatigue damage, which goes along well with the results presented in [29]. It should be mentioned that the results shown in Fig. 10 refer to one sample per condition and should therefore be treated with caution.

Next, the influence of water-induced fatigue damage on the specific surface area and pore size distribution of HCP samples was evaluated. The results of the BET surface area are depicted in Fig. 11a as a mean value. It can be seen that water-induced fatigue damage leads to a significant increase in specific pore surface area from 50.73 ± 10.57 m^2/g to 64.79 ± 6.60 m^2/g, which can be interpreted as damage in the microstructure.

The pore size distribution determined by gas adsorption and evaluated by the BJH method is shown in Fig. 11b. Unfortunately, the results showed a high scatter, with the one of the damaged samples being higher than that of the undamaged samples. The fatigue tests were halted in phase III of the strain curve, at which point the samples were close to complete damage. At this stage, the structure is highly unstable, and the damage pattern of different samples is differently advanced, which explains the high scatter of the damaged samples.

Despite the high scatter, a clear tendency can be observed: Water-induced fatigue damage leads to an increase in the pore volume. The most significant increase was visible for pore sizes lower than about 10 nm, which indicates that the greatest damage

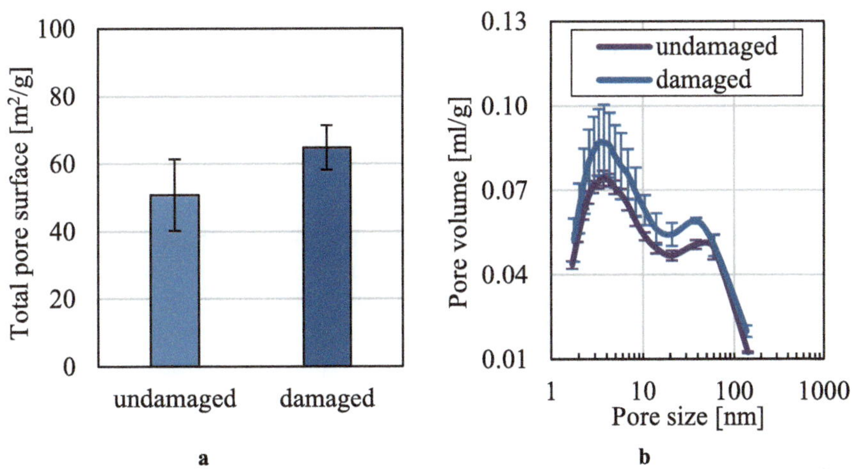

Fig. 11 **a** Pore size distribution obtained by gas adsorption was evaluated using the BJH method. **b** The BET surface area was measured for undamaged and damaged HCP samples stored underwater

occurs in this range. This result corresponds very well with observations obtained using the NMR techniques.

3.6 Impact of Water-Induced Damage on the Micromechanical Properties of HCP

Another important aspect to gain insight into the mechanisms of water-induced fatigue damage is understanding how this damage influences the micromechanical behaviour of HCP. Accordingly, micro- and nanoindentation tests were conducted on fatigue-loaded and reference HCP samples with different moisture contents.

The results of the microindentation, presented in [13], showed a clear trend: While the micromechanical properties–i.e. the HCP hardness H and indentation modulus M–of dry samples do not appear to be affected by fatigue loading, a significant decrease in the hardness H of wet HCP samples was observed. By contrast, no significant impact of water-induced fatigue damage on the indentation modulus M was seen. This can be traced back to the contribution of water to the load transfer, which may still play a role in the modulus of elasticity at this scale and, thus, mask the effects of moisture-induced fatigue damage (for more details, see [13]).

Using the microindentation technique, a volume of approximately 0.1 mm^3 – 0.2 mm^3 was probed. However, as shown by the NMR and gas adsorption results, water-induced fatigue damage probably occurs on a scale of a few nanometres. In order to evaluate the damage at this scale, the sub-micromechanical properties determined by nanoindentation (with a probed volume of about 1 μm^3) were analysed using a Gaussian Mixture Model approach coupled with a micromechanical model, as presented in [34].

The results of the Gaussian Mixture Model revealed the presence of three clusters representing three different phases in the HCP investigated. The focus of this work is on the two types of C-S-H phases: low-density (LD) with low packing density, and HD, with high packing density [35], as these two phases are primarily responsible for the macroscopic mechanical properties.

It is well-known that the microstructure of HCP is highly heterogeneous. Therefore, it was crucial to conduct multiple replicates to ensure reliable results, even though the method is very complex. Nanoindentation measurements were performed on a total of 23 HCP samples (11 samples at 100% RH, 10 samples at 65% RH and 2 samples at 90% RH) in this research with 625 indents per sample. In HCP samples, the distributions of LD-C-S-H, HD-C-S-H, and CH phases within an indentation area are typically similar, although they are strongly affected by the w/c ratio. Figure 12 illustrates the characteristic indentation modulus M, its probability density, and the corresponding deconvolution into three phases for a representative sample with a w/c ratio of 0.35.

Figure 13 illustrates the changes in the sub-micromechanical properties of LD- and HD-C-S-H phases due to fatigue loading. The results clearly show that fatigue

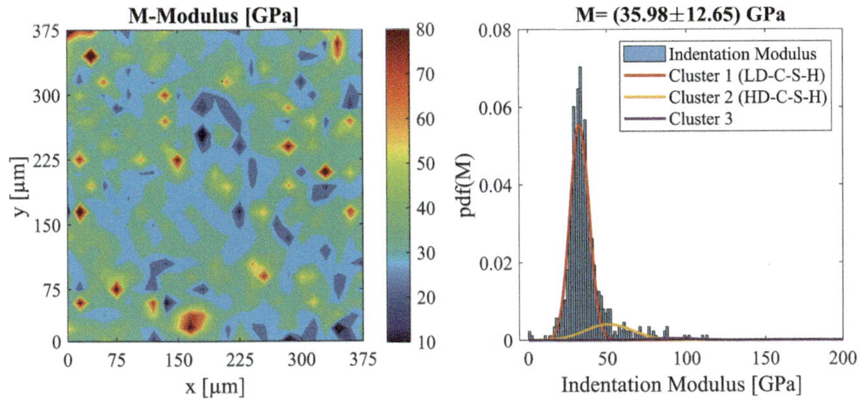

Fig. 12 **a** Characteristic spatial distribution of the indentation modulus M in [GPa] of an exemplarily undamaged HCP sample. **b** Probability density and deconvolution of indentation modulus M in three phases (LD-C-S-H, HD-C-S-H and a mixed phase of LD-C-S-H intertwined with CH) of an exemplarily undamaged HCP sample

loading has a higher, yet different, effect on the sub-micromechanical properties of both LD and HD-C-S-H phases, depending on the moisture content. It can be seen that fatigue loading leads to a decrease in the indentation modulus M and hardness H of the LD-C-S-H of the dry samples. However, this decrease seems to be less significant as the moisture content increases. Indeed, the water-saturated samples showed an increase in the indentation modulus M after fatigue loading, which was an unexpected result.

By contrast, the HD-C-S-H exhibited the completely opposite behaviour under fatigue loading. The indentation modulus M and hardness H decreased significantly for water-saturated samples due to fatigue loading, whereas dry samples showed an increase in the sub-micromechanical properties of the HD-C-S-H phases under fatigue loading.

In addition to the sub-micromechanical properties, the changes in the volume fraction of LD- and HD-C-S-H phases due to fatigue loading were analysed based on the Gaussian Mixture Model results. The latter are presented in [13] and also showed a different impact of fatigue damage on the content of LD- and HD-C-S-H phases depending on the moisture content. Fatigue loading for dry samples leads to an increase in the number of LD-C-S-H phases and a slight decrease in the number of HD-C-S-H phases, simultaneously. By contrast, the number of LD-C-S-H phases of wet samples decrease while the number of HD-C-S-H phases increase. The results of the samples stored at 90% RH were located between dry and wet samples, with a decrease in the volume fraction of LD-C-S-H phases and an increase in HD-C-S-H phases.

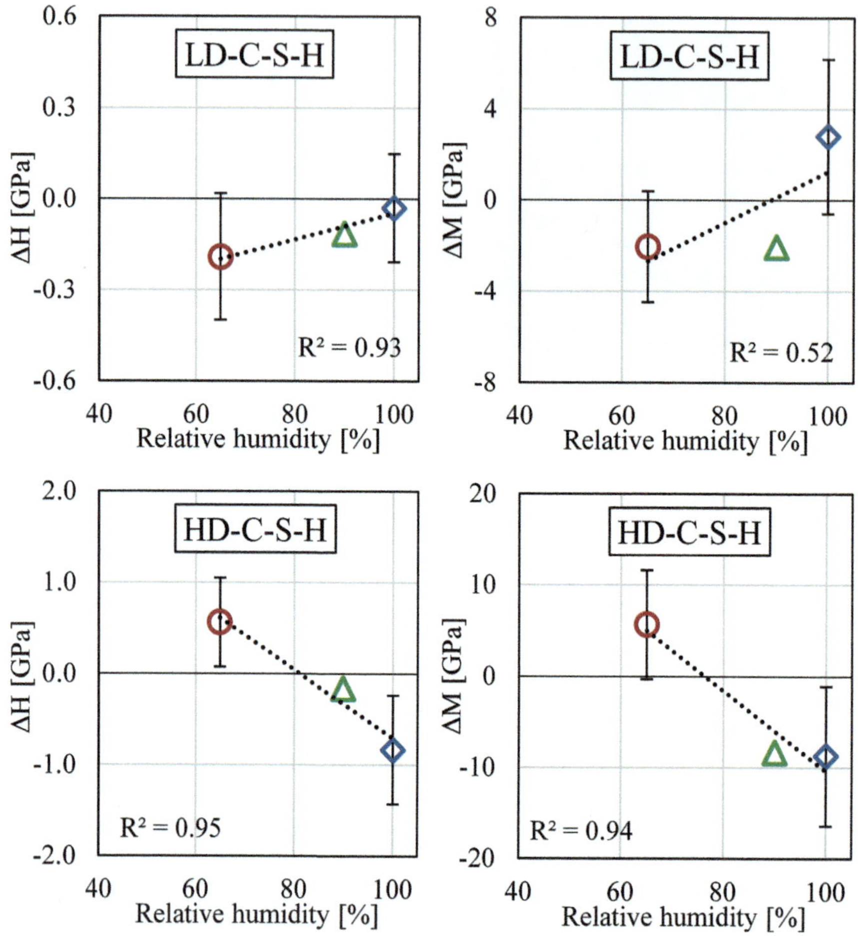

Fig. 13 Changes in sub-micromechanical properties of LD- and HD-C-S-H phases due to the fatigue loading (e.g. $\Delta H = H_{\text{undamgaed}} - H_{\text{damaged}}$) of samples stored at 65% RH, 90% RH and 100% RH, respectively. The fatigue tests were conducted with $S_{\max}$ of 0.70 and a load frequency f_t of 1 Hz. The tests were interpreted at phase III

3.7 *Numerical Results*

This section demonstrates the performance of the proposed fatigue failure theory applied to the HCP microstructure [21, 36]. Two- and three-dimensional examples are investigated to evaluate the experimental observations. The simulations are conducted under two distinct conditions, namely, dry and wet environments, each consisting of three phases: hydrated cement paste, clinker particles, and either dry or saturated voids, depending on the storage conditions (see Fig. 14). The material parameters were considered and applied based on [20, 21].

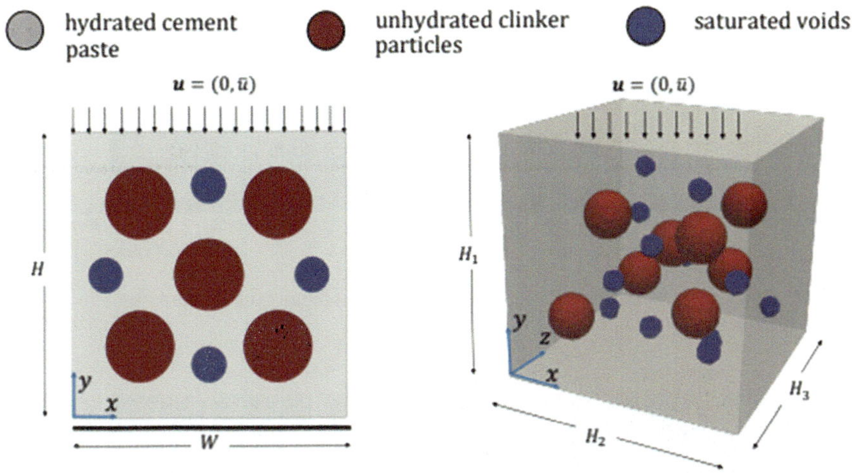

Fig. 14 Geometry and boundary conditions for **a** two- and **b** three-dimensional exemplary wet specimens consisting of HCP (matrix), unhydrated clinker particles (inclusions) and saturated pores (voids)

Most phase-field models primarily address tensile fractures, limiting their applicability to rock-like materials under compression and shear loads. Therefore, a phase-field driving force suitable for such materials with shear frictional characteristics is derived, and a three-dimensional explicit parallel phase-field model is introduced. Various crack driving forces presented in [37–40] are defined and analysed.

Firstly, a two-dimensional model under compression loading is considered to evaluate the proposed fatigue failure theory for concrete microstructures [36, 41]. The analysis includes force–displacement curves for dry and water-saturated specimens, as shown in Fig. 15. The results indicate that water-saturated specimens experience earlier damage than dry ones, confirming that fatigue resistance decreases with increasing loading cycles [21, 36].

The effect of the load frequency on the fatigue behaviour is studied for $f_t = 1$ Hz and 5 Hz. The force–displacement curves in Fig. 16 show that lower load frequencies

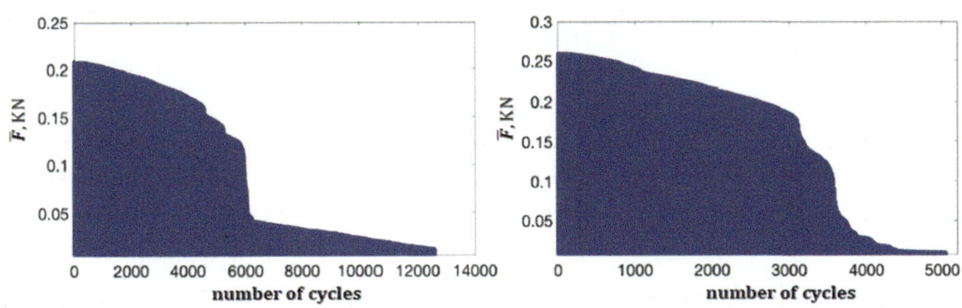

Fig. 15 Load–displacement curves for **a** dry and **b** water-saturated HCP specimens in a two-dimensional setting

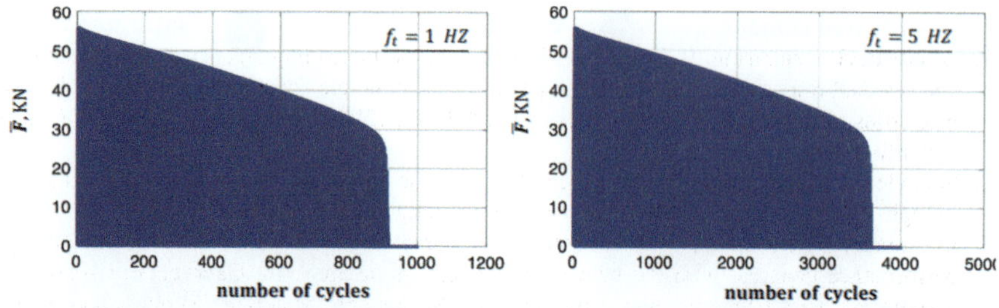

Fig. 16 Quantitative load–displacement curve due to the fatigue cyclic loading for two different frequencies: **a** $f_{\mathrm{t}} = 1$ Hz and **b** $f_{\mathrm{t}} = 5$ Hz

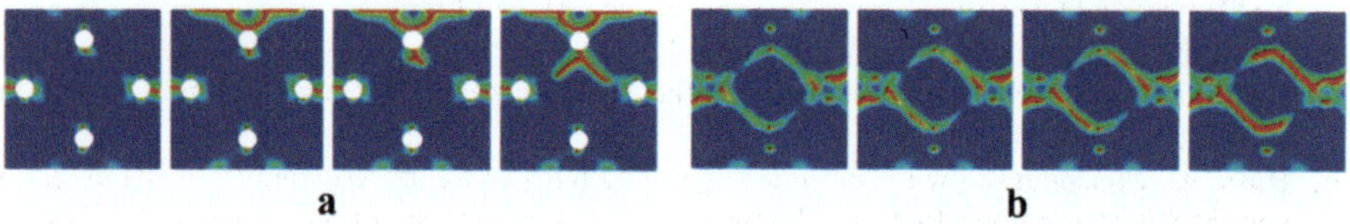

Fig. 17 The crack phase-field evolution for **a** dry and **b** water-saturated concrete in a two-dimensional setting

result in a significant reduction in the number of cycles to failure, accelerating crack propagation, consistent with experimental observations.

Figure 17 presents the computed regularised crack phase field for dry (Fig. 17a) and (Fig. 17b) saturated cases. Different crack driving forces lead to varying crack patterns, highlighting the importance of selecting an appropriate driving force to match experimental observations. Cracks in fully saturated media primarily nucleate around pores, where pressure acts as an additional driving factor.

Additionally, a boundary value problem is applied to a cube to represent the concrete microstructure in a three-dimensional setting (see Fig. 14b). The dimensions are set to 1 mm, with eight inclusions each of a radius of 0.1 mm and fourteen asymmetrically arranged pores each of a radius of 0.01 mm. The computed regularised crack phase-field for fully saturated cases (see Fig. 18) shows that cracks initially nucleate around pores due to pressure effects, before propagating into the HCP matrix, eventually leading to crack merging. This aligns well with the descriptive model presented in Sect. 3.3 and microstructural results.

4 Discussion and Conclusion

The underlying mechanism of water-induced fatigue damage was comprehensively investigated both experimentally and numerically in a joint research project at Leibniz University Hannover. High-strength concrete and HCP samples stored and tested

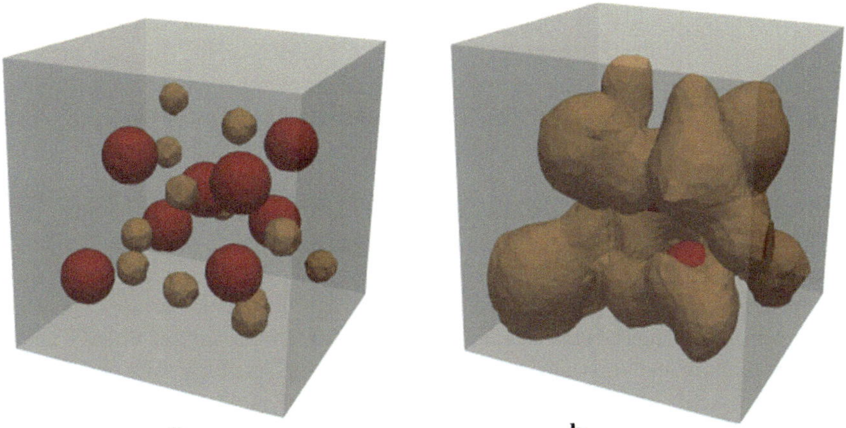

Fig. 18 Computed crack evolution at different stages for a three-dimensional setting at compression cyclic loading

under different environmental conditions were experimentally subjected to cyclical loading and analysed using a combination of macro- and microscopic methods.

The experimental results clearly showed that in fully saturated case, the water in the microstructure of concrete is substantially responsible for the accelerated water-induced fatigue damage. Additionally, indications were found that there is a moisture threshold in the range of about 85% RH (i.e. an internal concrete moisture content equivalent to a storage of the sample at 85% RH) beyond which the water-induced damage mechanisms in the concrete start to become dominant. Hence, water-induced damage should be considered for the design of fatigue-loaded structures in permanently humid environments.

Additionally, it was shown that the effect of water-induced fatigue damage increases massively with reducing load frequencies and relatively low stress levels, making this issue highly appropriate for design structures such as offshore wind turbines. Moreover, no significant impact of a high w/c ratio could be observed, indicating that large pores, such as capillary pores, do not play a major role in the water-induced damage mechanism. Additional influencing parameters, namely, the minimum stress level and pre-damage, were also investigated within the framework of the project. Unfortunately, no clear trends were observed for these parameters.

The investigations performed further showed that the water-induced fatigue damage can be distinguished from 'dry' fatigue damage based on the time of occurrence of the AE hits. Hits in the dry cases occur mainly in the ascending branch of the sinusoidal load curve, in the region between S_{max} and S_{mean}. By contrast, the majority of hits in wet cases shift from the ascending to the descending branch in the region between S_{mean} and S_{min}. This trend clearly indicates that water-induced fatigue damage occurs mainly during the unloading phase.

However, it should be noted that some hits in dry cases behave in terms of their time of occurrence, similarly to those in the wet cases, and vice versa. This may mean that both water-induced and "dry" fatigue damage occur simultaneously in concrete but with different intensities depending on the moisture content. Consequently, understanding and considering water-induced fatigue damage may be essential even for concrete subjected to dry conditions.

Additionally, this work presents a novel variational phase-field simulation for modelling water-induced fatigue damage mechanisms due to the cyclic loading in concrete [36, 41]. As a key feature, the fatigue-induced fracture was formulated based on the accumulation of the bulk energy in time, and, thus, a critical stress state will be drastically reduced. Numerical results confirmed that the fatigue effect results in the reduction of the crack resistance for the water-saturated case in comparison to the dry test.

Further microstructural investigations were conducted using NMR and gas adsorption techniques to gain insight into the underlying damage mechanisms and verify historically proposed damage theories in the literature. Regarding the NMR measurement, it was shown that storage at high RH mainly influences the filling degree of gel pores, which may imply conversely that there is a correlation between the filling degree of gel pores and water-induced fatigue damage, at least at such a low w/c ratio as that investigated in this work. Unfortunately, no critical degree of saturation of the gel pores could be identified.

Moreover, a water redistribution was observed due to cyclical loading, which occurs mainly between interlayer and gel pores (scale range < about 10 nm), confirming the hypotheses postulated by Tomann [14]. Interestingly, the results obtained also fit well with the recent findings of Wyrzykowski et al. [8], who investigated the water redistribution process in HCP samples in situ during creep loading. This makes it desirable to discuss and investigate in the future whether overlapping effects exist between creep and fatigue behaviour, especially at higher moisture contents.

The gas adsorption results showed an increase in porosity due to fatigue loading in the same size range as the water transport observed by NMR. This confirms that the water redistribution process on a scale of a few nanometers is responsible for the water-induced fatigue damage.

Micro- and nanoindentation techniques were used as an additional method to identify how water-induced fatigue damage manifests itself in the microstructure of HCP samples. Regarding the microindentation, no significant impact of dry fatigue damage on the micromechanical properties of HCP was observed. By contrast, a strong decrease in the indentation hardness was observed due to water-induced fatigue damage. However, no significant influence on the indentation modulus could be seen, which can be traced back to the contribution of water in the load transfer (for more details, see [13]).

As the findings from NMR and gas adsorption measurements narrow down the scale of damage in the microstructure of HCP to a few nanometres, it was important to access the sub-micromechanical properties at this low scale level. For this purpose, micromechanical properties obtained by nanoindentation were coupled with a micromechanical model, as described in [34]. The results showed that cyclical loading has different effects on the sub-micromechanical properties of LD- and HD-C-S-H phases depending on the moisture content. In the dry case, fatigue loading decreases the sub-micromechanical properties of LD-C-S-H phases, while the volume fraction of these phases increases. The sub-micromechanical properties of HD-C-S-H phases of these samples decrease with an increase in the volume fraction. In the wet case, fatigue loading increases the hardness and indentation modulus of LD-C-S-H phases, while the volume fraction of these phases decreases. Simultaneously, a strong decrease in the sub-micromechanical properties of HD-C-S-H phases was observed with an increase in their volume fraction.

Even though fatigue loading appears to have different effects on the sub-micromechanical properties of LD- and HD-C-S-H phases depending on the moisture content, a consistent scenario for the underlying mechanisms of dry and wet fatigue damage emerges. Both scenarios are discussed in detail in [13]. The scenario for mechanisms of wet fatigue damage can be summarised as follows:

LD-C-S-H phases: At a high moisture content, the pores in the LD-C-S-H phase are fully saturated with water. Subjected to compressive stress, the LD-C-S-H phases experience less compaction because water is nearly incompressible. This minimal compaction increases the mechanical stiffness of LD-C-S-H phases; however, it is probably insufficient to cause fragmentation or damage to these phases. Additionally, the water in the pores within the LD-C-S-H phases contributes to load transfer, which also helps to enhance their mechanical properties. Due to the compaction effect, the saturated pores within the LD-C-S-H phases are subjected to higher pressure. This may lead to water transport from the gel pores within the LD-C-S-H phases into the significantly smaller interlayer pore spaces within the HD-C-S-H phases, as demonstrated by the NMR results.

HD-C-S-H phases: The water transport from the LD-C-S-H phases causes an increase in the internal pressure within the HD-C-S-H phases. Due to this internal pressure, a differential stress is likely to build up within in the HD-C-S-H phases. However, this is probably suppressed by the compressive stress during cyclical loading, which prevents damage. As the compressive stress decreases during the unloading phase, the internal pressure also decreases; however, it does so at a slower rate, resulting in a phase shift between the external loading and internal pressure, as postulated by Tomann [14]. The remaining internal pressure may induce differential stress, which can eventually cause fractures in the HD-C-S-H phases, leading to the formation of new, smaller and mechanically weaker fragments. In concrete with high moisture content, this loosening may mainly occur during the unloading phase, which may explain why most of the AE hits were recorded in the descending branch of the load curve. The temporal allocation of the hits between S_{mean} and S_{min} implies that a reduction in compressive stress below the mean stress level S_{mean} is needed to eliminate the suppressed effect and induce the loosening effect. The creation of new and mechanically weaker fragments due to the loosening effect also explains the substantial decrease in the mechanical properties of the HD-C-S-H phases and the simultaneous increase in the particle volume fraction.

In order to prove the loosening effect of HD-C-S-H phases and, thus, the proposed damage scenario, the changes in the contribution of C-S-H phases in the total porosity were computed using the packing density of each phase, according to [42]. Firstly, the calculated contribution of the C-S-H phases' porosity was validated by comparing it with the surface area obtained through BET analysis, as this is a direct method for determining porosity. The results are shown in Fig. 19a. It can be seen that the calculated contribution of C-S-H porosity using the packing density, determined by nanoindentation techniques, correlates well and linearly with the BET surface area, which, in turn, correlates with the total porosity. Additionally, the water-saturated samples exhibited by far the highest porosity of C-S-H phases, followed by samples stored at 90% RH and 65% RH, respectively.

Next, the changes in the contribution of LD- and HD-C-S-H porosity caused by fatigue loading in wet HCP samples were evaluated. For this purpose, fatigue tests were carried out on wet samples and stopped at different damage stages: N = 0, 126, 631, and 1156 (for more details, see [13]). Due to the complexity and expense of this procedure, it was unfortunately not possible to conduct multiple replicates to quantify the precession of the results. Nevertheless, the results presented in Fig. 19b show a clear tendency: Fatigue loading leads to a decrease in the porosity of LD-C-S-H phases, which can be traced back to a compacting effect. Simultaneously, an increase in the porosity of HD-C-S-H phases was observed due to fatigue loading, which can be explained by loosening effects or, rather, damage. This confirms the damage scenario described above, which highlighted the loosening effect of the HD-C-S-H phases caused by internal pressure resulting from water transport due to cyclic loading as the primary factor responsible for water-induced fatigue damage.

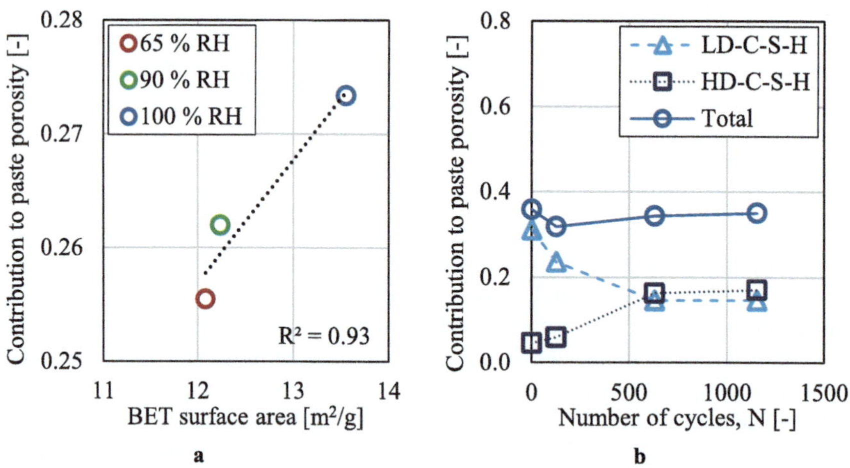

Fig. 19 **a** Comparison of the surface area obtained by BET and the contribution of C-S-H phases to the overall porosity of undamaged samples stored at 65% RH, 90% RH and 100% RH. **b** Contribution of LD- and HD-C-S-H phases to the overall porosity of HCP samples stored at 100% RH

Moreover, it can be seen from Fig. 19b that the total porosity of the C-S-H phases in phase I and at the beginning of phase II was massively dominated due to the compacting effect of LD-C-S-H phases, which led to a higher decrease in overall porosity. By contrast, the increase in the porosity of HD-C-S-H phases clearly has a significant impact on the total porosity and, consequently, on the fatigue behaviour in phase II and III. In these damage phases, the HD-C-S-H phases were loosened so far that they had even more porosity than the LD-C-S-H phases, which can be attributed to strong damage in the structure of the HD-C-S-H.

As the nanoindentation results were analysed using statistical methods, determining the exact proportion of LD-C-S-H phases that were compacted and potentially transformed into HD-C-S-H phases remains challenging. Additionally, quantifying the number of HD-C-S-H phases affected by loosening effects or damage is challenging. To address this, a chemical analysis of the C-S-H phases using techniques such as scanning electron microscopy, as described by Haist et al. [43], is required. A recent study [44], contributed to by the authors, employed both scanning and transmission electron microscopy to characterise fatigue damage in wet concrete samples. The results indicated that water-induced fatigue damage causes microstructural changes in the cement paste at the nanometer scale, which strongly supports the mechanisms proposed in the study at hand. However, further investigations are needed to validate these findings and correlate them with the micromechanical properties presented in this work.

In conclusion, combining different macro- and microscopic methods provided a conclusive picture of the water-induced fatigue damage. Whereby the nanoindentation methods stand out as innovative and suitable methods to detect fatigue damage evolution even at very low scales. However, a large number of investigations is needed in the future to deal with the large scatter of this method, which is very complex and time-consuming. A combination of nanoindentation methods and micro-computed tomography, as described in [45], which can be additionally combined with an artificial intelligence model, may be the solution in the future to cope with the complexity of the nanoindentation method. This can further provide new insights into how the described damage within HD-C-S-H phases accumulates into macroscopic fractures or cracks at later stages. Overall, in addition to the damage mechanisms, several questions remain open, such as the possible interaction between creep and fatigue effects, particularly at higher moisture levels, which should be investigated in future studies.

References

1. Muguruma, H., Watanabe, F.: On the low-cycle compressive fatigue behaviour of concrete under submerged conditions. In: Proceedings 27th Japan Congress on Materials Research, pp. 219–224. Kyoto, Japan (1984)
2. Hohberg, R.: Zum Ermüdungsverhalten von Beton [About the fatigue behaviour of concrete]. Doctoral dissertation, Technische Universität Berlin, Germany (2004). https://doi.org/10.14279/depositonce-815

3. Hümme, J., Lohaus, L.: Fatigue behaviour of high-strength grout in dry and wet environments. Paper presented at the International Wind Engineering Conference IWEC, Hannover, Germany (2014)
4. Oneschkow, N., Hümme, J., Lohaus, L.: Compressive fatigue behaviour of high-strength concrete in a dry and wet environment. Constr. Build. Mater. **262**, 119700 (2020). https://doi.org/10.1016/j.conbuildmat.2020.119700
5. Hümme, J.: Ermüdungsverhalten von hochfestem Beton unter Wasser [Fatigue behaviour of high-strength concrete under water]. Doctoral dissertation. Reports of the Institute of Building Materials Science, 18, Leibniz University Hannover, Germany (2018). https://doi.org/10.15488/3563
6. Sørensen, E.V., Westhof, L., Yde, E., Serednicki, A.: Fatigue life of high-performance grout for wind turbine grouted connections in wet or dry environment. Poster presented at the EWEA Offshore 2011, Amsterdam, the Netherlands (2011)
7. Brosge, S.: Beitrag zur Ermüdungsfestigkeit von Hochfestem Beton [Contribution to the fatigue strength of high-strength concrete]. Doctoral dissertation, Universität Leipzig, Germany (2001)
8. Wyrzykowski, M., Gajewicz-Jaromin, A.M., McDonald, P.J., Dunstan, D.J., Scrivener, K.L., Lura, P.: Water redistribution—microdiffusion in cement paste under mechanical loading evidenced by 1H NMR. J. Phys. Chem. C **123**, 16153–16163 (2019). https://doi.org/10.1021/acs.jpcc.9b02436
9. Abubakar Ali, M., Tomann, C., Aldakheel, F., Mahlbacher, M., Noii, N., Oneschkow, N., Drake, K.-H., Lohaus, L., Wriggers, P., Haist, M.: Influence of moisture content and wet environment on the fatigue behaviour of high-strength concrete. Materials **15**, 1025 (2022). https://doi.org/10.3390/ma15031025
10. Tomann, C., Oneschkow, N.: Influence of moisture content in the microstructure on the fatigue deterioration of high-strength concrete. Struct. Concr. **20**, 1204–1211 (2019). https://doi.org/10.1002/suco.201900023
11. German Institute for Standardization: DIN EN ISO 7500-1:2018-06. Metallic materials - Calibration and verification of static uniaxial testing machines—Part 1: Tension/compression testing machines - Calibration and verification of the force measuring system (German version of EN ISO 7500-1:2018). Beuth, Berlin (2018). https://doi.org/10.31030/2843057
12. German Institute for Standardization: DIN EN ISO 12571:2000-04. Hygrothermal performance of building materials and products—determination of hygroscopic sorption properties (ISO 12571:2013; German version EN ISO 12571:2013). Beuth, Berlin (2000)
13. Abubakar Ali, M.: Mechanisms of moisture-induced fatigue damage of concrete. Doctoral dissertation, Leibniz Universität Hannover, Germany (2025, accepted)
14. Tomann, C.: Wasserinduzierte Ermüdungsschädigung von Beton [Water-induced fatigue damage of concrete]. Doctoral dissertation. Reports of the Institute of Building Materials Science, 22, Leibniz Universität Hannover, Germany (2021). https://doi.org/10.15488/11314
15. Hardy, E.H.: NMR Methods for the Investigation of Structure and Transport. Springer, New York (2012). https://doi.org/10.1007/978-3-642-21628-2
16. Muller, A., Scrivener, K., Gajewicz, A., McDonald, P.: Densification of C-S-H measured by H NMR relaxometry. J. Phys. Chem. C **117**(1), 403–412 (2013). https://doi.org/10.1021/jp3102964
17. Miller, M., Bobko, C., Vandamme, M., Ulm, F.-J.: Surface roughness criteria for cement paste nanoindentation. Cem. Concr. Res. **38**, 467 (2008). https://doi.org/10.1016/j.cemconres.2007.11.014
18. Brunauer, S., Emmett, P.H., Teller, E.: Adsorption of gases in multimolecular layers. J. Am. Chem. Soc. **60**(2), 309–319 (1938). https://doi.org/10.1021/ja01269a023
19. Barrett, E.P., Joyner, L.G., Halenda, P.P.: The determination of pore volume and area distributions in porous substances. Computations from nitrogen isotherms. J. Am. Chem. Soc. **73**(1), 373–380 (1951). https://doi.org/10.1021/ja01145a126
20. Aldakheel, F.: A microscale model for concrete failure in poro-elasto-plastic media. Theor. Appl. Fract. Mech. **107**, 102517 (2020). https://doi.org/10.1016/j.tafmec.2020.102517

21. Noii, N., Khodadadian, A., Wick, T.: Bayesian inversion for anisotropic hydraulic phase-field fracture. Comput. Methods Appl. Mech. Eng. **386**, 114118 (2021). https://doi.org/10.1016/j.cma.2021.114118
22. Ulloa, J., Wambacq, J., Alessi, R., Degrande, G., François, S.: Phase-field modeling of fatigue coupled to cyclic plasticity in an energetic formulation. Comput. Methods Appl. Mech. Eng. **373**, 113473 (2020). https://doi.org/10.1016/j.cma.2020.113473
23. Carrara, P., Ambati, M., Alessi, R., De Lorenzis, L.: A framework to model the fatigue behavior of brittle materials based on a variational phase-field approach. Comput. Methods Appl. Mech. Eng. **361**, 112731 (2019). https://doi.org/10.1016/j.cma.2019.112731
24. Storm, J., Pise, M., Brands, D., Schröder, J., Kaliske, M.: A comparative study of micro-mechanical models for fiber pullout behavior of reinforced high-performance concrete. Eng. Fract. Mech. **243**, 107506 (2020). https://doi.org/10.1016/j.engfracmech.-2020.107506
25. Schreiber, C., Kuhn, C., Müller, R., Zohdi, T.I.: A phase-field modeling approach of cyclic fatigue crack growth. Int. J. Fract. **225**, 89–100 (2020). https://doi.org/10.1007/s10704-020-00468-w
26. fib—Fédération Internationale du Béton (ed.): fib Model Code for Concrete Structures (2020). Lausanne (2023). ISBN: 978-2-88394-176-2
27. Abubakar Ali, M., Oneschkow, N., Lohaus, L., Haist, M.: Influence of the load frequency and moisture content on fatigue deterioration of high-strength concrete. Ce/papers **6**(6), 849–855 (2022). https://doi.org/10.1002/cepa.2837
28. Schneider, S.: Frequenzabhängigkeit des Ermüdungswiderstandes von hochfestem Beton [Frequency dependence of the fatigue resistance of high-strength concrete]. Doctoral dissertation, Leibniz Universität Hannover, Germany (2020). https://doi.org/10.15488/11147
29. Abubakar Ali, M., Oneschkow, N., Lohaus, L., Haist, M.: Moisture-induced damage mechanisms in high-strength concrete due to compressive fatigue loading. In: Proceedings fib Symposium 2024, New Zealand (2024)
30. Everett, D.H.: Manual of symbols and terminology for physicochemical quantities and units: Appendix II—definitions, terminology, and symbols in colloid and surface chemistry (Part I). Pure Appl. Chem. 46, 71–90 (1976). https://doi.org/10.1351/pac197231040577
31. Muller, A., Scrivener, K., Gajewicz, A., McDonald, P.: Use of bench-top NMR to measure the density, composition, and desorption isotherm of C-S-H in cement paste. Micropor. Mesopor. Mater. **178**, 99–103 (2013). https://doi.org/10.1016/j.micromeso.2013.01.032
32. Korb, J.P.: NMR and nuclear spin relaxation of cement and concrete materials. Curr. Opin. Colloid Interface Sci. **14**, 192–202 (2009). https://doi.org/10.1016/j.-cocis.2008.10.004
33. Jennings, H.M., Kumar, A., Sant, G.: Quantitative discrimination of the nano-pore structure of cement paste during drying: new insights from water sorption isotherms. Cem. Concr. Res. **76**, 27–36 (2015). https://doi.org/10.1016/j.cemconres.2015.05.006
34. Vandamme, M., Ulm, F.-J.: Nanogranular origin of concrete creep. Proc. Natl. Acad. Sci. USA **106**(26), 10552 (2009). https://doi.org/10.1073/pnas.0901033106
35. Jennings, H.M.: Colloid model of C-S-H and implications for creep and shrinkage. Mater. Struct. **37**(1), 59–70 (2004). https://doi.org/10.1007/BF02481627
36. Noii, N., Khodadadian, A., Wick, T.: Bayesian inversion using global-local forward models applied to fracture propagation in porous media. Int. J. Multiscale Comput. Eng. **20**, 57–79 (2022). https://doi.org/10.1615/IntJMultCompEng.2022041735
37. Miehe, C., Schaenzel, L.-M., Ulmer, H.: Phase-field modeling of fracture in multiphysics problems. Part I. Balance of crack surface and failure criteria for brittle crack propagation in thermo-elastic solids. Comput. Methods Appl. Mech. Eng. 294, 449–485 (2015). https://doi.org/10.1016/j.cma.2014.11.016
38. Spetz, A., Denzer, R., Tudisco, E., Dahlblom, O.: A modified phase-field fracture model for simulation of mixed-mode brittle fractures and compressive cracks in porous rock. Rock Mech. Rock Eng. **54**(10), 5375–5388 (2021). https://doi.org/10.1007/s00603-021-02627-4
39. Zhuang, X., Zhou, S.: An experimental and numerical study on the influence of filling materials on double-crack propagation. Rock Mech. Rock Eng. **53**(12), 5571–5591 (2020). https://doi.org/10.1007/s00603-020-02220-1

40. Bilgen, C., Weinberg, K.: On the crack-driving force of phase-field models in linearized and finite elasticity. Comput. Methods Appl. Mech. Eng. **353**, 348–372 (2019). https://doi.org/10.1016/j.cma.2019.05.009
41. Noii, N., Fan, M., Wick, T., Jin, Y.: A quasi-monolithic phase-field description for orthotropic anisotropic fracture with adaptive mesh refinement and primal-dual active set method. Eng. Fract. Mech. **258**, 108060 (2021). https://doi.org/10.1016/j.engfracmech.2021.108060
42. Ulm, F.-J., Vandamme, M., Bobko, C., Alberto Ortega, J., Tai, K., Ortiz, C.: Statistical indentation techniques for hydrated nanocomposites: concrete, bone, and shale. J. Am. Ceram. Soc. **90**, 2677–2692 (2007). https://doi.org/10.1111/j.1551-2916.2007.02012.x
43. Haist, M., Divoux, T., Krakowiak, K.J., Skibsted, J., Pellenq, R.J.-M., Müller, H.S., Ulm, F.-J.: Creep in reactive colloidal gels: a nanomechanical study of cement hydrates. Phys. Rev. Res. **3**(4), 043127 (2021). https://doi.org/10.1103/PhysRevResearch.3.-043127
44. Engelhardt, M., Kalytta-Mewes, A., Volkmer, D., Lohmann, J., Ritter, M., Schaan, G., Schmidt-Döhl, F., Abubakar Ali, M., Basaldella, M., Haist, M., Kern, B., Lohaus, L., Oneschkow, N., Rozanski, C., Timmermann, T., Grosse, C.U., Birtel, V., Garrecht, H., Madadi, H., Markert, M., Steeb, H.: Micro- and nanostructural investigations of high and ultra-high performance concrete under fatigue. Int. J. Fatigue **199**, 109038 (2025). https://doi.org/10.1016/j.ijfatigue.2025.109038
45. Kim, J., Lim, J., Stephan, D., Park, K., Han, T.: Mechanical behavior comparison of single and multiple phase models for cement paste using micro-CT images and nanoindentation. Constr. Build. Mater. 342(A), 127938 (2022). https://doi.org/10.1016/j.conbuildmat.2022.127938

Investigation of Fatigue Damage of (Ultra) High-Performance Concrete at Varying Environmental Conditions

Hamid Madadi, Martin Markert, Veit Birtel, Harald Garrecht, and Holger Steeb

Abstract High-Performance Concrete (HPC) is gaining attention due to its exceptional compressive strength and durability. Understanding its mechanical behaviour under high-cycle fatigue loading is crucial for designing long-lasting structures. This study examines how factors like loading frequency, moisture content, and temperature affect the fatigue damage behaviour of HPC and Ultra-High-Performance Concrete (UHPC). Mechanical fatigue damage tests were conducted at different cyclic loading frequencies and upper stress levels to characterise strain evolution and its correlation with failure cycles. An evaluation method was developed to minimize variability in the relationship between failure cycles and applied load or temperature variations. High-frequency fatigue testing (20–200 Hz) simulated dynamic cyclic loading conditions. By stopping fatigue tests before complete failure, concrete samples were collected at specific damage phases and analysed using micro X-ray computed tomography (μXRCT) and dynamic mechanical analysis (DMA). This μXRCT-DMA characterization provides insights into local damage processes and the evolution of mechanical properties. Results highlight that moisture significantly impacts failure cycles, strain development, and temperature changes. The DMA and μXRCT effectively characterise stiffness degradation, while high-frequency fatigue testing enhances understanding of concrete behaviour under different frequency loads.

Keywords High-performance concrete (HPC) · Dynamic mechanical analysis (DMA) · Micro X-ray computed tomography (μXRCT) · High frequency fatigue test (HFFT)

H. Madadi · H. Steeb (✉)
Institute of Applied Mechanics (CE), University of Stuttgart, Stuttgart, Germany
e-mail: holger.steeb@mechbau.uni-stuttgart.de

M. Markert · V. Birtel · H. Garrecht
Material Testing Institute, University of Stuttgart, Stuttgart, Germany

H. Steeb
SC SimTech, University of Stuttgart, Stuttgart, Germany

N. Oneschkow and L. Lohaus (eds.), *Cyclic Deterioration of High-Performance Concrete*, https://doi.org/10.1007/978-3-032-19394-0_8

1 Introduction

The fatigue damage behaviour of complex building materials like concrete remains insufficiently understood. The heterogeneous microstructure and variations in material composition make HPC and UHPC a subject of ongoing research [1]. Although considered in design methods and regulations, fatigue damage is not adequately represented in standards [2]. An overview of fatigue damage for different concretes is provided by Hsu [3]. Most experimental studies use force-controlled methods with different loading conditions. It is important to distinguish between compression, tension, and alternating loading. Fatigue-related damage in concrete is defined differently; Kim and Kim [4] associate it with strains, while Gao and Hsu [5] use changes in stiffness. The exact damage processes remain under investigation.

Experimental analysis identified three phases [6, 7]. In phase I, no significant macroscopic crack growth occurs; damage is primarily due to local microstructural deformations and early microcracking in the cement paste. Phase II, which typically represents 80–90% of the total fatigue life, is characterised by progressive internal microcrack development, limited crack coalescence, and gradual surface damage accumulation [6]. Phase III (less than 10%) involves unstable crack growth leading to failure. Due to inhomogeneous initial conditions, strain distribution becomes uneven, accelerating damage. Stiffness reduction, often measured through sound wave propagation speed, serves as an indicator of fatigue damage [6]. Fatigue crack development occurs at macro-, meso-, and micro-scales [7].

Classical fatigue tests for concrete have been performed at low frequencies (1–10 Hz) due to hydraulic equipment limitations [8]. Most research has been conducted up to 5×10^6 load cycles, although real-world structures endure much higher cycles. Even small force pulses at high frequencies can initiate and propagate cracks, leading to failure [9]. High-frequency fatigue testing offers benefits and challenges. The high-power ultrasonic method, discussed by Bathias [10], mimics high deformation rates but requires considerable power and shows variable strain amplitudes. It shortens testing duration but involves complex design and resonance length changes with temperature. The electrodynamics shaker method, explored by Ghielmetti et al. [11], is cost-effective for testing up to 20 Hz but may face non-linear disturbances. The Single Gear-Tooth Bending Test Facility, examined by Stringer et al. [12], enables rapid cycle accumulation up to 1000 Hz but demands advanced equipment and high costs. Most studies focus on low to moderate frequencies, with high-frequency methods only recently explored. Research on UHPCs is limited, with Dong et al. [13] using a maximum frequency of 5 Hz. Current methods count cycles until failure without analyzing strain evolution [14]. Design principles in DIN EN 1991 [15] and *fib* Model Code 2010 [16] do not fully address frequency effects. Load frequencies between 1 and 10 Hz with stress levels $S_0 > 0.75$ increase maximum cycle numbers, but for test frequencies above 10 Hz, the influence remains unclear. Each frequency fatigue method reduces test times and simulates real-world conditions but presents complexity challenges. These advancements highlight the benefits of high-frequency testing in improving accuracy under real-world loading.

Environmental conditions during sample storage and testing are crucial for many structures. Hohberg [7] highlights the impact of moisture on fatigue failure cycles. Higher porosity concretes absorb more water, leading to greater variability in fatigue life. In contrast, denser concretes exhibit more consistent failure cycles. Despite the presence of water-filled pores, high-strength concrete generally demonstrates better resistance to cyclic loading due to its lower overall porosity and improved microstructural integrity. Scheydt [17] examined moisture-dependent fatigue damage in different concretes, revealing gel, microcapillary, and capillary pore distribution. High-strength concrete has fewer microcapillaries, with moisture movement primarily through gel pores, where effusion dominates over liquid transport.

Heating due to increased loading speeds is another factor. Elsmeier et al. [18] observed significant temperature evolution during cyclic loading (1–10 Hz). Increased temperature raises pore water pressure, causing tensile stresses. This effect is particularly relevant for high-strength concrete with its less-permeable pore structure. It remains uncertain whether the resulting mechanical stresses from higher pore water pressure alter damage behaviour and premature failure.

Recent advancements in micro X-ray computed tomography (μXRCT) have improved imaging for materials and geosciences. Bench-top μXRCT allows micrometre-resolution characterizations, while synchrotron-based μXRCT provides ultra-fast scanning with high photon fluxes [19]. X-ray transparent testing devices enable coupling hydro-chemo-mechanical experiments with X-ray tomography under "in-situ" stress and temperature conditions [20]. Despite these advancements, essential equipment for solid-state material testing, such as multi-axis devices and high-force testing ($F > 10$ kN) or high-test speeds ($f > 1$ Hz), remains limited to few facilities. While μXRCT has examined microstructural changes in concrete [21], combining it with high-frequency fatigue testing is crucial for understanding crack initiation, propagation, and failure mechanisms. This integration offers insights into mechanical property changes, damping behaviour, and crack morphology. Current models do not fully address the influence of loading frequency on UHPC fatigue performance, highlighting the need to develop and validate new fatigue models for high-frequency dynamic loading.

This chapter addresses knowledge gaps in HPC fatigue behaviour under varying frequencies, compressive loading, and environmental conditions. It aims to enhance concrete design for high-frequency dynamic loads, focusing on moisture content, test frequency, and temperature. Understanding the impact of these parameters on fatigue tests is significant important. Cyclic compressive loading tests combined with "in-situ" μXRCT will image crack formation under transient loading, revealing micrometer-wide cracks and their spatial spread. This links stiffness reduction (micro-crack formation) and dissipative effects (force–displacement phase shift) with moisture influence on effective stresses and load cycles. Correlating effective mechanical properties with changing morphology under changing environmental conditions remains a crucial first step.

2 Materials and Methods

In this research, various methods and experimental setups were developed to prepare, analyse, and test concrete. These methods are described in detail below. The HPCs and UHPCs used in this study are based on the reference mixtures from the Priority Programme SPP 2020 (see Sect. 2.1 "Concrete Composition" in the chapter by M. Basaldella et al., this volume) [21–23].

2.1 Sample Preparation and Application

In order to investigate the influence of concrete moisture, temperature and load type on the fatigue behaviour, two different test series were analysed. For this purpose, the influence of the concrete boundary conditions was analysed at three different concrete moisture contents and temperatures that are generated by self-heating during the fatigue tests. In these tests, an uniaxial load at a constant loading frequency of 10 Hz was applied (series 10 Hz). The influence of the spatial stress state on the concrete was investigated by additional uniaxial and triaxial tests (Series 2 Hz). For both series, a sinusoidal fatigue load of 2 Hz or 10 Hz was always applied in the longitudinal direction of the sample. In the triaxial tests, an additional constant confining pressure was applied to the surface of the concrete. Samples with two different concrete moisture contents were analysed. For both test series cylindrical samples with a diameter of 60 mm and a height of 180 mm were prepared. Three different storage conditions were used to study the influence of moisture. All samples were first stored underwater for 7 days. After day 7 the samples of Series UW-2, UW-10 and D-2 remained underwater, while the samples of Series C-10 and D-10 were stored in a climate chamber with 20 °C and 65% relative humidity. Series D-2 and D-10 were dried in an oven at 105 °C after day 56. The exact storage conditions and the resulting concrete moisture can be found in Table 1. To determine the moisture content of the concrete, twin samples were dried in an oven at 105 °C for at least 14 days. At the end of this drying period, no further mass loss could be detected within 24 h. Some of the chemically bound water can be expelled at 105 °C, so after 14 days of drying no further time was spent in the oven. The amount of water was then calculated by weighing the samples before and after oven drying. A residual water quantity of 0.1% was assumed for dried samples. All samples were cut to length, and the top surfaces were ground plane parallel, polished at an age of 7–28 days. Immediately before testing, the curved surface areas of the cylinders were sealed with a water vapor diffusion-tight aluminium composite foil (profiTHERM) to prevent the concrete from drying out during the test.

Also, the test samples used for part of the high-frequency fatigue testing are designed based on the material properties of UHPC, which exhibits high strength and durability. Consequently, smaller samples with a diameter of 1/4″ (6.35 mm) are investigated. This sample size also aligns with the limitations of the High Frequency

Table 1 Storage conditions and concrete moisture of the HPC conditions

Series	Day 1	Day 1–7	Days 8–56	Days + 56	Before testing	MC [m%]
D-2	Strip the formwork	UW	UW	105 °C	Wrap	~ 0.1
D-1			20 °C/65%	105 °C		~ 0.1
C-10			20 °C/65%	20 °C/65%		4.0
UW-2			UW	UW		5.2
UW-10			UW	UW		5.1

UW—under water, D—dry, C—climate chamber, MC—moisture content

Fatigue Testing (HFFT) setup, making it possible to perform μXRCT imaging in parallel while achieving the highest voxel resolution.

2.2 Test Setup and Experimental Programme

For the uniaxial fatigue tests at 10 Hz, a servo-hydraulic testing machine (Schenck Hydropuls PC 400 N) with a maximum force of 400 kN was used (Setup 1, Fig. 1a). For the uniaxial and triaxial tests at 2 Hz, a 4-column test frame from "Form + Test Seidner & Co. GmbH", designed for forces up to 1 MN, was used. This frame was fitted with a servo-hydraulic cylinder from Schenk with a maximum capacity of 630 kN (Setup 2, Fig. 1b). The only difference between the uniaxial and triaxial tests setup was the height of the load frame, which had to be raised for installation of the used triaxial test-cell. The installed triaxial test-cell (DBTA60-100-RT-Dyn from GL Test Systems GmbH) had a maximum operating pressure of 100 MPa at a maximum loading frequency of 5 Hz and a maximum operating temperature of 50 °C. The triaxial test-cell was kindly provided for the investigations by the Institute of Concrete Construction of the Leibniz University Hannover. Due to the large deformations of the test setup (up to ≈ 5 mm cylinder path) that occurred during the tests, only a test frequency of 2 Hz could be applied. An absolute pressure transducer from HBM type K-P3-MBP with a measuring range of up to 200 MPa was used to measure the confining pressure of the oil in the triaxial cell. The temperature at the centre of the surface samples, as well as the temperature of the test room and at two points in the triaxial cell (inside and outside), were measured using K-type thermocouples. To protect the sample from oil and to apply the confining pressure, the sample was wrapped in a protective rubber jacket. The installed instrumentation was approved for the high oil pressures. As the uniaxial load was limited to 630 kN, that refers to 250 MPa axial stress of the sample and the confining pressure has an influence on the compressive strength (Sect. 3.1), a confining pressure of 20 MPa was applied. The confining pressure was kept constant during the tests. The measurement technique and the triaxial cell that was used are shown in (Fig. 1c). To minimise unintended bending of the test sample due to imperfections in the plane-parallel top and bottom surfaces of the cylinders, a load transfer plate with a shell and a ball bearing was used for both setups.

During the tests, the applied forces, the number of load cycles, the deformation of the sample and the temperature development were measured. The deformation was measured using three inductive displacement transducers WA 1–3 (Typ HBM K-WA-U-010W) for the uniaxial setup. For the triaxial setup three Linear Variable Differential Transformer (internal LVDT of triaxial cell DBTA60-100-RT-Dyn) were used. In setup 1, 4 displacement transducers (WA1-) were installed because the load plate was not round but rectangular. The strain was calculated from the measured deformations under the assumption of constant strain over the height of the sample. All

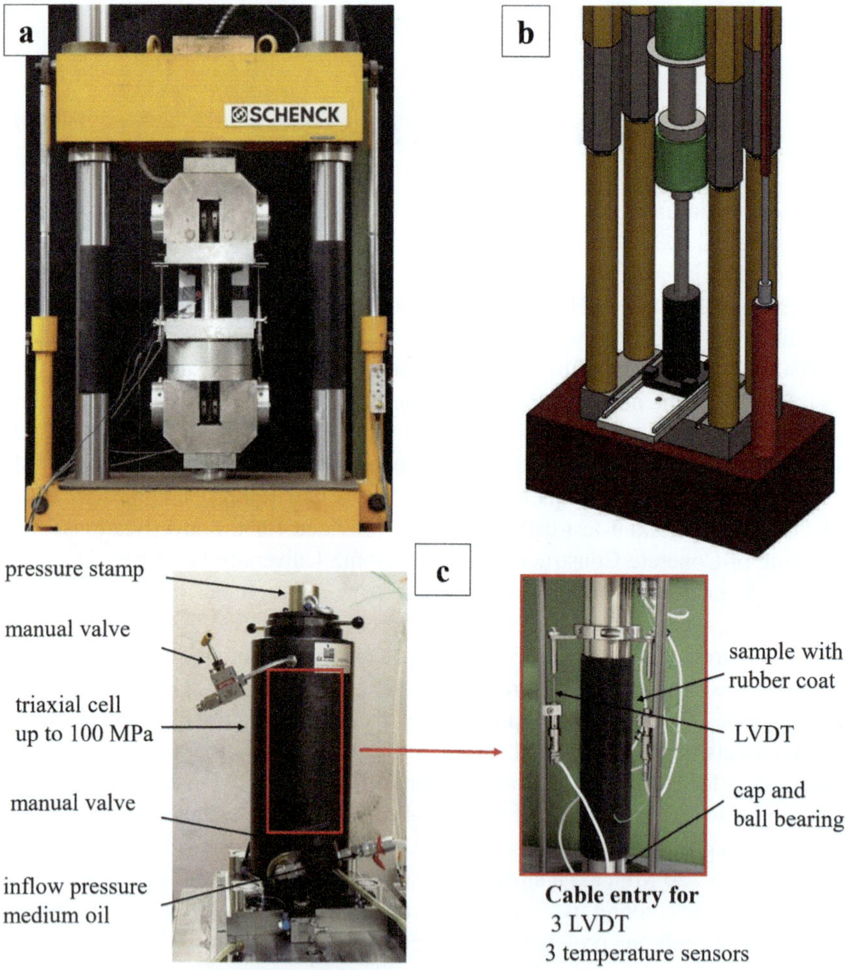

Fig. 1 **a** Test setup 1 fatigue test of Schenck PC 400 N, **b** 3D drawing of the experimental setup 2, **c** Triaxial cell used, measurement technology and internal installation of triaxial cell

Table 2 Overview of the test programme

Series	Storage condition	Loading type	S_o	S_u	Test frequency (Hz)
HPC-D-U-2	D	Uniaxial	0.80/0.85	0.05	2
HPC-UW-U-2	UW	Uniaxial	0.75/0.80	0.05	2
HPC-D-U-10	D	Uniaxial	0.70/0.75/ 0.80	0.05	10
HPC-C-U-10	C	Uniaxial	0.55/0.60/ 0.70/0.80	0.05	10
HPC-UW-U-10	UW	Uniaxial	0.50/0.55/ 0.60/0.70	0.05	10
HPC-D-T-2	D	Triaxial	0.80/0.85	0.05	2
HPC-UW-T-2	UW	Triaxial	0.70/0.75/ 0.80	0.05	2

UW—under water, D—dry, C—climate chamber

tests were carried out force controlled with a sinusoidal load at a constant frequency of 2 Hz or 10 Hz. The triaxial fatigue test setup allowed a maximum frequency of 2 Hz. Uniaxial tests were therefore also carried out at 2 Hz so that they could subsequently be compared with each other. The normalised minimum longitudinal total stress was defined as$S_u = \sigma_u/f_{c,cyl} = 0.05$, and the normalised maximum longitudinal total stress $S_o = \sigma_o/f_{c,cyl}$ was varied according to Table 2. When varying the upper stress level, it was necessary to ensure that the influence applied to all stress levels. In this case, $f_{c,cyl}$ represents the static compressive strength (Table 3), while σ_o/σ_u represents the ratio of the upper to the lower stress. All test series were analysed with regard to the number of cycles to failure, strain and temperature development. Tests with > 2.75 million load cycles were treated as runouts.

2.3 Dynamic Mechanical Analysis Setup

In order to characterise the mechanical properties and to investigate and quantify changes of the morphology of the material, harmonic experiments at small strains, denoted as DMA tests, are performed. DMA tests are standard methods used to characterise rate-dependent materials, such as viscoelastic materials. For brittle materials, DMA tests have been used in the past to characterise the morphology of porous and fluid-saturated porous rocks, like sedimentary sandstones. In contrast to rubber-like materials where tests have been performed in tension or torsion, DMA tests for rocks have been performed under compression [24].

Applying harmonic excitations, the overall viscoelastic response of the material can be characterised by an in-phase and an out-of-phase part, respectively, cf. [25]. Therefore, in the frequency domain, the constitutive behaviour of viscoelastic

Table 3 Compressive strengths of the different types of loading and storage conditions

HPC variation	HPC UW-U2	HPC D-U2	HPC UW-T2	HPC D-T2	HPC UW-U10	HPC C-U10	HPC D-U10
Storage condition	UW	D	UW	D	UW	C	D
Loading type	UAX	UAX	TAX	TAX	UAX	UAX	UAX
Confining pressure (MPa)	0	0	20	20	0	0	0
Static compressive strength ($f_{c,cyl}$) (MPa)	112.1	114.0	215.5	240.8	106.3–109.9	95.4–103.7	98.4–104.0
Standard deviation (MPa)	3.23	0.80	3.04	5.74	4.55–7.31	3.70–7.36	3.63–7.5

UW—under water, D—dry, C—climate chamber, UAX—uniaxial, TAX—triaxial

materials can be described by complex effective material properties like a complex Young's modulus ($E^* = E' + iE''$). The elastic part, represented by the real part of the complex modulus, is the storage modulus (E'). The imaginary part, is the loss modulus (E''). The loss factor (tan δ) arises from the phase shift (phase angle δ) between the applied stresses and the measured strains of the material. If the phase angle δ tends to zero, the material's response is purely elastic; if the phase angle δ tends to $\pi/2$, the material's response is purely viscous. For uniaxial compression or tension has been used following Eqs. (1–3):

$$\varepsilon(t) = \varepsilon_0 \sin(\omega t), \sigma(t) = \sigma_0 \sin(\omega t + \delta_E), \tag{1}$$

$$E' = \frac{\sigma_0}{\varepsilon_0}\cos(\delta_E), E'' = \frac{\sigma_0}{\varepsilon_0}\sin(\delta_E) \Rightarrow E^* = E' + iE'', \tag{2}$$

$$Q_E^{-1} = \tan\delta_E = \frac{E''}{E'}. \tag{3}$$

Furthermore, Poisson's ratio is directly obtained from the ratio of lateral strains (ε_q) to axial strains (ε_l). It can be calculated according to frequency as follows

$$v' = \frac{\varepsilon_q}{\varepsilon_l}\cos(\delta_v), v'' = \frac{\varepsilon_q}{\varepsilon_l}\sin(\delta_v) \Rightarrow v^* = v' + iv''. \tag{4}$$

The complex Poisson's ratio (v^*) consists again of a real part (v') and an imaginary part (v''). The loss factor ($Q_v{}^{-1} = \tan\delta_v$) can be determined as

$$Q_v^{-1} = \tan \delta_v = \frac{v''}{v'}. \tag{5}$$

Figure 2 provides a schematic view of the DMA setup. Detailed descriptions of this setup and other DMA configurations under compression used e.g. in rock physics are available in references [26, 27]. A static preload is applied to the setup using a static universal testing machine (Schenk-Trebel RM 100), and a metallic stamp is utilized to apply a static preload to the sample attached to the actuator. Harmonic force excitations are generated by producing low-amplitude sinusoidal voltage signals (0–10 V) using arbitrary waveform generators (Agilent 33500B). These signals are then amplified by a high-voltage linear amplifier (Physik Instrumente E-482), with a maximum amplification factor of 100. The amplified continuous sinusoidal signal is transmitted to a high-voltage piezoelectric actuator (Physik Instrumente P-235.2 s), capable of displacements up to 40 μm. The displacement of the actuator typically affects the input voltage in a highly nonlinear manner. A piezoelectric force sensor (HBM CFW/50KN) with high frequency sensitivity is integrated to directly measure the force. This sensor is installed in line, allowing for force measurement either directly from the sensor or through the strain gauge of the (calibrated) aluminium sample, particularly when high frequencies reduce the accuracy of the load cell of the universal testing machine (UTM). The samples are positioned at the centre of the unit. Each sample is equipped with a strategically placed strain gauge (HBM K-CXY3 T-Rosette) to transmit comprehensive strain responses throughout the testing process. Additionally, a PT-1000 temperature sensor is employed to monitor temperature evolution. The displacement of the piezoelectric actuator, forces, strains, and temperatures are recorded by a high-speed transient recorder (HBM-HBK GEN2tB and GN1640B) through a MATLAB script.

2.4 *High Frequency Fatigue Test Setup*

This section describes the setup of the high-frequency fatigue conducted using the HFFT method [26, 27], which builds on the DMA setup but is specifically adapted for fatigue testing at higher frequencies and amplitudes. While most components, such as the data acquisition system, sensors, and strain gauges, remain the same, the main differences lie in the actuator specifications and the loading conditions. Classical fatigue testing of concrete typically uses servo-hydraulic machines, limited to 10 Hz due to hydraulic constraints. In contrast, piezoelectric actuators offer faster response and higher dynamic stiffness, though with lower displacement. These properties make them suitable for generating cyclic loads at much higher frequencies. To enable high-frequency fatigue testing of UHPCs, the setup uses a high-voltage actuator ("Physik Instrumente P-035.90P") with displacements up to 180 μm and forces up to 30 kN. Sinusoidal signals are generated by the integrated "Physik Instrumente P-517" waveform generator and limited to 700 V for sustained operation. This allows frequencies between 20 and 200 Hz, well beyond DMA or hydraulic methods.

Fig. 2 Schematic of a HFFT and DMA setup

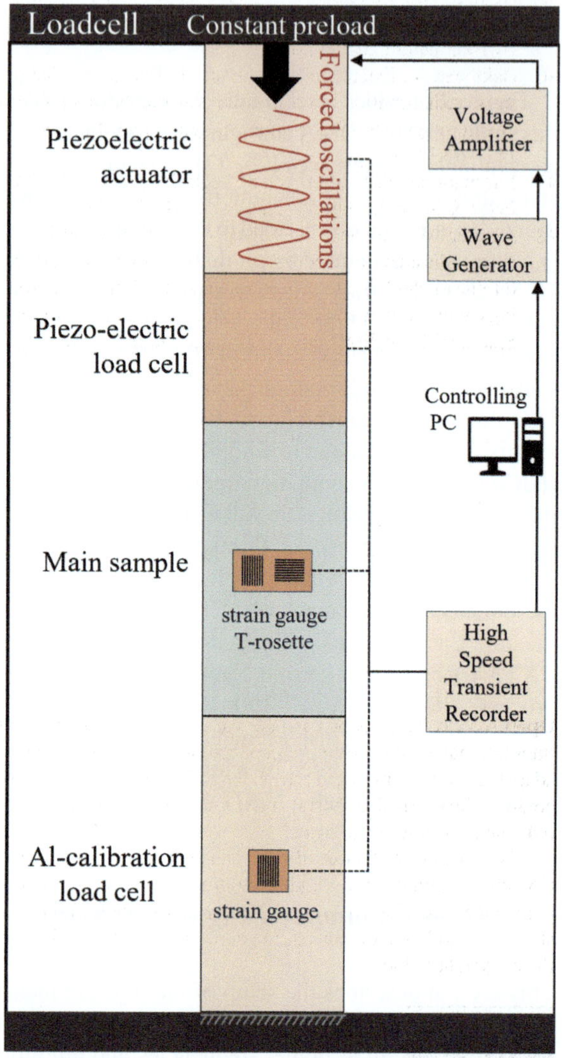

Compared to the DMA setup, which uses a smaller actuator and up to 200 V excitation, the HFFT setup is optimized for continuous cyclic loading. All other components are consistent with the DMA system. In summary, this configuration enables fatigue testing at higher frequencies with precise control. Care is taken to avoid resonance and manage actuator heating during extended tests.

3 Results

The results presented below detail the fatigue failure and characterization of mechanical properties of HPCs and UHPCs samples under different loading conditions, temperatures, and moisture levels. The experimental fatigue data were analysed using various characterization methods, such as μXRCT and DMA.

3.1 Static Compressive Strength

Prior to each fatigue test series, three cylindrical samples were evaluated for every concrete batch to ascertain the average static compressive strength ($f_{c,cyl,mean}$). This data was then used to calculate the maximum and minimum normalised stress (S_o and S_u), respectively. The resulting average compressive strength values and standard deviations for every concrete batch are presented in Table 3. The uniaxial static compressive strength for the dried (HPC-D-U-2) and wet concrete (HPC-UW-U-2) of series HPC-2 is almost identical at 114.0–112.1 MPa. For the uniaxial HPC-10 series, results from two batches for every storage condition are shown. Therefore, two values for static compressive strength and standard deviation are given. The static compressive strengths of the concrete samples with different moisture content are relatively similar to each other and are about 3–10 MPa below the HPC-2 series. This shows no major influence of the concrete moisture for the uniaxial loading on the static compressive strength. This is slightly different when comparing the triaxial static compressive strengths of the dry and the wet stored batches. The compressive strength of the dried samples (HPC-D-T-2) is 11.7% higher than that of the wet concrete (HPC-UW-T-2), with a value of 240.8 MPa compared to 215.5 MPa. A comparison of the uniaxial and triaxial compressive strengths of series HPC-2 at the same moisture content shows an increase of 112.3% for the dried concrete and 92.2% for the wet concrete. This effect is related to the confining pressure of 20 MPa in case of the triaxial tests. The compressive strength doubles from uniaxial to triaxial loading. The standard deviation of the HPC-UW-2 series is similar for both load types (3.23 and 3.04 MPa). For the HPC-D-2 series, on the one hand the standard deviation is very low for the uniaxial tests (0.80 MPa) and on the other hand very high for the triaxial tests (5.74 MPa). For the HPC-10 series, the standard deviation is much wider and ranges from 3.63 to 7.50 MPa.

3.2 Number of Cycles to Failure

The results of each uniaxial fatigue test with a test frequency of 10 Hz are shown in Fig. 3a. The results of all uniaxial and triaxial fatigue test with a frequency of 2 Hz are shown in Fig. 4a. The number of cycles to failure has been plotted on a logarithmic scale against the normalised maximum stress S_o on a linear scale. The results are plotted in blue (HPC-UW), in green (HPC-C) and black (HPC-D) depending on the moisture content. The uniaxial test results are plotted using circle and rectangles symbols, for the triaxial test results stars are used. Due to the wide dispersion of the individual values the mean number of cycles to failure was calculated for each stress level of each test series in Figs. 3b and 4b. In addition, a linear regression function was determined to describe the fatigue behaviour using the function $f = m \times \log(N) + b$ for each of the test series. In addition to the trendlines, the S–N curve according to *fib* Model Code 2010 [16] has been included in both figures. It can be seen that the results with the climate chamber storage (HPC-C-U-10) are in accordance with the *fib* Model Code 2010 [21]. The trendlines of the moist concretes (-UW) have a lower fatigue resistance than predicted by the *fib* Model Code 2010, regardless of the test frequency. The fatigue resistance of the dried concrete (-D) is higher for both test frequencies than predicted by *fib* Model Code 2010. The results under uniaxial loading at 10 Hz (Fig. 3) indicate clearly that moisture has a significant influence on the number of cycles to failure. The concrete HPC-UW-U-10 stored under water with 5.1% moisture content achieves significantly lower numbers of cycles to failure than the HPC-C-U-10 stored in the climate chamber with 4.0% moisture content or the dried concrete HPC-D-U-10. At an upper stress level of $S_o = 0.7$, the HPC-C achieves mean numbers of cycles to failure about more than 10 times higher mean numbers of cycles to failure than the HPC-UW, and the HPC-D even 100 times higher than the HPC-UW. If the test results of the uniaxial and triaxial tests with 2 Hz are compared, it is obvious that the trendlines of the same humidity are parallel to each other (Fig. 4b). In both cases, the trendlines of the triaxial load (dotted) are slightly to the right of the trendlines of the uniaxial load. This indicates that regardless of the moisture content, the results of the triaxial loading lead to nearly the same number of cycles to failure if the applied load is related to the normalised stress S_o. It should be noted, that the total fatigue load had to be doubled in the triaxial tests due to the doubling of the static compressive strength as a result of the additional confining pressure of 20 MPa. The influence of concrete moisture on the number of cycles of failure is clearly visible as the comparison of the trendlines of dried concrete (HPC-D) and wet concrete (HPC-UW) demonstrates. Both the comparison of the number of cycles of dry and wet concrete for uniaxial loads (blue and black rectangles) and triaxial loads (blue and black stars) show an increase in fatigue resistance at least by a factor of 10. Regardless of the type of loading, increasing moisture content in the concrete has a negative influence on the number of cycles to failure.

Whether the test frequency has an influence on the number of cycles in the range from 2 to 10 Hz is illustrated by the joint representation in Fig. 5. It is initially noticeable that the results with different test frequencies and the same concrete moisture are in the same scatter range compared to the S–N Model Code curve [21]. This applies to the individual values in Fig. 5a and can also be assumed for the distance of the trendlines compared to the Model Code curve in Fig. 5b. The influence of the test frequency on the number of cycles below 10 Hz is therefore considered to be rather low, as also found in other work [21, 23]. The influence of moisture has been already described in Figs. 3 and 4. It is also clearly visible in this illustration and can be

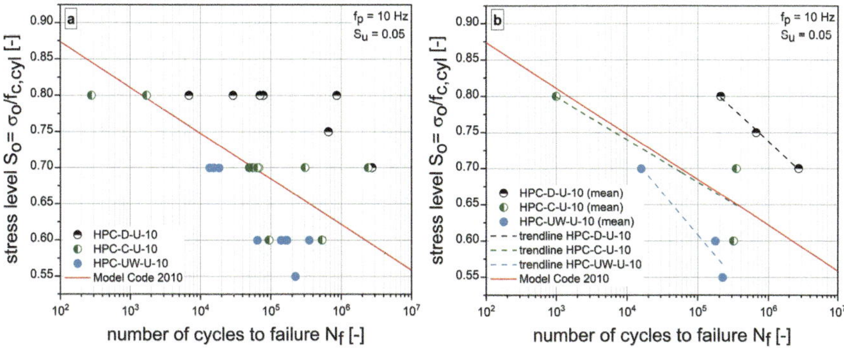

Fig. 3 **a** Number of cycles to failure of HPC-U-10, **b** Mean values of the number of cycles to failure and the corresponding trendlines of HPC-U-10

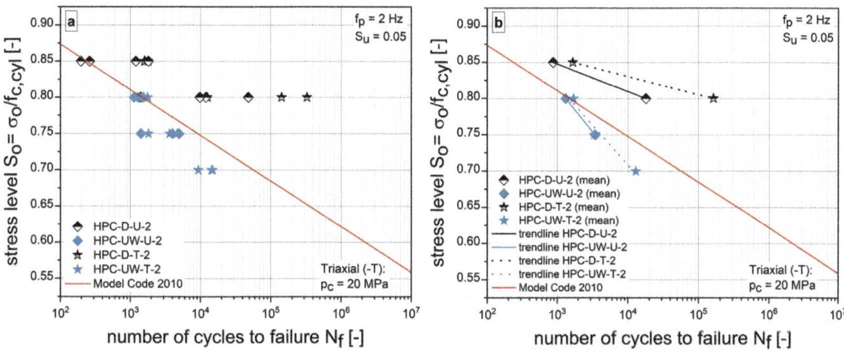

Fig. 4 **a** Number of cycles to failure of HPC-U-2 and HPC-T-2, **b** Mean values of the number of cycles to failure and the corresponding trendlines of HPC-U-2 and HPC-T-2

regarded as overriding effect. The results of the wet samples (blue symbols) achieve significantly lower numbers of cycles at both 2 and 10 Hz than the dry samples (black symbols).

3.3 *Strain Development as a Damage Indicator*

The strain development for concrete shows a characteristic S-shaped curve, which can be divided into three damage phases (Fig. 6a) depending on the applied stress state and the moisture content [28–30]. In general, strain development, i.e. stiffness decrease, is a very good indicator of damage [28]. In order to be able to better compare the different tests with each other, the gradient m of the strain in phase II was examined and compared (Eq. 6). For this purpose, the strain ($\Delta\varepsilon_o$) within 15%

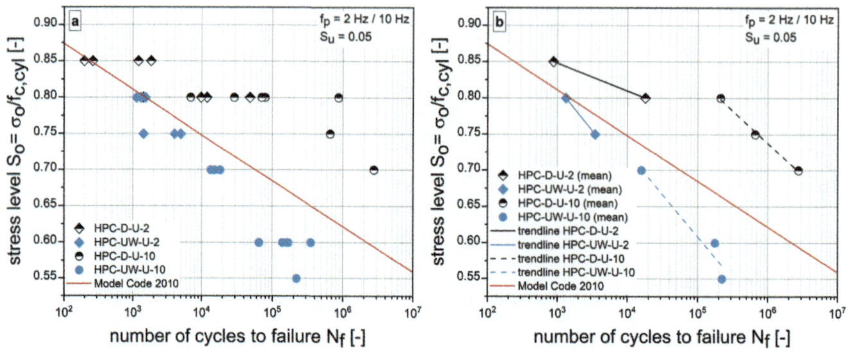

Fig. 5 **a** Number of cycles to failure of HPC-U-2 and HPC-U-10, **b** Mean values of the number of cycles to failure and the corresponding trendlines of HPC-U-2 and HPC-U-10

and 75% of the number of cycles to failure (ΔN) was analysed.

$$\text{grad m}(\varepsilon_o^{0.15-0.75}) = (\frac{\Delta\varepsilon_o^{0.15-0.75}}{\Delta N^{0.15-0.75}}) \tag{6}$$

The runout tests were not evaluated for this purpose. Figure 6b shows the gradients from Phase II together with the corresponding load cycles on a double logarithmic scale. It can be seen that the scatter range of the calculated gradient m is very small, although the tests results were very scattered in the plots used in Figs. 3, 4 and 5. In addition, moisture content seemed to have an influence to a smaller extent. While the difference between the HPC-UW-U-10 and the HPC-C-U-10 was barely noticeable, the difference to the HPC-D-U-10 was clear. For the same number of cycles to failure, the increase in strain per cycle is less for dry concrete than for wet concrete.

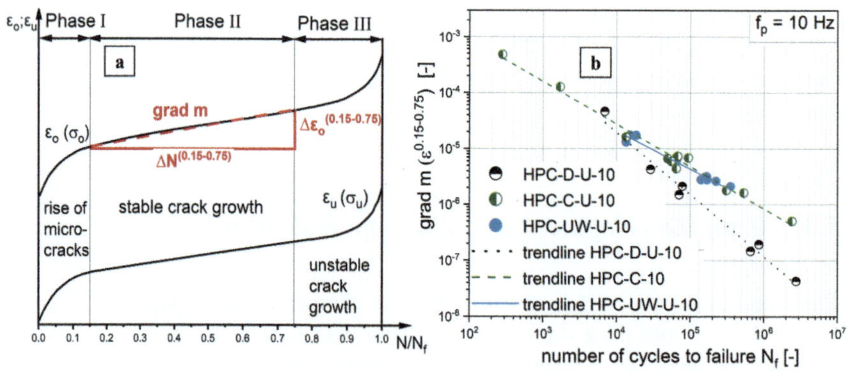

Fig. 6 **a** Schematic strain development and division into three phases, **b** Gradient of strain in damage phase II and the associated trendlines of HPC-U-10

3.4 Secant Modulus E_s as Damage Indicator

On the basis of the strain curves at maximum and minimum stress, the modulus of elasticity can be determined as the secant modulus (E_s) (Eq. 7). This curve can also be divided into three phases, see Fig. 7a. The development of stiffness in response to uniaxial and triaxial stress state and to both moisture contents at a top stress level of $S_o = 0.8$ is illustrated in Fig. 7a. For the uniaxially loaded sample, the curve of the wet concrete runs above the comparison curve of the dry concrete, for the triaxially loaded samples it is inverted. In both cases it can be seen, that the stiffness decreases $\Delta E_s = E_s(n = n_{max}) - Es(n = 1)$ for wet samples is larger than that for the dried samples. It can be observed that the decrease of stiffness for the two tests conducted in the triaxial cell (dashed line) is more extensive (ΔE_s(D) = 11.1 GPa, ΔE_s(UW) = 11.9 GPa) than for the uniaxial tests (ΔE_s(D) = 2.3 GPa, ΔE_s(UW) = 4.0 GPa).

$$E_s = (\frac{\sigma_o - \sigma_u}{\varepsilon_o - \varepsilon_u}) \tag{7}$$

This indicates that the stiffness drops at a much greater rate during the fatigue tests. Due to this effect and the well-known effect of large variation in the results of concrete fatigue tests, it is difficult to compare the individual tests with each other. Because the evaluation of the strain development in Sect. 3.3 was characterised by a low scattering, the gradient m of the secant modulus E_s in phase II was investigated and compared. Therefore, the gradient m of the secant modulus E_s was analysed within 15 and 75% of the number of cycles to failure, see Eq.

$$|m(E_s^{0.15-0.75})| = |(\frac{\Delta E_s^{0.15-0.75}}{\Delta N^{0.15-0.75}})| \tag{8}$$

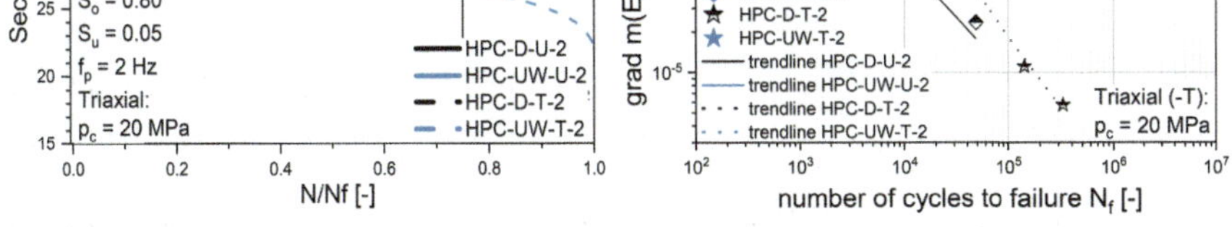

Fig. 7 **a** Secant modulus E_s at stress level $S_o = 0.8$ and $S_u = 0.05$ of HPC-U-2 and HPC-T-2, **b** Gradient of secant modulus E_s in damage phase II and the associated trendlines of HPC-U-2 and HPC-T-2

Figure 7b shows the results of the gradients $m(E_s^{0.15–0.75})$ of Phase II on a double logarithmic scale together with the corresponding number of cycles to failure. It can be observed that the gradients m are for each test very close to the calculated trendline, although there is a large range in the achieved number of cycles to failure. The results show that all four trendlines are almost parallel to each other. A comparison of the different types of loading at the same moisture content shows that the trendlines of the triaxial loading (dotted) are located above the trendlines of the uniaxial tests within one moisture content. From this, it can be concluded that for the same number of cycles to failure, the decrease in stiffness per cycle in phase II is greater for triaxial loading than for uniaxial loading. Because the results of the dried samples (black) have smaller values of the gradient $m(E_s^{0.15–0.75})$ than the wet concrete (blue) for the same type of loading it can be assumed that dry concrete has a lower loss of stiffness per cycle than wet concrete.

3.5 Temperature Development

For all samples, the temperature was measured on the surface of the concrete cylinder at half of the height of the sample. All samples heated up due to the mechanical loading during the fatigue tests. This self-heating is assumed to be a result of the internal friction [31]. In [32] it was shown that the temperature evolution is similar to the evolution of strains and therefore can be considered as an alternative indicator of damage evolution. It is also shown that moisture has a significant effect on the heating rate. It is obvious that temperature effects play a significant role in the damage processes in the material's microstructure and therefore affect the number of cycles to failure [31, 33]. Figure 8a shows the temperature increase $\Delta T = T(n_{\max}) - T(n = 0(in\ \text{K}))$ plotted over the number of cycles to failure for series HPC-10. The effect of moisture content of the sample can be clearly seen. It can be seen that the higher the number of cycles to failure achieved, the higher the self-heating of the sample. Looking at the trendlines, the trendlines of HPC-UW-U-10 with 5.1% moisture content and the trendline of HPC-C-U-10 with 4.0% moisture content match. The gradient of the trendlines is significantly larger than for the dried HPC-D-U-10 concrete. Temperature increases of more than 30 K are achieved for the wet concrete, whereas the maximum temperature increase for the dried concrete is more than 15 K. For the HPC-2 series with uniaxial and triaxial loading (Fig. 8b) it can be seen that the temperature increases of the wet samples (blue) were higher than those of the dry samples (black) for both types of loading. It can be observed that the triaxially-loaded samples (stars) heated up significantly more than the uniaxially-loaded samples (rectangles). A direct comparison of the load types of the wet samples shows that the samples of the HPC-UW-U series heated up by a maximum of approximately 5 K. For the same number of load cycles, the temperature increased by more than 25 K under triaxial loading conditions, which is 5 times higher than under uniaxial loading conditions. Because the thermal conductivity of all components in the triaxial cell are significantly better than that of air, it is assumed that the triaxial stress state

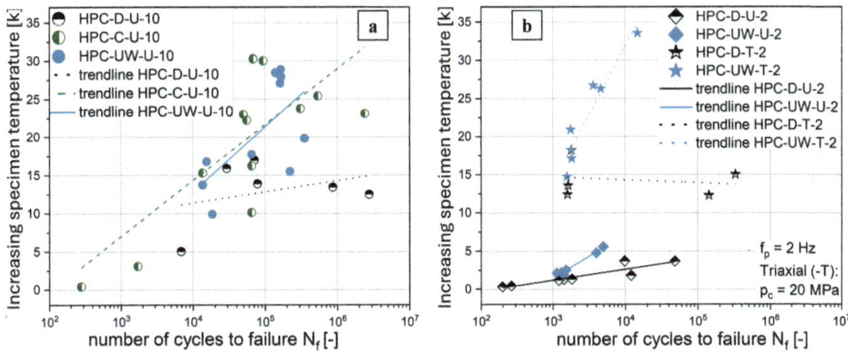

Fig. 8 Increase of the surface temperatures: **a** tests with a loading frequency of 10 Hz, **b** tests with a loading frequency of 2 Hz

leads to a significantly higher temperature increase. The results also show that the triaxial stress state results in significantly higher temperature increases compared to uniaxial loading conditions. However, the temperature increase observed in the dried samples is only about three to four times higher, from about 4 K to about 12–15 K. In general, samples with higher moisture content heated more than those with lower moisture content. In addition, the temperature increases were larger for triaxial loading scenarios than for uniaxial loading conditions. Temperatures generally increased with increasing number of cycles to failure. However, an exception to this was the case of the dried samples under triaxial loading where no significant increase was observed.

3.6 DMA—Dynamic Mechanical Analysis

In the following, the linear small-strain mechanical properties of HPC and UHPC are characterised in the frequency domain under different conditions, such as frequency-dependency, static pre-loads, amplitudes, before and after fatigue damage. The experimentally obtained results for Poisson's ratio and Young's modulus will be interpreted with an effective complex modulus and a loss factor (attenuation). The first test conducted using the DMA setup was to investigate the linear visco-elastic regime under harmonic (cyclic) loading. In Fig. 9, nine different amplitudes were applied, with each test series conducted at two different frequencies: 100 and 1000 Hz. The results show no significant change in Young's modulus and Poisson's ratio with increasing strain amplitude. A small change in Poisson's ratio with varying amplitude was observed, indicating that Poisson's ratio remains approximately constant with increasing amplitudes. Investigating the effective visco-elastic properties at different amplitudes is important to determine a strain regime for the subsequent frequency-dependent. Technically, an excitation voltage of the piezoelectric actuator was obtained from the performed amplitude-dependent. In this study, a (to-be amplified) voltage of 2.31 V was selected for the frequency-dependent.

According to Fig. 10, for HPC material, the DMA results show a (normal) dispersive behaviour. Young's modulus increases with higher frequencies. It is interesting to note that the effective properties are depending on the applied static pre-stress, i.e. Young's modulus decreases with higher axial compressive stress. This is due to the non-linear behaviour observed already in the previously discussed stress–strain curves of HPC. In other words, as strain increases, the slope of the stress–strain curve reduces, which means that Young's modulus, decreases as the pre-stress is increasing. Consequently, Young's modulus is lower under higher uniaxial compressive stress states. Additionally, between 400 and 600 Hz, disturbances caused by resonance are observed. In contrast to Young's modulus, Poisson's ratio shows a rather frequency-independent, i.e. non-dispersive behaviour. It should be noted that the dispersive behaviour of these mechanical quantities is independent from each other. Young's modulus is a typical modulus while Poisson's ratio, measured from two strain gauges mounted on the surface of the concrete sample, compares longitudinal and transversal harmonic strains. Regarding intrinsic material's attenuation captured by the loss factor, both, the phase angles of Young's modulus and the one of Poisson's ratio are small which is typical for less-attenuating materials like HPC or UHPC.

One of the main goals of this investigation was to simultaneously characterise changes in the mechanical properties and material behaviour during fatigue testing.

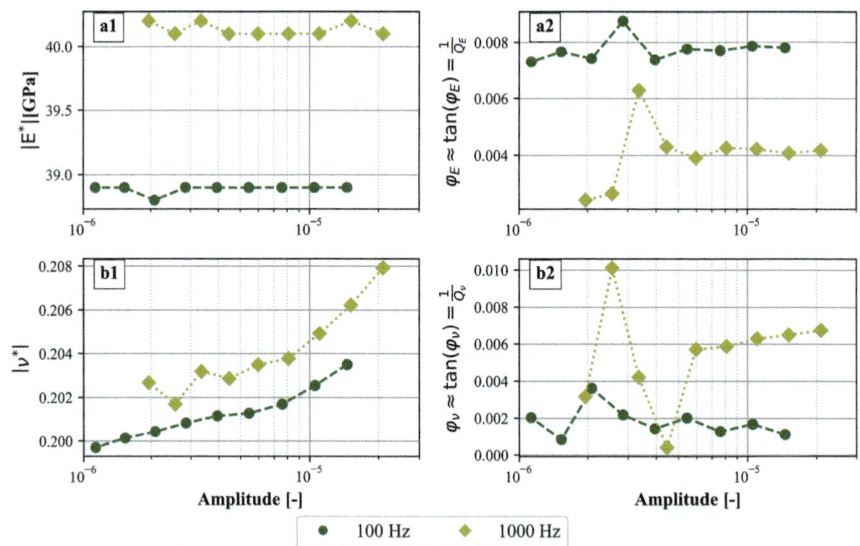

Fig. 9 Results of HPC based on amplitude-dependent: **a1** absolute value of Young's modulus and **a2** Young's modulus loss factor. Amplitude-dependent: **b1** absolute value of Poisson's ratio and **b2** Poisson's ratio loss factor

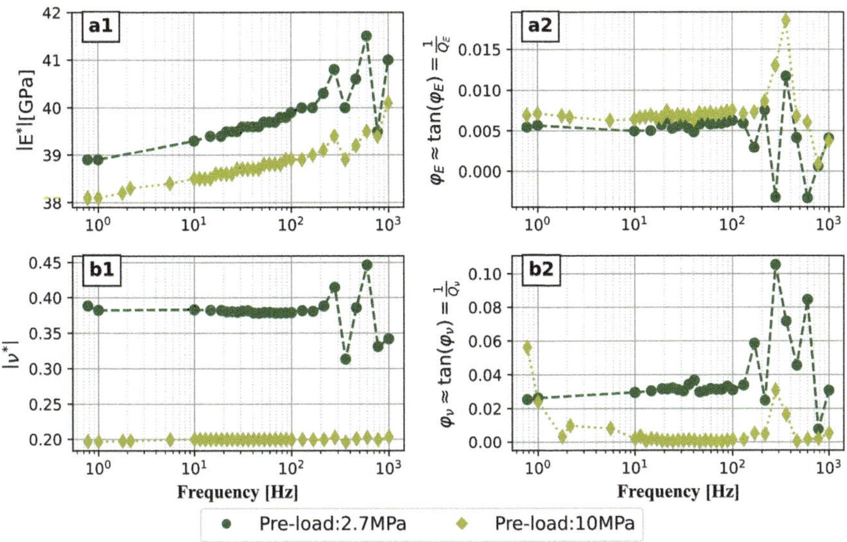

Fig. 10 Results of HPC based on frequency-dependent: **a1** absolute value of Young's modulus and, **a2** Young's modulus loss factor. Frequency-dependent: **b1** absolute value of Poisson's ratio and, **b2** Poisson's ratio loss factor

As (micro-)fractures develop after several loading cycles in a fatigue test, the material's effective behaviour also changes. DMA with its resolution of displacement amplitudes in the single micrometre regime is a very sensitive characterization technique which allows to measure such small changes of the inherent properties of the samples. Therefore, the obtained result could also be understood as the effective "fingerprint" of the material. For instance, in Fig. 11, the results of the UHPC series 9, analysed using the DMA method, are presented for both the (virgin) reference state and phase 3 of fatigue damage. The results show that in the reference (virgin) state, the stiffness of the sample is approximately 47 GPa decreasing to 45 GPa in the third phase. The level of intrinsic attenuation, i.e. the related loss factor associated with these phases also differs. Poisson's ratio remained nearly constant before and after the fatigue loading test. However, changes in Young's modulus and damping coefficients can indicate alterations in the mechanical properties of concrete and the formation of microcracks. Specifically, Young's modulus decreased by about 5%. Additionally, the trendline of the loss factor (attenuation) shifted in relation to Young's modulus and Poisson's ratio, with the peak of the loss factor (attenuation) curve moving to the left in both cases. A leftward shift in the peak suggests an increase in microcracks within the concrete sample. Overall, it can be concluded that the highly sensitive DMA method effectively identifies decreases in material's stiffness, the development of microcracks, changes in material behaviour under various loads such as frequency-dependent and pre-loads, and alterations in complex mechanical properties under fatigue tests.

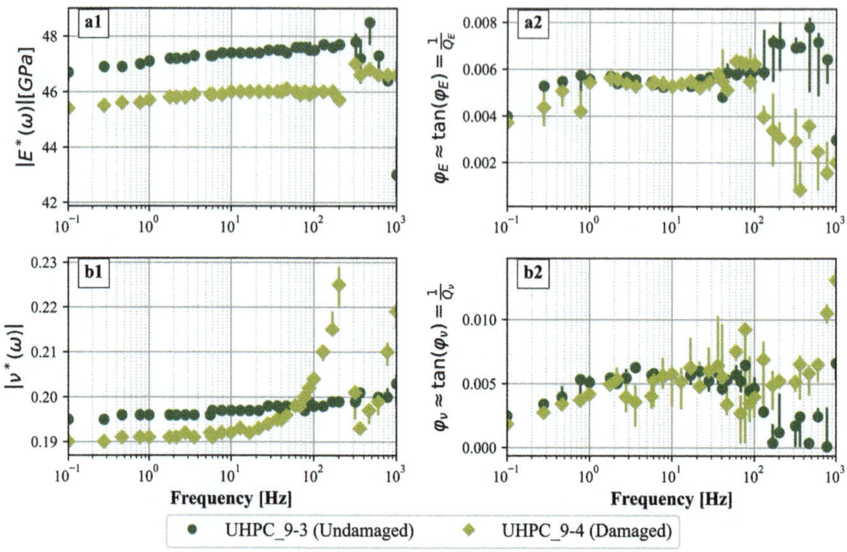

Fig. 11 Changes in the complex mechanical properties of UHPC in the virgin state and the third phase of the fatigue test based on: **a1** frequency dependence of the absolute value of the complex Young's modulus, **a2** attenuation coefficient/loss factor of Young's modulus, **b1** absolute value of the complex Poisson's ratio, and **b2** attenuation coefficient/loss factor of Poisson's ratio

Next, the UHPC series 9, in the 3rd phase of fatigue, was scanned using μXRCT [34, 35] in two different states. In the first state, the samples were scanned in their original size, which consisted of disks with a diameter of 60 mm and a height of 30 mm. As shown in Fig. 12, no microcracks were detected within the samples because the voxel size in this scanning mode was 25 μm, while microcracks exist on a micrometre scale. In the second state, cylindrical samples with a diameter of $1/4'' = 6.35$ mm were randomly extracted from within the disks and scanned with a voxel size of 2 μm. According to Fig. 12, microcracks were visible in various layers of the μXRCT images. Microcracks are marked in red circles. However, these microcracks had not yet developed into a fatigue crack network. This explains the small reduction in stiffness, as the macrocracks, which are formed from the coalescence of microcracks, had not yet developed.

3.7 High Frequency Fatigue Results

As results of the high frequency fatigue tests shows in Fig. 13a that the shear failure in HPC samples subjected to fatigue testing under loading frequencies ranging from 50 to 200 Hz. In these HPC samples, the initial crack begins at the top side and propagates at an angle of 60°. Due to the brittle nature of the samples, most of the fatigue test duration is spent on forming the initial crack. Once the initial crack forms, the crack

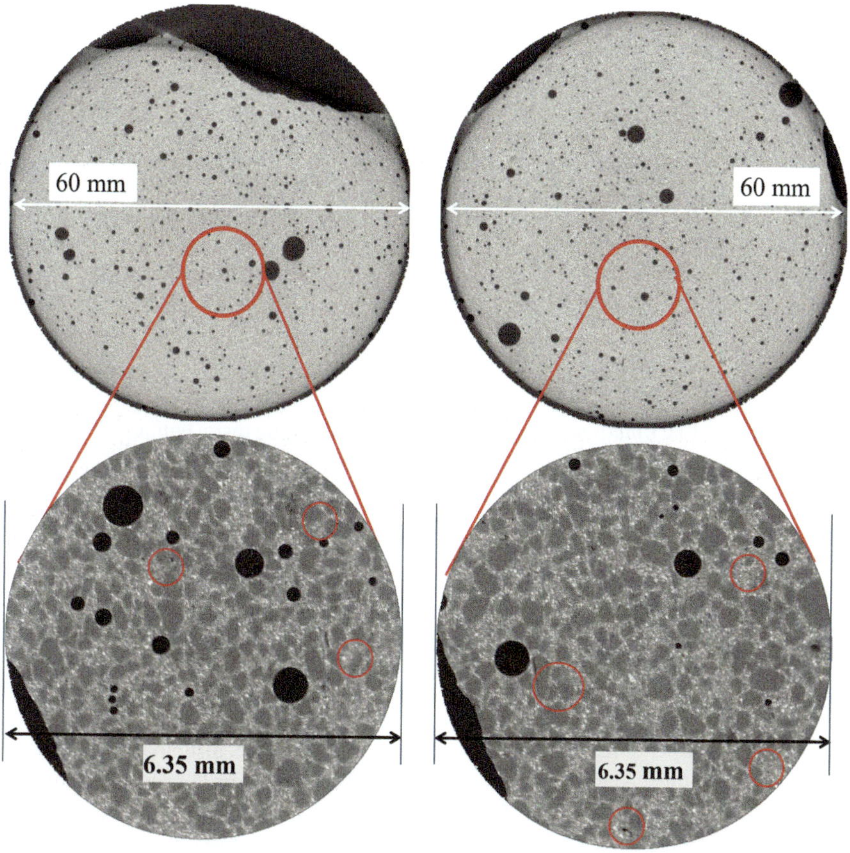

Fig. 12 Image of μXRCT scanning from two UHPC samples in two different scale

growth to complete failure of the sample occurs within a few seconds. Figure 13b presents results were arranged according to the number of cycles to failure relative to S_{max}. These results indicate that the number of cycles to failure decreases as the loading frequency increases. Consequently, with fewer cycles and higher loading frequencies, the time to failure is significantly reduced. At higher frequencies, the increased loading frequency allows less time for stress relaxation during individual loading cycles, causing cumulative stress effects on the sample. As a result, at lower frequencies, there is less residual stress per cycle, while at higher frequencies, the stress per cycle increases. The faster accumulation of stress at higher frequencies leads to failure in fewer cycles. In other words, as the loading frequency increases, the rate of energy input per unit time to the sample also increases. For instance, at a loading frequency of 20 Hz, the sample receives 44 J/s, while at 200 Hz, the energy rate increases to 450 J/s. This tenfold increase in the energy rate from 20 to 200 Hz corresponds to a reduction in the number of cycles to failure by approximately a factor of 10. Also, it was observed that as the frequency increases, the temperature of the sample also rises during the test. This temperature rise is another reason for reducing the loading frequency to delay failure, which aligns well with the findings of previous research [18]. Based on these results, the HFFT setup can be effectively used for conducting fatigue tests on concretes such as HPC and UHPC for high frequency.

The results of the fatigue tests on UHPC samples from series 9, conducted at different test frequencies of 50 Hz and 100 Hz, are presented in Fig. 14. During the fatigue tests, measurements of axial strain, lateral strain, and force were taken to assess stiffness reduction and changes in Poisson's ratio. At a frequency loading of 100 Hz, the number of cycles to failure is approximately 140,000, whereas at a frequency of 50 Hz, it increases to around 350,000 cycles. The experiment at 50 Hz took 115 min, while at 100 Hz, it lasted only about 23 min. For the UHPC samples, it was found that at a fatigue loading frequency of 50 Hz, the material's behaviour was like that observed almost in classical fatigue tests, behaving as a brittle material. Fatigue cracks developed during the third phase, leading to sample failure, with properties declining uniformly over time. However, when the fatigue loading frequency reaches 100 Hz, the material's behaviour changed. From the beginning, there were significant changes in both axial and lateral strains, causing a sharp reduction in stiffness right from the first phase of fatigue. This change does not indicate a shift from brittle to ductile behaviour; instead, it reflects a change in how microcracks open. At higher frequencies, as microcracks open, their tips are subjected to high-frequency loading, which prevents stress relief through relaxation, leading to continuous stress increase at the crack tips compared to lower frequencies. Also, the rate of crack growth and reduction in mechanical stiffness is affected by changes in the test temperature. At a loading frequency of 50 Hz, the sample's average (surface) temperature reaches around 40 °C. In contrast, at a loading frequency of 100 Hz, the average temperature of the sample during the fatigue test reaches approximately

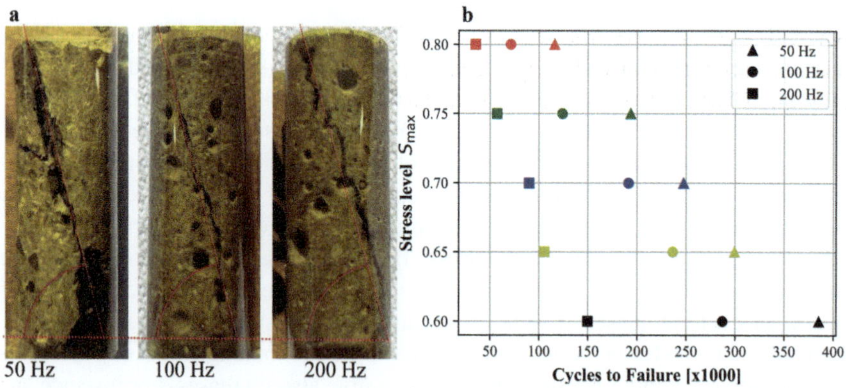

Fig. 13 **a** Final fatigue failure images and **b** number of cycles to fatigue failure for HPC samples under different loading frequencies (from [27], licensed under CC-BY 4.0)

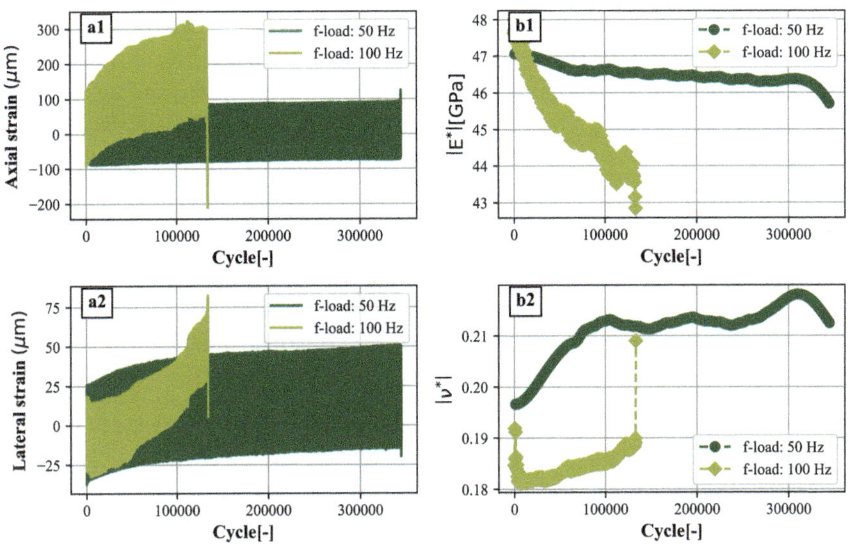

Fig. 14 Evolution of axial, lateral strains, Young's modulus, and Poisson's ratio in UHPC at high(er) loading frequencies of 50 and 100 Hz

50 °C. This higher temperature accelerates the failure of the sample. Another observation concerns changes in Poisson's ratio. At a loading frequency of 50 Hz, Poisson's ratio increases during the fatigue process and decreases just before final failure. This is because the axial strain remains relatively constant throughout the test but increases near failure, while the lateral strain increases steadily. However, at a frequency of 100 Hz, Poisson's ratio behaves differently: it decreases at the start of the test and then increases near failure. This occurs because the axial strain increases during the fatigue test.

Overall, increasing the fatigue loading frequency, coupled with an increase in the samples' surface temperature and a reduced relaxation time per cycle, leads to higher stress accumulation around microcracks. This significantly reduces the number of cycles needed to reach failure in the sample.

4 Summary and Conclusion

Based on results obtained from classical compression fatigue tests on concrete cylinder samples with a diameter of $d = 60$ mm, the influence of concrete moisture on the fatigue damage behaviour has been investigated at different frequencies and loading scenarios. The investigated HPC was subjected to uniaxial loading at 2 and 10 Hz, compared to triaxial loading at 2 Hz and a confining pressure of $P_c =$ 20 MPa. In addition, for reasons of material characterisation, small-format samples were examined using various experimental methods. For this purpose, the effectiveness of DMA and high-frequency fatigue testing setups in characterizing the mechanical behaviour of HPC and UHPC under various loading conditions and during fatigue testing has been studied. The results provided important insights into how loading frequency affects material stiffness and the formation of microcracks, suggesting that mechanical properties can change depending on the frequency. Additionally, a high-frequency fatigue testing, allowing fatigue tests to be conducted at frequencies much higher than the classical range of 1–10 Hz, extending up to 1000 Hz, and providing significant enhancements. Since the DMA method is non-destructive and operates within a small strain regime, it can be ideally combined with imaging techniques like μXRCT to monitor morphological changes during fatigue tests. To summarize, the following outcomes were observed:

- Concrete moisture has a significant effect on the fatigue resistance of concrete. Depending on the concrete moisture, the number of cycles to failure can be up to 100 times lower than for dry concrete at the same loading condition.
- At a test frequency of 10 Hz, dried concrete achieves higher numbers of cycles to failure as pronounced by the S–N curve of the *fib* Model Code 2010 Concrete stored under water achieves lower numbers of cycles to failure. Concrete stored in a climate-controlled environment (20 °C, 65% RH) achieves the pronounced values.
- A confining pressure of $P_c = 20$ MPa has a positive effect on the number of cycles to failure of dried concrete. For wet concrete, the number of cycles to failure under triaxial loading increased only slightly.
- It appears that the damaging effect of concrete moisture is a decisive influencing parameter for fatigue damage. The test frequency in the range of 2–10 Hz has no or only a very small influence on the fatigue damage in relation to the effect of concrete moisture.
- The increase in strain per cycle in damage phase II is less for dried concrete than for wet concrete.
- The loss of stiffness per cycle in phase II is more pronounced for triaxial stress states than for uniaxial stress states.
- The HPC samples heat up considerably due to the mechanical loading during the fatigue tests. The wet HPC samples heat up more than the dried samples.
- Microcracks inside concrete test samples were successfully detected using high-resolution μXRCT. Fatigue microcracks were monitored by comparing μXRCT data with (small) $d = 1/4''$ sample sizes adjusted to achieve the highest voxel resolution for observing the initiation of microcracks.
- A high sensitivity visco-elastic characterization of the material's damage state was achieved using the small-strain DMA method. It was shown that damage could be clearly detected in the measured values of Young's modulus through frequency sweeps.

- Young's modulus and Poisson's ratio were determined through DMA characterization during fatigue testing. The sequence of changes in mechanical properties from the start to the end of fatigue could serve as an indicator for detecting damage and eventual failure in materials.
- It was found that in HFFT, as the frequency of fatigue loading increases, the number of cycles until failure decreases. This is due to the rising temperature of the sample and the subsequent accumulation of stress resulting from the decreased relaxation time. A significant and linear reduction in stiffness was observed during the third phase of fatigue, especially at low frequencies. At higher frequencies, the reduction in stiffness was noted to become more pronounced at the initial stage of fatigue.

Associated data is published via the link: https://doi.org/10.18419/DARUS-4458.

References

1. Azmee, N.M., Shafiq, N.: Ultra-high performance concrete: from fundamental to applications. Case Stud. Constr. Mater.. **9**, 197 (2018). https://doi.org/10.1016/J.CSCM.2018.E00197
2. Kachkouch, F., Noberto, C., de Albuquerque Lima Babadopulos, L., Melo, A., Machado, A., Sebaibi, N., Boukhelf, F., El Mendili, Y.: Fatigue behavior of concrete: a literature review on the main relevant parameters. Constr. Build. Mater. **338**, 127510 (2022). https://doi.org/10.1016/J.CONBUILDMAT.2022.127510
3. Hsu, T.: Fatigue of plain concrete. J. Proc. **78**, 292–305 (1981). https://doi.org/10.14359/6927
4. Kim, J.K., Kim, Y.Y.: Experimental study of the fatigue behavior of high strength concrete. Cem. Concr. Res. **26**, 1513–1523 (1996). https://doi.org/10.1016/0008-8846(96)00151-2
5. Gao, L., Hsu, C.: Fatigue of concrete under uniaxial compression cyclic loading. Mater. J. **95**, 575–581 (1998). https://doi.org/10.14359/407
6. Thiele, M.: Experimentelle Untersuchung und Analyse der Schädigungsevolution in Beton unter hochzyklischen Ermüdungsbeanspruchungen. Doctoral thesis (2016). https://doi.org/10.14279/depositonce-4886
7. Hohberg, R.: Zum Ermüdungsverhalten von Beton. Doctoral thesis (2004). https://doi.org/10.14279/depositonce-815
8. Medeiros, A., Zhang, X., Ruiz, G., Yu, R.C., Velasco, M.: Effect of the loading frequency on the compressive fatigue behavior of plain and fiber reinforced concrete. Int. J. Fatigue **70**, 342–350 (2015). https://doi.org/10.1016/J.IJFATIGUE.2014.08.005
9. Brara, A., Klepaczko, J.R.: Fracture energy of concrete at high loading rates in tension. Int. J. Impact Eng **34**, 424–435 (2007). https://doi.org/10.1016/J.IJIMPENG.2005.10.004
10. Bathias, C.: Piezoelectric fatigue testing machines and devices. Int. J. Fatigue **28**, 1438–1445 (2006). https://doi.org/10.1016/J.IJFATIGUE.2005.09.020
11. Ghielmetti, C., Ghelichi, R., Guagliano, M., Ripamonti, F., Vezzù, S.: Development of a fatigue test machine for high frequency applications. Procedia Eng. **10**, 2892–2897 (2011). https://doi.org/10.1016/J.PROENG.2011.04.480
12. Stringer, D.B., Dykas, B.D., Laberge, K.E., Zakrajsek, A.J., Handschuh, R.F., Army, U.S.: A new high-speed, high-cycle, gear-tooth bending fatigue test capability (2011)
13. Dong, S., Wang, Y., Ashour, A., Han, B., Ou, J.: Uniaxial compressive fatigue behavior of ultra-high performance concrete reinforced with super-fine stainless wires. Int. J. Fatigue **142**, 105959 (2021). https://doi.org/10.1016/J.IJFATIGUE.2020.105959
14. Lohaus, L., Oneschkow, N., Wefer, M.: Design model for the fatigue behaviour of normal-strength, high-strength and ultra-high-strength concrete. Struct. Concr. **13**, 182–192 (2012). https://doi.org/10.1002/SUCO.201100054

15. DIN EN 1992-1-1/A1, Ausgabe 2015-03. https://www.baunormenlexikon.de/norm/din-en-1992-1-1-a1/b624bfe2-200f-4421-9d99-55323aa74b9f
16. fib model 2010: fib Model Code for Concrete Structures 2010. fib Model Code for Concrete Structures 2010. 1–402 (2013). https://doi.org/10.1002/9783433604090
17. Scheydt, J.C.: Mechanismen der Korrosion bei ultrahochfestem Beton. KIT Scientific Publishing (2014). https://doi.org/10.5445/KSP/1000036611
18. Elsmeier, K., Hümme, J., Oneschkow, N., Lohaus, L.: Prüftechnische Einflüsse auf das Ermüdungsverhalten hochfester feinkörniger Vergussbetone. Beton- und Stahlbetonbau. **111**, 233–240 (2016). https://doi.org/10.1002/BEST.201500065
19. Bultreys, T., De Boever, W., Cnudde, V.: Imaging and image-based fluid transport modeling at the pore scale in geological materials: a practical introduction to the current state-of-the-art. Earth Sci. Rev. **155**, 93–128 (2016). https://doi.org/10.1016/J.EARSCIREV.2016.02.001
20. Fusseis, F., Xiao, X., Schrank, C., De Carlo, F.: A brief guide to synchrotron radiation-based microtomography in (structural) geology and rock mechanics. J. Struct. Geol. **65**, 1–16 (2014). https://doi.org/10.1016/J.JSG.2014.02.005
21. Markert, M., Katzmann, J., Birtel, V., Garrecht, H., Steeb, H.: Investigation of the influence of moisture content on fatigue behaviour of HPC by using DMA and XRCT. Materials **2022**(15), 91 (2021). https://doi.org/10.3390/MA15010091
22. Basaldella, M., Jentsch, M., Oneschkow, N., Markert, M., Lohaus, L.: Compressive fatigue investigation on high-strength and ultra-high-strength concrete within the SPP 2020. Materials 2022 **15**, 3793 (2022). https://doi.org/10.3390/MA15113793
23. Tomann, C., Oneschkow, N.: Influence of moisture content in the microstructure on the fatigue deterioration of high-strength concrete. Struct. Concr. **20**, 1204–1211 (2019). https://doi.org/10.1002/SUCO.201900023
24. Fauser, D., Agudo, J.A.R., Madadi, H., Haeberle, J., Renner, J., Steeb, H.: Complex Poisson's ratio for viscoelastic materials: direct and indirect measurement methods and their correlation. Proceedings of the Royal Society A. **481** (2025). https://doi.org/10.1098/RSPA.2024.0543
25. Lakes, R.S.: Viscoelastic materials. Cambridge University Press (2009)
26. Madadi, H., Steeb, H.: High-speed fatigue testing of high-performance concretes and parallel frequency sweep characterization. PAMM. **23**, e202300262 (2023). https://doi.org/10.1002/PAMM.202300262
27. Madadi, H., Steeb, H.: The high cycle fatigue testing of high-performance concretes using high frequency excitation. PAMM. **22**, e202200221 (2023). https://doi.org/10.1002/PAMM.202200221
28. Oneschkow, N.: Analyse des Ermüdungsverhaltens von Beton anhand der Dehnungsentwicklung. Doctoral thesis, Reports of the Institute of Building Materials Science, 13, Leibniz University Hannover (2014).
29. Fatigue of concrete by constant and variable amplitude loading. Spec. Publ. **75**, 71–110 (1982)
30. Klausen, D.: Festigkeit und Schädigung von Beton bei häufig wiederholter Beanspruchung. Doctoral thesis. Technische Hochschule Darmstadt, Darmstadt (1978).
31. Deutscher, M., Tran, N.L., Scheerer, S.: Experimental investigations on the temperature increase of ultra-high performance concrete under fatigue loading. Appl. Sci. **9**, 4087 (2019)
32. Markert, M., Birtel, V., Garrecht, H.: Temperature and humidity induced damage processes in concrete due to pure compressive fatigue loading. In: Proceedings of the Fib Symposium, pp. 27–29, Krakow, Poland (2019)
33. Elsmeier, K.: Einfluss der Probekörpererwärmung auf den Ermüdungswiderstand von hochfestem Vergussbeton. ifB, Institut für Baustoffe, Leibniz Universität Hannover, Hannover (2019)
34. Ruf, M., Steeb, H.: An open, modular, and flexible micro X-ray computed tomography system for research. Rev. Sci. Instrum. **91** (2020). https://doi.org/10.1063/5.0019541
35. Ruf, M., Lee, D., Steeb, H.: A multifunctional mechanical testing stage for micro x-ray computed tomography. Rev. Sci. Instrum. **94** (2023). https://doi.org/10.1063/5.0153042

Effect of Microfibres on the Degradation in High-Performance Concrete Under Cyclic Loading

Vladislav Gudžulić, Niklas Schäfer, Günther Meschke, and Rolf Breitenbücher

Abstract This chapter presents results from investigations of the effect of steel and carbon microfibres on the degradation and fatigue of high-performance fibre-reinforced concrete (HPFRC) under monotonic and cyclic loading using experimental and numerical approaches. The experiments included investigations across multiple scales, from individual fibres to beams under various loading conditions. The addition of microfibre reinforcement has a nonnegligible effect on crack resistance and fatigue life, leading to smaller microcrack widths, limiting the absolute values of measured strain, and showing more ductility in phase 3 in the S–N diagram. However, the impact of fibres on the overall concrete lifetime remains inconclusive. Complementing the experimental work, a computational framework was developed to study the mesoscale behaviour of fibre-reinforced concrete in high fidelity. The framework was utilized to conduct CT-image-based numerical simulations of fibre-reinforced concrete specimens, realistically capturing crack initiation mechanisms and damage development under compressive loading. It also facilitated the exploration of the influence of different fibre orientation cases on the residual strength and unloading/reloading behaviour in fibre-reinforced concrete subjected to cyclic tensile loading. By combining experiments with numerical modelling, this research provides new insights into the influence of microfibres on the behaviour of HPC under monotonic as well as cyclic loading.

Keywords Steel and carbon microfibres · High-performance concrete · Cyclic loading · Cohesive crack · Discrete fibre model · CT-image based simulation

---

V. Gudžulić · G. Meschke (✉)
Institute for Structural Mechanics, Ruhr University Bochum, Bochum, Germany
e-mail: guenther.meschke@ruhr-uni-bochum.de

N. Schäfer · R. Breitenbücher
Institute for Building Materials, Ruhr University Bochum, Bochum, Germany

N. Oneschkow and L. Lohaus (eds.), *Cyclic Deterioration of High-Performance Concrete*, https://doi.org/10.1007/978-3-032-19394-0_9

1 Introduction

Even minor external mechanical influences can lead to local stresses and, consequently, changes in the microstructure of concrete due to its inhomogeneity. This is particularly relevant in the case of cyclic loading. Structural changes in the form of microcracks occur even at stresses that are significantly below the static strength of the concrete. As the number of load cycles increases, microcracks and the associated degradation of the microstructure also increase, which can lead to premature failure during cyclic loading [1].

Degradation of concrete is associated with microcrack initiation and microcrack growth within the concrete matrix. Crack coalescence and widening leads to progressive microcracking and eventually macrocracks, which are critical to the failure process. This involves a growing microcrack that forms in the interfacial transition zone (ITZ) [2]. According to [3], there exist two hypotheses regarding crack development under cyclic compressive loading. One hypothesis suggests that there is a failure of the bond between the aggregate and the cement paste matrix, leading to crack formation. The second hypothesis posits that pre-existing microcracks and their growth due to cyclic loading are responsible for fatigue failure.

The extent to which microfibres affect the microstructure and fatigue behaviour of high-performance concrete is still insufficiently understood. On one hand, fibres are attributed with a crack-bridging effect, while on the other hand, they are supposed to have a crack-initiating effect within the concrete matrix [4]. Conflicting evidence also exists regarding macro steel fibres: some studies report no improvement in fatigue resistance [5, 6], while others demonstrate improved fatigue behaviour [7, 8] or even reduced resistance under cyclic compressive loading [4]. The majority of research has focused on steel fibre-reinforced concretes in cyclic flexural and tensile tests, showing that fibre effects depend on factors such as fibre geometry, which influences load transfer and crack bridging [9–12]. For instance, microfibres are effective in controlling microcracks, while longer fibres are suitable for macrocrack control. Additionally, using other types of microfibres, such as carbon fibres, can enhance crack resistance, fatigue life, and energy absorption [13, 14].

While experimental studies have provided valuable insights into the fatigue behaviour of fibre-reinforced concretes, computational modelling has emerged as a powerful tool to complement these findings. Computational models, such as finite element analysis (FEA) and discrete element methods (DEM), can simulate the complex microstructural behaviour of concrete under monotonic and cyclic loading. In recent decades, substantial progress has been made in the numerical modelling of concrete fracture, with continuum damage models effectively capturing homogenized behaviour across various loading conditions [15–20].

However, accurately modelling concrete behaviour under load, characterized by numerous cracks and intricate interactions between microstructural components like fibres, aggregates, and the cement matrix, remains a significant challenge. To address these complexities, more detailed mesoscale models that explicitly simulate microstructure and crack development, though computationally demanding, provide valuable insights into fracture mechanics and the influence of fibre reinforcement [21–28].

This study presents a combined experimental and numerical investigation of high-performance fibre-reinforced concrete under various cyclic loading conditions. The experimental work systematically examines the impact of different fibre types and contents on concrete behaviour, supported by sophisticated computational models that explicitly resolve the fibre-reinforced concrete mesostructure. In the scope of this work, an effort was made to establish a virtual lab that allows for the integration of experimental data, such as concrete mesostructure obtained from CT-images, into the simulation environment. In this way, added value can be extracted from simulations that would be hard or impossible to obtain otherwise. The virtual lab enhances the accessibility and relevance of numerical simulations by grounding them in experimental data. The capabilities of the developed virtual lab were demonstrated by means of participation in numerical benchmarks organized within the Priority Programme SPP 2020, as well as in conducting numerical simulations of a number of experiments presented in this chapter. This research aims to provide a deeper understanding of the mechanisms driving the cracking and fatigue in fibre-reinforced concrete.

2 Materials

The investigations were carried out on high-performance concrete with and without microfibres, see Fig. 1. The reference high-performance concrete HPC (RH1) was utilized, as detailed in Sect. 2.1 "Concrete Composition" in the chapter by M. Basaldella et al., this volume. The compositions are labelled as follows: high-performance concrete HPC-08 equals HPC (RH1), with steel fibres HPC-08-SF, with carbon fibres HPC-08-C. The number indicate the maximum grain size in the composition. The fibre content for the HPC-08-SF was 1 percent by volume and for the HPC-08-C 0.5 percent by volume. The content of superplasticizer and stabilizer was retained for the HPC-08-SF. For HPC-08-C, the content was increased to improve the workability of the fresh concrete. The fibre properties are shown in Table 1.

The mean compressive strengths of the tested specimens were approximately 110 MPa for HPC-08 and -SF. The compressive strength of HPC-08-C was slightly lower at 100 MPa. The centric tensile strength was 6.4 ± 0.5 MPa for the HPC-08 and 5.8 ± 0.5 MPa for the HPC-08-SF. For the HPC-08 the flexural strength was 4.0 MPa, for the HPC-08-SF: 5.0 MPa and the HPC-08-C: 4.4 MPa as mean values. The lower flexural strength resulted from different storage conditions of the beams. The storage conditions are described in the experiment sections. The reference strengths for the cyclic tests were always determined on specimens with the same dimensions and boundary conditions as for the fatigue tests.

In addition to the influence of microfibres on fatigue behaviour and material degradation of high-performance concrete, investigations into the influence and presence

a b

Fig. 1 **a** Steel fibres, and **b** carbon fibres (Reproduced from the 20th fib Symposium Proceedings – ReConStruct - Resilient Concrete Structures, held in Christchurch from 11–13 November 2024, article "Influence of different microfibers on the flexural fatigue characteristics of high-strength concrete" page 225—"Fig. 2: steel fibers (left) and carbon fibers (right)" with permission from the International Federation for Structural Concrete (fib) [29])

Table 1 Properties of the microfibres

	Strength	Length	Ø	Density
	(MPa)	(mm)	(mm)	(kg/dm^3)
Steel (Dramix OL 6/0.16, Bekaert)	2600	6	0.16	7.85
Carbon (C C6-4.0/240-G100, Sigrafil)	4000	6	0.007	1.80

of aggregate on fatigue behaviour and material degradation were carried out as part of the research project. For this purpose, the reference composition was scaled down to a high-strength mortar (designated HPC-02) and a hardened cement paste (HPC-00) by gradually eliminating the aggregate. The results obtained are not part of this book chapter but are presented in detail in the PhD thesis of Schäfer [30].

3 Computational Modelling of Cracking Behaviour in Fibre-Reinforced Concrete

In this research, a computational model was developed to explicitly resolve the mesostructure of fibre-reinforced concrete. A comprehensive exposition of the modelling equations and discretization strategies is provided in [22, 31–34], while a brief summary is given here. The central component of the computational model is a penalty-based zero-thickness interface element with a cohesive-frictional traction–separation model, accounting for frictional sliding between crack surfaces and progressive crack closure. The key kinematic factor driving crack development is the crack opening (separation) $[[\mathbf{u}']]$, calculated at each integration point of the interface element.

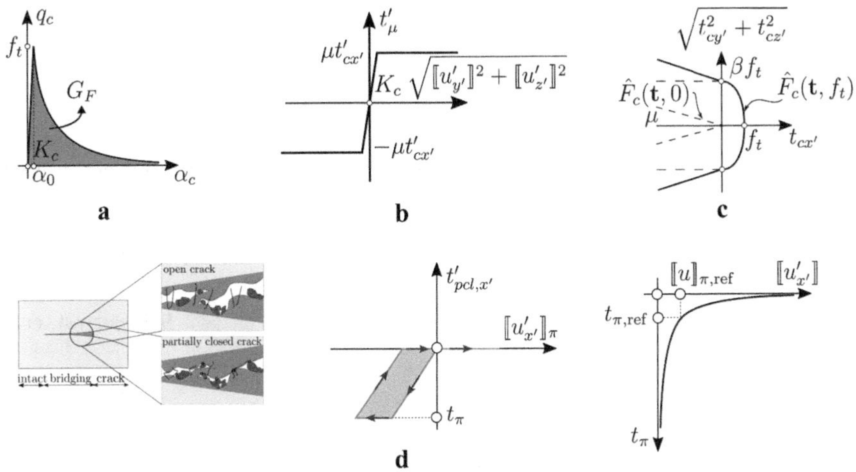

Fig. 2 Components of the cohesive-frictional discrete fracture model including, **a** residual strength evolution, **b** friction law, **c** strength envelope, and **d** components of the partial crack closure model (conceptualization, elementary hysteresis, and dependence of closure stress to crack opening) (from [31], licensed under CC-BY 4.0)

Figure 2 illustrates the model's key components and their dependencies on internal variables: a cohesive softening relation for material strength degradation during cracking, a frictional law for sliding of crack faces under compression, and a cohesive-frictional strength envelope that defines mixed-mode cracking behaviour. The final component, a sub-model for imperfect crack closure, models frictional sliding between "rough" asperities, essential for capturing the hysteretic cycles during cyclic loading. It should be noted that compression failure is not directly modelled but results from a combination of tensile and shear cracking. The model's capability to capture compression failure using fundamental assumptions of fracture mechanics provides valuable insights into complex fracture processes.

In Fig. 2, K_c denotes the penalty stiffness, $\beta = f_s/f_t$ is the ratio of shear to tensile strength, $\kappa = G_{F,II}/G_{F,I}$ is the ratio of mode II to mode I fracture energy, $d_c(\alpha_c)$ is the damage parameter which depends on the maximal experience crack opening α_c, $[\![\boldsymbol{u}'_{\pi}]\!]$ is the crack slip due to asperity sliding in the normal direction, and $\boldsymbol{t}_{\pi}$ is the crack closure stress.

In the model, fibres are discretely represented as lines using elastoplastic Timoshenko beam elements following J2 plasticity model. The fibres are coupled to the finite element mesh representing the cement/mortar matrix through a specially developed coupling strategy. Figure 3 provides a visual representation of the Timoshenko beam element, fibre discretization strategy, and the bond-slip law governing the fibre-matrix debonding process. The penalty-based bond-slip relation, adapted from [35], describes the progressive degradation of the fibre-matrix bond. It is characterized by six parameters: τ_f, τ_{max}, s_1, s_2, s_3, as well as the friction coefficient μ. The bond-slip relation is pressure-sensitive, accounting for the pressure p exerted by the fibre on

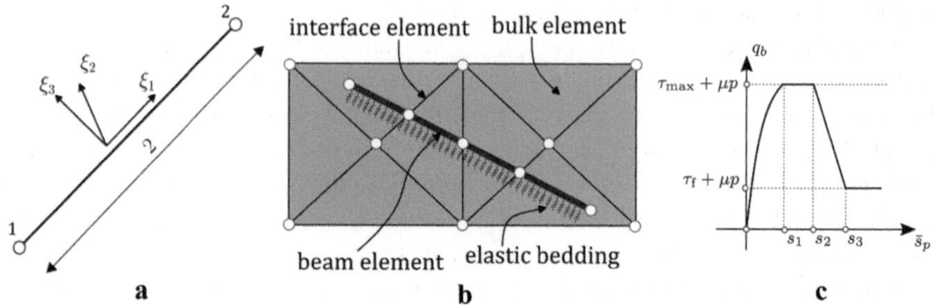

Fig. 3 Discrete fibre and bond models: **a** representation of a fibre element, **b** illustration of the finite element discretization of fibre and matrix mesh, **c** bond-slip relation (from [31], licensed under CC-BY 4.0)

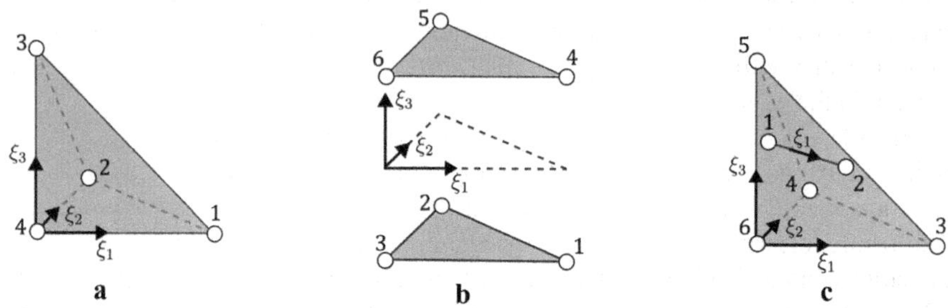

Fig. 4 Finite elements used for modeling **a** bulk material, **b** cracks, and **c** fibre-matrix bonds (from [31], licensed under CC-BY 4.0)

the matrix during crack bridging, which occurs as the fibre bends significantly during pullout (snubbing action).

Figure 4 summarizes the finite elements used in the computational strategy described above to discretize the bulk material, cracks, and fibre-matrix bond domains.

4 Experimental and Numerical Investigations

This section describes the experimental and numerical investigations conducted in the scope of this research. The investigations encompass a wide range of scales and loading conditions, providing valuable insights into constitutive behaviour of high-performance fibre-reinforced concrete.

4.1 Fibre Pullout Tests

First, experimental and numerical fibre pullout tests aimed at characterizing the behaviour of fibres as individual components within the complex microstructure of fibre-reinforced concretes and mortars are presented. These tests provide means for calibration and validation of the developed computational model of discrete fibres, which will be used in subsequent numerical simulations at the specimen scale.

4.1.1 Pullout of Single Inclined Fibre

The single fibre pullout experiments with an inclination of 45° performed by [36], serve to characterize the bond-slip behaviour of fibres used in the HPC-08-SF mixture, utilized in experiments on the specimen scale presented in Sect. 4.2 and 4.3 of this chapter. Since the Dramix OL 6/0.16 fibre used in the concrete mix was too short for pullout tests, it was decided to use a single long fibre with the same diameter and properties in the pullout tests.

Test Setup and Material Properties Fig. 5 shows the test setup with dimensions and the test rig. During specimen preparation, fibre ends were fixed in the formwork that was used to cast two mortar blocks with dimensions $20 \times 10 \times 8$ mm^3, 60 mm apart. The fibre embedment lengths were 8 mm on both sides.

The steel fibres properties used in the numerical simulation in this section are provided in Table 1, while the mortar-steel fibre bond properties identified by performing fibre pullout simulation are listed in Table 2.

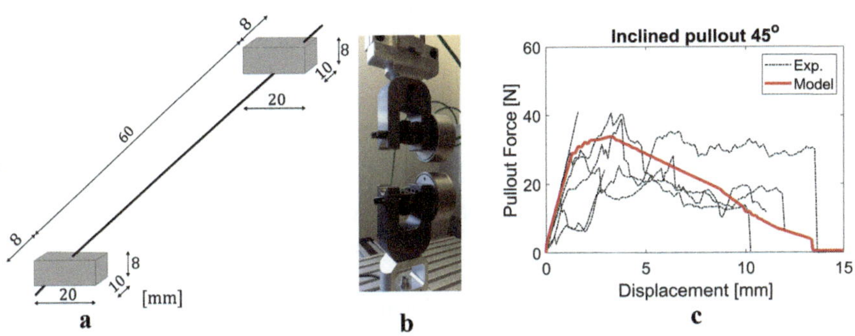

Fig. 5 Inclined fibre pullout test: **a** test setup with dimensions, **b** photo of the testing rig (Reproduced from the 16th fib Symposium Proceedings – Concrete Innovations in Materials, Design and Structures, held in Krakow from 27–29 May 2019, article "Pull-Out tests of carbon and high-strength steel microfibres in a cement-based matrix" page 1882—"Fig. 5: Test set-up for pull-out tests, left: for carbon fibre, middle and right: for steel fibre" with permission from the International Federation for Structural Concrete (fib) [36]), **c** comparison of experimental and numerical force–displacement curves

Table 2 Mortar-steel fibre bond parameters used in analysis of the inclined fibre pullout tests

Property	Unit	Mortar-steel fibre bond
τ_{max}	(MPa)	10.0
τ_f	(MPa)	7.0
s_1	(mm)	0.8
s_2	(mm)	5.0
s_3	(mm)	7.5
μ	(–)	0.6

Remark The fibre and bond properties determined in this calibration test are used in numerical simulations presented in Sects. 4.2 and 4.3.

Experimental and Numerical Pullout Results The experimental and simulated load–displacement response of the inclined fibre pullout test is presented in Fig. 5c. As mentioned, the simulation performed in this section serves as a calibration test to identify the bond parameters. Optimally, the bond parameter calibration should be performed on the pullout tests with 0° inclination and the validation on inclined pullout tests. However, using the 0° inclination results was not possible since the scatter was too high, and many tests failed very early and at extremely low pullout loads.

4.1.2 Pullout of Group of Fibres

The numerical analyses of fibre group pullouts performed in this section are a part of the numerical benchmark II organized within the Priority Programme SPP 2020 described in the chapter by J. Storm et al., this volume. In this subsection, the fibre scale is studied, and the model performance on ultra-high-performance fibre-reinforced concrete (UHPFRC) beams subjected to three-point cyclic bending is presented in Sect. 3.1 "Discrete Approach for Simulating Fracture in UHPFRC" in the chapter by J. Storm et al., this volume.

Test Setup and Material Properties This subsection briefly summarizes the setup and material properties used in fibre group pullout tests studied in this subsection. Two sets of fibre group pullout experiments under inclinations of 0° and 45° to the free surface were performed in the experimental campaign. Figure 6a illustrates the pullout test setup with 0° inclination, and the 45° inclination was set analogously. More details on the material and test setup can be found in [37, 38] and in the chapter by J.-P. Lanwer et al., this volume, and on numerical modelling in [31].

The elastoplastic material behaviour of steel fibres was characterized by uniaxial tension tests performed by [37], and they determined that some material properties differ significantly from those reported by fibre producers (see [37] for more details). The numerical simulation of a single fibre's uniaxial tensile behaviour was performed

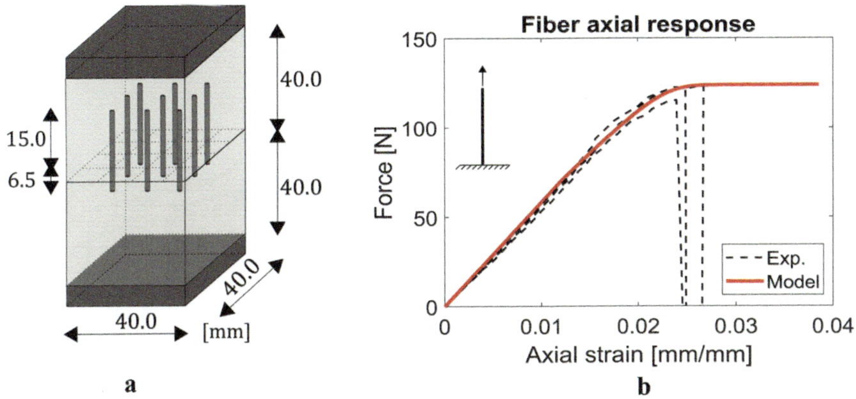

Fig. 6 **a** Fibre group pullout setup according to [37]. **b** Virtual material characterization: Fibre subjected to uniaxial tension. Experimental data from [38]

Table 3 Mortar-steel fibre bond properties used in analyses of the fibre group pullout tests

Property	Unit	Mortar-steel fibre bond
τ_{max}	(MPa)	9.0
τ_f	(MPa)	7.0
s_1	(mm)	0.7
s_2	(mm)	4.8
s_3	(mm)	6.0
μ	(–)	0.4

to characterize the material behaviour virtually and calibrate fibre material properties. The obtained results are presented in Fig. 6b.

The mortar-steel fibre bond parameter calibration is performed on fibre pullout tests with 0° inclination, and the identified bond parameters are listed in Table 3.

Numerical Simulations The simulations performed in this section serve primarily as calibration for the structural cyclic three-point beam bending simulations with explicitly modelled fibres performed in the chapter by J. Storm et al., this volume. The results shown in Fig. 7 demonstrate that the numerical model can capture the experimental behaviour well and replicate the main characteristics of the pullout behaviour.

4.2 *Cyclic Tensile Fatigue Tests and Crack Opening Tests*

Test Setup The cyclic tensile tests were carried out on prismatic ($40 \times 40 \times 160\,mm^3$) specimens with two notches (5 mm in width, 10 mm in depth) cut in the middle of

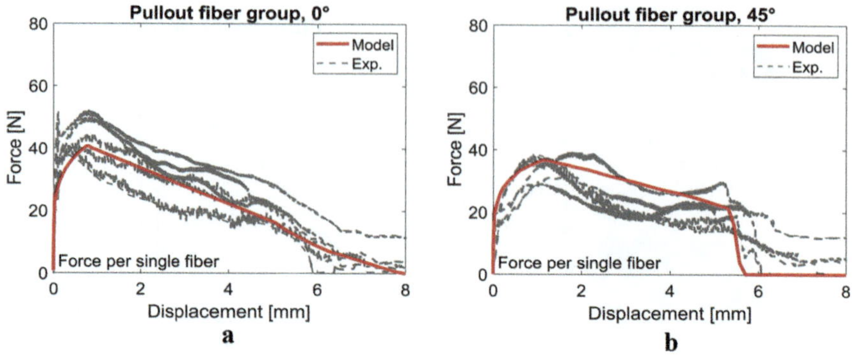

Fig. 7 Comparison between experimentally and numerically obtained pullout force–displacement responses: **a** 0°, and **b** 45° inclination. The force shown on the ordinate of the diagram is an average value carried by a single fibre

the specimens. The notches, introduced on two sides, define the weakest section, subsequently the breaking point. The cyclic tensile tests were carried out in a servo-hydraulic 100 kN tensile testing machine.

Method In force-controlled fatigue tests, the test specimens were subjected to tensile stresses according to sinusoidal load functions with $S_{\max} = 0.70$ and $S_{\min} = 0.35$. The cyclic tests were stopped after reaching 100,000 cycles. The displacement in the static tensile tests was measured by a clip-on-gage, whereas in the cyclic tensile tests, the displacements in the predefined breaking points (area of the notch) were measured using strain gauges with a length of 20 mm, see Fig. 8a.

To investigate crack closing and re-opening mechanisms, crack opening tests were conducted using notched prismatic specimens of the same dimensions as those in static and cyclic tests. Displacement transducers the plunger version were used to measure strain. The specimen was subjected to monotonic tensile loading until the first crack appeared, then unloaded to 10% of tensile strength to measure the deformation difference (Δl). Displacement was applied at the specimen's top end until the deformation at the notch reached Δl, followed by unloading to 10% of tensile strength. This process of crack opening and controlled loading continued until a crack width of CMOD + $10\Delta l$ was reached, with subsequent force-controlled unloading, see Fig. 8b [34].

Microscopic Investigations After $N = 0$, 1, 100, 1000 and 100,000 load cycles, one test specimen was chosen, from which the partial samples for microscopic examinations (amount, length, width, and position of the microcracks) were prepared as thick sections, see Fig. 8c. On these thick sections an examination field of 35×35 mm^2 was plotted. The thick sections were prepared after the specimens were subjected to load cycles, so it was possible to inspect the crack evolution [34].

Experimental Programme The cyclic tensile tests were carried out on the HPC-08 and HPC-08-SF in force-controlled fatigue tests with sinusoidal load functions. The

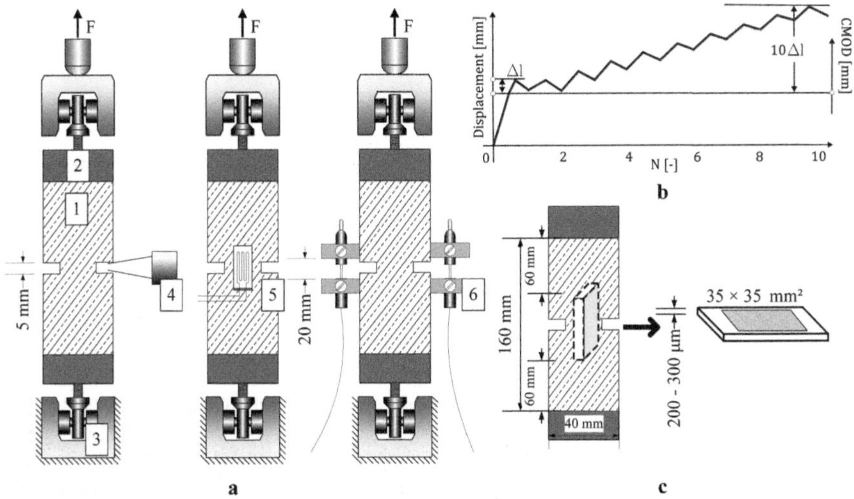

Fig. 8 **a** Test setup including the specimen (1), the two adhered test stamps (2), the lower/upper clamping in the machine (3), the clip-on-gage (4), strain gauge (5) and displacement transducer in plunger version (6), **b** Loading protocol used for tensile crack opening tests. **c** Preparation of thick sections (from [34], licensed under CC-BY 4.0)

crack opening tests were only conducted on the HPC-08-SF. For further information on experiments see also [34]. Above-described investigations also included additional cyclic tensile tests on the HPC-02 and HPC-00, which are described detailed in [39].

4.2.1 Experimental Analyses of Cyclic Tensile Fatigue Tests

Strains Fig. 9 shows exemplarily a comparison of the developments of strain at $S_{max} = 0.70$ and $S_{min} = 0.35$ for HPC-08 and HPC-08-SF. The strain of the HPC-08-SF grows only marginally in phase I and no noticeable increase in strain can be seen in phase II in comparison to the HPC-08. Due to the limited numbers of load cycles in the tests, phase III was not reached.

Microscopic Investigations The microscopic analyses showed that the number of cracks increased continuously with increasing number of cycles, see [34]. Most microcracks initiated in the interfacial transition zone (ITZ) in both HPC-08 and HPC-08-SF. In the case of HPC-08 wider and longer cracks and, thus, a larger total crack area was found. In HPC-08-SF, more cracks were found, but these cracks were smaller and shorter, i.e., more finely distributed. Thus, the total crack area was smaller than in case of the HPC-08. It is assumed that the bridging effect between the crack faces of the micro steel fibres leads to a finer distribution of microcracks under cyclic loading [34].

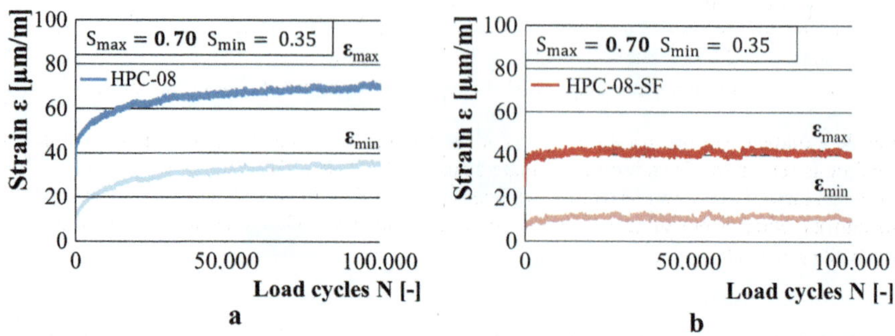

Fig. 9 Strain development at S_{min} and S_{max} of **a** HPC-08 and **b** HPC-08-SF (from [34], licensed under CC-BY 4.0)

4.2.2 Experimental and Numerical Analyses of Cyclic Crack-Opening Tests

This section utilizes the experimental virtual lab to investigate the mechanical behaviour of HPC-08-SF specimens under cyclic tensile loading. In addition to actual experiments, virtual experiments with different fibre orientation realizations that could stem from various casting procedures are studied. The fibre orientations are systematically varied to produce virtual experiments to investigate the effects of fibre orientation on post-peak ductility, and the unloading/reloading behaviour of notched prismatic specimens. The actual fibre orientation in the tested specimens is unknown.

Material Properties and Fibre Orientation Samples In numerical simulations, the concrete mesostructure, comprising aggregates, pores, and the mortar matrix, is not explicitly resolved due to the computational demands of high-resolution simulations on specimens of this size. However, fibres are explicitly resolved to consider the effects of different fibre orientations. The material properties of HPC-08 used in the simulation are listed in Table 4. Concrete properties not determined experimentally were adopted according to literature recommendations (see [31]).

The fibre and bond properties were previously determined through single fibre pullout tests presented in Sect. 4.1 and are listed in Tables 1 and 2, respectively.

In this study, four distinct simulations of the notched prismatic specimen were performed, each containing fibres sampled according to a different synthetically generated orientation distribution. The selected fibre orientation distributions are defined through second-order orientation tensors $\boldsymbol{a}$ (see [41, 42]). An orientation tensor is a compact tensorial representation of the fibre orientation function, which can be easily visualized using ellipsoidal surfaces. It is possible to reconstruct the entire fibre orientation function utilizing the procedure described in [31, 43] and use it to generate fibre samples with various orientations. The investigated cases are shown in Fig. 10, where the left column contains stereographic projections of the fibre orientations plotted on a unit hemisphere, the middle column shows the

Table 4 Concrete (HPC-08) properties used in numerical simulation

Property	Unit	Value
Young's modulus (E)	(MPa)	43,224.4[a]
Poisson's ratio (ν)	(–)	0.21[a]
Tensile strength (f_t)	(MPa)	5.9[a]
Shear strength ($f_s = \beta f_t$)	(MPa)	20.6
Fracture energy (G_F)	$(\frac{Nmm}{mm^2})$	0.133[b]
Fracture energy mode II ($G_{F,II} = \kappa G_F$)	$(\frac{Nmm}{mm^2})$	1.33
Friction coefficient (μ)	(–)	0.7
Reference closure stress ($t_{\pi,ref}$)	(MPa)	– 5.9
Reference crack opening ($\boldsymbol{u}_{\pi,ref}$)	(μm)	6.0

[a] Experimentally determined
[b] Estimated according to CEB-FIP Model Code 90 [40]

ellipsoidal representation of the orientation tensors, and the right column shows the actual discrete fibre realizations within the bounding box before the placement of fibres into the finite element model.

Experimental and Numerical Results Fig. 11 presents calculated responses of the notched prism specimens with four fibre orientation cases described above. On the left side of Fig. 11 are the experimentally measured and numerically calculated load–displacement curves, and on the right side are the snapshots of the fibres bridging the crack surface (in dark grey) at an opening of 0.1 mm, where the crack-bridging action of the fibres is the primary contributor to the residual strength of the HPC-08-SF specimen. Only the fibre elements with the Von-Mises stress in the cross-section greater than 40 MPa are shown in the figure.

Figures 11a–d show that once the crack is fully formed, indicated by a significant drop in force, fibres contribute to the residual strength. The alignment of fibres perpendicularly to the crack surface (Fig. 11a) yields the highest strength and residual carrying capacity. As the orientation becomes more isotropic (Figs. 11b–d), both peak and residual strengths decrease. The right column also shows that fewer fibres engage in crack bridging as their orientation deviates from perpendicular, reducing the likelihood of crack-fibre intersection. The 2-2-6 orientation case best matches the experimental results, suggesting it closely resembles the actual fibre orientations in the tested specimens. For further details on interpreting various fibre orientations, see [31].

A small hysteresis is observed in both the experiments and simulations, primarily due to the residual carrying capacity and energy dissipation from fibres. Figures 11a–d indicate that unfavourable fibre orientations result in a slight decrease in unloading stiffness and an increase in the area enclosed by hysteresis. Consequently, more fibres are inclined to the crack surface and undergo plasticisation in these cases. The

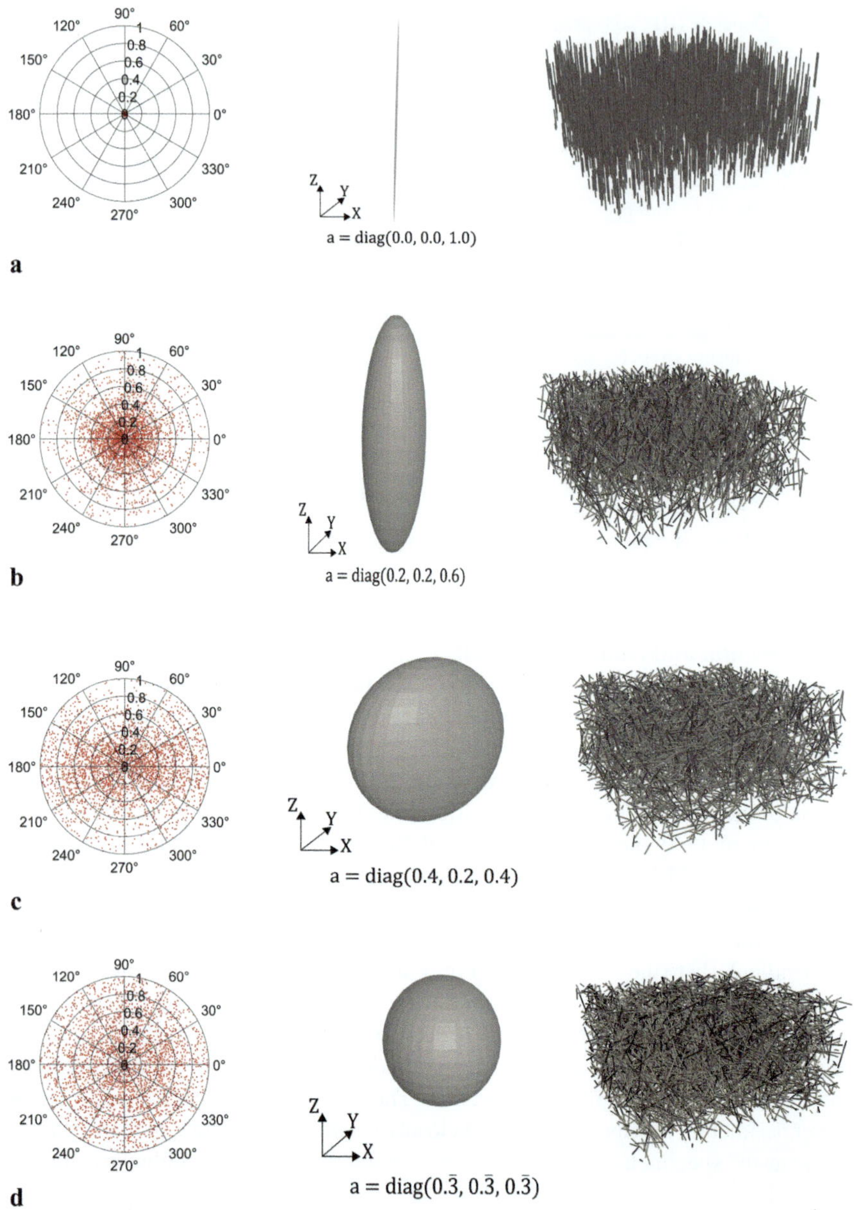

Fig. 10 Realizations of discrete fibre samples according to four distributions defined by orientation tensors: **a** aligned, **b** 2-2-6, **c** 4-2-4, **d** isotropic. Left column contains stereographic projections of the fibre orientations plotted on a unit hemisphere, middle column shows the ellipsoidal representation of the orientation tensors, and the right column shows the actual fibre realizations (from [31], licensed under CC-BY 4.0)

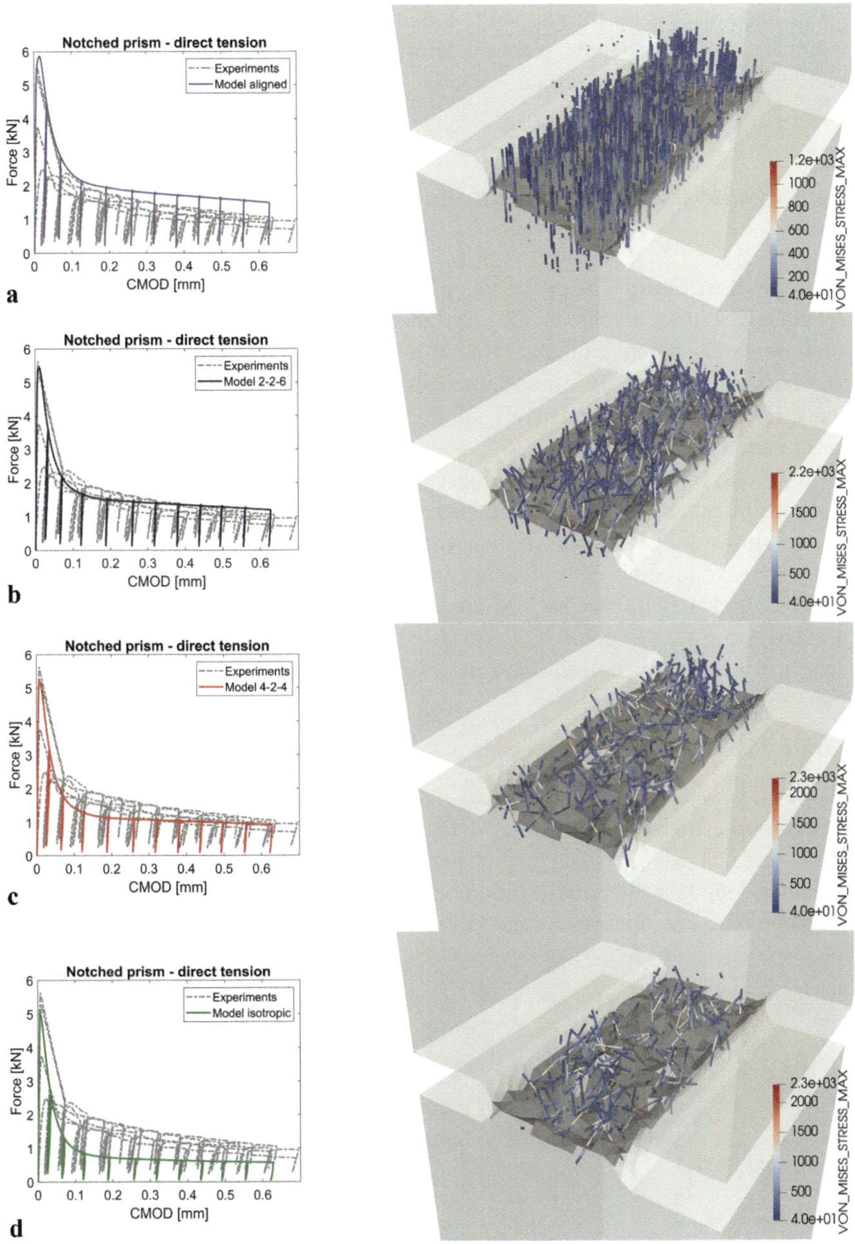

Fig. 11 Notched HPC-08-SF prismatic specimen under cyclic tensile loading. Comparison of load–displacement curves (left) and fibre elements actively bridging the crack (right) for the following test cases: **a** aligned; **b** 2-2-6; **c** 4-2-4; **d** isotropic (from [31], licensed under CC-BY 4.0)

unloading stiffness for the aligned, 2-2-6, and 4-2-4 orientations is somewhat overestimated, but the residual ("plastic") deformation upon unloading aligns reasonably well with the experimental data.

4.3 Cyclic Compressive Fatigue Tests

Multiple Test Setup The cyclic compressive tests were performed on cylindrical specimens with a height of 300 mm and a diameter of 100 mm for the high-performance concrete with and without microfibres. This means an *h/d* ratio of 3 and corresponds to a typical specimen shape for fatigue tests. The slenderness of the specimens implies a relatively unimpeded transverse strain in the centre of the specimen.

As cyclic tests are generally subject to a certain degree of scatter and therefore repeated tests are needed to ensure the reproducibility of the results, a multiple test set-up, shown in Fig. 12, was designed and built. This enables the simultaneous and congruent cyclic compressive loading of three test specimens. The test setup consists of three rigid steel frames, each equipped with a hydraulic differential cylinder with a maximum piston force of 1000 kN (force-controlled). The load is introduced via spherical bearing specifically designed for the multiple test setup with connected pressure plates [30, 44, 45].

Strain Measurement The strains in axial direction were continuously recorded by strain gauges with a length of 100 mm.

Microscopic Investigations After defined load cycles up to 5 million load cycles, thick sections were extracted, prepared and used for microscopic examinations (amount, length, width, and position of the microcracks). The thick sections were prepared with coloured resin (blue) to make the microcracks as visible as possible.

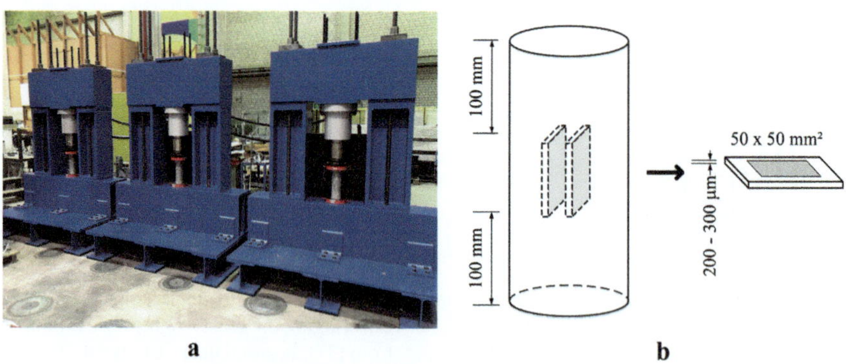

Fig. 12 Multiple test set-up for cyclic compressive tests (**a**) [30, 44, 45]. **b** Preparation of thick sections from the cylindrical specimens (d/h = 100/300 mm)

Two thick sections were taken out of the test specimen in the axial direction and an examination field of $50 \times 50\ mm^2$ was plotted on the thick sections, see Fig. 12b.

Experimental Programme Tests were conducted on high-performance concrete with and without microfibres (steel/carbon). The maximum stress level was varied ($S_{max} = 0.45$ and 0.70) while the minimum stress level ($S_{min} = 0.10$) remained constant. The tests used force-controlled sinusoidal load functions with a loading frequency of 5 Hz. Two batches per maximum stress level were produced and tested. The specimens, tested at around 28 days old, were stored under water until testing, resulting in high moisture content during cyclic loading.

Strains At $S_{max} = 0.45$, only low strains were observed, as expected, with no fatigue failure occurring up to the test stop ($N = 5$ million), see Fig. 13a. Phase 1 and the beginning of phase 2 showed similar characteristics across all compositions. Strains in fibre-reinforced concretes were slightly higher than in unreinforced high-performance concrete, likely due to increased matrix porosity from the fibres. At $S_{max} = 0.70$, a typical s-shaped curve with its 3 phases can be seen (Fig. 13b). Phase 3 of the high-performance fibre-reinforced concretes, especially HPC-08-SF, is significantly longer. This can be interpreted as a certain ductility under fatigue loading and indicates the failure of the concrete in comparison to a shorter phase 3 in HPC-08.

Microscopic Investigations Fig. 14 shows the number of microcracks and total crack area at $S_{max} = 0.70$, averaged from two thick sections up to just before specimen failure. The total crack area was normalized to a test area of $50 \times 50\ mm^2$. An HPC-08 specimen was removed after 80 k ($\log N = 4.9$) load cycles, while fibre-reinforced concrete specimens were removed after 15 k ($\log N = 4.2$) load cycles, corresponding to phase 3. HPC-08 already shows the largest microcrack area in the unloaded state. After 1000 cycles and just before failure, its microcrack area is significantly larger than that of HPC-08-SF and -C. HPC-08-C has the smallest total crack area, with no differences observed after 1000 cycles and just before failure.

HPC-08-C has the highest number of microcracks in the unloaded state. After 1000 load cycles, HPC-08-SF shows nearly double the microcracks, while HPC-08 sees

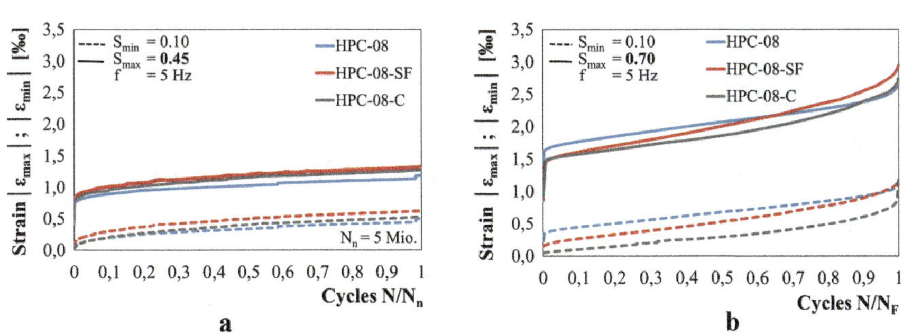

Fig. 13 Curves of averaged ($n = 3$) strains at: **a** $S_{max} = 0.45$ and **b** $S_{max} = 0.70$ [30, 45]

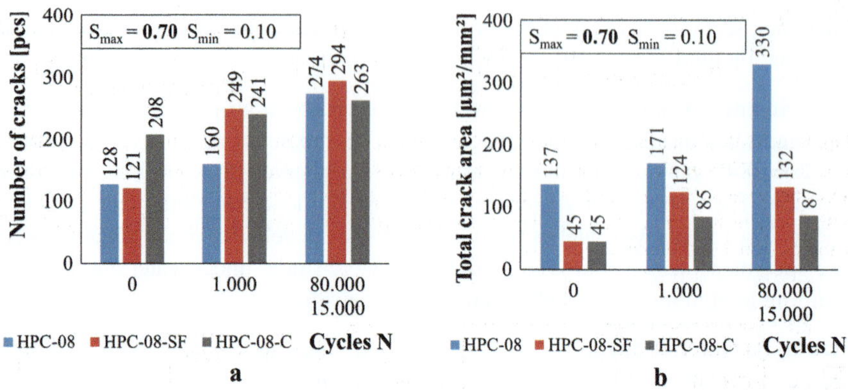

Fig. 14 **a** Number of cracks and **b** total crack area of HPC-08, -SF and -C at $S_{max} = 0.70$ [30, 45]

the greatest increase in number of microcracks in phase 3. Fibre-reinforced concretes consistently have smaller total crack areas due to their significantly narrower microcrack widths: HPC-08 averages 5–6 µm, HPC-08-SF 3–4 µm, and HPC-08-C 1–2 µm (see Fig. 15). Analysis shows that primarily new microcracks of similar width form, with no detectable widening of existing microcracks or length extension due to cyclic loading, observed for both $S_{max} = 0.70$ and $S_{max} = 0.45$.

Figure 16a illustrates the development of the number of cracks over loading cycles for investigated concretes. At $S_{max} = 0.70$, microcracks were oriented approximately 60–70% vertically (± 45°) and 30–40% horizontally (± 45°), both in the unloaded state and after cyclic loading. The percentage of vertically oriented cracks increased during cyclic loading, and newly formed cracks showed similar orientations. About 60–80% of microcracks formed in the interfacial transition zone (ITZ) after cyclic loading. At $S_{max} = 0.45$, the location of microcracks varied: in HPC-08-C, 30–40% were in aggregates, 60–70% in the ITZ, and less than 4% in the cement paste. In HPC-08-SF, 40–50% were in the cement paste, 30–40% in the ITZ, and 10–20% in aggregates, indicating fibres influence microcrack location at low stress levels [30]. Fewer than 5% of cracks were found in the fibre-matrix contact zone, suggesting it is not a crack-initiating weak point like the ITZ.

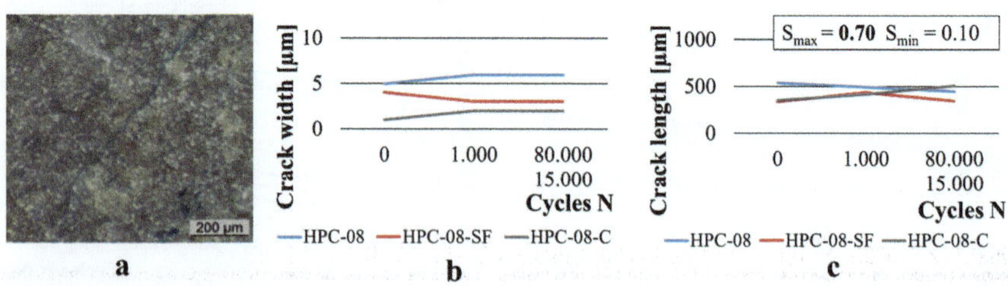

Fig. 15 **a** Example of a microcrack. **b** Crack width, and **c** crack length distributions over number of cycles [30]

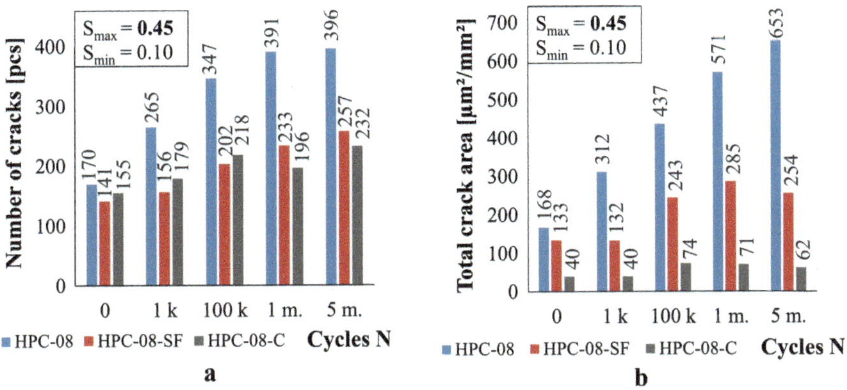

Fig. 16 Number of cracks and total crack area of HPC-08, -SF and -C at $S_{max} = 0.45$ [30]

Figure 16b shows the total crack area development over loading cycles for the investigated concretes. At $S_{max} = 0.45$, HPC-08-SF exhibits a significantly smaller total crack area that remains stable after 1 million cycles, with no further growth beyond a certain area. After 5 million cycles, HPC-08-SF's total crack area is about 2.5 times smaller than that of HPC-08. HPC-08-C has the smallest total crack area, remaining nearly constant after 100 k cycles and not increasing up to 5 million cycles; its microcrack area is about 10 times smaller than HPC-08's after 5 million cycles at $S_{max} = 0.45$. The width and length of microcracks are similar at both stress levels, indicating no influence of stress level on microcrack dimensions [30].

Number of Cycles to Failure Fig. 17 shows the mean cycles to failure ($n \geq 14$) for different microfibres. Fibre-reinforced compositions have similar cycles to failure, with $\log N = 4.13$ (HPC-08-SF) and $\log N = 4.09$ (HPC-08-C), indicating no influence of fibre type or content (HPC-08-SF: 1.0% vol., HPC-08-C: 0.5% vol.). In contrast, high-performance concrete without fibres (HPC-08) has a higher number of cycles to failure ($\log N = 4.82$). All specimens were stored and tested under the same conditions. The reduced cycles to failure in fibre-reinforced concrete are likely due to the introduction of fibres, which increase pores and defects, along with high moisture content.

4.4 Experimental-Virtual Lab: CT-Image-Based Mesoscale Analysis of a Cylinder Under Compressive Load

This section demonstrates the capabilities of the developed virtual lab framework to integrate the microstructure of an actual specimen obtained by means of CT-image into the simulation environment.

A cylindrical core (Ø 35.5 mm, height 40 mm) was cut from the central part of the top side of a standard HPC-08-SF specimen (Ø 100 mm, height 300 mm) to

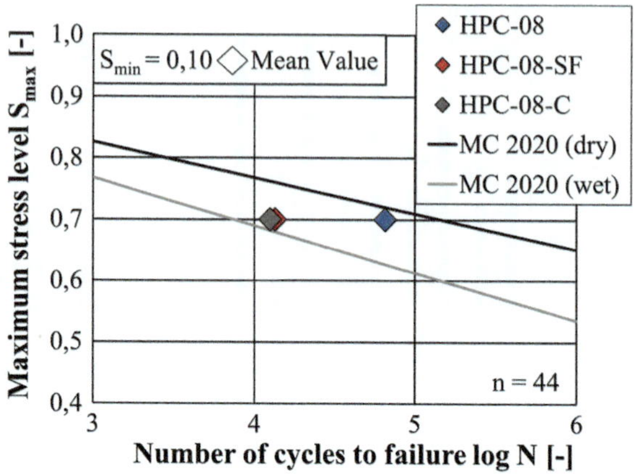

Fig. 17 Number of cycles to failure for HPC-08, -SF an -C [30, 45]

facilitate CT scanning of its microstructure, as shown in Fig. 18. Details on specimen preparation can be found in [31]. The specimen was then subjected to uniaxial compressive loading until failure. The material properties of mesoscale constituents are provided in Table 5. The properties that were not determined experimentally, were adopted from the literature (see [31, 46, 47]). The fibre and bond properties are given in Tables 1 and 2.

Generation of the Finite Element Model from CT-Image This subsection presents the methodology for converting voxel-based CT-image data into a finite element model.

The steps involved in image processing, segmentation, and mesh generation, are illustrated in Fig. 19, without delving into the algorithms, see [22].

The image processing performed in this work was done using the Image Processing Toolbox in MATLAB® and mesh generation tools used were the

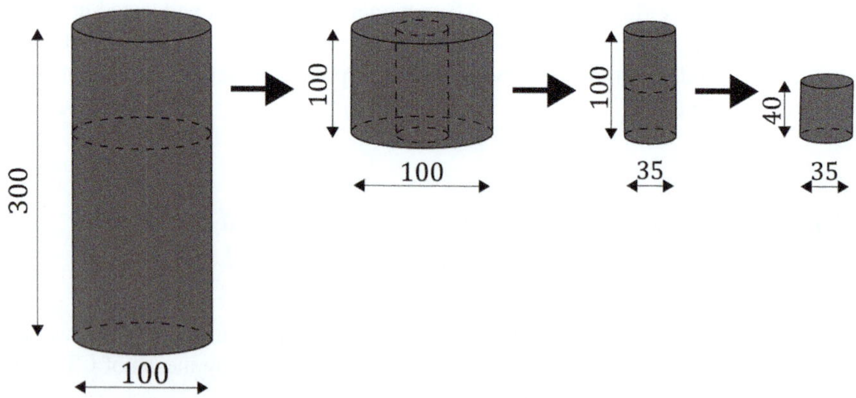

Fig. 18 Three-step cutting procedure to extract the small cylindrical core for CT scanning from HPC-08-SF specimen (from [31], licensed under CC-BY 4.0)

Table 5 Material properties of mesoscale constituents

Property	Unit	Aggregates	Mortar (HPC-02)	ITZ[a]
Young's modulus (E)	(MPa)	57,500.0[b]	39,183.0[c]	–
Poisson's ratio (ν)	(–)	0.18[b]	0.221[c]	–
Tensile strength (f_t)	(MPa)	14.0	5.9[c]	5.9
Shear strength ($f_s = \beta f_t$)	(MPa)	70.0	13.28	13.28
Fracture energy (G_F)	(N/mm)	0.05	0.05	0.05
Fracture energy mode II ($G_{F,II} = \kappa G_F$)	(N/mm)	0.25	0.25	0.25
Friction coefficient (μ)	(–)	0.7	0.7	0.7
Ref. closure stress ($t_{\pi,ref}$)	(MPa)	– 14.0	– 5.9	– 5.9
Ref. crack opening ($u_{\pi,ref}$)	(μm)	6.0	6.0	6.0

[a] Due to absence of experimental data, the ITZ properties were assumed same as for mortar matrix
[b] Determined according to Mori–Tanaka homogenization scheme (see [31] for details)
[c] Experimentally determined

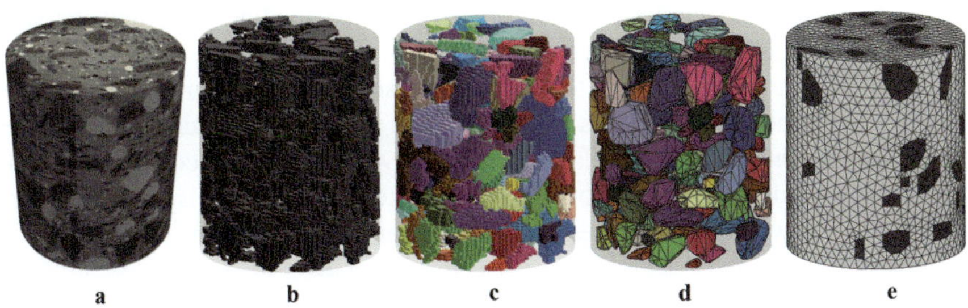

Fig. 19 Stages of generating the computational model from CT-image: **a** CT-image of an actual specimen with an adapted colour scheme for better visualization showing 4 phases: fibres (black), aggregates (dark grey), mortar matrix (grey), and air pores (light grey). **b** Preliminary segmentation of the aggregates. **c** Identification of separate aggregates. **d** Surface triangulation of individual aggregates. **e** Final tetrahedral mesh with two sets of elements representing aggregates (dark grey) and mortar (light grey) [22]

MATLAB® mesher and Gmsh®. Computed tomography (CT) scans of the cylindrical core were performed at Bundesanstalt für Materialforschung und -prüfung (BAM). Figures 20a and b shows a comparison of the cross-sections of the original CT-image and the obtained finite element mesh, indicating the level of resolved details. Figure 20c shows the final computational mesh of aggregates ($\geq$ 2 mm) and fibres.

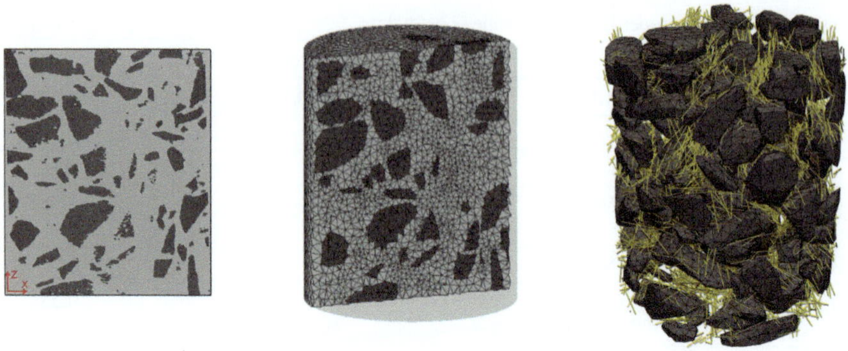

a b c

Fig. 20 A comparison between the segmented CT-image and the obtained mesh: **a** X–Z section of CT-image cut through the middle of the specimen, **b** cut through the generated mesoscale finite element model, **c** computational mesh of aggregates and fibres (from [31], licensed under CC-BY 4.0)

Experimental and Numerical Results Fig. 21a shows the comparison between the experimentally and numerically obtained load–displacement curves, as well as the failure patterns.

Due to surface irregularities, contact between the loading plate and the specimen was gradually established, causing a nonlinear loading path. The experimental diagram was shifted by 154 μm to align the intersection of experimental and numerical load–displacement diagrams with the point of where full contact was established. The numerical curve shows displacements up to 492 μm, as further convergence could not be achieved with the Newton–Raphson strategy. The load–displacement curve in Fig. 21 shows good agreement with the experimental curve, resulting from the

a b c

Fig. 21 CT-image based mesoscale analysis of HPC-08-SF cylinder under uniaxial compression: **a** comparison of experimental and numerical load–displacement curves; triangles refer to pre-peak, peak, and post-peak stages in Fig. 22, **b** photo of the specimen after test, and **c** snapshot of the fracture pattern in the last converged step of numerical analysis. The deformations in sub-figure, **c** are scaled fivefold (from [31], licensed under CC-BY 4.0)

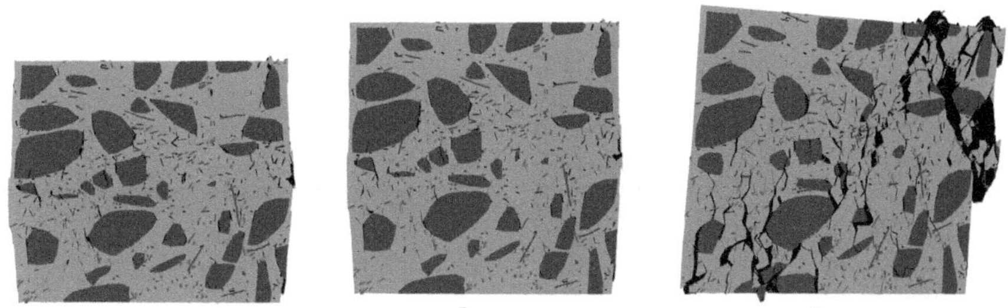

a b c

Fig. 22 Vertical cross-sections of the numerical model of the HPSFRC cylinder under uniaxial compression at different loading stages: **a** pre-peak, **b** peak, and **c** post-peak. Cracked interface elements are painted black for visibility. Fibres are dark yellow. Deformations in pre-peak, peak, and post-peak stages are scaled 50-fold, 25-fold, and tenfold, respectively (from [31], licensed under CC-BY 4.0)

complex interactions between mesoscale constituent and damage mechanisms. The peak load occurs at a displacement of 230 μm, with a strain of 5.81‰, indicating increased ductility due to short steel fibres.

Figure 21b and c display the experimental and numerical (at the last converged step) fracture patterns, showing diagonal crack band formation and significant edge spalling. Figures 22a–c show vertical cross-sections of the cylinder at pre-peak, peak, and post-peak stages of the numerical simulation, at load levels indicated in Fig. 21.

Damage initiates with multiple splitting cracks, often in the ITZ, due to deformation incompatibility between matrix and aggregates, and not ITZ weakness (same material properties were selected for the mortar and ITZ). As localization occurs, a shear band forms from merging splitting cracks (Fig. 22c), and some aggregates fracture, which is characteristic of high-performance concretes. Fibre bridging activates in larger cracks, leading to debonding and bending processes that slow down the crack growth and increase specimen ductility.

4.5 Cyclic Flexural Tests

Multiple Test Setup The cyclic tests were performed on beams with the dimensions 150 × 150 × 700 mm^3 (Fig. 23b). The cyclic tests were carried out in a specially developed multiple test setup, see Fig. 23a. With this setup, it is possible to test 6 beams with the same dimensions simultaneously. The setup is controlled via a central main hydraulic cylinder, which means that the same oil pressure is always present in the six individual test stations, see Fig. 23c. Two hydraulic cylinders apply the specified load to the test specimens at each of the six test stations.

Measurement The strains in the tension zone were continuously recorded by a strain gauge with a length of 150 mm. The degradation that occurs during cyclic loading

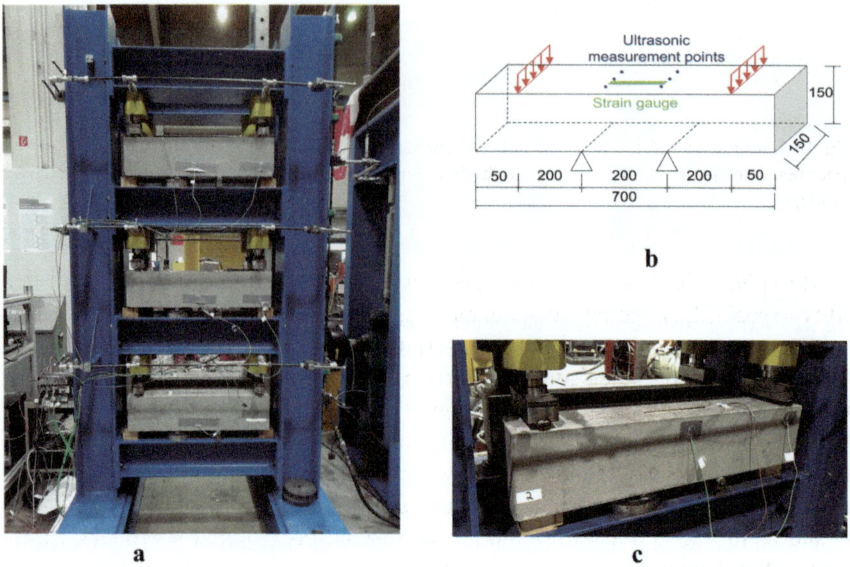

Fig. 23 **a** Multiple test setup for the flexural fatigue tests, **b** schematic test setup, and **c** individual test station (Reproduced from the 20th fib Symposium Proceedings—ReConStruct—Resilient Concrete Structures, held in Christchurch from 11–13 November 2024, article "Influence of different microfibers on the flexural fatigue characteristics of high-strength concrete" page 225—"Fig. 3: Multiple test setup for the flexural fatigue tests (left) and representation of the test beams and individual test station (right)." with permission from the International Federation for Structural Concrete (fib) [29])

was recorded using discontinuous ultrasonic (US) running time measurements. The dynamic modulus of elasticity can be calculated according to [48] based on the measured US running times, considering the density and the Poisson's ratio of the concrete tested.

Method The experiments were force-controlled, and the load applied as a sinusoidal load function with a constant load frequency of 5 Hz. First, the flexural strength was determined on at least three specimens of each series with the same dimensions as for the fatigue tests. From this, the stress levels $S_{max} = 0.45$ and $S_{max} = 0.70$ were determined. $S_{min} = 0.10$ was defined as a constant. The flexural fatigue tests were completed after 5 million cycles [29].

Experimental Programme Static and cyclic flexural tests were conducted on HPC-08, -SF and -C. The test specimens were produced in 6 batches. One batch was produced for each composition for the stress level $S_{max} = 0.45$ and $S_{max} = 0.70$. The beams were stored under water for up to 7 days and then in laboratory climate (20 ± 5 °C, 65 ± 5% r.h.) for up to 28 days and then tested, see also [29].

Strains Fig. 24 shows the strains of the analysed composition as mean values of 5 tested beams each. At $S_{max} = 0.45$ (Fig. 24a), the HPC-08 shows the highest strains

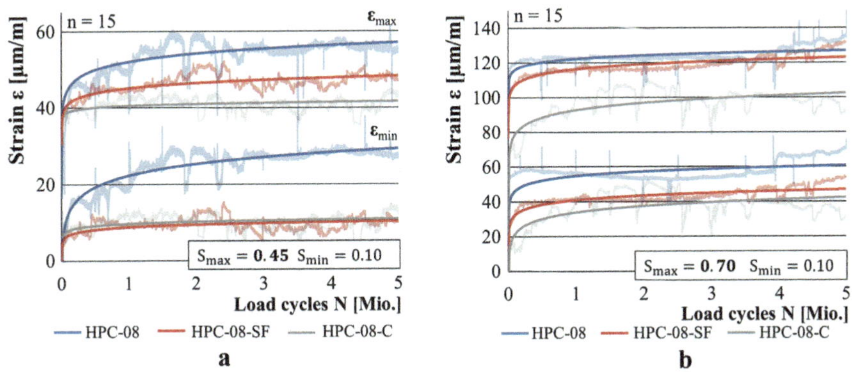

Fig. 24 Strain-cycle diagram for HPC-08, HPC-08-SF and HPC-08-C at **a** $S_{max} = 0.45$, and **b** $S_{max} = 0.70$, as mean values (Reproduced from the 20th fib Symposium Proceedings—ReConStruct—Resilient Concrete Structures, held in Christchurch from 11–13 November 2024, article "Influence of different microfibers on the flexural fatigue characteristics of high-strength concrete" page 229—"Fig. 7: Strain-cycle (S–N) diagram for HPC-08, HPC-08-SF and HPC-08-C at stress level Smax = 0.45 (**a**) and stress level Smax = 0.70 (**b**) as mean values of five tested beams" with permission from the International Federation for Structural Concrete (fib) [29])

and the strain of HPC-08 grew faster and showed a more significant increase in phase I of damage evolution. Comparatively, the strain of HPC-08C only grew marginally in phase I, and in phase II there was no noticeable increase in strain. The HPC-08-SF is in between but shows a more significant increase in strain in Phase II compared to the HPC-08-C. HPC-08 shows a steady increase in strain in phase II of damage evolution.

Due to the limited numbers of load cycles, phase III was not reached. At $S_{max} = 0.70$ (Fig. 24b) the expansions of the HPC-08 and HPC-08-SF come closer together. However, the HPC-08 also exhibits the greatest strains at $S_{max} = 0.70$. The HPC-08-C clearly shows low strains. The carbon fibres reduce the strains in the high-performance concrete most efficient.

Relative Dynamic Modulus of Elasticity Fig. 25 shows the development of the relative dynamic modulus of elasticity determined via the US measurements in the tensile zone. A significant drop in stiffness already occurs in the early phase of cyclic loading. While this initial drop in the beams at a stress level of $S_{max} = 0.45$ takes place within the first 500,000 load cycles (see Fig. 25a), the beams with a stress level of $S_{max} = 0.70$ suffer a much faster drop in stiffness within the first 100,000 cycles (Fig. 25b). After an initial decrease in stiffness (phase I), the development of stiffness stabilizes in phase II. The HPC-08 (4.3%) and HPC-08-SF (4.8%) exhibit the largest drop in stiffness.

At 2.5%, the HPC-08-C showed only half as great a decrease in the relative dynamic E-modulus. The tests at a stress level of $S_{max} = 0.70$ showed a decrease in stiffness of 10.4% (HPC-08), 10.5% (HPC-08-SF) and 5.5% (HPC-08-C) and thus similar proportions to $S_{max} = 0.45$. The HPC-08 and HPC-08-SF show a comparable

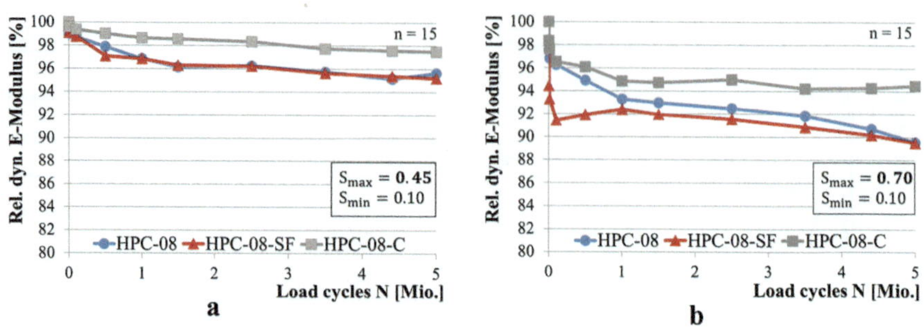

Fig. 25 Relative dynamic E-modulus at stress level, **a** $S_{\max} = 0.45$, and **b** $S_{\max} = 0.70$

stiffness decrease after 5 million load cycles at both stress levels and the HPC-08-C showed the lowest.

5 Conclusion

This work investigated the effects of steel and carbon microfibres over a wide range of experimental scenarios. In the scope of this work, a sophisticated computational model was developed and used to establish a virtual lab that facilitated the integration of synthetic or actual experimental data into the numerical simulation environment, complementing the experiments, especially in aspects that are hard or impossible to measure.

The investigations conducted in this research range from single fibre- to structural-element scales and offer valuable insights into interactions between individual constituents in high-performance concrete subjected to monotonic and cyclic loads. The experiments were supported by numerical simulations on each scale. The presented research has shown that the effect of fibre reinforcement on fatigue lifetime is non-negligible. Although, no definitive answer could be reached on whether adding fibres benefits or detriments fatigue resistance in cyclic tensile and flexural tests, many noteworthy findings emerged from this research. The key findings are summarized as follows.

Cyclic Tensile Tests The fatigue tests revealed that HPC-08-SF specimens exhibited a lower rate of strain increase, with strain stagnating after about 10,000 cycles, unlike the HPC-08 specimens. Microscopic analyses showed a continuous increase in number of cracks during cyclic loading. HPC-08 exhibited wider and longer cracks, leading to a larger total crack area, whereas HPC-08-SF displayed more numerous yet smaller and shorter cracks, resulting in a reduced total crack area. Displacement-controlled crack opening tests assessed the post-cracking behaviour of HPC-08-SF. Results indicated that adding short steel fibres did not substantially affect tensile strength but significantly enhanced ductility. The unloading stiffness after cracking was slightly lower than the initial stiffness but remained stable throughout the experiment, even for larger crack openings [34]. The cyclic tensile crack-opening virtual experiments showed that fibre orientation significantly impacts residual strength. The analysed virtual scenarios demonstrated that as the fibre orientation progresses from perfectly aligned to isotropic, fewer fibres get activated in the crack-bridging processes, and the specimens show reduced residual strength.

Cyclic Compressive Tests At a stress level of $S_{max} = 0.45$, low strains were observed, and no fatigue failure occurred by the end of the test. At $S_{max} = 0.70$, a characteristic S-shaped curve with three phases appeared. Phase 3 was notably longer for high-performance fibre-reinforced concretes, particularly for HPC-08-SF, indicating greater ductility under fatigue loading compared to the shorter phase 3 observed in HPC-08. Fibre-reinforced concretes consistently showed lower total microcrack surface areas. The evaluation revealed no widening or extension of existing microcracks due to cyclic loading; degradation was characterized by the formation of new microcracks at both stress levels. At $S_{max} = 0.70$, most microcracks formed in the ITZ. At $S_{max} = 0.45$, this varied by fibre type, with very few cracks ($< 5\%$) found in the fibre contact zone, indicating it is not a weak point like the ITZ. HPC-08-SF and -C exhibited fewer cycles to failure than HPC-08, as the addition of fibres increases number of pores and defects in the structure, which, along with high moisture content, can further reduce the number of cycles to failure.

Experimental-Virtual Lab The mesoscale analysis of the cylindrical specimen under compressive loading demonstrated the capabilities of CT-based mesostructure modelling. The numerical results emerging from the complex interactions between fibres, aggregates, and the mortar demonstrated a good agreement with experimental data. It was found that the cracks initiated in the ITZ region were due to deformation incompatibility, even though the material properties were assumed to be the same as those in the mortar matrix. The analysis emphasizes the importance of resolving mesostructure features to accurately capture the crack initiation and fracture processes that govern the behaviour of fibre-reinforced concrete.

Cyclic Flexural Tests The addition of fibres effectively limited strain at both investigated stress levels, with carbon fibres proving to be the most effective. At both stress levels ($S_{max} = 0.45$ and 0.70), the reduction in stiffness (relative dynamic modulus of elasticity) for HPC-08-C was only about half that of HPC-08 and HPC-08-SF. This smaller decrease in stiffness in HPC-08-C is attributed to less microcrack development. The developed numerical framework was not used for the high-cycle fatigue flexural experiments due to the significant computational intensity required to resolve the detailed fibre-reinforced concrete microstructure. Nevertheless, the model was successfully applied in numerical benchmark II of the Priority Programme SPP 2020, simulating cyclic three-point bending experiments (see the chapter by J. Storm et al., this volume), demonstrating a good agreement with experimental results.

The results and models developed in this research serve as a valuable link between experimental and numerical research, and also contribute to a better understanding of high-performance concrete behaviour. Utilizing the experimental–numerical framework, future work will focus on developing scale-bridging models to simulate fibre-reinforced concrete under fatigue loading at the structural level.

References

1. Ibuk, H.: Ermüdungsverhalten von Beton unter Druckschwellbelastung. PhD thesis, Ruhr-University Bochum, Germany (2008)
2. Kustermann, A.: Einflüsse auf die Bildung von Mikrorissen im Betongefüge. PhD thesis, Universität der Bundeswehr München, Germany (2005)
3. RILEM TC 36-RDL: Report by RILEM Technical Committee 36-RDL on Long term random dynamic loading of concrete structures. Matériaux et Constructions **1**, 1–28 (1984)
4. Lohaus, L., Oneschkow, N., Elsmeier, K., Hümme, J.: Ermüdungsverhalten von Hochleistungsbetonen in Windenergieanlagen. Bautechnik, Heft **8**, 533–541 (2012)
5. Yin, W., Hsu, T.: Fatigue behavior of steel fiber reinforced concrete in uniaxial and biaxial compression. ACI Mater. J. **92**(1), 71–81 (1995)
6. Cachim, P.B., Figueiras, J.A., Pereira, P.A.: Fatigue behavior of fiber-reinforced concrete in compression. Cem. Concr. Compos., Heft **2**, 211–217 (2002)
7. Paskova, T., Meyer, C.: Low-cycle fatigue of plain and fiber-reinforced concrete. ACI Mater. J. **94**(4), 273–286 (1997)
8. Medeiros, A., Zhang, X., Ruiz, G., Yu, R.C., Velasco, M.d.S.L.: Effect of the loading frequency on the compressive fatigue behavior of plain and fiber reinforced concrete. Int. J. Fatigue 342–350 (2015)
9. Chang, D.-I., Chai, W.-K.: Flexural fracture and fatigue behavior of steel-fiber-reinforced concrete structures. Nucl. Eng. Des. **1–2**, 201–207 (1995)
10. Naaman, A.E., Hammoud, H.: Fatigue characteristics of high performance fiber-reinforced concrete. Cem. Concr. Compos., Heft **5**, 353–363 (1998)
11. Kaur, G., Singh, S.P., Kaushik, S.: Flexural fatigue strength of steel fibre reinforced concrete containing blends of limestone powder and silica fume. Int. J. Emerg. Technol. Adv. Eng. **2**(6), 436–445 (2012)
12. Caggiano, S., Plizzari, G.A., Slowik, V.: Experimental investigations into the fatigue crack growth in concrete. In: Proceedings of the 3rd International Conference on Fracture Mechanics of Concrete and Concrete Structures, pp. 645–654, Gifu, Japan (1998)
13. Wang, W., Wu, S., Dai, H.: Fatigue behavior and life prediction of carbon fiber reinforced concrete under cyclic flexural loading. Mater. Sci. Eng., A **434**(1–2), 347–351 (2006)
14. Deng, Z.: The fracture and fatigue performance in flexure of carbon fiber reinforced concrete. Cement Concr. Compos. **1**, 131–140 (2005)
15. Grassl, P., Xenos, D., Nyström, U., Rempling, R., Gylltoft, K.: CDPM2: a damage-plasticity approach to modelling the failure of concrete. Int. J. Solids Struct. **50**(24), 3805–3816 (2013)
16. Jefferson, A.D., Mihai, I.C.: The simulation of crack opening–closing and aggregate interlock behaviour in finite element concrete models. Int. J. Numer. Meth. Engng. **104**(1), 48–78 (2015)
17. Richard, B., Ragueneau, F.: Continuum damage mechanics based model for quasi brittle materials subjected to cyclic loadings: formulation, numerical implementation and applications. Eng. Fract. Mech. **98**, 383–406 (2013)
18. Sellier, A., Casaux-Ginestet, G., Buffo-Lacarriere, L., Bourbon, X.: Orthotropic damage coupled with localized crack reclosure processing. Part I Const. Laws. Eng. Fract. Mech. **97**, 148–167 (2013)
19. Caner, F.C., Bažant, Z.P.: Microplane model M7 for plain concrete. I: formulation. J. Eng. Mech. **139**(12), 1714–1723 (2013)
20. Kirane, K., Bažant, Z.P.: Microplane damage model for fatigue of quasibrittle materials: subcritical crack growth, lifetime and residual strength. Int. J. Fatigue **70**, 93–105 (2015)

21. Snozzi, L., Molinari, J.F.: A cohesive element model for mixed mode loading with frictional contact capability. Int. J. Numer. Meth. Eng. **93**(5), 510–526 (2013)
22. Gudžulić, V., Daadouch, K., Meschke, G.: Modeling of damage processes in concrete under monotonic and cyclic loading. In: Proceedings of the 11th International Conference on Fracture Mechanics of Concrete and Concrete Structures, Bangalore, India (2023)
23. Cusatis, G., Schauffert, E.: A. Cohesive crack analysis of size effect. Eng. Fract. Mech. **76**(14), 2163–2173 (2009)
24. Cusatis, G., Pelessone, D., Mencarelli, A.: Lattice discrete particle model (LDPM) for failure behavior of concrete I: theory. Cement Concr. Compos. **33**(9), 881–890 (2011)
25. Bolander, J.E., Saito, S.: Discrete modeling of short-fiber reinforcement in cementitious composites. Adv. Cem. Based Mater. **6**(3–4), 76–86 (1997)
26. Kang, J., Kim, K., Lim, Y.M., Bolander, J.E.: Modeling of fiber-reinforced cement composites: discrete representation of fiber pullout. Int. J. Solids Struct. **51**(10), 1970–1979 (2014)
27. Schauffert, E.A., Cusatis, G.: Lattice discrete particle model for fiber-reinforced concrete I: theory. J. Eng. Mech. **138**(7), 826–833 (2012)
28. Yang, Z., Zhang, X., Wang, Z., Li, Q.M.: A 3D cohesive-frictional coupled interface model for mesoscale simulation of steel fibre-reinforced concrete. Int. J. Numer. Meth. Eng. **125**(2), e7370 (2024)
29. Schäfer, N., Breitenbücher, R.: Influence of different microfibers on the flexural fatigue characteristics of high-strength concrete. In Proceedings of the fib Symposium ReConStruct, 11–13 November 2024, Christchurch, New Zealand (2024)
30. Schäfer, N.: Ermüdungsverhalten von hochfestem Beton – Einfluss der Gesteinskörnung und Fasern, PhD thesis, Ruhr-University Bochum, Germany (2025)
31. Gudžulić, V.: Virtual Lab for Mechanical Testing of Steel- and Steel-Fiber Reinforced Concrete, PhD thesis, Ruhr-University Bochum, Germany (2024)
32. Gudžulić, V., Neu, G. E., Meschke G.: Numerical analysis of plain and fiber reinforced concrete structures during cyclic loading: influence of frictional sliding and crack roughness. PAMM **20**(1), e202000167 (2021)
33. Gudžulić, V., Meschke G.: Multi-level approach for modelling the post-cracking response of steel fibre reinforced concrete under monotonic and cyclic loading. PAMM **21**(1), e202100194 (2021)
34. Schäfer, N., Gudžulić, V., Breitenbücher, R., Meschke, G.: Experimental and numerical investigations on high performance SFRC: cyclic tensile loading and fatigue. Materials **14**, 7593 (2021). https://doi.org/10.3390/ma14247593
35. International Federation for Structural Concrete (fib): fib Model Code for Concrete Structures, Ernst & Sohn (2013)
36. Schäfer, N., Breitenbücher, R.: Pull-Out tests of carbon and high-strength steel microfibres in a cement-based matrix. In: Proceedings of the fib Symposium in Krakow, pp. 27–29, Krakow, Poland (2019)
37. Lanwer, J.P., Oettel, V., Empelmann, M., Höper, S., Kowalsky, U., Dinkler, D.: Bond behavior of micro steel fibers embedded in ultra-high performance concrete subjected to monotonic and cyclic loading. Struct. Concr. **20**(4), 1243–1253 (2019)
38. Oettel, V., Lanwer, J. P., Empelmann, M.: Auszugverhalten von Mikrostahlfasern aus UHPC unter monoton steigender und zyklischer Belastung. Bauingenieur **96**(1,2) (2021)
39. Schäfer, N., Gudžulic, V., Timothy, J.J., Breitenbücher, R., Meschke, G.: Fatigue behavior of HPC and FRC under cyclic tensile loading: experiments and modeling. Struct. Concr. **20**(4), 1265–1278 (2019)
40. Comite Euro-International Du Beton: CEB-FIP Model Code 1990, Bulletin d'Information, Lausanne, Switzerland (1990)
41. Advani, S.G., Tucker, C.L.: The use of tensors to describe and predict fiber orientation in short fiber composites. J. Rheol. **31**(8), 751–784 (1987)
42. Gudžulić, V., Dang, T.S., Meschke, G.: Computational modeling of fiber flow during casting of fresh concrete. Comput. Mech. **63**(6), 1111–1129 (2019)

43. Reinold, J., Gudžulić, V., Meschke, G.: Computational modeling of fiber orientation during 3D-concrete-printing. Comput. Mech. **71**(6), 1205–1225 (2023)
44. Schäfer, N., R. Breitenbücher, R.: Influence of aggregate on fatigue behavior of high-performance concrete under cyclic loading. In: Proceedings of the fib Symposium, pp. 781–787, Shanghai, China (2020)
45. Schäfer, N., Breitenbücher, R.: Influence of different microfibers on the compressive fatigue and microcracking behavior of high-strength concrete. In: Mechtcherine, V., Signorini, C., Junger, D. (Eds.) Transforming Construction: Advances in Fiber Reinforced Concrete, RILEM Bookseries, pp. 414–422. Springer Nature Switzerland, Cham (2024)
46. Van Mier, J.G.M.: Concrete Fracture: A Multiscale APPROACH. CRC Press, Boca-Raton (2012)
47. Oneschkow, N., Timmermann, T.: Influence of the composition of high-strength concrete and mortar on the compressive fatigue behaviour. Mater. Struct. **55**(2), 83 (2022)
48. Breitenbücher, R., Kunz, S.: Auswirkungen dynamischer Beanspruchungen in Beton von Fahrbahndecken auf potentielle Gefügeschädigungen. Forschung Straßenbau und Straßenverkehrstechnik, Heft 1124 (2017)

Effects of Steel-Fibres on the Degradation of High-Performance Concrete Subjected to Fatigue Loading—Testing and Modelling

Gregor Gebuhr, Mangesh Pise, Dominik Brands, Steffen Anders, and Jörg Schröder

Abstract This contribution aims to describe and model the fatigue behaviour of steel fibre-reinforced concretes under bending loading by combining experimental and numerical approaches on different scales. Initially, single fibre pull-out tests were conducted with different fibre types and inclinations, both with and without hooked-ends. Based on the results, an ellipsoidal unit cell is constructed to simulate the material behaviour in preferred fibre orientation. Then, low- and high-cycle bending tests on notched beams with varying fibre contents and compressive strengths were performed. It is shown that the deterioration development in these tests progresses similarly regarding load-levels and loading history. However, significant differences in deterioration development occurred between the compressive strengths and fibre contents tested. A phenomenological material model for the beam tests is constructed which is not only capable to predict the load to crack-mouth-opening behaviour for different steel fibre contents, but also the deterioration in terms of residual stiffness for low-cycle tests. Finally, the results suggest the conclusion that deformation controlled low-cycle tests are a suitable means to estimate high-cycle fatigue behaviour. Furthermore, constant deterioration development in the second phase of damage provides a suitable basis for estimations of failure and can be applied in numerical cycle-jump approaches.

Keywords (Ultra-)high-performance concrete · Steel fibres · Phase-field fracture modelling · Multi-level fatigue loading · Cycle jump · Experimental virtual lab

---

G. Gebuhr (✉) · S. Anders
Institute for Structural Engineering, University of Wuppertal, Wuppertal, Germany
e-mail: gebuhr@uni-wuppertal.de

M. Pise · D. Brands · J. Schröder
Institute of Mechanics, University of Duisburg-Essen, Essen, Germany

N. Oneschkow and L. Lohaus (eds.), *Cyclic Deterioration of High-Performance Concrete*, https://doi.org/10.1007/978-3-032-19394-0_10

1 Introduction

Concrete is one of the most widely used materials in civil engineering construction. In recent decades, the growing demand for higher strength and durability has driven the development of high-performance concrete (HPC) and ultra-high-performance concrete (UHPC). However, as the material's strength increases, its ductility tends to decrease. Consequently, the investigation of concrete's fatigue behaviour becomes crucial. For example, the use of these concretes in structures such as bridges and precast wind turbine towers requires high resistance to fatigue-induced degradation. In general, the design of components using the extremely brittle HPCs is not feasible without adding fibres. The complex composition of HPC and UHPC, along with the mechanical and physical properties of their components and the interactions between them, plays a critical role in determining their overall material behaviour.

This contribution focuses on the quantification of the effects of different types and contents of steel fibres on the damage process at the mesoscale. It aims to develop a phenomenological material model that numerically represents the fibres within the concrete matrix. The influence of fibre reinforcement under different cyclic load scenarios is analysed experimentally in order to understand the damage mechanisms taking place on different scales. Here, experimental data is used to develop and calibrate a phenomenological material model, enabling realistic and accurate predictions of deterioration in fibre-reinforced HPC, particularly under cyclic loading conditions.

To be able to investigate damage development in cracked sections, three different types of tests were performed. On the mesoscale, pull-out behaviour of single steel fibres was investigated. For the macroscale, deformation-controlled low-cycle tests and force-controlled high-cycle tests were conducted. Here, the effects of different loading histories were looked into. The results suggest that different damage indicators are firstly applicable to differentiate damage development with respect to the fibre content. Secondly, the indicators are suitable to quantitatively describe damage development in low-cycle tests and transfer it to time consuming high-cycle tests.

On this basis, a systematic step-by-step approach was used for the development, calibration, and validation of a phenomenological material model. The first step involved the numerical calibration of a micro-mechanical model based on the load-displacement characteristics observed in single-fibre pull-out tests. Subsequently, the calibrated macro-mechanical model is used to analyse the macroscopic behaviour of ellipsoidal representative volume elements (RVEs) of both pure HPC and reinforced HPC through virtual experiments. Subsequently, the efficiency of the phenomenological material model is evaluated by comparing the macroscopic responses obtained from the virtual experiments and from macroscopic boundary value problems (BVPs) under similar loading conditions. In the next step, two distinct continuous step-wise linearly approximated degradation functions were implemented to model the asymmetric tension–compression behaviour of HPC. The accuracy of the numerical model is validated by analysing the stiffness degradation in HPC beams with varying fibre contents and orientations, characterised by the orientation distribution function (ODF), and comparing the numerical results with experimental data.

Table 1 Concrete and fibres, overview

	HPC[a]	UHPC[a]
Concrete compressive strength	110 N/mm^2	160 N/mm^2
Steel fibres type	Bekaert 3D 65/60 Hooked-end macro fibres	Stratec Weidacon FM 0.19/13 Smooth micro fibres
Fibre contents	0/23/57/115 kg/m^3 0/0.3/0.75/1.5 vol.-%	0/57/115 kg/m^3 0/0.75/1.5 vol.-%

[a]Information regarding the composition can be found in Sect. 2.1 "Concrete Composition" in chapter by M. Basaldella et al., this volume

2 Materials and Concrete Compositions

The experiments described here were performed on one high-strength and one ultra-high-strength concrete mix. Both were the reference concretes of the SPP 2020 as defined in Sect. 2.1 "Concrete Composition" in the chapter by M. Basaldella et al., this volume. Furthermore, two different types of fibres were used. Depending on the maximum grain size, hooked-end macro steel fibres with a length of 60 mm and a diameter of 0.92 mm were chosen for the HPC. For the UHPC, smooth micro steel fibres with a length of 13 mm and a diameter of 0.19 mm were applied. The contents of steel fibres were also varied in a wide range as can be seen in Table 1. The investigations focused on the change in mechanical fatigue properties due to the addition of steel fibres. The fibres were kindly provided by Bekaert and Stratec, respectively.

3 Single Fibre Pull-Out Test—Characterization and Modelling

3.1 Experimental Characterization

The basic approach was to experimentally analyse the behaviour of high-performance concretes with steel fibres on several levels and to describe it, focusing on the usability for the calibration of numerical models. The lowest level of investigation was the behaviour of individual fibres, which were quasi-statically pulled out of the concrete matrix under a range of boundary conditions such as embedded length and different inclinations. For comprehensive descriptions of the various parameters, such as embedded length and fibre inclination, and test results, see [1–3].

For the experimental evaluation of the load-bearing behaviour of the macro fibres, Fig. 1a shows the force-slip curve of fibres embedded vertically 20 mm (or approximately one third of the fibre length) into the concrete matrix as an example. As depicted, experiments were conducted on specimens with cut end hooks in addition to pull-out tests on hooked-end fibres. This approach enables an evaluation in terms

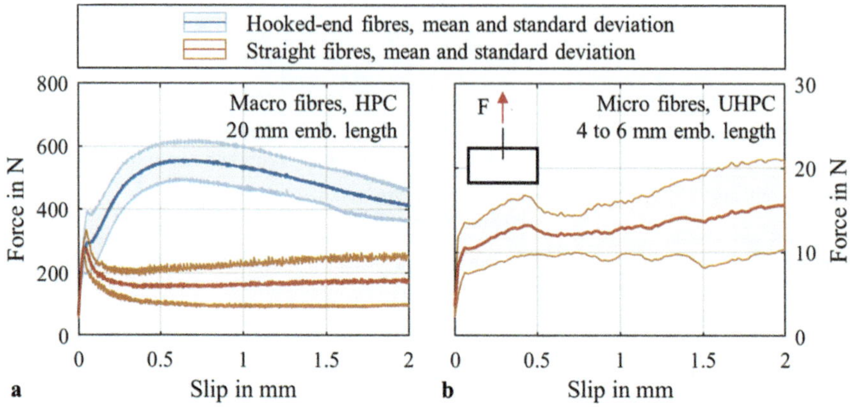

Fig. 1 Force-slip curves of vertically embedded **a** macro fibres with hooked-ends (blue) and cut end hooks (red) in HPC, cf. [1], and **b** straight micro fibres in UHPC

of the load-bearing behaviour prior to and after the overcoming of the adhesive fibre-matrix bond at about 0.05 to 0.1 mm of slip. Subsequent to this point the mechanical end anchorage is activated, leading to an increase in maximum force for hooked-end fibres.

In principle, it becomes apparent that the maximum pull-out forces of the hooked-end macro fibres were only reached at slips of 0.6–0.8 mm. Given that this range is only relevant for the ultimate limit state of structural components and also represents extremely large damage values for modelling purposes, the force-slip development up to this point is of particular interest. The embedded fibre length has a greater overall influence on the force-slip curve before the maximum pull-out force is reached than at higher slip values. This is discussed in more detail in [4]. The test results were used as basis for a comparative study between several modelling approaches within the SPP 2020, see [5].

Figure 1b depicts the mean force-slip curve and standard deviation for micro fibres embedded vertically 4–6 mm into the concrete. For these specimens, comparable force-slip curves in terms of residual forces were reached during pull-out despite larger scattering of individual test results. The maximum pull-out forces for the micro fibres reached about 20 N, approximately a tenth of the values of the macro fibres. However, a larger number of fibres is present in concretes with micro fibres.

3.2 Phase-Field Model for Fracture in Fibre-Reinforced HPC

This section presents the formulation of the micro-mechanical model, which is calibrated based on experimental results from the single fibre pull-out tests. Virtual experiments are then conducted using the calibrated parameters, serving as the foundational basis for the development of a phenomenological material model. For the small strain elastic–plastic phase-field approximation of fracture the displacement field $\boldsymbol{u}(\boldsymbol{x}, t)$ and the phase-field parameter q, are considered, cf. [6]. The phase-field parameter $q \in [0, 1]$ characterizing undamaged $q = 0$ and the fully damaged $q = 1$ state of the material governs the regularised crack surface $\Gamma_l(q)$, i.e.,

$$\Gamma_l(q) = \int_{\mathcal{B}} \gamma(q, \nabla q)\mathrm{d}v \text{ with } \gamma(q, \nabla q) = \tfrac{1}{2l}q^2 + \tfrac{l}{2}\|\nabla q\|^2, \tag{1}$$

using the crack surface density function per unit volume of the solid $\gamma(q, \nabla q)$. The gradient of phase-field parameter q is denoted as ∇q, l is a length scale parameter.

The free energy function ψ is constructed as,

$$\psi(\boldsymbol{\varepsilon}, \boldsymbol{\varepsilon}^{\mathrm{p}}, \alpha, q, \nabla q) = \psi^{\mathrm{ep}}(\boldsymbol{\varepsilon}^{\mathrm{e}}, \alpha, q) + \psi^{\mathrm{c}} - g(q, m)\psi^{\mathrm{c}} + 2\tfrac{\psi^{\mathrm{c}}}{\zeta} l\gamma(q, \nabla q), \tag{2}$$

which depends on the specific critical fracture energy ψ^{c}, the equivalent plastic strain α, the degradation function $g(q, m) = (1 - q)^m$. The parameters m and ζ control the speed of damage evolution and the stress softening in the post critical region, respectively, cf. [5, 7–10]. Elastic strain tensor $\boldsymbol{\varepsilon}^{\mathrm{e}}$ depends on the plastic strain tensor $\boldsymbol{\varepsilon}^{\mathrm{p}}$ and the total strain tensor $\boldsymbol{\varepsilon}$. The relation is defined as $\boldsymbol{\varepsilon}^{\mathrm{e}} := \boldsymbol{\varepsilon} - \boldsymbol{\varepsilon}^{\mathrm{p}}$. An elastic energy ψ^{e} and a plastic energy ψ^{p} are parts of the elastic–plastic energy function ψ^{ep}, which reads

$$\psi^{\mathrm{ep}}(\boldsymbol{\varepsilon}^{\mathrm{e}}) = \psi^{\mathrm{e}}(\boldsymbol{\varepsilon}^{\mathrm{e}}) + \psi^{\mathrm{p}}(\alpha). \tag{3}$$

Therein, the elastic energy function ψ^{e}

$$\psi^{\mathrm{e}}(\boldsymbol{\varepsilon}^{\mathrm{e}}, q) = g(q, m)\psi_0^{\mathrm{e}+}(\boldsymbol{\varepsilon}^{\mathrm{e}}) + \psi_0^{\mathrm{e}-}(\boldsymbol{\varepsilon}^{\mathrm{e}}), \tag{4}$$

additively split using a positive part $\psi_0^{\mathrm{e}+}(\boldsymbol{\varepsilon}^{\mathrm{e}})$ and a negative part $\psi_0^{\mathrm{e}-}(\boldsymbol{\varepsilon}^{\mathrm{e}})$ of the reference energy function $\psi_0^{\mathrm{e}}(\boldsymbol{\varepsilon}^{\mathrm{e}})$, cf. [11], i.e.

$$\psi_0^{\mathrm{e}+}(\boldsymbol{\varepsilon}^{\mathrm{e}}) = \kappa\langle \mathrm{tr}[\boldsymbol{\varepsilon}^{\mathrm{e}}]\rangle_+^2/2 + \mu\|\mathrm{dev}\boldsymbol{\varepsilon}^{\mathrm{e}}\|^2 \text{ and } \psi_0^{\mathrm{e}-}(\boldsymbol{\varepsilon}^{\mathrm{e}}) = \kappa\langle \mathrm{tr}[\boldsymbol{\varepsilon}^{\mathrm{e}}]\rangle_-^2/2, \tag{5}$$

which depend on the shear modulus μ and the bulk modulus κ. Macaulay's notation describes the function $\langle x\rangle_\pm = 1/2(x \pm |x|)$. The considered plastic energy ψ^{p} reads

$$\psi^{\mathrm{p}}(\alpha) = g(q, m)\psi_0^{\mathrm{p}}(\alpha) \text{ where } \psi_0^{\mathrm{p}}(\alpha) = y_0\alpha + \tfrac{1}{2}h\alpha^2, \tag{6}$$

where y_0 and h denote the yield stress and hardening parameter, respectively.

The energy density function ψ is reformulated as

$$\psi = (1 - q)^m\left[\psi_0^{\mathrm{e}+} + \psi_0^{\mathrm{p}} - \psi^{\mathrm{c}}\right] + \psi_0^{\mathrm{e}-} + \psi^{\mathrm{c}} + 2\tfrac{\psi^{\mathrm{c}}}{\zeta} l[\tfrac{1}{2l}q^2 + \tfrac{l}{2}\|\nabla q\|^2]. \tag{7}$$

The equation of the stress tensor is

$$\boldsymbol{\sigma} := (1-q)^m \underbrace{\left[\kappa\langle\mathrm{tr}\boldsymbol{\varepsilon}^{\mathrm{e}}\rangle_+\mathbf{I} + 2\mu\,\mathrm{dev}\boldsymbol{\varepsilon}^{\mathrm{e}}\right]}_{\boldsymbol{\sigma}_0^+} + \underbrace{\left[\kappa\langle\mathrm{tr}\boldsymbol{\varepsilon}^{\mathrm{e}}\rangle_-\mathbf{I}\right]}_{\boldsymbol{\sigma}_0^-}, \tag{8}$$

where positive $\boldsymbol{\sigma}_0^+$ and negative $\boldsymbol{\sigma}_0^-$ are used as parts of the effective stress tensor $\boldsymbol{\sigma}_0 = \boldsymbol{\sigma}_0^+ + \boldsymbol{\sigma}_0^-$. The second-order identity tensor is denoted by the symbol $\mathbf{I}$. The value of the parameter $m = 2$ is chosen in formulating the governing equation for the phase-field parameter, which is

$$q - l^2\,\mathrm{Div}\big[\nabla q\big] - (1-q)\mathcal{H} = 0. \tag{9}$$

This choice ensures that the phase-field parameter q is bounded within the interval $q \in [0, 1]$. The irreversibility of fracture is ensured by a local history field $\mathcal{H}$, i.e.,

$$\mathcal{H} := \max_{\tilde{t}\in[0,t]} \quad \mathcal{H}_0\big(\mathbf{x}, \tilde{t}\big) \geq 0 \text{ where } \mathcal{H}_0 = \zeta\left\langle \frac{\psi_0^{\mathrm{e}+}(\boldsymbol{\varepsilon}^{\mathrm{e}})}{\psi^{\mathrm{c}}} + \frac{\psi_0^{\mathrm{p}}(\alpha)}{\psi^{\mathrm{c}}} - 1\right\rangle. \tag{10}$$

The quantity $\mathcal{H}_0$ is a dimensionless crack driving state function. For the loading–unloading process an update of the local history field at the time t_{n+1} is given as

$$\mathcal{H} = \begin{cases} \mathcal{H}_{0,\mathrm{n+1}} & \text{for } \mathcal{H}_{0,\mathrm{n+1}} > \mathcal{H}_{0,\mathrm{n}}\ , \\ \mathcal{H}_{0,\mathrm{n}} & \text{otherwise.} \end{cases} \tag{11}$$

To capture the distinct behaviour of concrete in tension and in compression, two different critical fracture energies are considered: $\psi_{\mathrm{t}}^{\mathrm{c}}$ for tension and $\psi_{\mathrm{c}}^{\mathrm{c}}$ for compression. They are differentiated based on the sign of the first invariant of the strain tensor, i.e.

$$\psi^{\mathrm{c}} = \begin{cases} \psi_{\mathrm{t}}^{\mathrm{c}} & \text{for } tr\,\varepsilon \geq 0\ , \\ \psi_{\mathrm{c}}^{\mathrm{c}} & \text{otherwise.} \end{cases} \tag{12}$$

The Drucker-Prager yield criterion, which accurately predicts the nonlinear behaviour of concrete in tension and compression, cf. [12], is used and it reads

$$\phi\big(\boldsymbol{\sigma}_0, \kappa_{\mathrm{p}}\big) = \tfrac{1}{\sqrt{2}}||\mathrm{dev}\boldsymbol{\sigma}_0|| + \beta_{\mathrm{p}}\mathrm{tr}\boldsymbol{\sigma}_0 - (y_0 + h\alpha). \tag{13}$$

For $\beta_{\mathrm{p}} = 0$, the von Mises yield criterion, see [12], can be used to capture the nonlinear behaviour of steel fibres. The presented model is implemented in the framework of the Finite Element Method. The staggered scheme documented in [6] and an adopted algorithm used in [13] are used to solve the weak forms of the balance of linear momentum using the stress tensor, see Eq. 8, and the governing equation for phase-field parameter, see Eq. 9, with incrementally decoupled updates, for details see [14, 15]. The implemented integration algorithm for the Drucker-Prager plasticity and the consistent elasto-plastic tangent moduli are documented in [14].

3.3 Numerical Simulations of Single Steel Fibre Pull-Out Tests

The fibre pull-out tests are simulated numerically to analyse the influence of the single embedded fibre on the material behaviour of HPC. In this procedure, calibration and validation of the presented numerical model in the Sect. 3.2 is done by comparing the numerical with experimental data taken from [1].

The boundary value problem which is based on the experimental pull-out test setup, taken from [5, 15], is shown in Fig. 2. The material parameters of steel fibres and HPC are taken from [1, 10]. The pull-out test with a fibre embedded length of $l_f = 20$ mm was simulated to calibrate the material parameters for the interface zone, see orange area shown in the Fig. 2. For the validation of the numerical model, a steel-fibre pull-out test with an embedded length of $l_f = 30$mm was simulated using the calibrated parameters. The numerical results, i.e. load–displacement diagrams and the distributions plots are shown in Fig. 3. The validation shows a good agreement with the experimental data. The efficiency and limitations of the presented model are discussed in-depth in [5, 15] (Fig. 4).

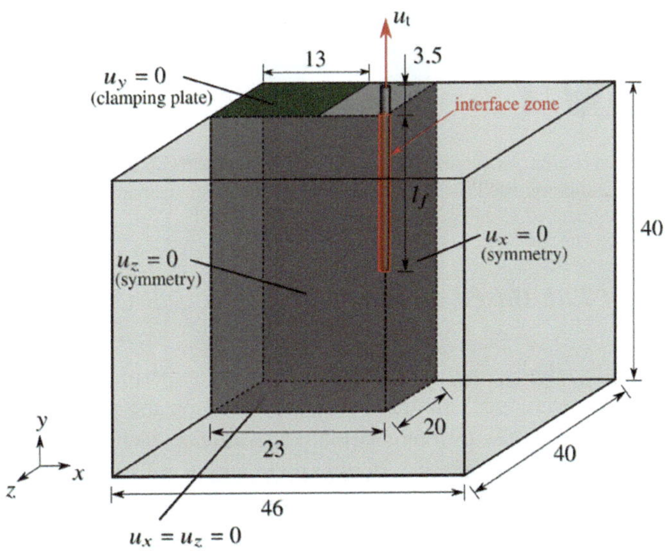

Fig. 2 Boundary value problem for steel-fibre pull-out tests based on the experimental assembly, taken from [5, 15]. All dimensions are in mm

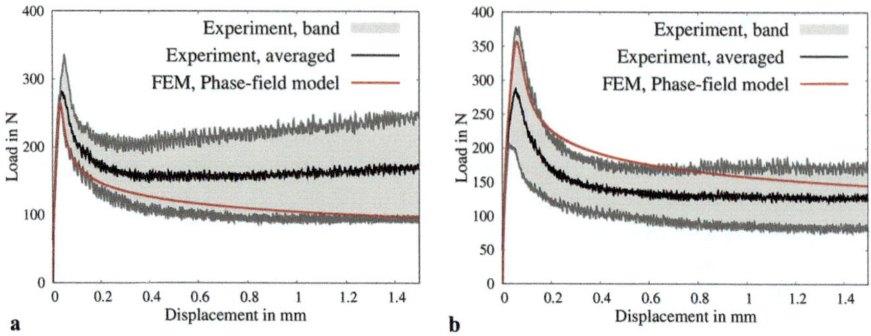

Fig. 3 Comparison of load–displacement diagrams of simulations with the experimental averaged curves and the bands characterizing the scattering of the experimental results, taken from [11], **a** $l_f = 20\,\text{mm}$ and **b** $l_f = 30\,\text{mm}$ embedded length, cf. [5, 15]

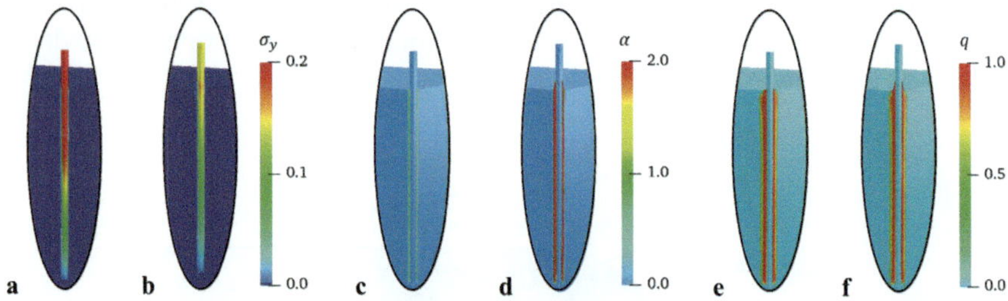

Fig. 4 Simulation results of steel fibre pull-out test for $l_f = 20\,\text{mm}$: stresses σ_y (in GPa), equivalent plastic strains α and phase-field parameter q in **a**, **c**, **e** at the displacement of $u_t = 0.08$ mm and **b**, **d**, **f** at the displacement of $u_t = 0.8$ mm, respectively, cf. [5, 15]

3.4 Virtual-Lab Based on Ellipsoidal Unit Cell Calculation

The single-scale microscopic simulations of virtual experiments were conducted using the micro-mechanical model of fracture based on the phase-field approach, presented in Sect. 3.2. Six different virtual experiments, each varying in boundary conditions, are simulated. These virtual experiments were based on an ellipsoidal unit cell of HPC, which includes the resolution of fibres and serves as a representative volume element (RVE), as illustrated in Fig. 5a. The ellipsoidal RVE, which characterises the material behaviour along the preferred fibre direction, is constructed with the consideration that fibres occupy 0.3% of the total volume of the unit cell, corresponding to a fibre content of 23 kg/m^3 in reinforced HPC, see [16–18]. The heterogeneity in the concrete matrix is accounted for by applying random numerical perturbations of up to 20% to the material parameters of the HPC matrix, as detailed in [15, 19, 20]. A macroscopic homogeneous strain state is considered and applied to the microscopic boundary value problem based on the ellipsoidal RVE, see

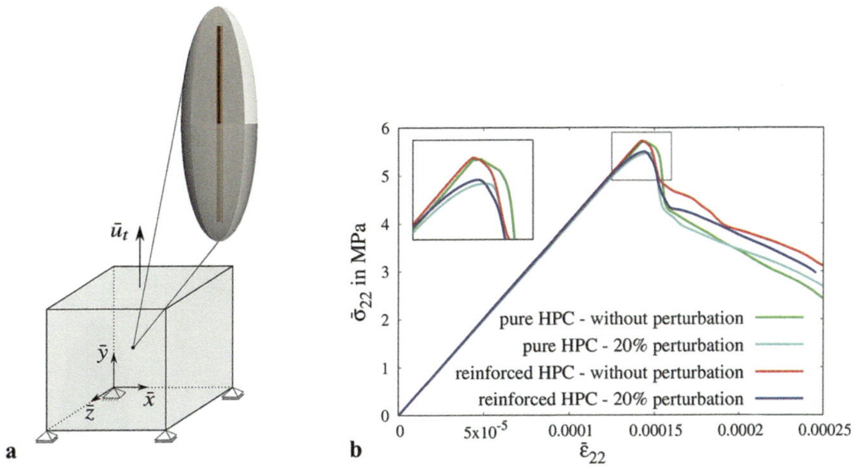

Fig. 5 Virtual experiment I: uniaxial tensile test without transverse stresses, **a** boundary value problem (RVE), **b** comparison of macroscopic stress–strain characteristic using pure and reinforced ellipsoidal RVE without perturbation and with 20% perturbation (from [15], licensed under CC-BY 4.0)

Figs. 5a, b. The macroscopic quantities are determined using the volumetric averages of the microscopic quantities over the RVE and plotted to study the effective material behaviour of fibre-reinforced HPC, see Fig. 5c. Details on the construction of the RVE and the numerical homogenization procedure are documented in [15].

Figure 5 and 6 illustrate an exemplary boundary value problem and the simulation results of the virtual experiment, which considers a uniaxial tensile test along preferred fibre direction with vanishing transverse stresses (only $\overline{\sigma}_{22} \neq 0$). Four simulations were conducted using different types of ellipsoidal RVEs: ellipsoidal RVEs of pure HPC with and without perturbation and ellipsoidal RVEs of reinforced HPC with and without perturbation of HPC parameters. For more details for all types of the virtual experiments, see [15].

4 Bending Beam Tests—Damage Evolution in Low-Cycle Fatigue

4.1 Setup and Procedure of Experiments

Based on the calibration of the numerical model for the ellipsoidal RVE and the simulation of the pull-out behaviour of individual fibres, the investigation of larger structures follows. Initially, quasi-static flexural tensile swelling tests were carried out on notched three-point bending beams in deformation control.

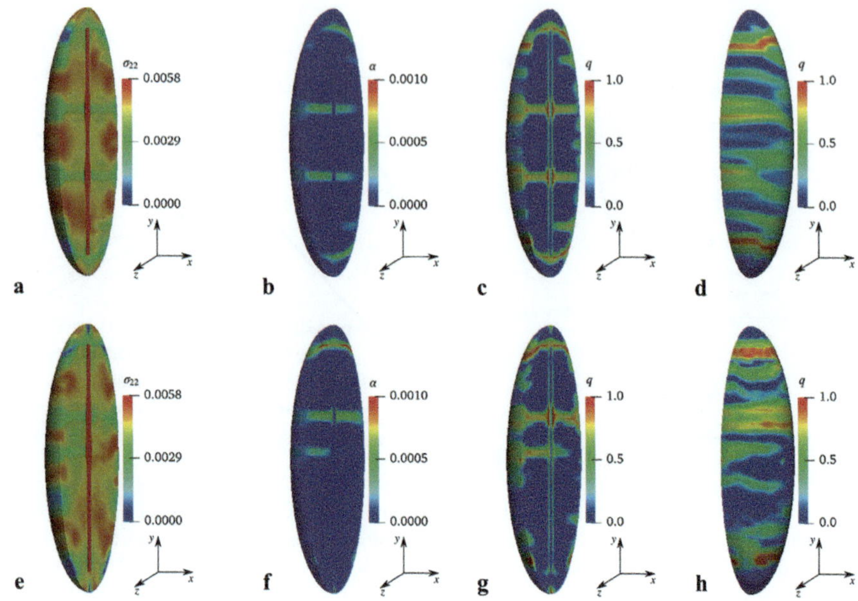

Fig. 6 Virtual experiment I: uniaxial tensile test without transverse stresses using ellipsoidal RVE of reinforced HPC, **a–d** HPC parameters without perturbation at $\overline{\sigma}_{22} = 0.00450$ GPa and **e–h** with 20% perturbation at $\overline{\sigma}_{22} = 0.00417$ GPa: distribution of **a**, **e** microscopic stress σ_{22} in GPa, **b**, **f** microscopic equivalent plastic strain α and microscopic phase-field parameter σ_{22} **c**, **g** cross-sectional view and **d**, **h** front view at macroscopic strain $\overline{\varepsilon}_{22} = 0.175$ ‰ (from [15], licensed under CC-BY 4.0)

All tests were conducted on HPC and UHPC bending beams with dimensions of 150 × 150 × 700 mm. As described in Sect. 2, the HPC beams contained 0/23/57/115 kg/m^3 macro steel fibres, equivalent to 0/0.3/0.75/1.5% by volume. The UHPC beams were tested with 0/57/115 kg/m^3 of micro steel fibres, representing 0/0.75/1.5% by volume.

The test procedure, which was based on EN 14651 [21], was augmented by unloading–reloading cycles in order to be able to comprehensively describe the changes in residual material properties during the transition of the material from the uncracked to the cracked state and the subsequent activation of the fibre load-bearing capacity. The test setup and procedure are shown schematically in Fig. 7.

To ensure clear data for both the behaviour of the concrete matrix before and after the formation and opening of the macro crack and the associated redistribution of tensile forces to the steel fibres, the first 0.05 mm of crack opening were investigated with a particularly fine resolution of one unloading cycle every 0.01 mm of crack growth. For the investigation of the behaviour with very large crack openings of up to 4 mm, an additional influence due to too many unloading cycles was to be avoided. Therefore, once a crack opening of 0.05 mm was reached, larger increases in crack openings were applied before starting the next unloading cycle.

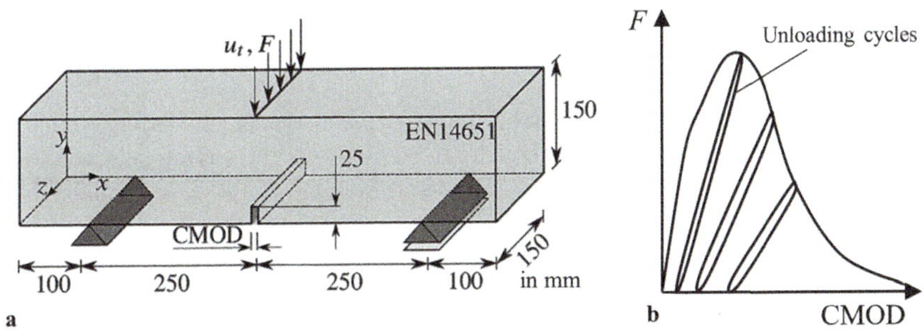

Fig. 7 Schematic depiction of **a** the test set-up according to EN 14651 [21] (from [4], licensed under CC-BY 4.0) and **b** test procedure with unloading cycles

4.2 Damage Indicators—Definition and Discussion

One major objective of this research was to identify and validate suitable macroscopic damage indicators in order to describe and subsequently model the damage process. This allows for a more accurate description of the degradation of the material properties, in addition to modelling only the envelope curve. As presented in [4] and [22], several indicators are comprehensively taken into account for this purpose, as illustrated in Fig. 8.

The main indicators analysed for the i-th load cycle are the plastic fraction of the crack opening $CMOD_{pl,i}$, the residual stiffness s_i and the damage energy $W_{pl,i}$. All indicators are set as functions depending on the total crack opening $CMOD_{tot}$ at the point of unloading. The results demonstrate that all damage indicators are suitable for resolving differences in degradation behaviour as functions of fibre content. Out of these indicators, the residual stiffness proved particularly suitable for calibration of numerical models. As the secant modulus s_i of any given hysteresis loop i it equates to

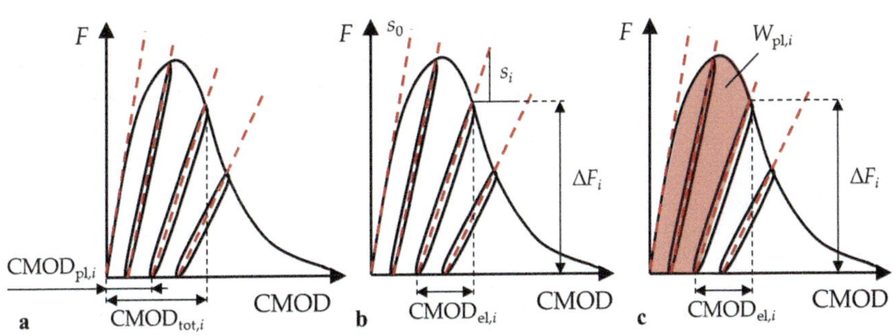

Fig. 8 Definition of damage indicators **a** plastic fraction of crack-mouth opening displacement $CMOD_{pl}$, **b** residual stiffness s and **c** damage energy W_{pl}, depicted for the second unloading cycle (from [4], licensed under CC-BY 4.0)

$$s_i = \frac{\Delta F_i}{\text{CMOD}_{\text{tot},i} - \text{CMOD}_{\text{pl},i}}, \tag{14}$$

where ΔF_i corresponds to the difference between the load at the unloading point and the minimum load of the unloading cycle. In addition, for the numerical modelling, a dimensionless relative damage value is derived from the stiffness by normalization to the initial stiffness s_0, i.e.

$$q_i = \frac{s_0 - s_i}{s_0}. \tag{15}$$

The direct formulation of a general damage parameter for the numerical approach rendered stiffness a widely employed damage indicator. In addition, while the data shows considerable scatter, as is usual for such tests, the stiffness development provides good selectivity between different types of concrete and fibre contents. The results can also be transferred to the behaviour of the concretes in high-cycle fatigue tests, which is further discussed in Sect. 6.1. Figure 9 shows an example of the development of stiffness in HPC depending on the fibre content.

Several different phases can be identified in the stiffness curve. The first load cycle provides the initial stiffness, which is approx. 1500 kN/mm for the HPC specimens, regardless of the fibre content. This constant stiffness at the beginning can be explained by the constant stiffness of the concrete matrix for all concrete compositions. The effects of fibres become relevant after the initiation of the crack. Starting from a crack opening of approx. 0.02 mm, the concrete matrix begins to degrade rapidly as the fibres start to be activated. Up to a crack opening of about 0.1 mm, the progression arranges according to the fibre content. Generally, higher fibre contents lead to higher residual stiffness. For larger crack openings from approx. 1 mm onwards, this behaviour weakens and the individual curves converge again. Strain-hardening behaviour is observable for both HPC and UHPC for all specimens with a fibre content of 57 and 115 kg/m^3 after reaching the limit of proportionality.

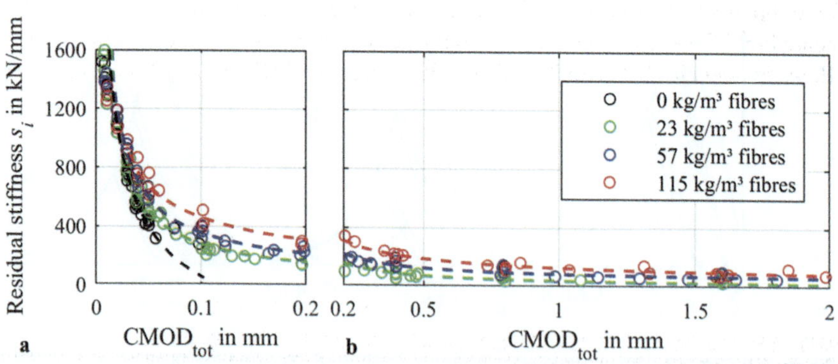

Fig. 9 Development of residual stiffness, HPC: **a** zoomed view of initial crack formation and **b** view of development for large crack openings, (adapted from [4], licensed under BY-CC 4.0)

4.3 Phenomenological Material Model for Fibre-Reinforced HPC

4.3.1 Constitutive Framework

To represent the heterogeneity in the steel fibre-reinforced HPC, an additive structure of a macroscopic stored energy function per unit volume ψ is formulated using energy functions ψ^{HPC} for the HPC phase and ψ^{F} for the embedded fibres, i.e.,

$$\psi = \mathrm{v}^{\mathrm{HPC}}\psi^{\mathrm{HPC}}\left(\boldsymbol{\varepsilon}, \boldsymbol{\varepsilon}^{\mathrm{p,HPC}}, q, \nabla q, \alpha^{\mathrm{HPC}}\right) + \mathrm{v}^{\mathrm{F}}\psi^{\mathrm{F}}\left(\boldsymbol{\varepsilon}, \mathrm{M}, \alpha^{\mathrm{F}}\right). \tag{16}$$

As observed in the experiments, fracture occurs only in the HPC phase, which is described by the phase-field parameter $q \in [0, 1]$. The volume fraction $\mathrm{v}^{\mathrm{HPC}}$ of the HPC phase and the volume fraction v^{F} of fibres are conserved by the condition $\mathrm{v}^{\mathrm{HPC}} = 1 - \mathrm{v}^{\mathrm{F}}$. The total macroscopic strain tensor $\boldsymbol{\varepsilon}$ contains the elastic strains $\boldsymbol{\varepsilon}^{\mathrm{e,HPC}}$ and the plastic strains $\boldsymbol{\varepsilon}^{\mathrm{p,HPC}}$ in HPC phase, i.e. $\boldsymbol{\varepsilon}^{\mathrm{e,HPC}} := \boldsymbol{\varepsilon} - \boldsymbol{\varepsilon}^{\mathrm{p,HPC}}$. An energy function ψ^{HPC} for the HPC phase is constructed using the Eqs. 1 and 2, cf. [4, 14, 15],

$$\psi^{\mathrm{HPC}} = g\left[\psi_0^{\mathrm{e+,HPC}} + \psi_0^{\mathrm{p,HPC}} - \psi^{\mathrm{c,HPC}}\right] + \psi_0^{\mathrm{e-,HPC}} + \psi^{\mathrm{c,HPC}} + \frac{2\psi^{c}l}{\zeta}\gamma(q, \nabla q), \tag{17}$$

where specific critical fracture energy $\psi^{\mathrm{c,HPC}} > 0$ controls the crack threshold in the HPC phase. Therein, the positive $\psi_0^{\mathrm{e+,HPC}}\left(\boldsymbol{\varepsilon}^{\mathrm{e,HPC}}\right)$ and negative $\psi_0^{\mathrm{e-,HPC}}\left(\boldsymbol{\varepsilon}^{\mathrm{e,HPC}}\right)$ reference elastic energy functions are, respectively,

$$\begin{aligned} \psi_0^{\mathrm{e+,HPC}}\left(\boldsymbol{\varepsilon}^{\mathrm{e,HPC}}\right) &= \kappa\langle \mathrm{tr}[\boldsymbol{\varepsilon}^{\mathrm{e,HPC}}]\rangle_+^2/2 + \mu\|\mathrm{dev}\boldsymbol{\varepsilon}^{\mathrm{e,HPC}}\|^2, \\ \psi_0^{\mathrm{e-,HPC}}\left(\boldsymbol{\varepsilon}^{\mathrm{e,HPC}}\right) &= \kappa\langle \mathrm{tr}[\boldsymbol{\varepsilon}^{\mathrm{e,HPC}}]\rangle_-^2/2, \end{aligned} \tag{18}$$

formulated using the Lamé coefficient μ and the bulk modulus κ for the HPC phase and the Macaulay's notation, i.e. $\langle x\rangle_\pm = 1/2(x \pm |x|)$. A plastic energy $\psi^{\mathrm{p,HPC}}$, which depends on the equivalent plastic strain α^{HPC}, yield stress y_0^{HPC} and hardening parameter h^{HPC} for HPC phase, reads

$$\psi^{\mathrm{p,HPC}}\left(\alpha^{\mathrm{HPC}}\right) = g(q)[y_0^{\mathrm{HPC}}\alpha^{\mathrm{HPC}} + \tfrac{1}{2}h^{\mathrm{HPC}}(\alpha^{\mathrm{HPC}})^2]. \tag{19}$$

An energy function ψ^{F} is constructed using the total strain tensor $e^{\mathrm{F}} = \boldsymbol{\varepsilon} : \boldsymbol{M}$ for fibres. It reads.

$$\psi^{\mathrm{F}}\left(\boldsymbol{\varepsilon}, \boldsymbol{M}, e^{\mathrm{p,F}}, \alpha^{\mathrm{F}}\right) = \psi^{\mathrm{e,F}}\left(\boldsymbol{\varepsilon}, \boldsymbol{M}, e^{\mathrm{p,F}}\right) + \psi^{\mathrm{p,F}}\left(\alpha^{\mathrm{F}}\right) \quad \text{where} \quad e^{\mathrm{e,F}} = e^{\mathrm{F}} - e^{\mathrm{p,F}}. \tag{20}$$

wherein $e^{\mathrm{e,F}}$ and $e^{\mathrm{p,F}}$ are the one-dimensional elastic and plastic strains along the fibre direction, respectively. Here, the structural tensor $\boldsymbol{M} := \boldsymbol{a} \otimes \boldsymbol{a}$ is used, cf. [23], which facilitates the description of the behaviour of the embedded steel fibre aligned in the preferred direction $\boldsymbol{a}$ with $||\boldsymbol{a}|| = 1$, see [15]. An elastic energy function $\psi^{\mathrm{e,F}}$ and a plastic energy function $\psi^{\mathrm{p,F}}$ for fibres are, respectively,

$$\psi^{\mathrm{e,F}}\left(\boldsymbol{\varepsilon}, \boldsymbol{M}, e^{\mathrm{p,F}}\right) = \frac{1}{2} E^{\mathrm{F}}\left(e^{\mathrm{F}} - e^{\mathrm{p,F}}\right)^2, \quad \text{and} \quad \psi^{\mathrm{p,F}}\left(\alpha^{\mathrm{F}}\right) = y_0^{\mathrm{F}} \alpha^{\mathrm{F}} + \frac{1}{2} h^{\mathrm{F}}\left(\alpha^{\mathrm{F}}\right)^2, \tag{21}$$

where the elastic moduli E^{F}, initial yield stress y_0^{F}, equivalent plastic strain α^{F} and hardening parameter h^{F} for fibres are considered. The total stress tensor is computed using Eq. 16, i.e.

$$\boldsymbol{\sigma} := \mathrm{v}^{\mathrm{HPC}} \boldsymbol{\sigma}^{\mathrm{HPC}} + \mathrm{v}^{\mathrm{F}} \boldsymbol{\sigma}^{\mathrm{F}}. \tag{22}$$

The stress tensor $\boldsymbol{\sigma}^{\mathrm{HPC}}$ for the HPC phase is formulated as

$$\boldsymbol{\sigma}^{\mathrm{HPC}} = g(q) \underbrace{\left[\mathrm{tr}\boldsymbol{\sigma}_0^{+,\mathrm{HPC}} \mathbf{I} + \mathrm{dev}\boldsymbol{\sigma}_0^{\mathrm{HPC}}\right]}_{\boldsymbol{\sigma}_0^{+,\mathrm{HPC}}} + \underbrace{\left[\mathrm{tr}\boldsymbol{\sigma}_0^{+,\mathrm{HPC}} \mathbf{I}\right]}_{\boldsymbol{\sigma}_0^{-,\mathrm{HPC}}}, \tag{23}$$

using the given quantities, i.e. $\mathrm{tr}\,\boldsymbol{\sigma}_0^{+,\mathrm{HPC}} = \kappa\langle\mathrm{tr}\boldsymbol{\varepsilon}^{\mathrm{e,HPC}}\rangle_+$, $\mathrm{dev}\boldsymbol{\sigma}_0^{\mathrm{HPC}} = 2\mu\,\mathrm{dev}\boldsymbol{\varepsilon}^{\mathrm{e,HPC}}$, and $\mathrm{tr}\,\boldsymbol{\sigma}_0^{-,\mathrm{HPC}} = \kappa\langle\mathrm{tr}\,\boldsymbol{\varepsilon}^{\mathrm{e,HPC}}\rangle_-$. Therein, $\boldsymbol{\sigma}_0^{\mathrm{HPC}} = \boldsymbol{\sigma}_0^{+,\mathrm{HPC}} + \boldsymbol{\sigma}_0^{-,\mathrm{HPC}}$ is the effective stress tensor for the HPC phase. The fibre phase stress tensor $\boldsymbol{\sigma}^{\mathrm{F}}$ is formulated as

$$\boldsymbol{\sigma}^{\mathrm{F}} := \sigma^{\mathrm{F}} \boldsymbol{a} \otimes \boldsymbol{a} \quad \text{with} \quad \sigma^{\mathrm{F}} = E^{\mathrm{F}}\left(e^{\mathrm{F}} - e^{\mathrm{p,F}}\right). \tag{24}$$

The governing equation for the phase-field parameter q is computed similarly as in Eq. 9. It reads

$$q - l^2\,\mathrm{Div}\,(\nabla q) - (1-q)\mathcal{H}^{\mathrm{HPC}} = 0, \tag{25}$$

using a local history field $\mathcal{H}^{\mathrm{HPC}}$ with a dimensionless crack driving state function $\mathcal{H}_0^{\mathrm{HPC}}$ for the HPC phase, i.e.

$$\mathcal{H}^{\mathrm{HPC}} := \max_{s\in[0,t]} \mathcal{H}_0^{\mathrm{HPC}}(\mathbf{x}, t) \geq 0 \quad \text{with} \quad \mathcal{H}_0^{\mathrm{HPC}} = \zeta\left\langle \frac{\psi_0^{\mathrm{e+,HPC}} + \psi_0^{\mathrm{p,HPC}}}{\psi^{\mathrm{c,HPC}}} - 1 \right\rangle. \tag{26}$$

The Drucker-Prager yield criterion is used to describe the non-linear behaviour in the HPC phase. The yield criterion $\phi_{\mathrm{p}}^{\mathrm{HPC}}$ and a plastic potential $\phi_{\mathrm{n}}^{\mathrm{HPC}}$ are given, see [12], as

$$\phi_{\mathrm{p}}^{\mathrm{HPC}} = \frac{1}{\sqrt{2}}\left|\left|\mathrm{dev}\boldsymbol{\sigma}_0^{\mathrm{HPC}}\right|\right| - \beta_{\mathrm{p}}\mathrm{tr}\boldsymbol{\sigma}_0^{\mathrm{HPC}} - y_0^{\mathrm{HPC}} - h^{\mathrm{HPC}}\alpha^{\mathrm{HPC}}, \tag{27}$$

$$\text{and} \quad \phi_{\mathrm{n}}^{\mathrm{HPC}} = \frac{1}{\sqrt{2}}\left|\left|\mathrm{dev}\boldsymbol{\sigma}_0^{\mathrm{HPC}}\right|\right| - \beta_{\mathrm{n}}\mathrm{tr}\boldsymbol{\sigma}_0^{\mathrm{HPC}}, \tag{28}$$

where β_{p} and β_{n} are the material parameters. The one-dimensional von Mises yield criterion is used to capture non-linear behaviour along the preferred fibre direction, cf. [24, 25], i.e.

$$\phi^{\mathrm{F}}\left(\sigma^{\mathrm{F}}, \kappa_{\mathrm{p}}^{\mathrm{F}}\right) = \left|\sigma^{\mathrm{F}}\right| - \kappa_{\mathrm{p}}^{\mathrm{F}} \quad \text{with} \quad \kappa_{\mathrm{p}}^{\mathrm{F}} := \partial_{\alpha^{\mathrm{F}}}\psi^{\mathrm{p,F}} = y_0^{\mathrm{F}} + h^{\mathrm{F}}\alpha^{\mathrm{F}}. \tag{29}$$

Note that in the described formulation of the phenomenological material model, plastic flow occurs independently in both the HPC phase and the fibres. For the details on the overall consistent elasto-plastic tangent modulus, algorithm integration for both used plasticity criteria, and implementation of the presented formulation in the framework of the Finite Element Method, please refer to [14, 15].

4.3.2 Calibration and Validation Using Unit Cell Calculations

The phenomenological material model is calibrated by conducting single-scale simulations of the macroscopic boundary value problems (BVPs) based on a cuboid made of both pure and reinforced HPC. In these simulations, the same homogeneous boundary conditions applied to the RVE for virtual experiments, see Sect. 3.4, are used. For details on all types of macroscopic BVPs, please refer to [15]. To achieve a fibre content of 23 kg/m^3 in the reinforced HPC, a volume fraction of $\mathrm{v}^{\mathrm{F}} = 0.003$ is chosen for the cuboid.

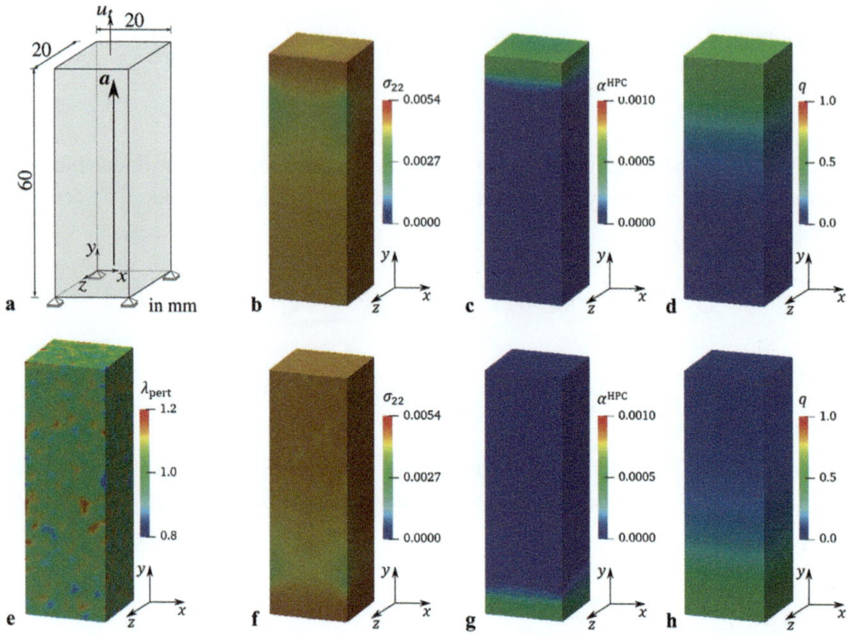

Fig. 10 Phenomenological material model applied to **a** macroscopic BVP—uniaxial tensile test with vanishing transverse stresses, **b–d** HPC parameters without perturbation at $\overline{\sigma}_{22} = 0.00435$ GPa and **e–h** with 20% perturbation at $\overline{\sigma}_{22} = 0.00432$ GPa: distribution of (**e**) perturbation parameter λ_{pert}, **b**, **f** total stress σ_{22} in GPa, **c**, **g** equivalent plastic strain α^{HPC} in HPC phase and **d**, **h** phase-field parameter q in HPC phase, at macroscopic strain $\overline{\varepsilon}_{22} = 0.175\text{‰}$ (from [15], licensed under CC-BY 4.0)

The preferred fibre direction a within the cuboid, which is aligned parallel to the y-axis, the material parameters for HPC phase and fibre and considered heterogeneity using 20% perturbations of HPC parameters are maintained consistent with that in the RVE, as shown in Fig. 10a. The boundary conditions and the geometry of a cuboid used for the simulations of the macroscopic BVP for a uniaxial tensile test with vanishing transverse stresses are shown in Fig. 10a. The resulting macroscopic stress–strain characteristics obtained from simulations of virtual experiments and macroscopic BVPs are compared for calibration and validation of the phenomenological material model, see Fig. 11a and b. Figure 10 shows the simulation results of macroscopic BVP using cuboid of reinforced HPC with and without perturbation of HPC parameters, for details see [15, 20].

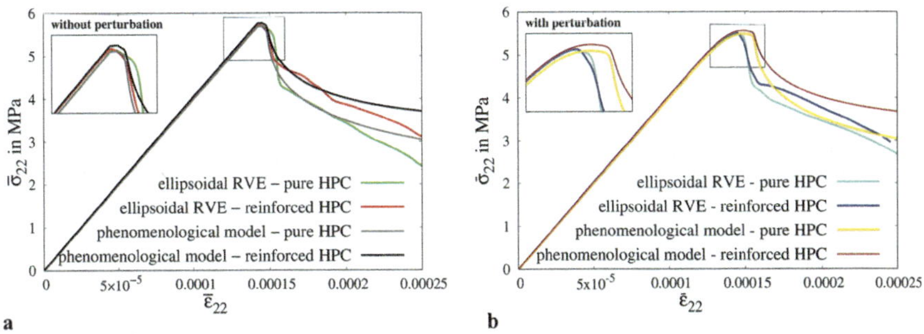

Fig. 11 Macroscopic BVP—uniaxial tensile test with vanishing transverse stresses: comparison of macroscopic stress–strain characteristic to results of virtual experiment, shown in Fig. 5, using ellipsoidal RVE, **a** without perturbation and **b** with 20% perturbation of HPC parameters (from [15], licensed under CC-BY 4.0)

5 Modelling of Failure Behaviour of HPCs During Low-Cycle Fatigue

To understand and further model the influence of fibres added in the HPC, the three-point bending beam tests at low-cycles were simulated using the phenomenological material model presented in Sect. 4.3. Two different continuous step-wise linearly approximated degradation functions for HPC in tension $g^+(q)$ and in compression $g^+(q)$, introduced in [14], are further implemented to model the distinct tension–compression failure behaviour of concrete.

The support points used for the interpolation of these degradation functions are calibrated by simulating the uniaxial tensile and compression tests as well as the three-point bending beam test for pure HPC, corresponding simulation results are documented in [14]. The initial guess of the value of the phase-field parameter q for the particular loading–unloading cycle determined using the measured residual stiffness in the experiments of the three-point bending test. This approach is explained in more detail in [4] and [14]. The geometry and the boundary conditions used for the three-point bending beam test with an example of a superimposed ODF are designed as per the European Standard EN 14651, see Fig. 7.

To understand the influence of fibres on failure behaviour of HPC, the three-point bending beam tests at low-cycle for reinforced HPC with different fibre distributions and contents were simulated. At first, the reinforced HPC beams with three different fibre contents of 23/57/115 kg/m^3 and the preferred fibre direction along x-axis corresponding to the beam axis were used for the simulations. Figure 12 shows the results of a three-point bending beam test at low-cycle of reinforced HPC with fibre content of 23 kg/m^3 and a preferred fibre direction along the x-axis.

The results are validated by comparing the degradation of stiffness, see Sect. 4.2 for the calculation, at each loading cycle in the numerical and experimental load-CMOD curves, see Fig. 12h. The results show good agreement with the experimental data, but the crack bridging effect could not be captured efficiently.

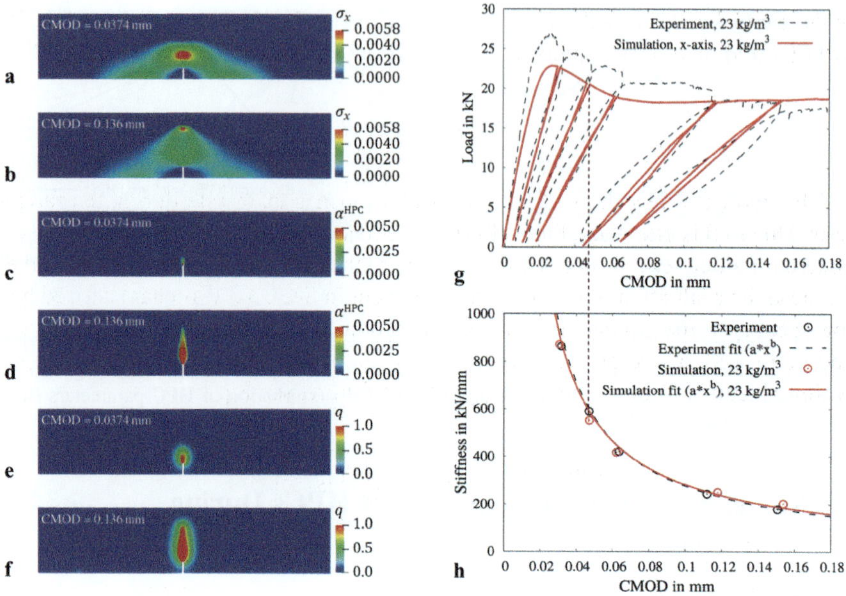

Fig. 12 Three-point bending beam test at low-cycle of fibre-reinforced HPC with fibre content of 23 kg/m^3 and a preferred fibre direction along x-axis: distribution of stresses σ_x (in GPa) in x-direction, the equivalent plastic strain α^{HPC} for HPC phase and the phase-field parameter q for HPC phase in **a, c, e** at CMOD = 0.0374 mm and in **b, d, f** at CMOD = 0.136 mm, respectively. Comparison of experimental data and simulated results: **g** load-CMOD diagram and **h** calculated and interpolated residual stiffness-CMOD diagram, taken from [4, 26, 27]

To incorporate different distributions and orientations of fibres, ODF using 5 and 24 preferred orientations lying in the x–y plane are constructed and implemented, see Fig. 13a and b, for details see [4, 26, 27].

The experimental load-CMOD curves, represented by the grey scatter bands, are compared with the numerical load-CMOD curves obtained from simulations of both a pure HPC beam and a fibre-reinforced HPC beam. These simulations consider various fibre distributions with fibre contents of 23 kg/m^3, 57 kg/m^3 and 115 kg/m^3, as shown in Fig. 14a–c. Figure 14d demonstrates the effect of increasing fibre content, which results in an increase in additional stiffness for the simulations considering a preferred fibre orientation along the x-axis.

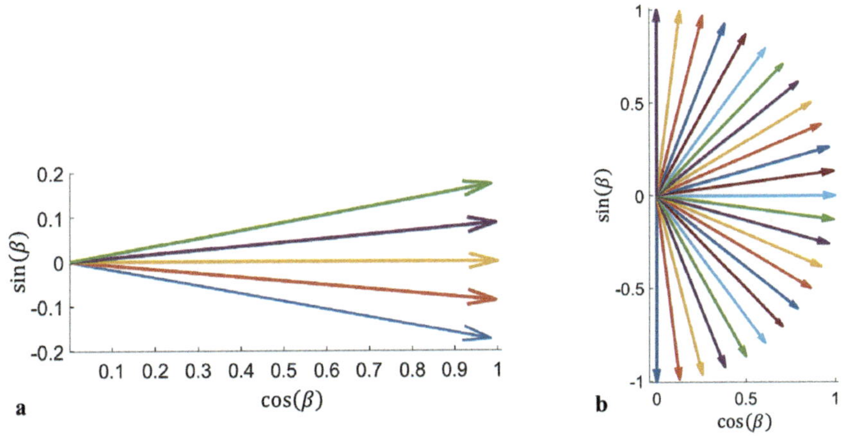

Fig. 13 Three-point bending beam cyclic test: orientation distribution functions (ODFs) for the reinforced steel fibres: **a** 5 preferred orientations between the values of angle β, i.e. $-10°$ to $10°$ and **b** 24 preferred orientations the values of angle β, i.e. $-90°$ to $90°$ (nearly isotropic), taken from [4, 26, 27]

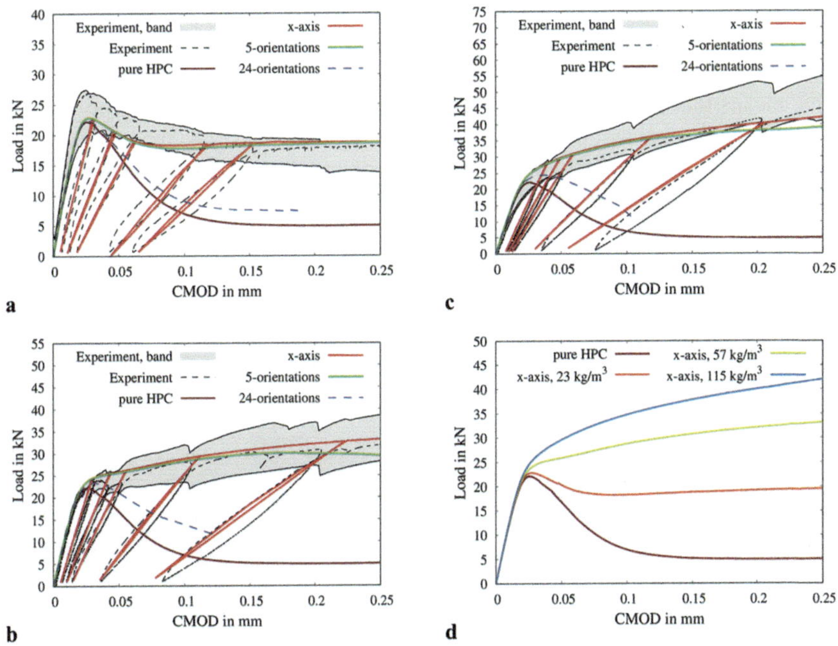

Fig. 14 Comparison of load-CMOD diagram for experimental data and simulation results: Three-point bending beam test of fibre-reinforced HPC with fibre content of **a** 23 kg/m^3, **b** 57 kg/m^3, **c** 115 kg/m^3 and **d** comparison of results using different fibre contents, taken from [4, 26, 27]

6 High-Cycle Fatigue and Lifetime Prediction of Bending Beams

6.1 *Damage Indicator Transfer Between Low- and High-Cycle*

Based on the indicators for damage development in low-cycle tests described in Sect. 4.2, their transferability to high-cycle fatigue tests was analysed. It was furthermore investigated whether the development of damage indicators in low-cycle and high-cycle is quantitatively comparable. Test specimens and test set-up corresponded to the previously described low-cycle tests. Various combinations of upper and lower load levels were applied for fatigue testing. Both test series with constant (single block) load levels and the influence of different load sequences, i.e. successive loads with different upper and lower load levels, were investigated. In the context of this chapter, results presented in [28] will be discussed in particular. Due to the qualitatively comparable results of the UHPC samples to HPC samples with regard to all damage indicators analysed, this contribution focuses on results of HPC samples.

Figure 15a and b provide a schematic overview of the analogy between the definitions of damage indicators for low-cycle tests presented in Sect. 4.2 and high-cycle tests. It has to be noted that the load levels are referred to the initial crack-force F_{init}. This means that for specimen containing super-critical fibre amounts, i.e. showing strain hardening behaviour after the initiation of the macro-crack, the load levels can exceed 100%. The overall good correlation in the CMOD-dependent development of the residual stiffness as shown exemplarily in Fig. 16 implies that the fatigue tests, which are by far more demanding in terms of both testing time and simulation complexity, can be well described using quasi-static low-cycle tests. While all indicators described show good selectivity with regard to the fibre content when applied to high-cycle tests, the development of residual stiffness in particular proves to be promising for a data-driven degradation description due to this correlation.

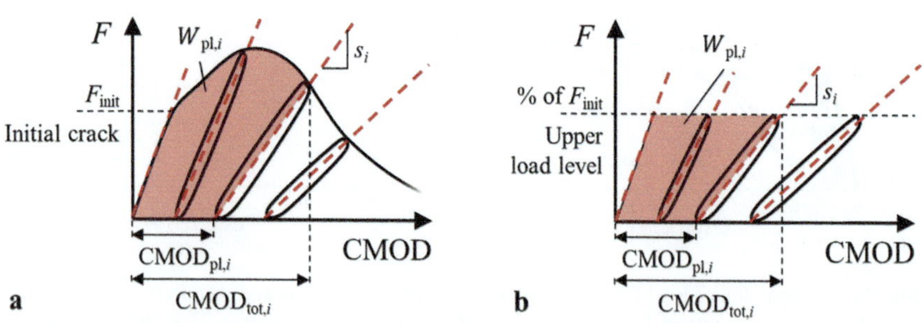

Fig. 15 Concept of transferability of damage indicators CMOD plastic $(\text{CMOD}_{\text{pl},i})$, residual stiffness (s_i) and damage energy $(W_{\text{pl},i})$ between **a** low-cycle and **b** high-cycle fatigue tests, reproduced from [28] (page 1328, Fig. 1)

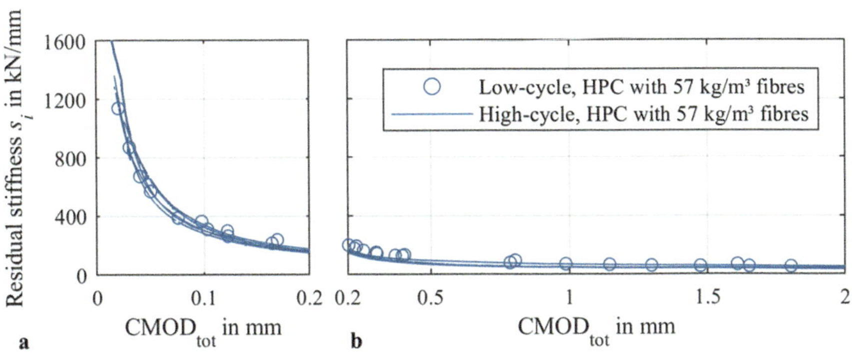

Fig. 16 Comparison of residual stiffness (s_i) as damage indicator in low-cycle (circles) and high-cycle (lines) tests, exemplarily for HPC specimens with 57 kg/m^3 fibres; **a** zoomed view of initial crack formation and **b** view of development for large crack openings

For the stiffness development, the magnitude of effects such as load level or loading history is much smaller compared to the influence of the fibre contents on the fatigue curve. This also applies to the influence of upper and lower load levels on the general degradation curve, both when applied unchanging as well as consecutively altered in block-wise testing. To illustrate this, Fig. 17a and b show the influence of different constant upper load levels applied as single block as well as tests with a block-wise increasing upper load level. Figure 17c and d give an exemplary comparison of the stiffness development of HPC specimens depending on the fibre content. Both depictions are described in more detail in [28].

As illustrated Fig. 17, the influence of the upper load level, both constant and increasing in blocks, is superseded by the influence of the fibres. In [28], it is shown that this is the case for all damage indicators discussed above as well as for HPC and UHPC. Fibre-free reference specimens reached a CMOD of approx. 0.05 mm and then failed immediately after the formation of the macro-crack. HPC specimens with a fibre content of 23 kg/m^3, which showed strain-softening behaviour in the previously described low-cycle tests, also failed at crack openings of about 0.1 to 0.15 mm. Generally, the shape of the curve is well comparable to the curve of fibre-free specimens. The difference in overall fatigue life for specimens with 23 kg/m^3 is due to the upper load level being coupled to the initial crack load F_{init} from low-cycle tests, which meant that larger crack openings could not be achieved in force-controlled high-cycle tests. The basic characteristics for all tests shown correspond to the behaviour described in Sect. 4.2 for quasi-static low-cycle tests. Especially for the range from 0.05 to approx. 1 mm crack opening, where the initial crack forms and load bearing mechanism changes from the matrix to the fibre, there is a clear differentiation of the development of residual stiffness according to fibre content.

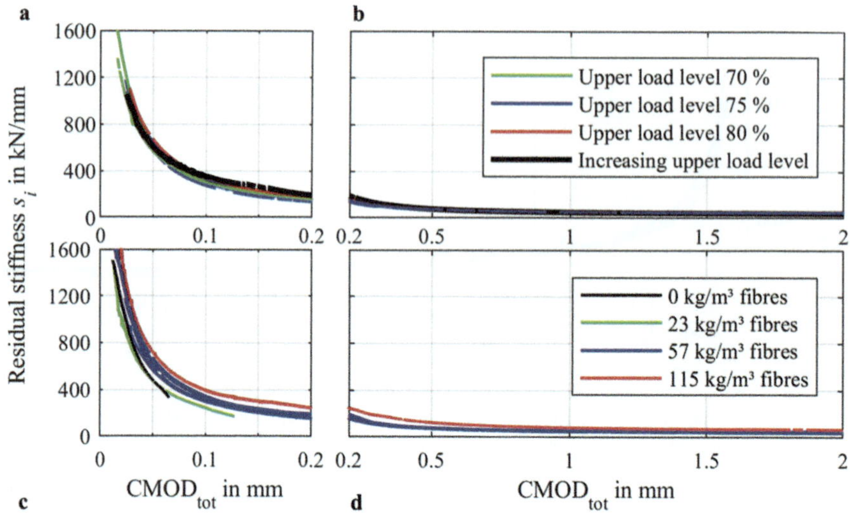

Fig. 17 Comparison of the influence of **a, b** upper load level and **c, d** fibre content on residual stiffness development (s_i); **a, c** zoomed view of initial crack formation and **b, d** view of development for large crack openings, reproduced from [28] (page 1329–1330, Figs. 4 and 5)

6.2 *Secondary Creep Phase Analysis*

While it is possible to carry out a limited number of load cycles experimentally in reasonable time, as in the case of the low-cycle tests described in Sect. 4.2, considerable computing time must be allocated for their simulation, depending on the complexity of the numerical model used. Although high-cycle tests can still be realised experimentally, they can no longer be explicitly simulated cycle by cycle. This results in a need for approaches to reduce computational effort. A possible basis for such a reduction are cycle jump approaches, as described e.g. in [29].

Although the representation of the CMOD to stiffness analogy considered in the previous Sect. 6.1 enables transferability between low-cycle and high-cycle tests, direct information on the number of load cycles up to fatigue is not included. Especially for long-term fatigue tests with low stress levels as well as run-out specimens which did not fail, it is important to know when the test has reached fatigue phase II, or secondary creep phase, as displayed in Fig. 18. This phase is known to provide a linear increase in damage per load cycle. Its gradient is furthermore known to be correlated to the number of cycles to failure, see e.g. [30].

Figure 18a schematically shows the well-known S-shaped three-phase progression of fatigue damage development to final failure. While the first and third phases are non-linear, the progression of damage development in phase II is approximately linear. In fatigue tests, this second phase accounts for between 70 and up to 90% of the total number of cycles, depending e.g. on the compressive strength of the concrete, see [31].

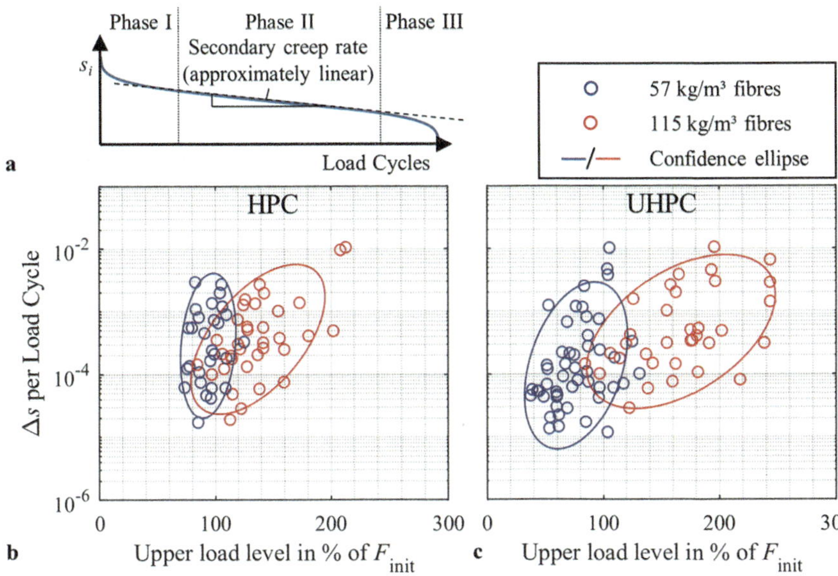

Fig. 18 Secondary creep phase analysis of stiffness deterioration rate Δs as **a** generalised schematic depiction, results for **b** HPC and **c** UHPC in high-cycle fatigue tests comprising different loading histories, cf. [32, 33]

Since the damage accumulation per cycle is constant in this range, areas of the curve can be extrapolated with reasonable precision. This results in enormous possible savings potential in the simulation of explicit load cycles.

In [32, 33], an approach is outlined that enables fatigue phase II to be identified even in ongoing high-cycle fatigue tests without knowing the point of final failure. The algorithm has been continuously developed since then in order to further improve the forecasting capacities.

Based on this, a comparison of the stiffness development in the secondary creep phase for HPC and UHPC samples as taken from [33] is shown in Fig. 18b and c respectively. Each point displayed indicates the gradient of the second creep phase, either of a single-stage test or one loading block in a multi-stage test. In addition, confidence ellipses are given as two-dimensional representations of the standard deviation of the tests. Only tests or blocks of tests with residual stiffness values lower than 500 kN/mm at the start of the secondary creep phase are included. From a structure-oriented perspective, this starting residual stiffness means that a macro crack is fully initiated and thus only the deterioration of fibre capacity as load-bearing mechanism is displayed. Additionally, tests that did not reach a stable secondary creep phase, i.e. due to imminent failure are excluded as well. Due to these limitations, no samples with 0 or 23 kg/m^3 fibres are included here, as their load-bearing capacity in the high-cycle tests is not sufficient to withstand at residual stiffnesses of less than 500 kN/mm over longer periods of time.

Evidently, an increased fibre content facilitates higher loading levels at a consistent degradation rate. This corresponds to the higher residual stiffness reported in Sect. 6.2. In addition, the applicable load levels for UHPC appear to be higher and the associated secondary creep rates lower compared to HPC. This is due to the difference in mass per individual fibre between the macro fibres used in HPC with a diameter of 0.92 mm and the micro fibres used in UHPC with a diameter of 0.19 mm. This means that the number of fibres bridging the crack and providing load-bearing capacity is significantly higher for UHPC. A quantitative differentiation of different loading histories using statistical approaches does not lead to noticeably different results. Therefore, the depiction represents all tests, including many different loading histories. This fact also explains parts of the scatter that is visible in Fig. 18. However, the results provide a possibility to predict the gradient of deterioration depending on the load level.

6.3 Definition of a Stiffness-Based Failure Criterion for Lifetime Prediction

As shown, the residual stiffness exhibits comparable degradation rates with respect to CMOD between high-cycle fatigue and low-cycle tests for comparable fibre contents, independent of upper load levels applied. This correlation allows the definition of a boundary criterion for the prediction of failure based on stiffness degradation. By comparing the residual stiffness observed during fatigue tests with stiffness degradation curves derived from low-cycle tests, it is possible to formulate a failure criterion that incorporates both, the residual stiffness at the final low-cycle load cycles and the corresponding relative load level at that stage. This approach is graphically outlined in the following Fig. 19.

Despite the scattering, there is a failure limit threshold for the stiffness achievable in fatigue up to failure for each fibre content. This threshold is found to be between 50 kN/mm and 100 kN/mm, provided the load-bearing capacity being high enough to reach this level in force-controlled fatigue tests. Here, this is particularly the case for concrete mixtures with high fibre contents exhibiting strain hardening behaviour. Due to the tests being run in a force-controlled manner, specimen with concrete mixtures with low fibre contents resulting in strain softening behaviour reach the envelope of the quasi-static tests shortly after crack initiation and thus fail in a few load cycles. Still, the envelop curves derived from the stiffness degradation in low-cycle tests can serve as failure criterion.

In Fig. 19, hollow markers represent residual stiffnesses in deformation controlled low-cycle tests, whereas filled markers represent failure in force-controlled high-cycle tests. The dashed lines indicate the stiffness-based failure curve derived from low-cycle tests (comparable to the stiffness degradation as displayed in Fig. 16. Failure of high-cycle fatigue tests lines up well at this given line. Furthermore, the solid red lines in Fig. 19) exemplarily show successive single load-steps of a

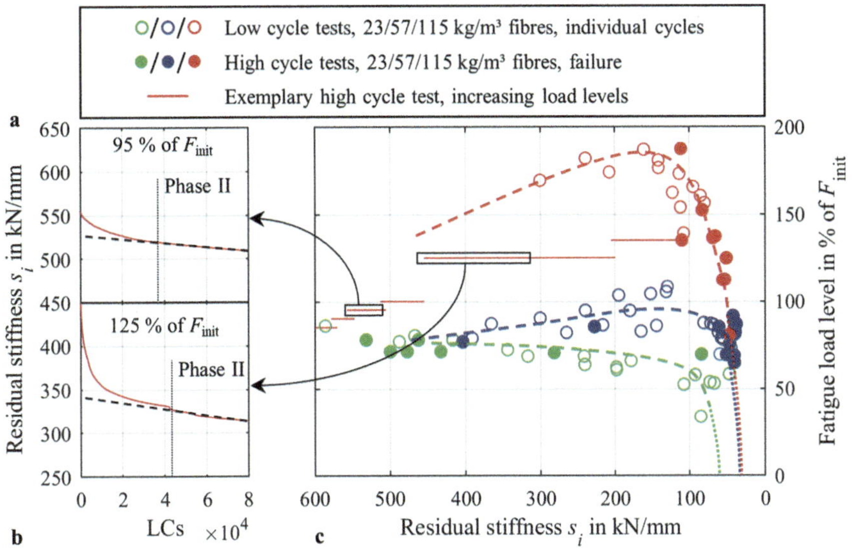

Fig. 19 Phase II stiffness degradation for an exemplary HPC specimen with 115 kg/m^3 fibres subjected to **a** 95% and **b** 125% of F_{init} and **c** a global failure criterion based on multiple low- and high-cycle tests

multi-step high-cycle fatigue tests with block-wise increasing upper load level. The development of stiffness of two load levels is shown in detail in the left part of Fig. 19. As could be expected, the lower load level of 95% of the first-crack load F_{init} shows a higher residual stiffness of about 525 kN/mm and a smaller gradient of stiffness degradation in fatigue phase II compared to the higher load level of 125% of the first-crack load. The gradients of the secondary creep phase correlates directly with the number of load cycles to failure. Here, the damage gradient in the second phase changes from − 2.2 kN/mm per 10^4 load cycles at 95% of F_{init} to −4.0 kN/mm per 10^4 load cycles at 125%. These gradients can finally be extrapolated against the stiffness-based failure curves derived from the low-cycle tests. Combining the findings presented in here and in Sect. 6.2, suitable lengths for cycle jumps based on experimental data can be derived, which can be used to reduce numerical effort in high-cycle simulations.

7 Conclusions

This chapter presents a summary of the principal findings of a comprehensive study on a scale-bridging methodology to combine short term experiments and numerical models to predict high-cycle fatigue of HPC and UHPC.

Initially, a combination of extensively tested and calibrated parameters from fibre pull-out tests, along with the virtual experiments based on the ellipsoidal RVEs incorporating structural heterogeneity, are used to achieve a realistic representation of steel fibre-reinforced HPC in the preferred fibre direction. These simulations results are used to develop and calibrate the phenomenological material model through the simulations of macroscopic BVPs.

At the macroscale, flexural beam tests were conducted and simulated using a phenomenological material model to further analyse the material behaviour. Both experimental and numerical approaches were designed to characterise and model the degradation in fatigue, taking into account different concretes, i.e. HPC and UHPC, as well as different types and quantities of steel fibres. It is shown, that different damage indicators are firstly applicable to differentiate damage development with respect to the fibre content. Secondly, the indicators are suitable to quantitatively describe damage development in quasi-static low-cycle tests while being transferable to time consuming high-cycle tests. However, focus is put on the residual stiffness as it is easily applicable to calibrate the numerical models when normalised into a dimensionless damage parameter.

For the phenomenological material model on the macroscale, an additive formulation of energy functions and the ODF for the fibres allows for the simulation of failure in both plain and fibre-reinforced HPC across different fibre contents, distributions and orientations with minimal adjustments. Moreover, the introduction of two distinct step-wise linear degradation functions enables precise control over stress degradation, based on the damage parameter's evolution, allowing for an accurate representation of the degradation behaviour of cracked sections in HPC and UHPC. Based on this, comparisons between experimental and numerical load-CMOD curves for the three-point bending beam tests at low-cycles with varying fibre contents further confirm the model's ability to replicate experimental responses—not only the load-CMOD curve, but also stiffness degradation in unloading–reloading cycles.

Regarding high-cycle fatigue, it is shown that the fibre content significantly affects the stiffness degradation function. In contrast, the upper load level and different histories of block-wise loading does not affect damage development to a statistically significant level. These statements do apply for HPC, UHPC and the types and contents of fibres tested.

A general procedure based on the experimental results is outlined to determine the onset of the linear secondary creep phase based solely on the previous cycles. Having in mind that the gradient of deterioration in the linear phase II correlates to the number of cycles to failure, this approach facilitates accurate failure prediction and can be used to implement cycle-jumping techniques in numerical models generally. Having further in mind that damage development in low-cycle fatigue and high-cycle fatigue are quantitatively comparable, a degradation curve for stiffness reduction can be derived from low-cycle tests. This curve can be used as a failure criterion against which damage development in high-cycle fatigue can be calculated to predict either damage accumulation for a given number of cycles (cycle-jump) or to predict final failure. Future research should incorporate full 3D simulations of failure and realistic fibre distributions, which are crucial for advancing a more comprehensive understanding of the material's behaviour. Additionally, the next step is to elaborate the ability of the presented numerical model to simulate high cycle fatigue tests of reinforced HPC using the cycle jump technique.

References

1. Gebuhr, G., Pise, M., Sarhil, M., Anders, S., Brands, D., Schröder, J.: Analysis and evaluation of the pull-out behavior of hooked steel fibers embedded in high and ultra-high performance concrete for calibration of numerical models. Struct. Concr. **20**, 1254–1264 (2019)
2. Gebuhr, G., Pise, M., Sarhil, M., Anders, S., Brands, D., Schröder, J.: Description and analyses of pull-out behaviour of hooked steel fibres embedded in high performance concrete using phase-field modeling. In: Derkowski, W., et al. (eds.) Concrete Innovations in Materials, Design and Structures, 16th fib Symposium Proceedings. FIB CEB-FIP (2019)
3. Gudžulić, V., Neu, G.E., Gebuhr, G., Meschke, G., Anders, S.: Numerisches Mehrebenen-Modell für Stahlfaserbeton: Von der Faserzur Bauteilebene. Betonund Stahlbetonbau **115**(2), 146–157 (2020). https://doi.org/10.1002/best.201900067
4. Gebuhr, G., Pise, M., Anders, S., Brands, D., Schröder, J.: Damage evolution of steel fiber reinforced high performance concrete in low-cycle flexural fatigue: numerical modeling and experimental validation. Materials **15**(3), 1179 (2022). https://doi.org/10.3390/ma15031179
5. Storm, J., Pise, M., Brands, D., Schröder, J., Kaliske, M.: A comparative study of micro-mechanical models for fiber pullout behavior of reinforced high performance concrete. Eng. Fract. Mech. **243**, 107506 (2021). https://doi.org/10.1016/j.engfracmech.2020.107506
6. Miehe, C., Hofacker, M., Welschinger, F.: A phase field model for rate-independent crack propagation: robust algorithmic implementation based on operator splits. Comput. Methods Appl. Mech. Eng. **199**, 2765–2778 (2010)
7. Kuhn, C., Schlüter, A., Müller, R.: On degradation functions in phase field fracture models. Comput. Mater. Sci. **108**, 374–384 (2015)
8. Borden, M.J., Hughes, T., Landis, C.M., Anvari, A., Lee, I.J.: A phase-field formulation for fracture in ductile materials: finite deformation balance law derivation, plastic degradation, and stress triaxiality effects. Int. J. Numer. Meth. Eng. **312**, 130–166 (2016)
9. Miehe, C., Aldakheel, F., Raina, A.: Phase field modeling of ductile fracture at finite strains: a variational gradient-extended plasticity-damage theory. Int. J. Plast **84**, 1–32 (2016)
10. Pise, M., Brands, D., Sarhil, M., Schröder, J., Gebuhr, G., Anders, S.: Numerical calibration of elasto-plastic phase-field modeling of fracture for experimental pullout tests of single steel fibres embedded in high-performance concrete. In: Zingoni, A. (ed.) Advances in Engineering Materials, Structures and Systems: Innovations, Mechanics and Applications: Proceedings of the 7th International Conference on Structural Engineering, Mechanics and Computation (SEMC 2019), Cape Town, South Africa, pp. 1391–1396. Taylor & Francis Group, London (2019)
11. Amor, H., Marigo, J.J., Maurini, C.: Regularized formulation of the variational brittle fracture with unilateral contact: numerical experiments. J. Mech. Phys. Solids **57**, 1209–1229 (2009)
12. Drucker, D.C., Prager, W.: Soil mechanics and plastic analysis or limit design. Q. Appl. Math. **10**, 157–165 (1952)
13. Ambati, M., Gerasimov, T., de Lorenzis, L.: Phase-field modeling of ductile fracture. Comput. Mech. **55**, 1017–1040 (2015)
14. Schröder, J., Pise, M., Brands, D., Gebuhr, G., Anders, S.: Phase-field modeling of fracture in high performance concrete during low-cycle fatigue: numerical calibration and experimental validation. Comput. Methods Appl. Mech. Eng. **398**, 115181 (2022). https://doi.org/10.3390/ma15031179

15. Pise, M., Brands, D., Schröder, J.: Development and calibration of a phenomenological material model for steel-fiber-reinforced high-performance concrete based on unit cell calculations. Materials (Basel, Switzerland) **17**(10) (2024). https://doi.org/10.3390/ma17102247
16. Zeman, J.: Analysis of Composite Materials with Random Microstructure, University of Prague (2003)
17. Saavedra Flores, E.I., de Souza Neto, E.A.: Remarks on symmetry conditions in computational homogenisation problems. Int. J. Comput.-Aided Eng. Softw. **27**, 551–575 (2010)
18. von Hoegen, Y., Hellebrand, S., Scheunemann, L., Schröder, J.: On the realization of periodic boundary conditions for hexagonal unit cells. Finite Elem. Anal. Des. **229**, 104067 (2024). https://doi.org/10.1016/j.finel.2023.104067
19. Pise, M., Brands, D., Schröder, J., Gebuhr, G., Anders, S.: Macroscopic model for steel fiber reinforced high performance concrete based on unit cell calculations. Proc. Appl. Math. Mech. **21**, e202100180 (2021)
20. Pise, M., Brands, D., Schröder, J., Gebuhr, G., Anders, S.: Macroscopic model based on application of representative volume element for steel fiber reinforced high performance concrete. In: Zingoni, A. (ed.) Current Perspectives and New Directions in Mechanics, Modelling and Design of Structural Systems: Proceedings of the Eighth International Conference on Structural Engineering, Mechanics and Computation (SEMC 2022), Cape Town, South Africa, pp. 1300–1306. CRC Press, Oxon, United Kingdom (2022)
21. DIN EN 14651:2007-12: Test Method for Metallic Fibre Concrete—Measuring the Flexural Tensile Strength (Limit and Proportionality (LOP), Residual); German version EN 14651:2005+A1:2007. German Institute for Standardisation, Berlin, Germany (2007)
22. Gebuhr, G., Pise, M., Sarhil, M., Anders, S., Brands, D., Schröder, J.: Deterioration development of steel fibre reinforced high performance concrete in low-cycle fatigue. In: Zingoni, A. (ed.) Advances in Engineering Materials, Structures and Systems: Innovations, Mechanics and Applications: Proceedings of the 7th International Conference on Structural Engineering, Mechanics and Computation (SEMC 2019), Cape Town, South Africa, pp. 1444–1449. Taylor & Francis Group, London (2019)
23. Boehler, J.P.: Applications of Tensor Functions in Solid Mechanics, vol. 292. Springer, Vienna (1987)
24. von Mises, R.: Mechanik der festen Körper im plastisch-deformablen Zustand. Nachrichten von der Gesellschaft der Wissenschaften zu Göttingen, Mathematisch-Physikalische Klasse, pp. 582–592 (1913)
25. Simo, J.C., Hughes, T.: Computational Inelasticity. Interdisciplinary Applied Mathematics. Springer, New York (2000)
26. Pise, M., Brands, D., Schröder, J., Gebuhr, G., Anders, S.: Phenomenological material model for damage in steel-fiber reinforced high performance concrete during low cycle fatigue. Proc. Appl. Math. Mech. **22**, e202200236 (2022)
27. Pise, M., Schröder, J., Brands, D., Gebuhr, G., Anders, S.: Phase-field modeling for fracture in steel fiber reinforced high performance concrete during low-cycle fatigue. Int. J. Plasticity Damage Fracture **1**, 1–11 (2023)
28. Gebuhr, G., Anders, S., Pise, M., Brands, D., Schröder, J.: Damage development of steel fibre reinforced high performance concrete in high cycle fatigue tests. In: Zingoni, A. (ed.) Current Perspectives and New Directions in Mechanics, Modelling and Design of Structural Systems: Proceedings of the Eigth International Conference on Structural Engineering, Mechanics and Computation (SEMC 2022), Cape Town, South Africa, pp. 1327–1332. CRC Press, Oxon, United Kingdom (2022)
29. Seleš, K., Aldakheel, F., Tonković, Z., Sorić, J., Wriggers, P.: A general phase-field model for fatigue failure in brittle and ductile solids. Comput. Mech. **67**(5), 1431–1452 (2021). https://doi.org/10.1007/s00466-021-01996-5
30. Cornelissen, H., Reinhardt, H.W.: Uniaxial tensile fatigue failure of concrete under constant-amplitude and programme loading. Mag. Concr. Res. **36**(129), 216–226 (1984). https://doi.org/10.1680/macr.1984.36.129.216

31. Mena-Alonso, Á.: Flexural Fatigue of High-Strength Plain and Fiber-Reinforced Concrete: Influence of Mesostructure and Study of Size Effect, Universidad de Burgos. Departamento de Ingeniería Civil (2023)
32. Gebuhr, G., Anders, S., Pise, M., Brands, D., Schröder, J.: Approach for detection of fatigue phases using the example of high-performance concrete. In: Chandra Kishen, J.M., et al. (eds.) Proceedings of the 11th International Conference on Fracture Mechanics of Concrete and Concrete Structures (FraMCoS XI) (2023). https://doi.org/10.21012/FC11.092394
33. Gebuhr, G., Mena-Alonso, Á., Pise, M., Anders, S., Vicente, M., Brands, D., Schröder, J.: Predictive power of crack opening rate for flexural fatigue life of steel fiber-reinforced high and ultra-high performance concrete. In: Chandra Kishen, J.M., et al (eds.) Proceedings of the 11th International Conference on Fracture Mechanics of Concrete and Concrete Structures (FraMCoS XI) (2023). https://doi.org/10.21012/FC11.092370

High-Resolution Electron Microscopic Investigations on the Fatigue Behaviour of Ultra-High-Performance Concrete and Modelling with Bonded-Particle Model

Sebastian Rybczynski, Gunnar Schaan, Frank Schmidt-Döhl, Martin Ritter, and Maksym Dosta

Abstract Given the increasing demand for sustainable and lightweight concrete structures, the application of ultra-high-performance concrete (UHPC) has gained significant relevance. However, in contrast to static loading conditions, cyclic loads are more prevalent in real-world scenarios. Despite the superior mechanical properties of UHPC, it may exhibit a different resistance to cyclic loading. Consequently, this study focuses on both experimental and numerical analyses to investigate the primary mechanisms responsible for crack initiation at the meso-, micro-, and nanoscale. Following mechanical fatigue testing, microstructural alterations were characterized using scanning electron microscopy (SEM) and transmission electron microscopy (TEM). To extend these findings to the mesoscale, a novel rheological model was developed. The results suggest that fatigue damage originates from nanoscale transformations of ettringite, combined with a densification of the cementitious matrix. Moreover, an increased content of unhydrated cement clinker within the matrix enhances fatigue resistance. At the mesoscale, stress concentrations around aggregate particles intensify with increasing load cycles, propagating into the binder matrix and contributing to further material degradation.

Keywords Rheological fatigue-model · Crack propagation · Ultra-high-performance concrete · Ettringite transformation · Bonded-particle model · Electron microscopy

S. Rybczynski (✉) · F. Schmidt-Döhl
Institute of Materials, Physics and Chemistry of Buildings, Hamburg University of Technology, Hamburg, Germany
e-mail: sebastian.rybczynski@tuhh.de

G. Schaan · M. Ritter
Electron Microscopy Unit, Hamburg University of Technology, Hamburg, Germany

M. Dosta
Institute of Solids Process Engineering and Particle Technology, Hamburg University of Technology, Hamburg, Germany

N. Oneschkow and L. Lohaus (eds.), *Cyclic Deterioration of High-Performance Concrete*, https://doi.org/10.1007/978-3-032-19394-0_11

1 Introduction

Cement-based structures such as bridges or the foundations of wind turbines are more frequently subjected to cyclic than static loads and, thus, to fatigue-related damage. Under cyclic loading, these structures will undergo plastic deformation, even though the peak loads applied are below the short-term strength measured under quasi-static conditions. As advanced materials like ultra-high-performance concrete (UHPC) are increasingly utilised, current research is concentrating on identifying the primary factors responsible for crack initiation. UHPC is a highly brittle material with a dense, high-strength composition. Its remarkable mechanical performance is achieved through the application of the maximum packing density theory [1, 2], high strength aggregates, the use of silica fume and an extremely low water-to-cement ratio (w/c). Due to the low w/c ratio, the degree of cement hydration in UHPC is relatively low, around 30%, meaning that the remaining unhydrated cement contributes as a physical filler. Additionally, the pozzolanic activity of silica fume promoting the formation of calcium silicate hydrate (C-S-H) enhances the interfacial transition zone (ITZ) and reduces porosity within the binder matrix [3, 4]. Consequently, crack initiation from pores or weaker ITZ regions, common in ordinary concrete, is less likely to occur in UHPC. Recent studies have focused on identifying inhomogeneities that play a role in crack initiation, particularly under cyclic loading.

For several decades, scanning electron microscopy (SEM) and transmission electron microscopy (TEM) have been used to study cementitious materials [5–9], but the influence of fatigue on the microstructure, especially in the binder phase, has received little attention. In this work, an investigation of microstructural changes down to the nanoscale is aimed by utilizing SEM and scanning TEM, with lamella samples prepared through focused ion beam (FIB) techniques. While energy-dispersive X-ray spectroscopy (EDS) has historically been applied to cement-based materials to analyse the local chemical composition of the binder phase [10–12], its use is extended to identifying nanoscale objects.

On the numerical side, the bonded-particle model (BPM) was used, an extension of the discrete element method (DEM), where particles are connected by solid bonds [13, 14]. This approach allows for the modelling of both porous [15] and coherent material structures, incorporating single or multiple components. The mechanical behaviour of the material is represented through models for particle interaction and solid bonds. The flexibility to adjust or break these bonds during the simulation enables the representation of local structural changes or material fracture. While BPM has been increasingly applied to plain concrete analysis in 2D and 3D simulations of static loading [16–22], its application to UHPC remains unexplored, and only a few studies have adapted it to cyclic loading scenarios [23–25].

This chapter aims to demonstrate the combined application of experimental and numerical methods to refine and validate a novel fatigue model for UHPC. By identifying critical weaknesses in the microstructure, a contribution is hoped to be made to the development of cementitious materials with enhanced resistance to fatigue.

The first part focuses on the mechanical characterization, providing an overview of the materials and their properties. This is followed by a detailed examination of fatigue-damaged samples using SEM and TEM techniques, presented in Sect. 3. Subsequently, the numerical framework is introduced, where a new rheological model for fatigue analysis is developed and explained. Building on this, the results of the low-cycle and high-cycle fatigue simulations are discussed in detail. Finally, the paper concludes with a summary of the key findings and the main conclusions drawn from the study.

2 Mechanical Characterisation

2.1 *Specimen Preparation*

The main experimental investigations are based on cylindrical test specimens with the dimensions *h/d* = 180/60 mm, made from the UHPC reference material called "RU1" and described in Sect. 2.1 "Concrete Composition" in the chapter by M. Basaldella et al., this volume. Based on the RU1 mixture, additional binder specimens were produced in which the aggregate (H33 quartz sand) was omitted. In addition, UHPC specimens with supplementary basalt and granite (5/8 mm, Glensanda granite and Oelberg basalt) were produced as part of exploratory tests. The added amounts were set to 10 vol.-% and 30 vol.-%. It is important to note that this involved a pure addition of the rock materials rather than a replacement of the H33 aggregate. The volume proportions of all original RU1 components were reduced accordingly to maintain the ratio of the starting materials. To avoid influences from different particle size distributions of the additional aggregates, the individual grading curves were also aligned. This allowed a direct comparison of the specimens. In subsequent sections, these specimens will be abbreviated as "BAS10/BAS30" and "GRA10/GRA30". All specimens were stripped after 24 h and then stored in water until a minimum age of 56 d, since the compressive strength of UHPC still increases considerably after a standard hydration time of 28 d [26]. This procedure ensures a high degree of hydration and a nearly constant compressive strength.

2.2 *Static Compression Tests*

To investigate the mechanical behaviour of UHPC, binder and specimens with additional coarse aggregate, uniaxial static compression tests were carried out. The compression tests were carried out strain-controlled with a load speed of 0.2 mm/min on three specimens of each material. Three high-precision inductive displacement transducers were used to capture the longitudinal strain during testing with a sampling rate of 30 Hz. The mean values of the mechanical properties are shown in

Table 1 Mechanical properties of UHPC, binder with additional coarse aggregate (from [27], licensed under CC-BY 4.0)

Material	Strength (MPa)	Young's modulus (GPa)
UHPC	196.8 ± 3.1	48.0 ± 1.0
Binder	164.6 ± 7.1	38.6 ± 2.0
BAS10	176.5 ± 8.5	51.8 ± 1.0
BAS30	169.8 ± 9.0	56.5 ± 1.1
GRA10	153.7 ± 5.0	47.1 ± 1.0
GRA30	151.6 ± 8.2	48.1 ± 1.0

Table 1. Both the strength and stiffness of the binder specimens are lower than for UHPC specimens, while the breakage strain is higher. This is related to the missing aggregate grains in the mixture.

The additional coarse aggregate (basalt and granite) leads to a reduction in static strength, but results in an increase in stiffness. This can be explained by its different grain sizes. While H33 aggregates range from 0.125 mm to 0.5 mm and contribute to the optimized packing density of UHPC, the coarse aggregate has a grain size between 5 mm and 8 mm, introducing voids into the finely tuned reference material. Another aspect is the potential weakening of the interfacial transition zone between the coarse aggregate and the surrounding binder. This weakening could be due to the reduced amount of cement and microsilica relative to the surface area of the aggregate, which may lead to a decrease in overall strength.

2.3 Cyclic Compression Tests

After static tests, uniaxial cyclic compression tests were carried out force-controlled with a triangular load regime at 1 Hz. The upper and lower load level (S_u, S_l) corresponded to 80% and 5%, respectively, of the mean value of the static strength at a minimum age of 56 d, while the longitudinal strain was measured by three high-precision inductive displacement transducers with a sampling rate of 30 Hz.

All concrete mixtures show a significant scatter in the number of load cycles until failure. Additionally, it is evident that, on average, the binder specimens exhibit the highest fatigue resistance after UHPC, as shown in Fig. 1. Considering the binder mixture as the common basis for all compositions, it is observed that any addition of aggregate resulted in a reduction of cycles to failure. In this context, both the type and size of the aggregate appear to be crucial. For the H33 quartz aggregate, which consists of single crystals with a maximum grain size of 0.5 mm, it represents a very fine aggregate and was incorporated into the packing density optimization. This suggests that the quartz is fully embedded in the binder matrix, preventing unplanned defects due to a locally insufficient amount of hydrated binder. Nonetheless, the addition of H33 could result in variable force redistribution within the structure, leading to a more heterogeneous stress distribution due to differing local stiffnesses

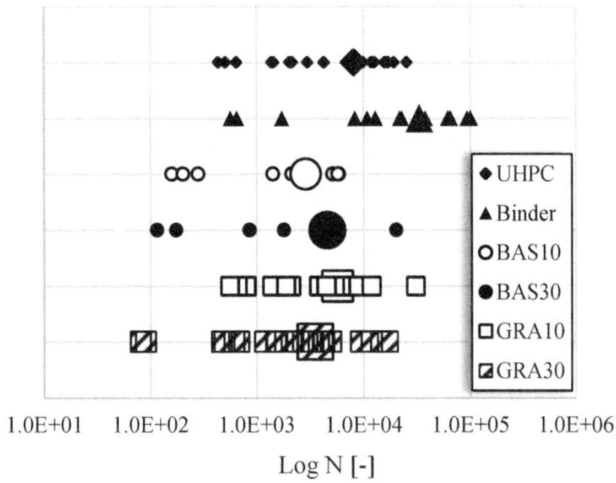

Fig. 1 Comparative representation of the number of cycles to failure of all analyzed UHPC mixtures. Small symbols = individual values, large symbols = mean values (from [27], licensed under CC-BY 4.0)

between the binder and aggregate. However, H33 remains a beneficial aggregate due to its single crystalline nature, which lacks internal grain boundaries that could act as additional weaknesses during fatigue. Although the high strength and stiffness of such single crystals contribute to increased static strength in UHPC, they may also induce adverse effects under cyclic loading.

In addition to the fatigue tests performed until specimen failure, phase-specific tests were also conducted. The primary goal was to generate various fatigue states (phases I-III) within the cylinders, which were subsequently used for microscopic examinations. Fatigue stages I and II were approximated at predefined cycle numbers. Phase III tests were generated using a strain criterion based on all available tests.

Regarding the influence of coarse aggregate, the negative effects on the number of load cycles sustained are intensified by the increased aggregate size. On the one hand, this results in a markedly more heterogeneous stress distribution within the concrete matrix. On the other hand, a reduced proportion of cement and microsilica relative to the surface area of the aggregate may lead to a higher incidence of voids around the aggregate particles. Additionally, from a mineralogical point of view, the used aggregates are of different types. Basalt is characterized by its fine grain size and appears macroscopically uniform, whereas granite consists of fewer, larger crystals on the millimetre scale, resulting in a macroscopically more heterogeneous texture. This could potentially explain why the GRA30 mixture especially exhibits, on average, lower fatigue resistance.

All tested mixtures exhibit a fatigue behaviour characteristic of concrete under cyclic loading, displaying an S-shaped fatigue curve. This curve is defined by micro-crack growth in phase I, continued by crack propagation in phase II, and unstable crack growth with significant strain increases leading to ultimate failure in phase III. The curves shown in Fig. 2 represent typical damage progressions for the various mixtures, with strain plotted against the relative number of load cycles (S-N curves). Additionally, phase transitions at 3% and 97% of the specimen lifetime, as identified in the literature, are indicated as vertical lines. A good agreement among all

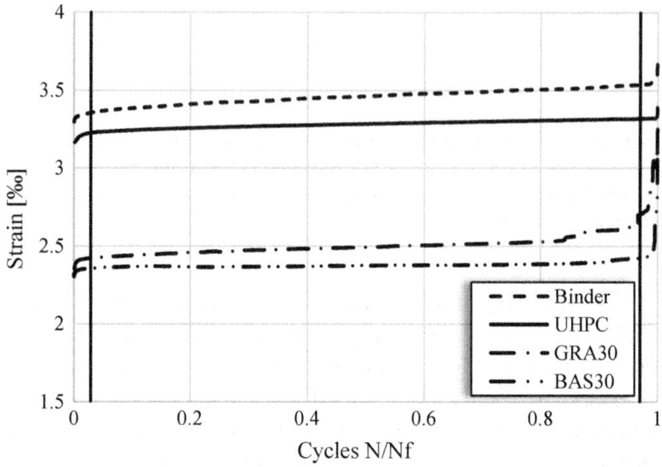

Fig. 2 Exemplary illustration of S-N curves for binder, UHPC, GRA30 and BAS30 (from [27], licensed under CC-BY 4.0)

mixtures is observed for the phase I-II transition, while for the phase II-III transition, the binder and UHPC mixtures show an even shorter phase III. This is likely due to the very fine-grained reference composition, which exhibits crack growth at very small scales up to failure, resulting in minimal macroscopic strain accumulation.

The varying levels of strain are attributable to the mechanical properties of the respective compositions and correspond to the values obtained from static tests. Notably, the GRA30 mixture exhibits an earlier onset of strain increase. Such deviations in the strain curve have been frequently observed for both GRA30 and GRA10. Initially, this might suggest a displacement issue with the measurement equipment, a known problem with inductive displacement transducers during cyclic tests. However, this effect is observed in many curves, with varying degrees of prominence. Additionally, displacement measurement equipment typically results in significant deviations that render the entire measurement immediately invalid, which is not the case here. This observation may be related to the aforementioned coarseness of the granite.

3 Electron Microscopy Investigations

3.1 Sample Preparation

As suggested previously, specimens from various stages of fatigue were investigated on the micro- and nanometre scale using a variety of methods of electron microscopy, among other techniques.

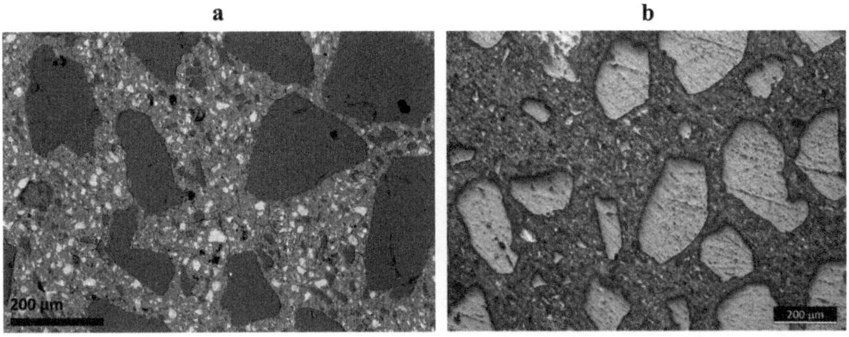

Fig. 3 **a** SEM image of a damaged UHPC sample with cracks caused by shrinkage or fatigue, **b** Reflected light microscope image of an undamaged UHPC sample (from [27], licensed under CC-BY 4.0)

In the case of pristine specimens or specimens in early stages of fatigue damage, platelets of ca. 1 cm × 1 cm × 0.3 cm were cut from the original cylinders. In the case of after-failure specimens, suitable fragments of a comparable size were selected. These were finely ground with silicon nitride grinding papers of various grain sizes down to 8.4 µm (FEPA P2500 grit) to ensure flat and smooth surfaces for electron microscopy investigations.

Scanning electron microscopy was utilized for a general characterization of the whole sample surface, especially concerning finding suitable sites to extract and prepare samples for transmission electron microscopy (TEM). Backscattered electrons (BSE) were primarily used for image formation because their intensity correlates with the mean atomic number of atoms present in individual sample regions. A quick classification of sample regions and the components of the material is possible.

The aggregate as well as the quartz powder, both consisting of pure silicon dioxide, appear dark, the cement paste matrix mainly consisting of C-S-H phase appears bright, while unhydrated residual cement clinker appears very bright. Cracks as well as small compaction pores appear almost completely black because practically no BSE can escape from them (see Fig. 3a). A comparative image of reflected light microscopy is shown in Fig. 3b.

3.2 Investigation of Aggregates, Unhydrated Cement Clinker, and the Interfacial Transition Zone Using Scanning Electron Microscopy

SEM survey images display that the quartz sand aggregate H33 consists entirely of single-crystalline, crack-free individual quartz grains, regardless of the stage of fatigue damage. The common hypothesis of the aggregate being the preferred site of

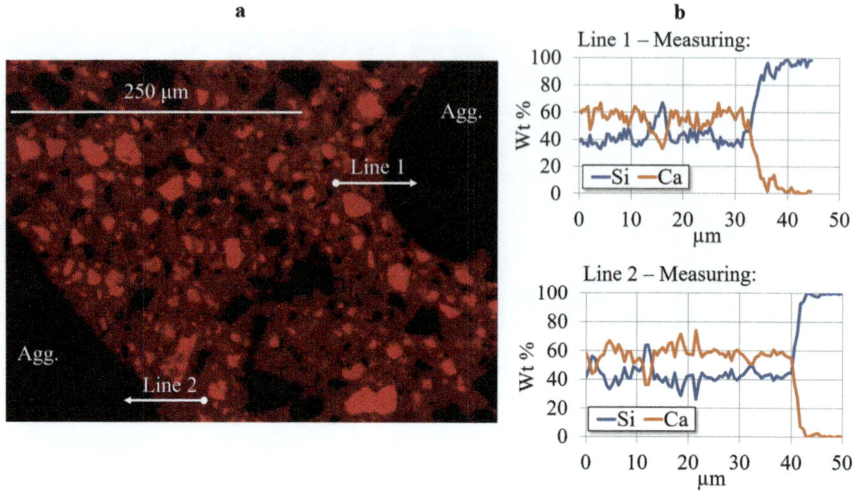

Fig. 4 **a** SEM–EDX overview of the interface between aggregate and binder, **b** Line scans of Si and Ca concentration with a length of 50 μm [28]. Reprinted with permission from Rybczynski, S., Dosta, M., Schaan, G., Ritter, M., Schmidt-Döhl, F. (2020), Structural Concrete, 23(1), 548–563.

crack initiation in UHPC due to the high strength of the cement paste matrix cannot be corroborated in the case of fine-grained, high-strength aggregate.

In addition, energy-dispersive X-ray spectroscopy (EDS) in SEM was employed to characterize the interfacial transition zone (ITZ) of aggregate grains (see Fig. 4). In ordinary concrete, platelet-shaped calcium hydroxide crystals preferentially agglomerate in the neighbourhood of coarse aggregate grains, decreasing the strength of the affected region [29].

In UHPC samples, however, higher concentrations of calcium in the neighbourhood of H33 aggregate grains could not be observed. Any possible concentration gradient is greatly outweighed by local maxima or minima due to unhydrated cement clinker or quartz powder grains, respectively. It can be concluded that any amount of calcium hydroxide formed intermediately will quickly react with silica fume to form C-S-H phases, compensating for any loss of strength in the ITZ.

In addition to the EDX analyses, an image segmentation analysis of the binder matrix components was conducted using SEM data and employing filters from [30] and algorithms from [31]. The aim of this investigation was to explore a potential correlation between the lifetime of a specimen and its content of unhydrated cement. The study involved examining fragments from three UHPC and three binder specimens, each with varying load cycles until failure. BSE-SEM images are processed in a multi-step method described in [32].

The fraction of unhydrated cement clinker in the cement paste matrix can be determined and correlated with the final lifetime of the specimen, showing a positive correlation. This applies to the reference mixture of UHPC as well as, to a lesser

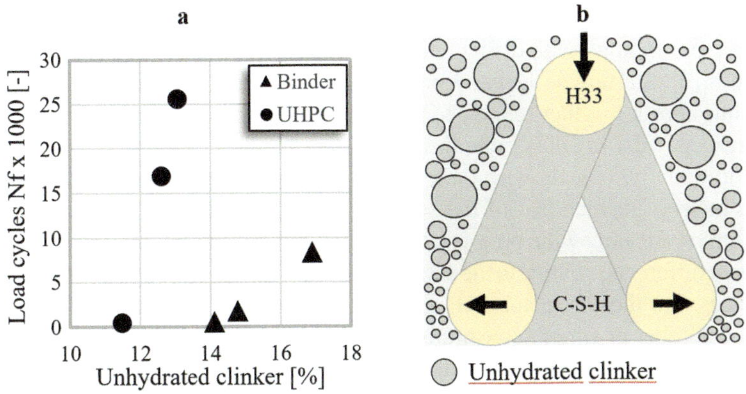

Fig. 5 **a** Correlation between the proportion of unhydrated cement clinker and the load cycles until failure for three UHPC and binder specimens each, **b** two-dimensional conceptual model of the blocking effect of unhydrated cement particles in the transverse direction (from [27], licensed under CC-BY 4.0)

extent, to a pure binder mixture without H33 aggregate (see Fig. 5a). Due to the low w/c value of UHPC, numerous unhydrated clinker grains remain in the concrete and can be considered a fine-grained, high-strength additive. Regarding the mentioned positive correlation, it can be hypothesized that the higher content of unhydrated cement contributes to an increase in strength, which would, in turn, reduce the stress level on the specimens during fatigue loading. Conversely, unhydrated cement particles may act to inhibit internal transverse tensile stresses at the microscale, ultimately resulting in less damage to the structure in the transverse direction. Figure 5b illustrates this conceptual model.

3.3 Characterization of Microstructural Changes Caused by Fatigue Damage

Platelets of 10 μm × 10 μm × 1.5 μm were extracted from suitable sites around the interface of aggregate and cement paste using Focused Ion Beam techniques (FIB). These platelets were thinned to a thickness of 0.1 μm to ensure electron transparency for TEM investigations. Material structures on a subnanometer scale can be resolved and imaged using TEM due to the higher energy of incident electrons compared to SEM and the usage of ultrathin samples. High-Angle Annular Dark-Field (HAADF) imaging was primarily used which, like BSE imaging in SEM, is based on the scattering of electrons. Thus, it exhibits mass-thickness contrast: sample regions that are thin or porous or that contain a large fraction of light elements appear dark and vice versa.

In samples from pristine specimens, the C-S-H phases exhibit a fibrous, loosely packed morphology with distinct porosity that has been established in literature for decades [9]. Starting at the early onset of fatigue, the C-S-H phases are compacted and thus appear brighter relative to non-porous components surrounding them, such as quartz powder or silica fume than in pristine samples (see Fig. 6).

Additionally, objects with dark image contrast form in the compacted cement paste matrix. Independent of the stage of fatigue damage, they have a size of ca. 200 nm × 30 nm; with progressing fatigue damage, their number rises and the contrast difference relative to the surrounding material increases. Dark image contrast indicates that these structural changes either consist of a material with low mean atomic number or simply constitute hollows in the material.

Upon long-term electron irradiation of the sample, the structural changes often either increase or decrease in size due to the energy transferred from the electron

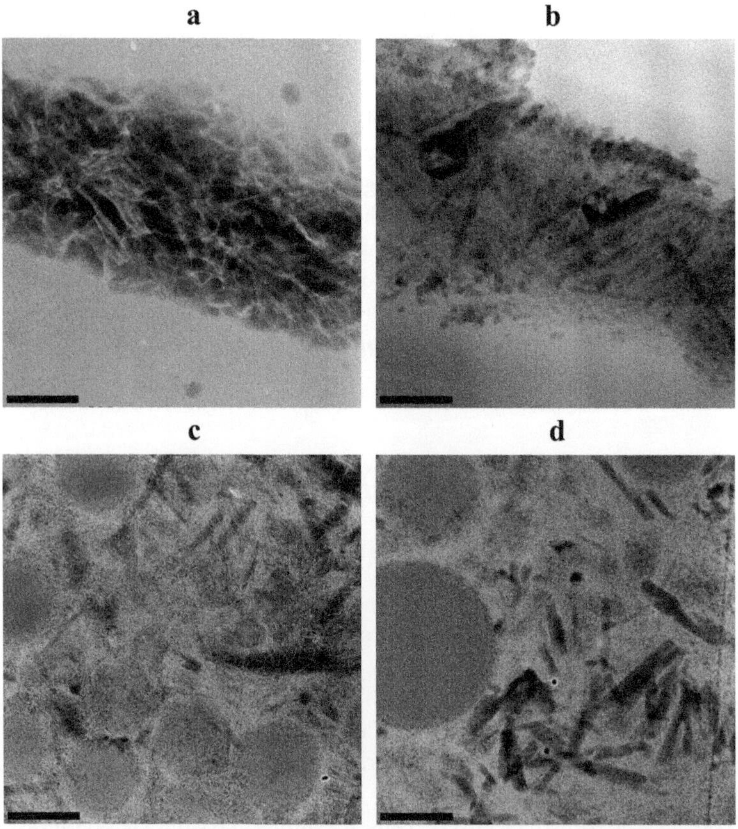

Fig. 6 HAADF-STEM micrographs of C-S-H phases in various stages of UHPC fatigue. **a** Pristine sample; **b** stage I fatigue; **c** stage II fatigue; **d** stage III fatigue. Scale markers are 200 nm [33]. Reprinted with permission from Schaan et al., © 2020 John Wiley & Sons Ltd.

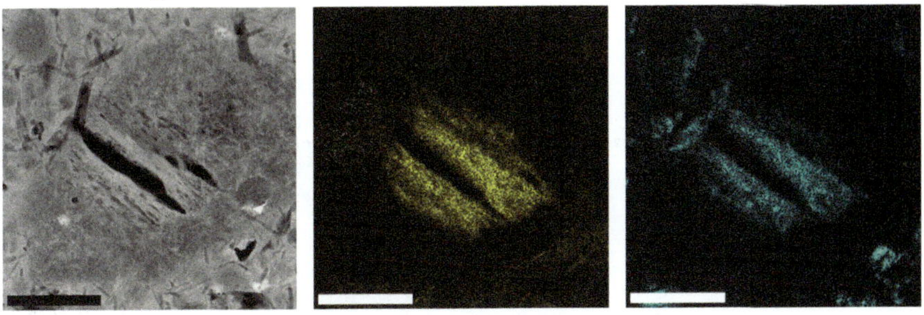

Fig. 7 **a** HAADF-STEM micrograph maps in the neighbourhood of a nanocrack in after-failure UHPC RU1. **b** EDS elemental distribution map of aluminium; **c** EDS elemental distribution map of sulphur. Scale markers are 500 nm (from [32], licensed under CC-BY 4.0)

beam. Chemical analysis using energy-dispersive X-ray spectroscopy (EDS) shows strongly increased concentrations of aluminium and sulphur in the vicinity of structural changes. This tendency is especially strong in the neighbourhood of nanocracks (see Fig. 7).

In combination, these phenomena—needle-like shape, tendency to dissolve or crystallize upon changing environmental parameters, high content of aluminium and sulphur—indicate a relation with the formation or transformation of ettringite on the nanoscale. Ettringite ($Ca_6Al_2(SO_4)_3(OH)_{12}$. 26 H_2O), as a reaction product of the cement phase C_3A (3 CaO. Al_2O_3) and sulphate carrier, plays a crucial role in the early stages of cement paste hardening. A delayed formation of ettringite after the hardening of the material, however, can result in expansion phenomena or damage [34, 35].

To minimize or completely inhibit ettringite transformation, a variation of the UHPC reference mixture using a sulphate-resistant SR0 portland cement instead of the original SR3 cement was produced. While the SR0 cement only has a C_3A content of 0.86% - compared to 2.61% in the SR3 cement -, its $C_4(A,F)$ content is considerably higher at 19.09% compared to 14.46%. The deciding factor is the content of Al_2O_3 equivalent, which is only slightly lower in the SR0 cement at 3.71% compared to 4.02%.

The mean static compressive strength of specimens fabricated using SR0 compared is ca. 25% lower than using the reference mixture at ca. 150 MPa; the mean lifetime during cyclic loading experiments is comparable at 7206 load cycles.

TEM investigations display a formation of structural changes and nanocracks even in samples fabricated using SR0 cement (see Fig. 8). Like in SR3 cement specimens, the neighbourhood of structural changes and nanocracks exhibits significantly higher concentration of aluminium and sulphur, indicating the presence of ettringite. At the same time, the concentration of iron, which can substitute for aluminium in ettringite, is increased as well in these regions.

Using a highly sulphate-resistant cement with low C_3A content does not immediately result in the inhibition of nanoscale ettringite transformation and an increase in

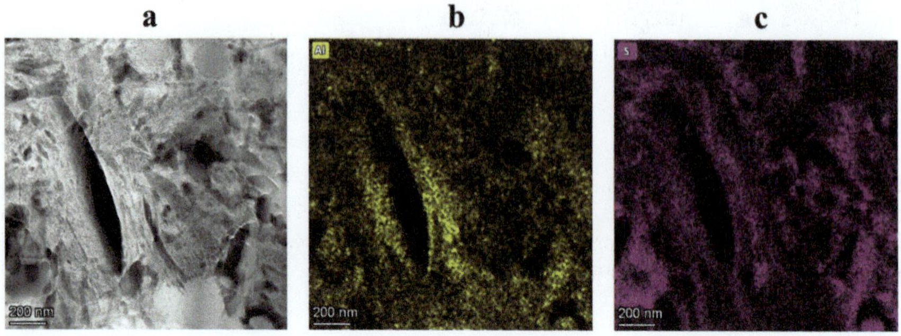

Fig. 8 **a** HAADF-STEM micrograph in the neighbourhood of a nanocrack in after failure UHPC fabricated with SR0 cement. **b** EDS elemental distribution map of aluminium; **c** EDS elemental distribution map of sulphur. Scale markers are 200 nm

lifetime during cyclic loading experiments. It might be necessary to utilize a specially designed cement with a very low content of Al_2O_3 equivalent, which will likely affect the hardening of the fresh concrete considerably.

4 Numerical Investigations

Numerical simulations were conducted using the bonded-particle model (BPM) within the open-source discrete element method (DEM) framework MUSEN [36], which is available for download from [41]. The computations were performed on a system equipped with an Intel Xeon Gold 5118 CPU and an NVIDIA A100 GPU, utilizing GPU-based force calculations and motion integration for enhanced performance. Generally, DEM models granular materials as assemblies of individual particles, governed by Newton's second law and interacting through specific contact models [37]. Modern advancements in DEM have expanded its application to non-spherical particles and multiphase flow modelling [38]. BPM extends DEM by connecting particles with solid bonds [13, 14], allowing for the simulation of cohesive material structures and internal fracturing processes. Key interactions include interparticle contacts, particle-wall contacts, and forces within solid bonds.

4.1 Contact Models and Structural UHPC Model

Given the nearly ideal elastic behaviour of UHPC and binder materials, elastic-based models were used for all three interaction types. A linear elastic model was utilized for interparticle interactions, while the Hertz-Mindlin model was applied for particle-wall contacts. A detailed description of these models can be found in [27]. For the bond model, a newly developed approach was implemented, incorporating

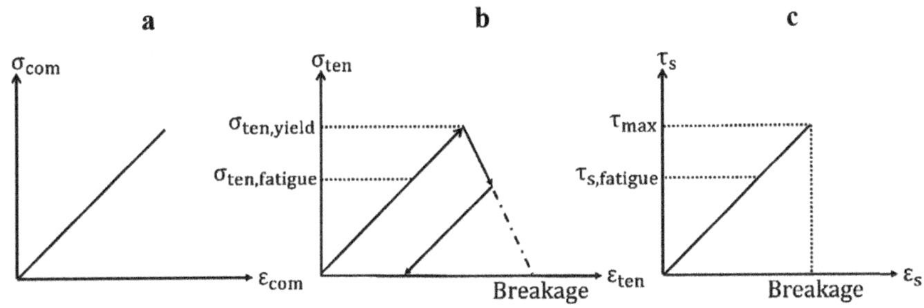

Fig. 9 Bond response in different load modes: **a** Compression; **b** tension; **c** shear (from [32], licensed under CC-BY 4.0)

both elastic and plastic deformations along with fatigue-induced damage. The fundamental equations governing the purely elastic components, which are also applicable in our model, can be found in [13].

Assuming that concrete typically fails due to internal transverse tensile stresses, the primary fatigue damage mechanisms were integrated into the tensile response of the model formulation. To capture fatigue degradation resulting from shear stresses, a fatigue damage component was also incorporated in shear mode. Fig. 9 illustrates the rheological formulations for each of the three loading conditions. A detailed description of the model equations can be found in [32].

In the case of fatigue loading, crack propagation occurs across multiple scales and does not necessarily result in plastic deformation. Therefore, modelling fatigue degradation solely through plastic behaviour is insufficient. Instead, crack growth occurs concurrently at the meso, micro, and nanoscale, and this multiscale phenomenon must be taken into account. To represent this behaviour on the mesoscale, a fatigue damage criterion was introduced, where the bond diameter progressively reduces as fatigue damage accumulates. This approach allows the representation of micro- and nanoscale crack formation without requiring the immediate breakage or removal of the bond from the simulation. Fig. 10 illustrates the relationship between small-scale damage mechanisms and their mesoscale implementation in the bond model.

In the numerical investigation of UHPC's fatigue behaviour, a three-phase structural model was employed, incorporating two particle types (binder and aggregate) and three bond types (binder–binder, binder–aggregate, and aggregate–aggregate). These bonds correspond to the primary phases of concrete (binder, interfacial transition zone (ITZ), and aggregate), facilitating a detailed analysis of the UHPC structure. The modelled dimensions represent a real concrete volume of around $h/d = 6/2$ mm, as detailed in [28]. To account for realistic non-spherical shapes of aggregate grains, reconstructed 3D models from X-ray computed tomography (μCT) were used.

SEM investigations did not reveal a notably weaker ITZ, likely due to the optimized packing density of the UHPC composition. Consequently, ITZ and binder bond parameters were assumed to be identical, simplifying the calibration process to the binder and aggregate phases. Calibration of binder parameters relied on uniaxial

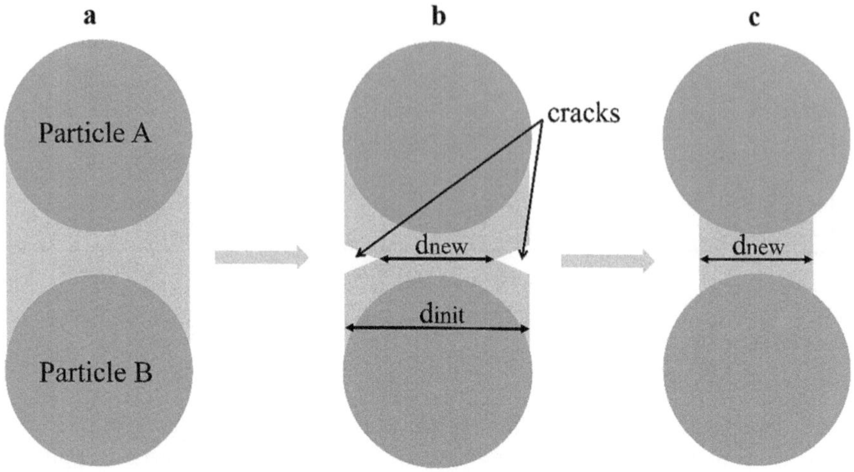

Fig. 10 Bond damage caused by a small-scale crack and its implementation: **a** Undamaged bond connection of two primary particles; **b** A fatigue crack manifests in a reduction of the load-bearing cross-cut of a meso bond; **c** Representation of a resulting fatigue crack in the simulation (from [32], licensed under CC-BY 4.0)

compression test results from Table 1, focusing on static simulations. Since experimental cracks did not penetrate individual grains, only the aggregate stiffness was calibrated, with bond strength set to an infinite value. Aggregate parameter calibration was performed on cylindrical specimens under uniaxial compression.

Generally, calibration involved a gradient-based trial and error approach, iterating with different parameters to minimize discrepancies between experimental and numerical results. The slope of the stress-strain curve was used to fine-tune Young's modulus of the bonds, while bond strength parameters were adjusted based on macroscopic fracture strength. Initial parameter estimates were derived using a linearization technique, specifically the direct stiffness method, as detailed in [39]. A detailed description of the algorithms used for structural generation and the whole calibration process can be found in [28].

4.2 Explicit Low-Cycle Fatigue Simulations

In the case of low-cycle simulations, each loading and unloading cycle is explicitly computed. Given that these simulations are computationally and temporally intensive, taking approximately 22 hours per cycle, the structural UHPC model was slightly scaled down. The reduced model can represent dimensions of approximately $h/d = 4.2/1.4$ mm relative to a real laboratory specimen.

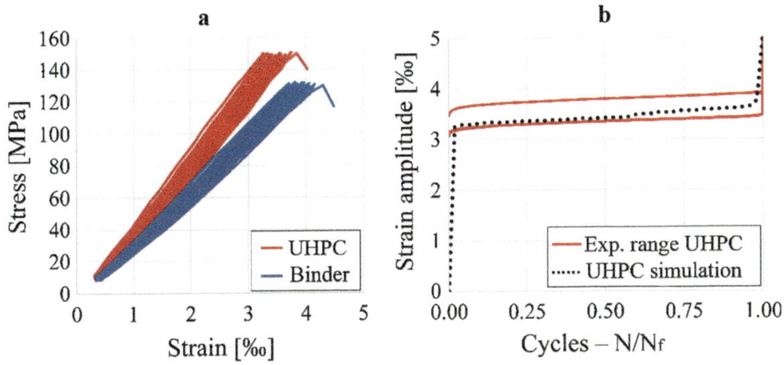

Fig. 11 Macroscopic fatigue simulation results of UHPC using the proposed bond formulation: **a** Hysteresis loops obtained from fatigue simulations for UHPC and binder; **b** Comparison of mean S-N curve from UHPC simulations with experiments (from [32], licensed under CC-BY 4.0)

Both the binder and UHPC simulations exhibit the main characteristics of decreasing stiffness and increasing strain over the material's lifetime, as shown in Fig. 11a. Furthermore, the typical S-shaped fatigue behaviour aligns well with the range of experimental fatigue tests, validating the model as representative and suitable for further analysis of fatigue-induced damage (see Fig. 11b).

To identify potential crack initiation areas, the average stresses on the binder particles over the material's lifetime were analysed. This qualitative analysis was conducted for various fatigue stages ($N = 1$; $N/Nf = 0.3, 0.6, 0.9$) at the upper load level, as shown in Fig. 12. The results indicated that, in the first cycle, higher-stress regions of the UHPC binder predominantly concentrate around the aggregate, while areas with less aggregate are subjected to lower stress. The stress peaks around the aggregate particles promote localized material degradation, which, despite the optimized interfacial transition zone (represented in the model by identical bond parameters used for the binder), was identified as a crack-initiating factor.

Further analysis of the higher-stress zones revealed significant propagation of stress concentrations into the surrounding matrix with increasing load cycles. This propagation can be attributed to recurring stress peaks, which initially damage or break directly adjacent bonds upon first loading. Since the material bridges between aggregate and matrix, such as C-S-H phases, vary in length, this leads to activation or redistribution of forces to more distant bonds. These bonds, which initially contributed less to the matrix-aggregate interaction, become activated due to the damage.

The pronounced material degradation around the aggregate particles was further investigated through a quantitative analysis of broken bonds. The analysis revealed that significant damage to the interfacial transition zone occurs particularly during phase I of fatigue, until a first equilibrium state of initial damage is reached, as shown in Fig. 13a and c. This primary damage is followed by uniform damage growth, which leads to a noticeable increase in bond failure at around 75-80% of the material's lifetime. This occurs even though macroscopic strain amplitudes do not

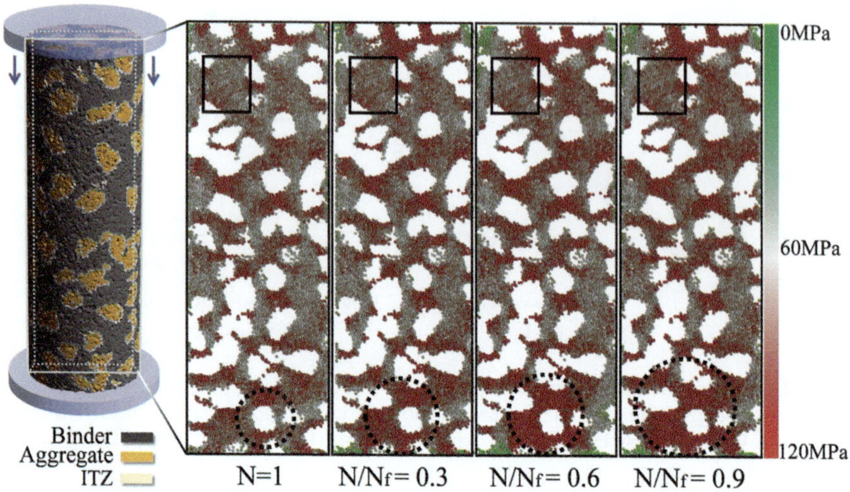

Fig. 12 Visualization of stresses on primary binder particles at the upper load level at different fatigue stages; stresses are analysed in the loading direction (from [32], licensed under CC-BY 4.0)

yet show signs of imminent fracture or increasing damage, as depicted in Fig. 13b and d. This observation suggests that the ITZ plays a critical role in initiating the final, unstable damage growth characteristic of phase III.

4.3 Cycle-Jump Method and High-Cycle Fatigue Simulations

The Cycle-Jump Method represents an approach that facilitates extended calculation processes through partial simulations. In traditional DEM/BPM simulations, the movements and interactions of individual particles are explicitly computed in small, discrete time steps as described and used for the low-cycle fatigue simulations. However, simulating a large number of macroscopic cycles is impractical due to the extensive computational time and storage requirements. The Cycle-Jump Method addresses this by reducing the number of cycles simulated explicitly. It achieves this by extrapolating the fatigue damage for each bond to a later time point, which is then used as the basis for further explicit calculations. This approach allows a reduction in simulation time without significantly impacting the overall system behaviour. The developed Cycle-Jump algorithm is based on the analysis and extrapolation of all bond states at defined simulation points, and is detailed with reference to the flowchart presented in Fig. 14. A detailed description of all algorithms, the calibration process and the equations underlying the Cycle-Jump Method can be found in [27].

Given the consistency achieved between explicit and extrapolated simulations, the parameter set for the cycle-jump method is considered representative. To account for the scattering observed in experimental fatigue tests, REM investigations into the

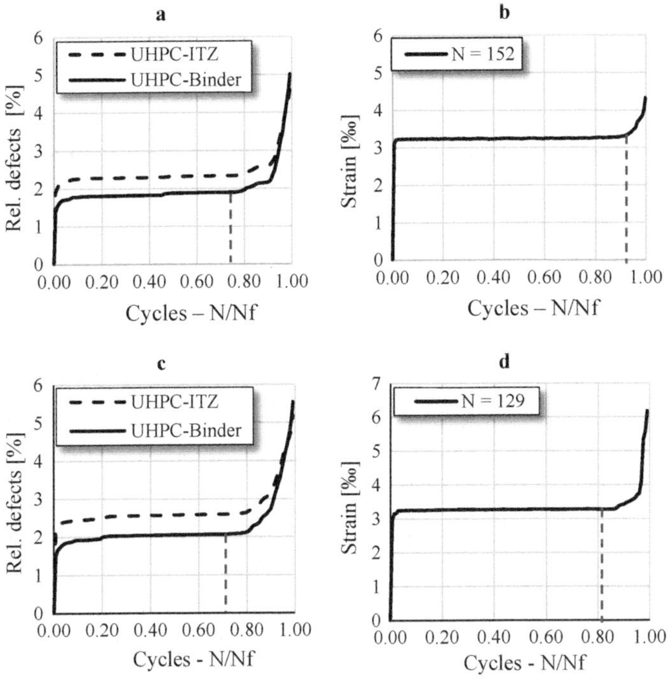

Fig. 13 Exemplary representation of the S-N curves with the corresponding proportions of defects for two low-cycle simulations (N = 152 and N = 129). The initial increase in the curves is highlighted in red (from [27], licensed under CC-BY 4.0)

effect of unhydrated cement on the fatigue behaviour of UHPC were utilized (see Fig. 5a). These investigations indicate that an increased content of unhydrated cement clinker positively impacts the load cycles endured in experimental fatigue tests.

In the simulation, a varying proportion of unhydrated cement clinker is represented by implementing partial deactivation of fatigue damage for a specific part of bonds. This approach aims to replicate the mentioned blocking effect of unhydrated cement clinker (see the conceptual model at Fig. 5b). A random allocation algorithm is used to assign different proportions to bonds. The resulting load cycle counts from 152 to 16,362 cycles. The individual values are available in [27]. However, an increase in the fraction of non-fatigue-prone bonds is associated with a rise in the number of load cycles the material can withstand. This variation highlights less of a quantitative relationship with unhydrated cement clinker and more of a qualitative correlation with the model concept shown in Fig. 5b.

It should be noted that the fatigue process involves a complex interaction of various damage mechanisms and cannot be reduced to a single factor. Nevertheless, it can be demonstrated that improving fatigue resistance occurs when individual bonds are less susceptible to fatigue due to blocking effects. A comparison of the experimentally observed scattering in UHPC lifetime with simulation results further reveals parallels in the range of variability (see Fig. 15).

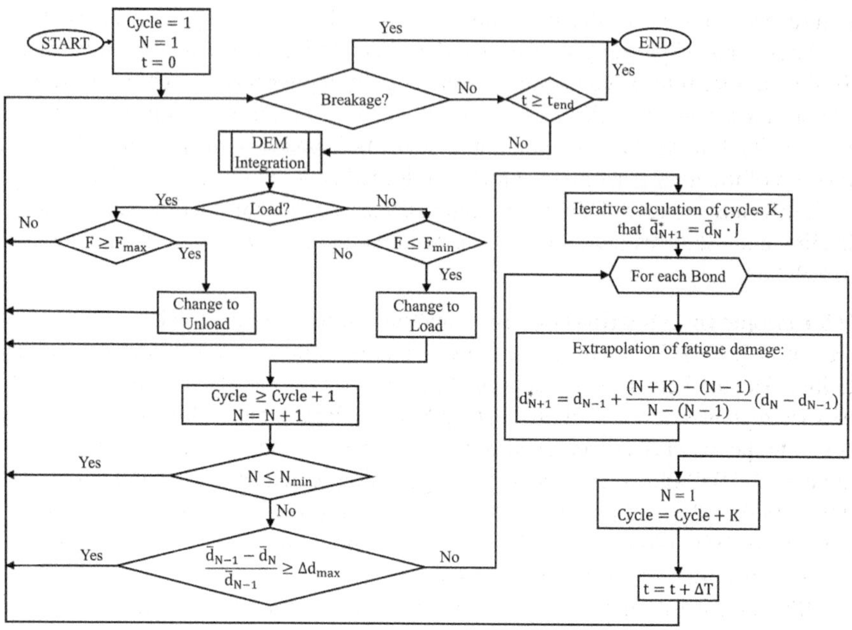

Fig. 14 Flowchart of the developed Cycle-Jump algorithm (from [27], licensed under CC-BY 4.0)

Fig. 15 Comparison of experimental fatigue tests on UHPC with high-cycle fatigue simulations featuring varying proportions of fatigue-prone bonds (from [27], licensed under CC-BY 4.0)

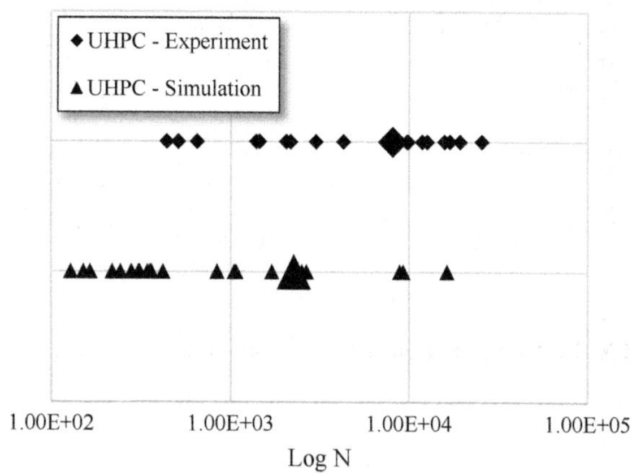

Building on the insights from the explicit low-cycle simulations, a comprehensive investigation of material degradation was conducted through a quantitative analysis of broken binder and ITZ bonds. Two high-cycle simulations are presented as representative examples: one with N = 16,362 cycles (Fig. 16a, b) and another with N = 8885 cycles (Fig. 16c and d). The x-axis is scaled to separately display both the cycle jumps and the explicit simulation segments, ensuring a clearer representation.

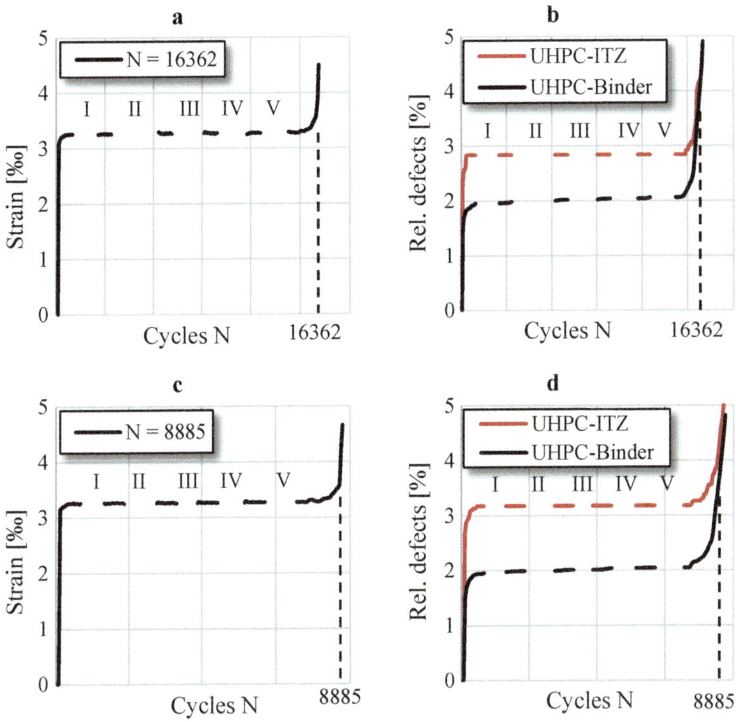

Fig. 16 **a**, **c** S-N curves and **b**, **c** corresponding relative defects for two high-cycle simulations (N = 16,362 and N = 8885) (from [27], licensed under CC-BY 4.0)

It is consistently observed that, in phase I of high-cycle fatigue, significant damage occurs in the ITZ until an initial equilibrium state is reached. This primary damage is followed by uniform and minimal damage growth in phase II. Unlike the low-cycle simulations, no definitive conclusion can be drawn about whether a rise in damage occurs at 80% of the total cycles due to the extrapolation phase. However, high-cycle simulations suggest that similar damage patterns develop within the structure, with stress peaks and the newly mapped distribution of fatigue-prone bonds having a significant impact on the number of load cycles the material can withstand. Additionally, the increasingly shorter jump distances indicate that fatigue damage intensifies with prolonged loading duration, suggesting that the extent of damage increases in parallel with the duration and intensity of local stress.

5 Conclusion

This chapter provides a comprehensive experimental and numerical investigation into the fatigue behaviour of ultra-high-performance concrete, employing both scanning and transmission electron microscopy alongside a newly developed bond formulation and cycle-jump technique. Structural inhomogeneities, such as local stiffness variations and ettringite transformation, are identified as potential factors contributing to crack initiation. Similar to these inhomogeneities, material degradation occurs across multiple scales—meso, micro, and nano.

The key findings from this study are as follows:

- Microscopic analysis confirmed that fatigue-induced damage extends to the nanoscale, where ettringite undergoes a constant local crystallization and dissolution. Therefore, to accurately simulate degradation mechanisms under cyclic loading, nanoscale changes must be considered, even when modelling at the mesoscale. The use of sulphate-resistant cement (SR) with low or no calcium aluminate content could not mitigate delayed ettringite formation.
- A higher proportion of unhydrated cement clinker within the binder matrix enhances local strength caused by a blocking effect, resulting in an extended fatigue life of the specimens under cyclic loading.
- Stress concentrations due to significant stiffness variations lead to increased degradation in the binder surrounding aggregate particles during fatigue. This promotes crack initiation within the interfacial transition zone and encourages crack propagation into the surrounding matrix. Reducing the stiffness of the aggregate and employing finer particles could reduce localized stress peaks and improve fatigue resistance.

References

1. Arora, A., Aguayo, M., Hansen, H., Castro, C., Federspiel, E., Mobasher, B., Neithalath, N.: Microstructural packing-and rheology-based binder selection and characterization for ultra-high performance concrete (UHPC). Cem. Concr. Res. **103**, 179–190 (2018). https://doi.org/10.1016/j.cemconres.2017.10.013
2. Wang, X., Yu, R., Song, Q., Shui, Z., Liu, Z., Wu, S., Hou, D.: Optimized design of ultra-high performance concrete (UHPC) with a high wet packing density. Cem. Concr. Res. **126**, 105921 (2019). https://doi.org/10.1016/j.cemconres.2019.105921
3. Alkaysi, M., El-Tawil, S., Liu, Z., Hansen, W.: Effects of silica powder and cement type on durability of ultra high performance concrete (UHPC). Cem. Concr. Compos. **66**, 47–56 (2016). https://doi.org/10.1016/j.cemconcomp.2015.11.005
4. Abdulkareem, O.M., Fraj, A.B., Bouasker, M., Khelidj, A.: Mixture design and early age investigations of more sustainable UHPC. Constr. Build. Mater. **163**, 235–246 (2018). https://doi.org/10.1016/j.conbuildmat.2017.12.107
5. Groves, G.W.: TEM studies of cement hydration. Res. Soc. Symp. Proc. **85**, 3–12 (1987). https://doi.org/10.1557/PROC-85-3

6. Richardson, I.G., Groves, G.W.: Microstructure and microanalysis of hardened ordinary Portland cement pastes. J. Mater. Sci. **28**, 265–277 (1993). https://doi.org/10.1007/BF00349061
7. Diamond, S.: Aspects of concrete porosity revisited. Cem. Concr. Res. **29**, 1181–1188 (1999). https://doi.org/10.1016/S0008-8846(99)00122-2
8. Gatty, L., Bonnamy, S., Felessoufi, A., Clinard, C., Richard, P., van Damme, H.: A transmission electron microscopy study of interfaces and matrix homogeneity in ultra-high-performance cement-based materials. J. Mater. Sci. **36**, 4013–4026 (2001). https://doi.org/10.1023/A:1017938725748
9. Wenzel, O., Schwotzer, M., Müller, E., Chakravadhanula, V.S.K., Scherer, T., Gerdes, A.: Investigating the pore structure of the calcium silicate hydrate phase. Mater. Charact. **133**, 133–137 (2017). https://doi.org/10.1016/j.matchar.2017.09.035
10. Diamond, S.: The microstructure of cement paste and concrete—a visual primer. Cem. Concr. Compos. **26**, 919–933 (2004). https://doi.org/10.1016/j.cemconcomp.2004.02.028
11. Wei, F.Y., Lan, X.H., Lv, Y.N., Xu, Z.: Effect of pozzolanic reaction products on alka-li-silica reaction. J. Wuhan Univ. Technol. Mater. Sci. Ed. **21**, 168–171 (2006). https://doi.org/10.1007/BF02840910
12. Sakalli, Y., Trettin, R.: Investigation of C3S hydration mechanism by transmission electron microscope (TEM) with integrated SuperXTM EDS system. J. Microsc. **267**, 81–88 (2017). https://doi.org/10.1111/jmi.12549
13. Potyondy, D.O., Cundall, P.A.: A bonded-particle model for rock. Int. J. Rock Mech. Min. Sci. **41**, 1329–1364 (2004). https://doi.org/10.1016/j.ijrmms.2004.09.011
14. Dosta, M., Dale, S., Antonyuk, S., Wassgren, C., Heinrich, S.: Numerical and experimental analysis of influence of granule microstructure on its compression breakage. Powder Technol. **299**, 87–97 (2016). https://doi.org/10.1016/j.powtec.2016.05.005
15. Dosta, M., Bistreck, K., Skorych, V., Schneider, G.A.: Mesh-free micromechanical modeling of inverse opal structures. Int. J. Mech. Sci. **204**, 106577 (2021). https://doi.org/10.1016/j.ijmecsci.2021.106577
16. Cusatis, G., Mencarelli, A., Pelessone, D., Baylot, J.: Lattice discrete particle model (LDPM) for failure behavior of concrete. II: calibration and validation. Cem. Concr. Compos. **33**, 891–905 (2011). https://doi.org/10.1016/j.cemconcomp.2011.02.010
17. Donze, F., Magnier, S., Daudeville, L., Mariotti, C., Davenne, L.: Numerical study of compressive behavior of concrete at high strain rates. J. Eng. Mech. **125**, 1154–1163 (1999). https://doi.org/10.1061/(ASCE)0733-9399(1999)125:10(1154)
18. Hentz, S., Donze, F., Daudeville, L.: Identification and validation of a discrete element model for concrete. J. Eng. Mech. **130**, 709–719 (2004). https://doi.org/10.1061/(ASCE)0733-9399(2004)130:6(709)
19. Poinard, C., Piotrowska, E., Malecot, Y., Daudeville, L., Landis, E.: Compression triaxial behavior of concrete: the role of the mesostructure by analysis of X-ray tomographic images. Eur. J. Environ. Civ. Eng. **16**, 115–136 (2012). https://doi.org/10.1080/19648189.2012.682458
20. Kim, H., Buttlar, W.G.: Discrete fracture modeling of asphalt concrete. Int. J. Solids Struct. **46**, 2593–2604 (2009). https://doi.org/10.1016/j.ijsolstr.2009.02.006
21. Nitka, M., Tejchman, J.: Modelling of concrete behavior in uniaxial compression and tension with DEM. Granul. Matter **17**, 145–164 (2015). 10.1007/s10035-015-0546-4
22. Skarzynski, L., Nitka, M., Tejchman, J.: Modelling of concrete fracture at aggregate level using FEM and DEM based on X-ray μCT images of internal structure. Eng. Fract. Mech. **147**, 13–35 (2015). https://doi.org/10.1016/j.engfracmech.2015.08.010
23. Sinaie, S., Ngo, T.D., Nguyen, V.P.: A discrete element model of concrete for cyclic loading. Comput. Struct. **196**, 173–185 (2018). https://doi.org/10.1016/j.compstruc.2017.11.014
24. Nguyen, N.H.T., Bui, H.H., Kodikara, J., Arooran, S., Darve, F.: A discrete element modelling approach for fatigue damage growth in cemented materials. Int. J. Plast. **112**, 68–88 (2018). https://doi.org/10.1016/j.ijplas.2018.08.007
25. Song, Z., Konietzky, H., Herbst, M.: Three-dimensional particle model based numerical simulation on multi-level compressive cyclic loading of concrete. Constr. Build. Mater. **225**, 661–677 (2019). https://doi.org/10.1016/j.conbuildmat.2019.07.260

26. Deutscher, M., Tran, N.L., Scheerer, S.: Experimental investigation of load-induced temperature development in UHPC subjected to cyclic loading. In: Proceedings of the 6th International fib Congress, Krakow, Poland, 27–29 May 2019
27. Rybczynski, S.: Ermüdungsinduzierte Schädigung von UHPC—Experimentelle Ergebnisse und BPM-Modellierung. Dissertation, TU Hamburg (2024). https://doi.org/10.15480/882.13203
28. Rybczynski, S., Dosta, M., Schaan, G., Ritter, M., Schmidt-Döhl, F.: Numerical study on the mechanical behavior of ultrahigh performance concrete using a three-phase discrete element model. Struct. Concr. **23**(1), 548–563 (2020). https://doi.org/10.1002/suco.202000435
29. Scrivener, K., Crumbie, A.K., Laugesen, P.: The Interfacial Transition Zone (ITZ) between cement paste and aggregate in concrete. Interface Sci. **12**, 411–421 (2004). https://doi.org/10.1023/B:INTS.0000042339.92990.4c
30. Buades, A., Coll, B., Morel, J.-M.: Non-local means denoising. Image Process 208–212 (2011). https://doi.org/10.5201/ipol.2011.bcm_nlm
31. Chen, L., Shan, W.B., Peng, L.: Identification of concrete aggregates using K-means clustering and level set method. Struct. **34**, 2069–2076 (2021). https://doi.org/10.1016/j.istruc.2021.08.048
32. Rybczynski, S., Schaan, G., Dosta, M., Ritter, M., Schmidt-Döhl, F.: Discrete element modeling and electron microscopy investigation of fatigue-induced microstructural changes in ultra-high-performance concrete. Materials **14**(21), 6337 (2021). https://doi.org/10.3390/ma14216337
33. Schaan, G., Rybczynski, S., Schmidt-Döhl, F., Ritter, M.: Imaging Microsc. **2**, 25–27 (2020)
34. Stark, J., Bollmann, K.: Delayed ettringite formation in concrete. Nordic Concr. Res.-Publ. **23**, 4–28 (2000)
35. Zhang, Y.L., Pan, Y.F., Zhang, D.K.A.: A literature review on delayed ettringite formation: mechanism, affect factors, and suppressing methods. Mag. Concr. Res. **73**, 325–342 (2021). https://doi.org/10.1680/jmacr.20.00268
36. Dosta, M., Skorych, V.: MUSEN: an open-source framework for GPU-accelerated DEM simulations. SoftwareX **12**, 100618 (2020). https://doi.org/10.1016/j.softx.2020.100618
37. Cundall, P.A., Strack, O.D.L.: A discrete numerical model for granular assemblies. Geotechnique **29**, 47–65 (1979). https://doi.org/10.1680/geot.1979.29.1.47
38. Kieckhefen, P., Pietsch, S., Dosta, M., Heinrich, S.: Possibilities and limits of computational fluid dynamics–discrete element method simulations in process engineering: a review of recent advancements and future trends. Annu. Rev. Chem. Biomol. Eng. **11**, 397–422 (2020). https://doi.org/10.1146/annurev-chembioeng-110519-075414
39. Jarolin, K., Dosta, M.: Linearization-based methods for the calibration of bonded-particle models. Comput. Part. Mech. **8**, 511–523 (2020). https://doi.org/10.1007/s40571-020-00348-z
40. Ramdohr, P.: The Ore Minerals and Their Intergrowths, 3rd edn. Pergamon Press, Oxford (1969)
41. GitHub—Msolids/Musen. Available online: https://github.com/msolids/musen

Influence of Load-Induced Temperature Fields on the Fatigue Behaviour of UHPC Subjected to High Frequency Compression Loading

Silke Scheerer, Melchior Thieme, and Ngoc Linh Tran

Abstract An experimental study on load-induced temperature fields generated in cyclic compression tests was carried out to close an existing knowledge gap on the material behaviour of ultra-high-performance concrete (UHPC). The main result of the experimental part is an analytical model with which the misleading influence of specimen heating on the results of such fatigue tests can be eliminated. The theoretical work was based on the microplane model M4 for concrete. The model was calibrated on laboratory tests on the investigated UHPC. The results of the numerical simulations confirmed the influence of temperature on the fatigue strength.

Keywords Ultra-high-performance concrete (UHPC) · Temperature fields · Cyclic compression test · Analytical model

1 Intension

Saving resources and reducing CO_2 emissions are two of the most important challenges of our time. Construction-related emissions are a significant proportion of global emissions of greenhouse gases [1]. In order to reduce the amount of concrete, components must be designed according to the force flow and materials should be adapted to specific requirements. This results, for example, in (i) customized special

S. Scheerer (✉)
Institute of Concrete Structures, TUD Dresden University of Technology, Dresden, Germany
e-mail: silke.scheerer@tu-dresden.de

M. Thieme
Konstruktiver Ingenieurbau—Schwerpunkt Massivbau, Hochschule Magdeburg-Stendal, Magdeburg, Germany

Former: Institute of Concrete Structures, TUD Dresden University of Technology, Dresden, Germany

N. L. Tran
Vietnamese-German University, Ben Cat City, Vietnam

Former: Institute of Concrete Structures, TU Darmstadt, Darmstadt, Germany

N. Oneschkow and L. Lohaus (eds.), *Cyclic Deterioration of High-Performance Concrete*, https://doi.org/10.1007/978-3-032-19394-0_12

concretes whose properties have often not yet been researched in every detail, and (ii) in lightweight, filigree structures, which are often exposed to higher fatigue loads than massive, conventional steel-reinforced concrete structures.

Ultra-high-performance concretes (UHPCs) are predestined for the realization of filigree structures. However, studies on their fatigue behaviour under high-frequency compression loading, see for instance [2–6], led to the hypothesis that load-induced temperature fields have a non-negligible influence on their damage process and the fatigue strength determined in experiments. Temperature increases in the range of 20–50 K are no exception. Particularly for UHPC, systematic investigations into the effect of load-induced temperature fields on the degradation of concrete under fatigue loading were still outstanding and, therefore, the focus of the research presented here. In parallel, research on the heating behaviour of high-performance concrete (HPC) and grouted concrete was carried out in [7, 8] and in the chapter by H. Madadi et al., this volume, but without overlaps.

Regarding material modelling, there was a deficit of appropriate material models for concrete due to the complex mechanical behaviour of cohesive friction material. In particular, load-induced temperature stresses and their effects on the damage process and fatigue strength could not yet be modelled sufficiently well. The microplane concept by Bažant et al. [9] was considered to have great potential for realistically capturing the diverse mechanical phenomena. For the numerical investigations, this approach served as the basis for the development of a new material model to describe the fatigue behaviour of UHPC under high-frequency compression loading.

2 Experimental Investigations

2.1 Materials, Specimen Manufacturing and Test Programme

The experimental results presented below were published in more detail in [10–17] and in Melchior Thieme's (neé Deutscher) doctoral dissertation [18].

Materials The reference material of the Priority Programme SPP 2020—UHPC 1—and a similarly composed UHPC with a larger maximum grain size—UHPC 2—were selected for the main tests. Table 1 gives an overview of the mixtures and of basic material values for both UHPC. For more details see [18] and Sect. 2.1 "Concrete Composition" in the chapter by M. Basaldella et al., this volume. Additional tests were carried out on a high-performance concrete (HPC), a normal-strength concrete (NC), and a high-strength mortar. However, the compositions and results regarding these three materials are not a central part of this report and will be only mentioned in passing.

Specimens The concrete cylinders for the cyclic compression tests had a height h of 180 mm and a diameter d of 60 mm and, therefore, a slenderness ratio of $d/h = 1/3$. This geometry allows good comparability with other research results (including tests for creating Wöhler curves of *fib* Model Code 2010 [19]).

Table 1 Composition and standard strength parameters of UHPCs 1/2, for more details see for instance [18]

	UHPC 1	UHPC 2
Components in kg/m^3		
CEM I 52,5R - SR3 (na)	759	667
Microsilica (powder)	169	182
Quarz flour W12 I W3	198 I –	333 I 134
Fine sand H33	971	363
Basalt 2/8	–	612
Water	188	162
Superplasticizer	24	23
Water-binder (w/b) ratio	0.20	0.19
Characteristics of hardened concrete in MPa (mean values from three specimens each)		
Compressive strength f_{cm} [a]	162.7	167.2
Elastic modulus E_{cm} [a]	47,000	50,500
Reference compressive strength $f_{cm,90}$ [b]	184.5	196.5

[a]Determined on 28 days old cylindrical specimens (height h = 300 mm, diameter d = 150 mm), [b]Determined on 90 days old cylindrical specimens (height h = 180 mm, diameter d = 60 mm)

The specimens were manufactured and tested in the Otto Mohr Laboratory of TU Dresden, with the exception of the HPC specimens, which were produced and tested within the SPP 2020 project of Steeb/Garrecht (see the chapter by H. Madadi et al., this volume). Using five formwork panels each consisting of nine 200 mm long plastic pipes, Fig. 1a), 45 specimens were produced per batch. Four batches were concreted for UHPC 1, and one batch each for the other concretes and the mortar. After concreting, the specimens remained covered in the formwork for 24 h. Further storage was carried out according to [20] in a water bath until the 7th day and then in a climatic chamber at 20 °C and a relative humidity of 65% until further processing and testing. The storage conditions were modified only for selected supplementary tests. Before the load tests, the specimens were sawn to the required length of 180 mm.

Test programme The aim was to detect the heating effect primarily in UHPC and to identify the influence of different parameters on test-related load-induced specimen heating. For this purpose, mainly the concrete composition and age, the load range and the frequency f_p were varied in the cyclic compression tests (Table 2) on three cylinders for each test configuration. Minimum and maximum stress levels $S_{c,min}$ and $S_{c,max}$ were defined as a percentage of the static compressive strength f_{cm} of the tested reference specimens.

Supplementary cyclic compression tests on HPC, NC, and mortar specimens were carried out. Also, deviating storage or test conditions—for example prevention of drying out or cooling, compression tests at different temperatures—were taken into

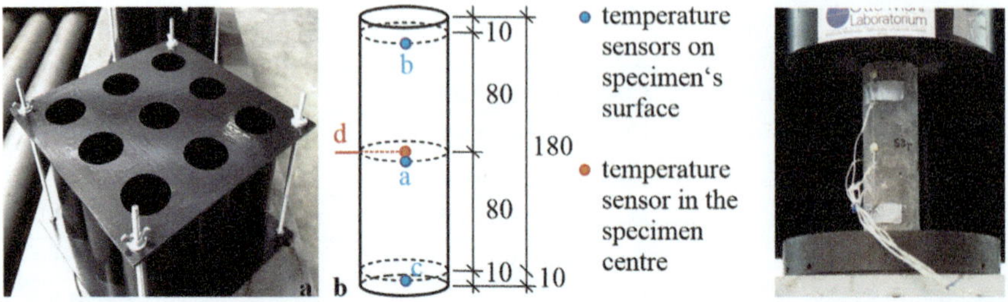

Fig. 1 Specimens for cyclic tests: **a** formwork, **b** arrangement of temperature sensors (configuration 1), **c** UHPC specimen with applied temperature measurement sensors before the cyclic test (photos: Melchior Thieme, graphic: Silke Scheerer)

Table 2 Overview of main tests on UHPCs conducted in the project; more details are given, e.g., in [18]

Material	Minimum stress level $S_{c,min}$ [–]	Maximum stress level $S_{c,max}$ [–]	Frequency f_p in Hz	Specimen age in d	Number of specimens [–]
UHPC 1	0.1	0.7	3 | 7.5 | 10 | 12.5 | 20 | 25 | 30	28 … 206	12 × 3
		0.6 | 0.8	3 | 10 | 20	89 … 177	6 × 3
		0.65 | 0.75	10	91 … 140	2 × 3
	0.05	0.65 | 0.7 | 0.75	10	93 … 120	4 × 3
	0.15	0.65 | 0.7 | 0.75	10	111 … 143	3 × 3
UHPC 2	0.1	0.6	3 | 10 | 20	93 … 216	9 × 3
		0.7		91 … 210	
		0.8		92 … 210	

account. In addition, the influence of pre-damage of test specimens on the fatigue behaviour was considered. Selected tests were conducted at the Versuchstechnische Einrichtung (VTE) of Bauhaus-Universität Weimar (test parameters: $S_{c,min} = 0.1$, $S_{c,max} = 0.7, f_p = 10$, 20, 30 Hz). In addition, some results of MPA Stuttgart were also included in the evaluation.

2.2 Test Conduction, Measuring Devices and Data Evaluation

The cyclic tests were carried out in servo-hydraulic testing machines. At the beginning, the scheduled mean stress level $S_{c,M}$ was applied. Then, the sinusoidal cyclic loading between the maximum and minimum stress level was started from this stress level. The cyclic load was applied until the specimen failed or until two million load cycles (LC) were applied. Specimens without failure up to this number of load cycles (so-called 'runout specimens') were finally subjected to a static compressive strength test.

During the fatigue tests, the applied force and frequency, the machine path, the specimen temperature at several positions, the temperature of surrounding and of the load application plates as well as specimen's strains were continuously recorded.

Three measurement configurations were used for the three tests per parameter combination [10]:

- during every test: temperature on three positions on the specimen's surface (Fig. 1b),
- configuration 1: temperature in the specimen centre,
- configuration 2: strains in vertical direction (three strain gauges evenly distributed around the circumference at half sample height),
- configuration 3: strains in vertical and circumferential direction (with one cross-like strain gauge at half sample height).

The temperature difference was the most interesting information for the evaluation. The test durations varied greatly depending on the parameter combination and could last up to several days. The fluctuating ambient temperature during a test was subtracted from the temperature measured on the specimens. To reduce the amount of data, only every 10th recorded dataset was considered.

An important characteristic value is the temperature rate or heating rate. It indicates the amount of temperature change generated during one load cycle or a certain time interval. It was calculated by deriving the raw data of the temperature curve. Due to the noise of the measured values, the derivation sometimes resulted in large scatter, which is why the data were smoothed using a moving average as a function of the load frequency f_{p}. Trend lines were generated from the smoothed data for further evaluation, details are given in [18].

2.3 Test Results

Temperature development in the specimen Non-uniform heating of specimens during cyclic compression tests has been reported for instance in [8, 21, 22]. The recorded temperature is usually highest at about half the height (inside) the specimen or at the point of fatigue failure, which was confirmed in our tests. Figure 2a shows a typical temperature development of cyclic loaded specimen, measured at four different positions. The specimen temperature increases steadily until specimen failure. Considering the varying slope, the shape of the curve is basically similar to the typical three-phase progression of strain increase over the number of load cycles: strong strain increase in phase I (microcrack formation, stress redistribution), phase II with moderate curve increase (stable crack growth), and phase III with rapidly

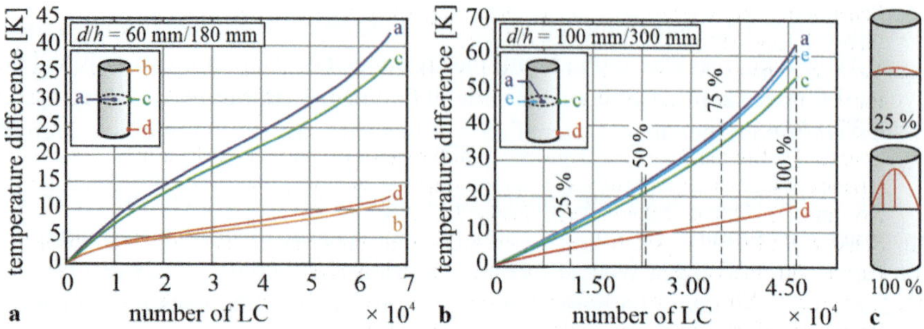

Fig. 2 Examples for specimen heating during cyclic compression tests ($S_{c,min}/S_{c,max} = 0.1/0.7$, $f_p = 10$ Hz): temperature change at the surface and inside of a specimen **a** with $d/h = 60$ mm/180 mm and **b** with $d/h = 100$ mm/300 mm, **c** temperature gradient curves at 25 and 100% of the test duration (specimen with $d/h = 100$ mm/300 mm) (based on [18], published under CC-BY-NC-ND 4.0, with permission from Melchior Thieme, modified)

increasing degradation until failure. The heating is significantly greater halfway up the specimen than near to the upper and lower edges.

The gradient between the specimen's surface and its centre is also clearly visible. To determine the gradient more precisely, an additional temperature sensor was embedded at the quarter point of the cross-section in a larger cylinder with $d/h = 100/300$ mm. The recorded non-linear course of the temperature field inside the specimen is shown in Fig. 2c. More details can be found in [18].

Parameter study In the following, only an excerpt of the whole parameter study is presented. For more details see [10, 13, 14, 16, 18]. The influence of the maximum stress level $S_{c,max}$, the frequency f_p, and of the concrete's maximum grain size on the fatigue strength was researched by 30 fatigue tests. The minimum stress level $S_{c,min}$ was set to 0.1. Nine UHPC 1 specimens each were tested at three different maximum load levels $S_{c,max} = 0.6, 0.7, 0.8$ at three frequencies $f_p = 3, 10, 20$ Hz. The diagrams in Fig. 3 show the heating at the specimens' surface at half the specimen height over the test duration (represented by the number of load cycles). For the runout specimens, there is an interruption on the x-axis from around 200,000 load cycles to just before the upper limit of 2 million LC.

A temperature increase was recorded in every test, but the characteristics of the curves indicate a strong dependency on the test conditions. The diagram in Fig. 3a shows results for the 3 Hz tests. With increasing $S_{c,max}$, the maximum temperature reached and the slope of the curves increased. At 0.6 $S_{c,max}$, a maximum temperature difference of 2 to 5 K was recorded which remained constant until termination after applying 2 million LC without specimen failure. Specimens tested with $S_{c,max} = 0.7$ showed a maximum temperature increase of 8 K after 20,000 to 25,000 load cycles. Then, the temperature remained almost constant until a small decrease was observed shortly before the end of the tests. One of the three tested cylinders did not withstand 2 million LC. At a load level of 0.8 $S_{c,max}$, the samples failed while the temperature continued to rise. In two of three cases, the curves show a nearly linear slope until an early specimen failure. With rising frequency (see Fig. 3b and c), the load-induced

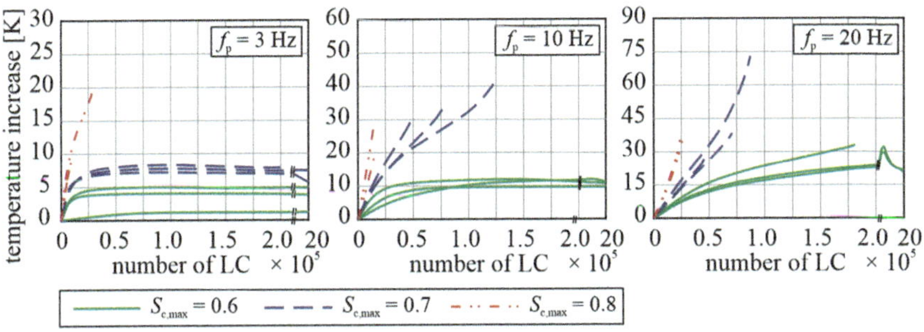

Fig. 3 Temperature development of UHPC 1 specimens for $S_{c,min} = 0.1$, three different maximum stress levels $S_{c,max} = 0.6$, 0.7, and 0.8, and three load frequencies $f_p = 3$, 10, and 20 Hz (published in [10], licensed under CC-BY 4.0, modified)

heating of the specimens increased, and, in general, the specimens failed after less load cycles. The series with $S_{c,max} = 0.6$ and $f_p = 20$ Hz shows a special feature. In two of three tests, the temperature increased over 500,000 LC and dropped then by about 12 K. From 1.25 to 2 million load cycles, the temperature remained constant. One specimen failed after 180,000 LC.

To research the influence of the maximum grain size, UHPC 2 cylinders were tested with $S_{c,min}/S_{c,max} = 0.1/0.7$ and $f_p = 10$ Hz [10]. All specimens failed during the load-induced heating process. The generated temperatures were similar for both UHPC. However, the temperature increase-load cycle curves of UHPC 1 (1 mm maximum grain size) always showed a steeper rise and therefore a lower number of endured LC than the curves determined for UHPC 2 (8 mm maximum grain size).

For the visualization of the heating up speed in regard of the number of LC or time, the heating rate is shown in Fig. 4. The displayed curves belong to the test specimens that reached the middle temperature value of the respective parameter combination. The fastest heating rate of 10^{-3} K/LC, Fig. 4a, was recorded during the 3 Hz test directly after the beginning of the cyclic loading. Compared to the other configurations, at 3 Hz the amount of applied load per LC was the highest. Consequently, more damage-inducing energy was introduced into the specimen, resulting in a higher heating per cycle. The temperature increase decreased rapidly afterwards. After 1500 and 4150 LC, the 10 Hz and the 20 Hz curves were crossed, respective. The heating speed has already halved after only 0.25% of the total number of applied LC. After 75,000 LC (less than 4% of the total number of LC), no further heating up was recorded.

In the 10 Hz test, the temperature development was again most at the beginning, but 20% less than for the 3 Hz specimen. The curve also dropped, but remained clearly above zero. The temperature generation never came to a standstill. After about 2/3 of the test duration, the heating speed increased again and reached nearly the level from the beginning at the point of failure. Based on Fig. 4b, the heating rate and the strain rate development can be related to each other. The highest heating rates were recorded in the phases I and III when increased deterioration takes place. In phase II

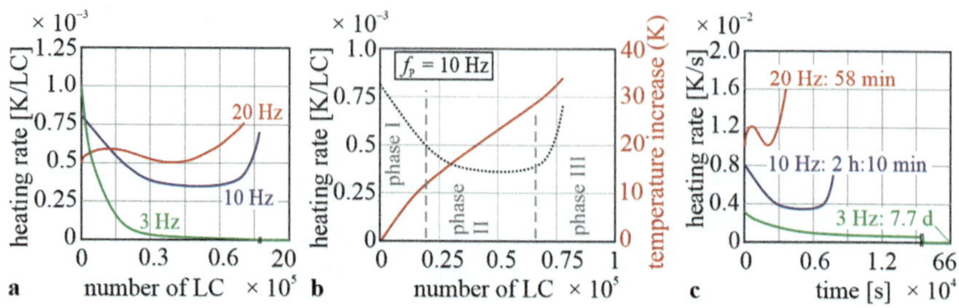

Fig. 4 Temperature development of UHPC 1 specimens tested with $S_{c,min}/S_{c,max} = 0.1/0.7$; **a** heating per LC for the three load frequencies of 3, 10, and 20 Hz, **b** relation between heating rate and temperature difference, **c** heating up per second ([10, 13], licensed under CC-BY 4.0, modified)

with continuous widening of cracks, the temperature generation almost stopped as no real (micro-)structural changes occurred. In the 20 Hz test, the observed heating rate did not fall below the value at the beginning of the cyclic loading. Shortly before failure, the highest heating rate was recorded with $0.76 \cdot 10^{-3}$ K/LC. In Fig. 4c, the heating rate is displayed over time. The curves' characteristics are similar to the curves in Fig. 4a. A high frequency f_p results in a high and small frequency in a low heating rate but without direct proportionality between the two values.

To summarize that part of the study, it can be pointed out, that all three discussed parameters – frequency f_p, maximum load level $S_{c,max}$ and maximum grain size – influence significantly the load-induced temperature rise. Whereas the maximum temperature reached depends strongly on the frequency, the other two parameters mainly influence the heating rate. Table 3 shows the summarized results for the various influencing parameters.

Special investigations into the heating process and effects on the mechanical properties Some supplementary tests mainly on UHPC 1 specimens were carried out to improve knowledge of the heating process itself, for instance regarding the influence of an insulation or a pre-damage on temperature release and increase. The basic configuration was $f_p = 10$ Hz, and $S_{c,min}/S_{c.max} = 0.1/0.7$. Measurement configuration 1 was used [13, 16, 17]. The study required several consecutive loading scenarios for individual specimens. At first, three specimens (1 T, 2 T, and 3 T) were heated up in an oven to 50 °C, then clamped in the test machine between the steel load introduction plates, analogous to the fatigue tests, for recording the cooling. Due to the direct contact to the steel, fastest cooling was observed at top and bottom of the cylinders. At a 4th specimen (4 T), insulation layers were placed between concrete and steel plates. Now, all sensors recorded cooling in the same magnitude and speed.

Next, specimens 1 T–3 T were cyclic loaded to temperature increases of 15, 25, and 30 K, respectively. The different required numbers of LC resulted in different pre-damage grades. Then, cooling was recorded again, showing very similar results, Fig. 5a, but no systematic change in heat dissipation due to previous damage. The temperature release changed depending on the temperature difference to the

Table 3 Summary of parameter study with effect of increasing the respective parameter; more details are given in [18]

Increasing parameter	Load cycles until failure	Temperature rate	Maximum temperature difference
Load range			
$S_{c,max}$	Strongly decreases	Decreases	Increase up to ~0.7, then decrease due to premature failure
$S_{c,min}$	Increases	Decreases	Decreases
$S_{c,M}$	Decreases	Constant	Decreases [a]
Frequency f_P	Decreases	Constant at the beginning of the test, then increasing	Increases up to 20 Hz with decreasing frequency influence
Maximum grain size	Decreases (temperature-dependent)	Decreases	Decreases
Compressive strength	No influence	Increases (primarily dependent on packing density, not on strength)	
Pre-damage	– [b]	Increases, temperature release is not affected	– [b]
Specimen age			
≤ 90 d	Increases	Decreases	Decreases
> 90 d	Constant	Constant	Constant

[a] dependent on number of LC until failure, [b] not specified because of insufficient data

surrounding. The number of load cycles had no influence. The relationship between temperature release rate T_{rate} and absolute specimen temperature T_{is} was linear.

Finally, the specimens 1 T–3 T were cyclically loaded until failure to assess changes in temperature increases depending on pre-damage. It was observed that a

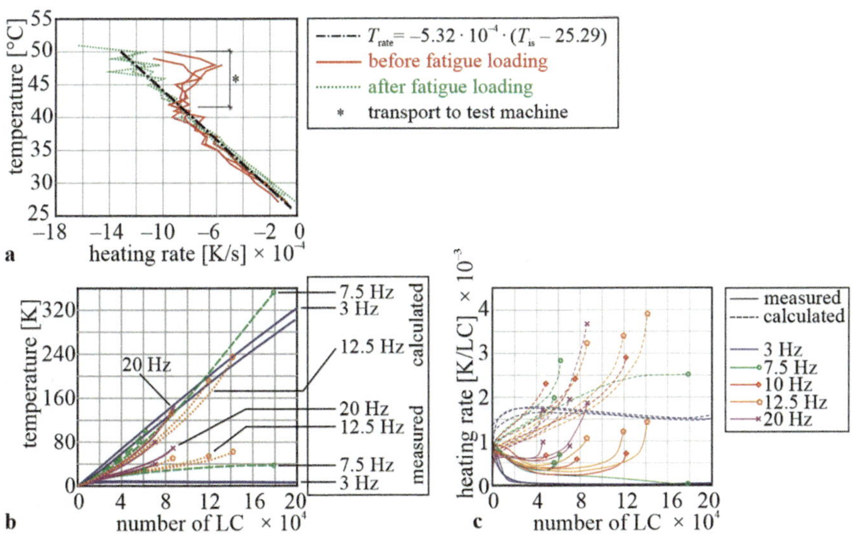

Fig. 5 **a** Cooling of UHPC 1 specimens after different pre-damage scenarios (with 25.29 °C as mean ambient temperature T_{surr} during the lab tests), **b** comparison of measured and calculated temperature development and **c** heating rate (data published in [16], modified)

longer cyclic preload and, thus, a higher heating of the specimen in a first loading phase (after specimen cooling) led to a significantly higher heating rate during renewed fatigue loading. The reason for this can be assumed to be the higher pre-damage in the concrete and, thus, the greater existing internal friction potential. If the temperature release is independent of the degree of damage, the heating up of a specimen can be calculated fictitiously as if cooling not takes place, resulting in a temperature rate without cooling. By integrating this rate over the load cycles, the summarized temperature generated over the test duration is obtained. This algorithm was applied to all UHPC 1 tests. Figure 5b shows examples of the temperatures generated without cooling, and Fig. 5c the corresponding heating rates. Basically, the heating rate increased over the entire course of the experiments, and the temperature increase was disproportionately high. The measured decrease of the slope of the temperature curves was therefore only generated by the simultaneous increase in temperature release.

It is interesting that the measured temperature curves of the 3 Hz tests did not show any significant heating. So, the actual heating is frequency-independent, only the possible cooling per LC during slower tests prevented strong heating. For the experimentally verification of this hypothesis, tests were carried out at 3 Hz and $S_{c,min}/S_{c,max} = 0.1/0.7$ on thermally insulated specimens. Figure 6a and b allow a comparison. The insulated specimens warmed up steadily by 23.2 K on average until failure. Without insulation, the temperature difference in the concrete was always below 10 K, and there was no failure within 2 million LC. The insulated specimens failed at ~ 50,000 LC. The calculated temperatures were not reached, as cooling took place via the steel plates.

To research the aspect of the temperature-dependent compressive strength, another supplementary test series was conducted on UHPC 1 and HPC specimens together with the SPP 2020 project of Steeb/Garrecht (see the chapter by H. Madadi et al., this volume) [17]. Strength losses on NC or HPC specimens are well documented, see for example [23], but knowledge on UHPC was not available so far. For UHPC 1, Fig. 7 shows a decrease of the uniaxial strength with increasing specimen temperature. A trend line and expected values for NC and HPC according to *fib* MC 2010 [19] are added.

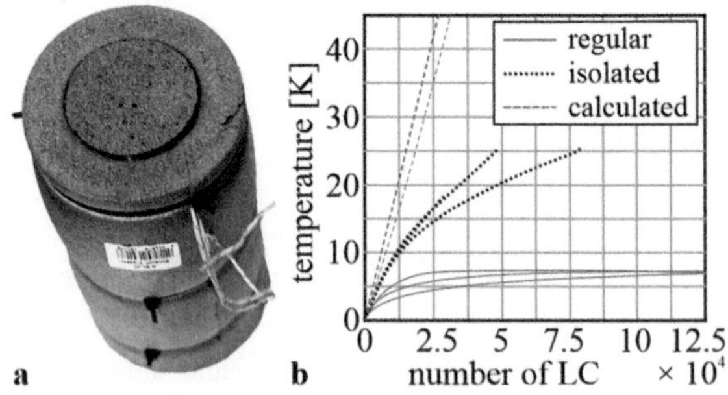

Fig. 6 **a** Insulated specimen, **b** heat development of UHPC 1 specimens without and with thermal insulation ($f_p =$ 3 Hz, $S_{c,min}/S_{c,max} = 0.1/0.7$) (published in [16], modified)

Fig. 7 Temperature-dependent relative static compressive strength of UHPC 1 (data published in [15], licensed under CC-BY 4.0, modified)

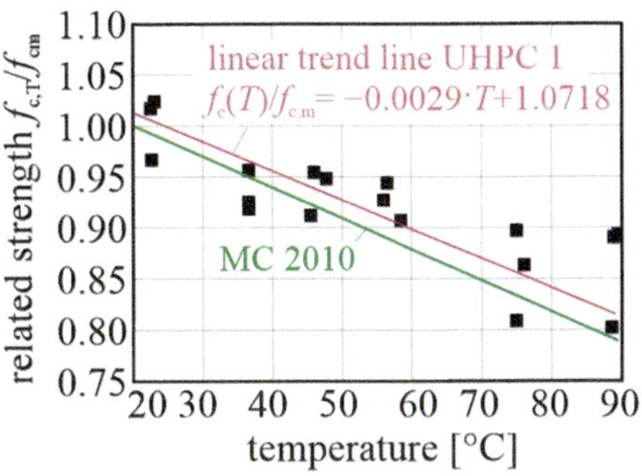

3 Analytical Model

The aim of the developing of an analytical model was to use temperature measurement data to subsequently correct results from laboratory tests with regard to a possible temperature-induced influence on fatigue strength [15]. This intension differed from previous approaches [7, 8]. The concept does not generate any significant additional effort in test execution and does not contain any strict limitations on test speed.

The basic idea was to relate the temperature-dependent static compressive strength to the fatigue curves. Based on the linear equation for the strength loss (Fig. 7), the related compressive strength, which changes (continuously) as a function of the observed concrete heating, can be determined for the fatigue tests (examples are displayed in Fig. 8a). Figure 8b shows the resulting change of the load range for the selected tests. The minimum stress level $S_{c,min}$ remains almost constant. The maximum stress level $S_{c,max}$ and the stress amplitude increased up to 10% when a high temperature increase was recorded. This increases the temperature rate and can also cause an earlier failure.

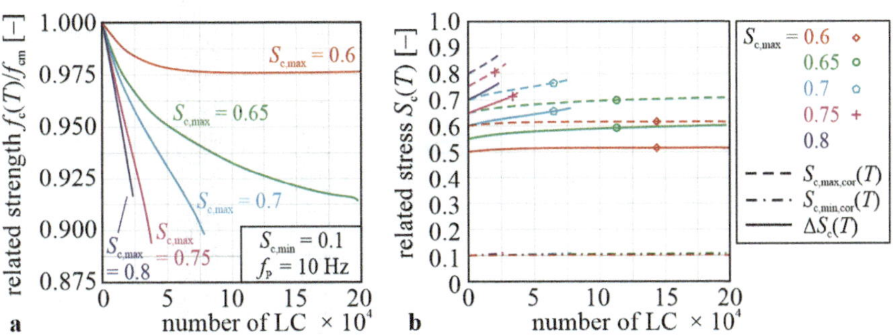

Fig. 8 **a** Decreasing concrete compressive strengths in fatigue tests dependent on the specimen heating occurred, **b** load range change during the cyclic tests as a function of temperature (data published in [15], licensed under CC-BY 4.0, modified)

Results with changing maximum stress levels $S_{c.max}$ cannot be included in Wöhler diagrams. So, it was necessary to determine a constant load range, which causes an equivalent fatigue damage determinable solely with a measured maximum temperature value. All fixed material parameters are included in the temperature-dependent strength reduction or the different maximum temperature reached. It is therefore sufficient to consider the initial load range in addition to T_{max}. In the tests, T_{max} was generally reached directly before failure or test termination. However, the curves in Fig. 8b show that the resulting maximum stress cycles max $\Delta S_c(T)$ are only degressive for load ranges up to 55% and exhibit the maximum value over a long time. With higher load ranges, T_{max} is only reached immediately before failure, and the associated load range is only present over a comparably short duration. Therefore, a correction value β_{cor} was added to consider this temporal dimension of T_{max}. The lower limit of 1.0 ensures that the calculated temperature does not exceed the measured one. Finally, the corrected maximum stress level $S_{c,max,cor}(T)$ can be calculated as follows:

$$S_{c,\max,cor}(T) = S_{c,\max} + \frac{\frac{S_{c,\max}}{f_c(T_{\max})/f_{cm}} - S_{c,\max}}{1 + 10 \cdot (\beta_{cor} - 1)} \tag{1}$$

$$\text{with}: \frac{f_c(T_{\max})}{f_{cm}} = 1.0718 - 0.0029 \cdot T_{\max}$$

$$\beta_{cor} = \frac{S_{c,\max} - S_{c,\min}}{0.55} \geq 1.0$$

The method was applied to all UHPC 1 tests and resulted in an adjustment as shown in Fig. 9 for different minimum stress levels of $S_{c,min}$. Regardless of the load range, the values with strong heating that laid originally below the Wöhler curve are corrected, as the maximum stress and load range changed the most during the tests.

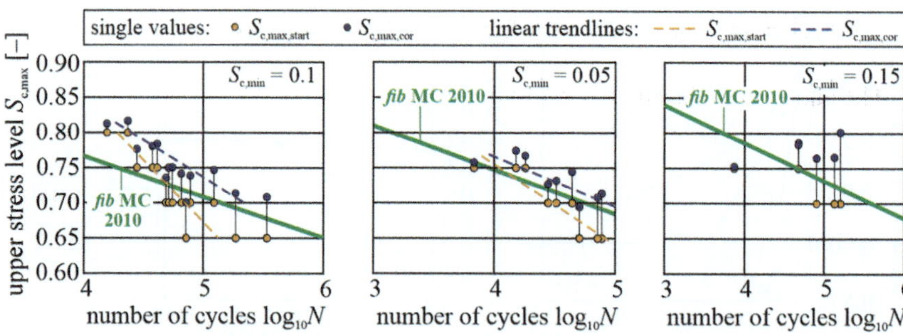

Fig. 9 Effect of the temperature-dependent correction of the load range on the mapping of the test values in *S*–*N* diagrams (tests on UHPC 1 with different levels of $S_{c,min}$) ([15], licensed under CC-BY 4.0, modified)

The method was also applied to test results on HPC from the SPP 2020 project of Steeb/Garrecht (see the chapter by H. Madadi et al., this volume) [18] with very good results.

For all verifications carried out, the temperature correction resulted in values approaching the Wöhler curves according to *fib* MC 2010. An overestimation by the analytical method can be ruled out, because the achieved LC numbers in the tests on specimens without significant heating ($f_\mathrm{p} = 1$ to 3 Hz) were higher than the corrected values at the same load range.

The proposed analytical method enables material testing at high frequencies with the common test procedure and independent of concrete strength or composition. Results of previous tests can be corrected retrospectively.

4 Numerical Investigations

The aim of the numerical investigations carried out at TU Darmstadt was to explain the influencing parameters obtained from the experimental work and to identify further influencing parameters, for example the size effect. The numerical approach used for the simulation of the fatigue behaviour of concrete is based on the macro level, in which only a few parameters are required. It is assumed that the heating resulting from cyclic loading significantly influences the damage process in UHPC. On the one hand, the temperature gradient causes residual stresses in the concrete and local stress increases, which lead to a lower fatigue strength of the specimen. On the other hand, the absolute temperature influences the concrete properties [8].

In particular, a degradation of the mechanical properties is to be expected. A thermo-mechanically coupled model was developed and implemented in the FE environment ATENA Science version 5.4.1q (Cervenka Company) for the analysis of the behaviour of concrete under cyclic compression loading with consideration of load-induced temperature fields. The calculation was carried out in a closed process with a thermal and a mechanical module. The results generated in one module are the input parameters for the analysis with the other. For the calibration of the model, the recalculation of the tests, and for further investigations, a constant element size was selected in order to eliminate its possible influence on the results. The element size was selected in the order of magnitude of the maximum aggregate size, but not larger than 10% of the diameter of the cylinder specimens. In this project, a maximum element size of 6 mm was used for both cylinders with diameters of 60 mm and 150 mm.

4.1 Mechanism of Heat Generation

Concrete under load shows elastic and plastic strains. The load-induced heat as a type of dissipation energy is caused by plastic work in the concrete. This was confirmed experimentally, whereby the rate of heat generation correlated very well with the development of plastic concrete strain in three phases (see Fig. 4b). The plastic strains occur in the weak points of the concrete specimen, one part in the contact zone between binder and aggregates (interfacial transition zone, ITZ) and another part in the binder. The crack distribution is mainly determined by the size of the aggregates, see for example [24]. The working hypothesis was that the heat is generated by internal friction between binder and aggregates (ITZ). The heat generated increases when the total area of the cracks in the concrete and the friction path increase. The latter corresponds to a larger stress range. By analysing the area of the contact zone as a function of the grain diameter D (in mm), the energy conversion fraction of the plastic work in concrete β_{D} can be formulated according to Eq. (2), where β_1 is the reference value corresponding to the energy conversion ratio into heat for a concrete with a grain diameter $D = 1$ mm. A detailed formulation can be found in [25].

$$\beta_D = \frac{\beta_1}{D} \tag{2}$$

In concrete, the aggregates generally correspond to a specific grading curve. Therefore, the equivalent grain diameter D_{ef} should be used, see Eq. (3). Here, $v_{\mathrm{a,Di}}$ is the relative volume of the aggregates with the diameter D_{i} in the concrete.

$$\frac{1}{D_{ef}} = \sum_{i=1}^{n} \frac{v_{a,Di}}{D_i} \tag{3}$$

Equation (2) was confirmed by test results for HPC taken from [5] in [26]. For the investigated UHPC 1, Eq. (3) was used to determine D_{ef}. The comparison of concretes with different maximum grain diameters resulted in a heat generation rate twice as high for UHPC 1 with $D_{\mathrm{max}} = 1$ mm ($D_{\mathrm{ef}} = 0.8$ mm) as for UHPC 2 with $D_{\mathrm{max}} = 8$ mm ($D_{\mathrm{ef}} = 1.6$ mm). This calculation result was confirmed by the tests presented in Sect. 2.

Based on the test results [27], the heat generation rate $\dot{q}_v$ without regard to cooling can be described as a function of the acting stress σ and the plastic strain rate $\dot{\varepsilon}_p$ as follows:

$$\dot{q}_v = C \cdot \frac{\sigma \cdot \dot{\varepsilon}_p}{D} \tag{4}$$

Here, C (in mm) is an empirical parameter that is constant for a specific concrete, which is determined by comparing the enclosed area of the experimentally recorded hysteresis of the stress–strain curve, Fig. 10a, and the calculated thermal energy. C has a value of 220 for the investigated concretes and for predominantly normal stresses, which was the case in our tests. In general, C can be determined separately for each stress and distortion component in the microplane model. It should be noted that the enclosed areas of the measured hysteresis of the stress–strain curves of UHPC 1 and UHPC 2 were almost the same, but the two concretes showed different

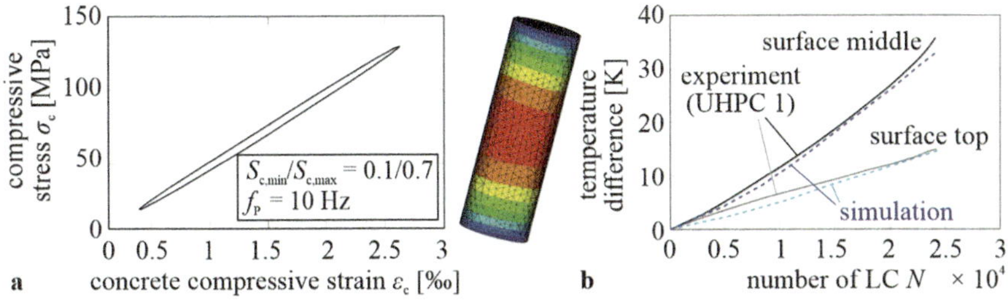

Fig. 10 **a** Hysteresis of a stress–strain curve and **b** numerical results of the heating in a concrete body·due to cyclic compression load

heat generation due to the different equivalent grain diameters, see [18]. With the presented description of the heat generation rate, the development of the temperature in a concrete body was simulated. The numerical recalculation of tests, see Fig. 10b for specimen UHPC 1–80-20b, showed very good agreement.

4.2 Material Model to Describe the Fatigue Behaviour and Influence of a Load-Induced Temperature Field

The M4 microplane model developed by Bažant et al. [9] and already calibrated for NCs [28] was initially used to describe the fatigue behaviour of the investigated concretes. The M4 model was calibrated based on the static and cyclic laboratory tests on UHPC. The results of the fatigue tests were used to determine the critical uniaxial concrete strain in the failure state [26], which was also used to calibrate the model. Due to the very high expected numbers of load cycles, the damage rate in the numerical simulation was adjusted using a variant of the jump cycle technique to reduce the calculation time. Here, the interval of plastic strain was selected and the corresponding equivalent heat generation rate for a representative hysteresis was calculated according to Eq. (4). This approach is suitable since the decisive impact on the fatigue strength in the given case is the restraint effect due to load-induced temperature fields. Calculation results after calibration of the material model for UHPC under static and cyclic loads are shown in Fig. 11.

The simulation results of the fatigue behaviour of UHPC under consideration of load-induced temperature fields confirmed the influence of temperature on the fatigue strength. For the fatigue tests carried out in the laboratory with 60/180 mm cylinders, a load range $S_{c,min}/S_{c,max}$ = 0.1/0.7 and a frequency f_P = 10 Hz, the simulation showed a decrease in the total number of load cycles of around 5% compared to the case without considering a temperature field, which corresponds to a very low load frequency (see Fig. 12a). For cylinder specimens with a larger diameter, the difference was significantly greater. In the steady state, it can be deduced from [25], that the difference between internal and external temperature can be mapped with a

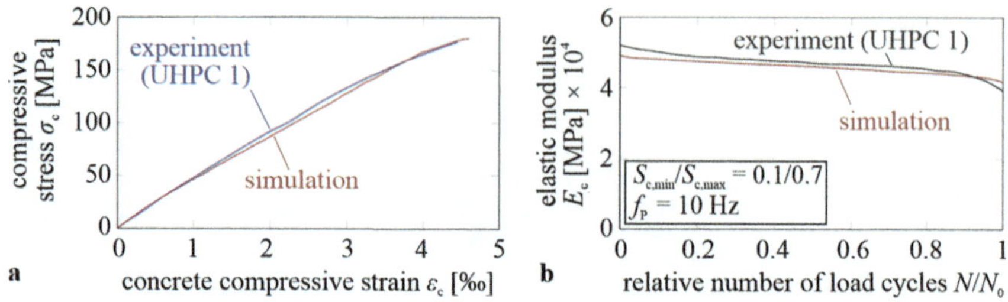

Fig. 11 Calibration of the material model for UHPC 1; **a** static and **b** numerical cyclic loading

quadratic function of the diameter of the concrete cylinder. The temperature gradient determines the residual stresses in the concrete specimens and thus the concrete fatigue strength. According to the numerical calculations, the reduction in the service life of a cylinder with $d = 150$ mm due to load-induced temperature will be around 16%.

The variation of the concrete compressive strength strongly influences the fatigue strength, especially at high compression loads. Figure 12b shows an example of a cylinder with $d/h = 60/180$ mm, constant absolute load amplitudes $\sigma_{c,o}$ and $\sigma_{c,u}$ and varying concrete compressive strength. This explains the large scatter of fatigue strength of concrete under cyclic loadings.

The calculation model can explain the source of the load-induced temperature and its influence on the concrete fatigue behaviour well, as well as the influencing parameters, like, for instance, the grain size or the duration of a load cycle. The model can be extended in order taking into account other influencing parameters such as moisture and pores in the concrete, or other types of concrete, for example fibre-reinforced concretes.

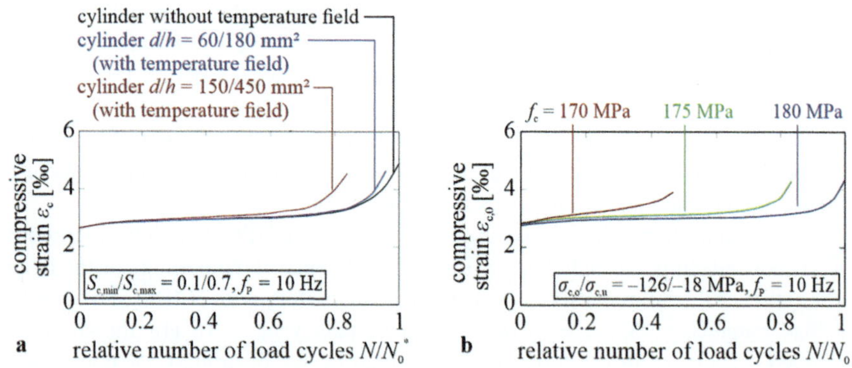

Fig. 12 Influence of load-induced temperature fields on the fatigue strength; variation of **a** specimen size (N_0^*: LC until failure without consideration of a temperature field) and **b** of concrete compressive strength

5 Conclusions

A comprehensive parameter study provided an overview of the key factors influencing the heating process of UHPC specimens in cyclic compression tests. The concrete's fatigue strength is primarily dependent on the maximum stress level $S_{c,max}$ and the heating per load cycle. The test speed and, therefore indirectly the load frequency, only influences the heating by determining the time period which is available for temperature dissipation during the relief phase of a load cycle. In cyclic compression tests in the laboratory, the heating up of a specimen results in a reduction of the number of load cycles until failure. This aspect as a result of the test conditions can be eliminated from the determined values using the proposed analytical model.

Based on the M4 microplane model by Bažant et al. [9], a thermo-mechanically coupled model with the introduction of mechanism of heat generation in concrete due to repeated loading was developed and implemented in the FE environment ATENA for simulating the behaviour of UHPC under cyclic compression loading. The calculation showed good agreement with the test results and is suitable to perform extended parameter studies.

References

1. United Nations Environment Programme: 2020 Global Status Report for Buildings and Construction: Towards a Zero-emission, Efficient and Resilient Buildings and Construction Sector. Nairobi. https://globalabc.org/news/launched-2020-global-status-report-buildings-and-construction. Accessed 13 September 2022
2. Anders, S.: Betontechnologische Einflüsse auf das Tragverhalten von Grouted Joints. Dissertation, LU Hannover (2007)
3. Fitik, B.: Ermüdungsverhalten von ultrahochfestem Beton (UHPC) bei zyklischen Beanspruchungen im Druck-Zug-Wechselbereich. Dissertation, TU München (2012)
4. Hohberg, R.: Zum Ermüdungsverhalten von Beton. Dissertation, TU Berlin (2004)
5. Elsmeier, K., Hümme, J., Oneschkow, N., Lohaus, L.: Prüftechnische Einflüsse auf das Ermüdungsverhalten hochfester feinkörniger Vergussbetone. Beton- und Stahlbetonbau **111**(4), 233–240 (2016). https://doi.org/10.1002/best.201500065
6. V. d. Haar, C., Hümme, J., Marx, S., Lohaus, L.: Untersuchungen zum Ermüdungsverhalten eines höherfesten Normalbetons. Beton- und Stahlbetonbau **110**(10), 699–709 (2015). https://doi.org/10.1002/best.201500034
7. Schneider, S.: Frequenzabhängigkeit des Ermüdungswiderstandes von hochfestem Beton. Dissertation, LU Hannover (2021)
8. Elsmeier, K.: Einfluss der Probekörpererwärmung auf das Ermüdungsverhalten von hochfestem Vergussbeton. Dissertation, LU Hannover (2019)
9. Bažant, Z.P., Caner, F.C., Carol, I., Adley, M.D., Akers, S.A.: Microplane model M4 for concrete. I: Formulation with work-conjugate deviatoric stress. J. Eng. Mech. **126**(9), 944–953 (2000). https://doi.org/10.1061/(ASCE)0733-9399(2000)126:9(944)
10. Deutscher, M., Tran, N.L., Scheerer, S.: Experimental investigations on the temperature increase of ultra-high performance concrete under fatigue loading. Appl. Sci. **9**(19), 4087 (2019). https://doi.org/10.3390/app9194087
11. Deutscher, M., Tran, N.L., Scheerer, S.: Experimental investigation of load-induced temperature development in UHPC subjected to cyclic loading. In: Derkowski, W., Gwozdziewicz, P.,

Hojdys, L., Krajewski, P., Pantak, M. (eds.) CONCRETE – Innovations in Materials, Design and Structures—Proceedings of the fib Symposium 2019, Krakow, 27–29 May 2019. Thelem, pp. 1904–1911 (2019)

12. Deutscher, M., Tran, N.L., Scheerer, S.: Experimental investigation of load-induced increase of temperature in UHPC. In: Zingoni, A. (ed.) Advances in Engineering Materials, Structures and Systems: Innovations, Mechanics and Applications—Proceedings of SEMC 2019, Cape Town, 2–4 September 2019, pp. 1450–1455. CRC Press/Taylor & Francis (2019). https://doi.org/10.1201/9780429426506
13. Deutscher, M., Tran, N.L., Scheerer, S.: Experimental investigations on temperature generation and release of ultra-high performance concrete during fatigue tests. Appl. Sci. **10**(17), 5845 (2020). https://doi.org/10.3390/app10175845
14. Deutscher, M., Markert, M., Tran, N.L., Scheerer, S.: Influence of the compressive strength of concrete on the temperature increase due to cyclic loading. In: Zhao, B., Lu, X. (eds.) Concrete Structures for Resilient Society—Proceedings of the fib Symposium 2020, online & Shanghai, pp. 773–780. 22–24 November 2020 (2020)
15. Deutscher, M.: Consideration of the heating of high-performance concretes during cyclic tests in the evaluation of results. Appl. Mech. **2**, 766–780 (2021). https://doi.org/10.3390/applmech2040044
16. Deutscher, M., Scheerer, S.: Heating rate with regard to temperature release of UHPC under cyclic compressive loading. CEND **3**, 143–152 (2021). https://doi.org/10.1002/cend.202100020
17. Deutscher, M., Markert, M., Scheerer, S.: Influence of temperature on the compressive strength of high performance and ultra-high performance concretes. Struct. Concr. **23**(4), 2381–2390 (2022). https://doi.org/10.1002/suco.202100153
18. Deutscher, M.: Berücksichtigung von Temperaturfeldern bei Ermüdungsversuchen an UHPC. Dissertation, TU Dresden. Schriftenreihe "Konstruktiver Ingenieurbau Dresden – KID", Heft 68. (2023). https://nbn-resolving.org/urn:nbn:de:bsz:14-qucosa2-840056. License: CC-BY-NC-ND 4.0
19. fib (ed.): fib Model Code for Concrete Structures 2010. Lausanne (2012)
20. Marx, S.: Empfehlungen zur Vereinheitlichung von Ermüdungsversuchen an Druckschwellbeanspruchtem Beton. DAfStb PA Ermüdung, LU Hannover (2013)
21. Elsmeier, K.: Influence of temperature on the fatigue behaviour of concrete. In: Stang, H., Braestrup, M. (eds.) Proceeding of International fib Congress, Copenhagen, 18–20 May 2015 (2015), pp. 325/326 (full paper: 9 pages).
22. Bode, M.: Energetische Schädigungsanalyse von ermüdungsbeanspruchtem Beton. Dissertation, LU Hannover (2020)
23. Budelmann H.: Zum Einfluß erhöhter Temperatur auf Festigkeit und Verformung von Beton mit unterschiedlichen Feuchtegehalten. Dissertation, TU Braunschweig (1987)
24. Thiele, M.: Experimentelle Untersuchung und Analyse der Schädigungsevolution in Beton unter hoch-zyklischen Ermüdungsbeanspruchungen. Dissertation, TU Berlin (2016)
25. Tran, N.L., Deutscher M., Scheerer, S.: An investigation of the heat generated during cyclic loading of ultra-high performance concrete. In: Eberhardsteiner, J., Schöberl, M. (eds.) Proc. of 90th Annual Meeting of the Int. Association of Applied Mathematics and Mechanics (GAMM), Vienna, 18–22 February 2019. PAMM **19**(1), e201900469 (2019). https://doi.org/10.1002/pamm.201900469
26. Tran, N.L., Deutscher, M., Scheerer, S.: Analysis of load-induced temperature development in UHPC under cyclic compressive loading. In: Zingoni, A. (ed.) Advances in Engineering Materials, Structures and Systems: Innovations, Mechanics and Applications—Proc. of SEMC 2019, Cape Town, 2–4 September 2019, pp. 1397–1402. London, CRC Press/Taylor & Francis (2019). https://doi.org/10.1201/9780429426506
27. Schneider, S., Marx, S.: Investigation on the influence of loading frequency on the fatigue resistance of a high strength concrete. In: Foster, S., Gilbert, R.I., Mendis, P., Al-Mahaidi, R., Millar, D. (eds.) Better, Smarter, Stronger—Proceedings of the 5th International fib Congress, pp. 595–606. Melbourne, 7–11 October 2018 (2018)

28. Caner, F.C., Bažant, Z.P.: Microplane model M4 for concrete. II: algorithm and calibration. J. Eng. Mech. **126**(9), 954–961 (2000). https://doi.org/10.1061/(ASCE)0733-9399(2000)126:9(954)

Degradation Processes in Ultra-High-Performance Fibre-Reinforced Concrete (UHPFRC) Under Cyclic Tensile Loading

Jan-Paul Lanwer and Martin Empelmann

Abstract The contribution contains the results of a fundamental investigation strategy to reveal the degradation processes of ultra-high-performance fibre-reinforced concrete (UHPFRC). First, the degradation of plain fibres and UHPC is investigated, followed by investigations of bond between fibres and UHPC. Final tests were conducted under cyclic loading with a 3D fibre distribution. The results show that fibres rupture in cyclic loaded UHPFRC. This was supported by experimental data of pullout tests and large-scale beam tests. The fibre rupture did occur for fibres that show a fibre alignment.

Keywords Ultra high performance · Fibre Reinforced Concrete · Post cracking tensile strength · Fatigue under tensile loading

1 Introduction

Modern structures demand high-performance materials efficiently utilizing its mechanical properties. UHPFRC is excelling in durability and mechanics. While plain UHPC is prone to brittle failure, added fibres give UHPFRC a post-crack tensile strength. This could lead in lessening the need for intricate bar installation.

Building with UHPFRC, elements tend to possess a slimmer and lighter profile compared to parts manufactured from conventional concrete [1]. The reduced dead weight or a less favourable cyclic load-to-dead weight ratio leads to heightened susceptibility to fatigue. As a result, special consideration must be given to cyclic behaviour. Presently, research of UHPFRC is primarily focussed on the macroscale [2–5] involving UHPFRC with 3D fibre distribution. Initial findings suggest that the fibres within UHPFRC exhibit a notable resistance to cyclic loading. However, due to

J.-P. Lanwer (✉) · M. Empelmann
Institute of Building Materials, Concrete Construction and Fire Safety, Division of Concrete Construction, Technische Universität Braunschweig, Braunschweig, Germany
e-mail: jan.paul.lanwer@t-online.de

N. Oneschkow and L. Lohaus (eds.), *Cyclic Deterioration of High-Performance Concrete*, https://doi.org/10.1007/978-3-032-19394-0_13

limited data available, the precise course of load-bearing behaviour for fibres within UHPFRC components under cyclic loading remains unclear.

2 Experimental Investigations

2.1 Structure of the Test Programme

Figure 1 shows the experimental programme.

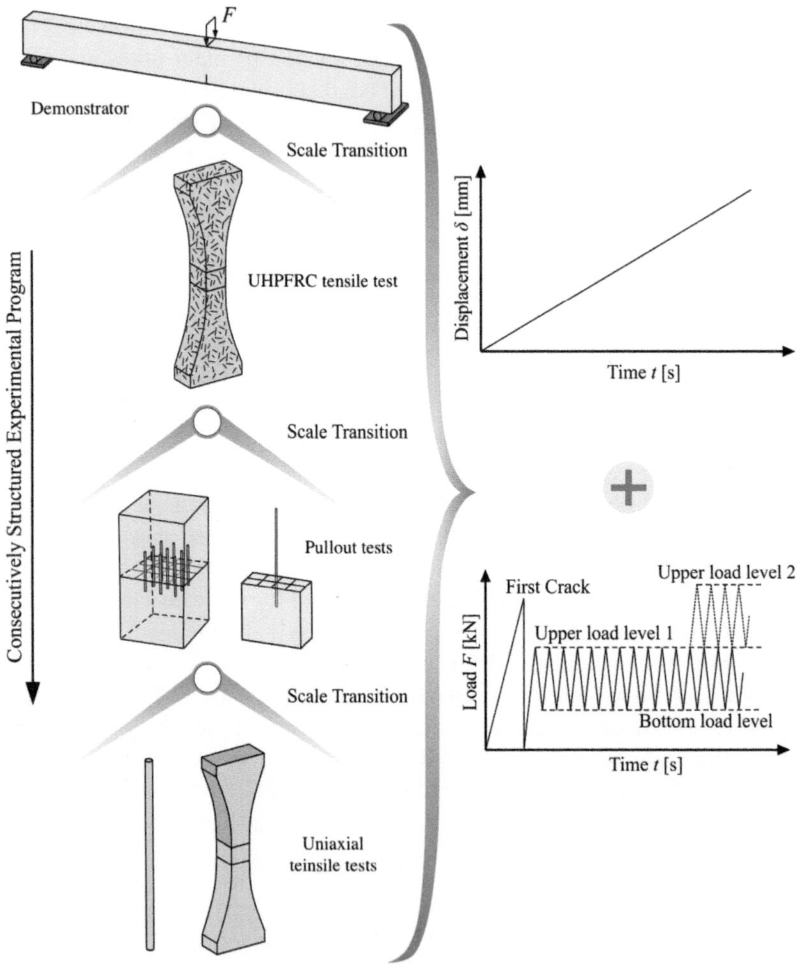

Fig. 1 Consecutively structured test programme and loading situations (see [6], licensed under CC-BY 4.0)

The programme is systematically graded containing experimental investigations in a range of small-scale specimens and large-scale demonstrators. All test series (with the exception of the pullout tests on individual fibres) were carried out under quasi-static and cyclic tensile loading. All tests under cyclic loading were carried out with a constant load-time function as single-stage tests. Selected specimens were also tested under dual-stage cyclic loading to capture the sequence effect. The tests under quasi-static load were also used as reference tests and to determine the input values for the tests under cyclic loading.

2.2 Bare Steel Fibre and Plain UHPC

The experimental programme of the bare steel fibre contained 15 monotonic and 26 cyclic loaded tensile tests. The test programme of the plain UHPC contained 12 monotonic tests and 13 cyclic loaded tensile tests. From both programmes, the material properties (tensile and compressive strength, young's modulus) of the steel fibre and the UHPC has been determined. Furthermore, S/N-approaches for both materials had been set up. Detailed information for both materials were published in [6–8].

In addition to the above-mentioned tests, the influences of a higher loading frequency were investigated for the plain UHPC and a dual-stage loading was investigated for both, the plain UHPC and the bare steel fibre. Figure 2 shows the S/N-curve of the plain UHPC with results of testing with a frequency of 4.0 Hz and approx. 50 Hz. An arrow pointing to the right implies that no failure of the specimen occurred.

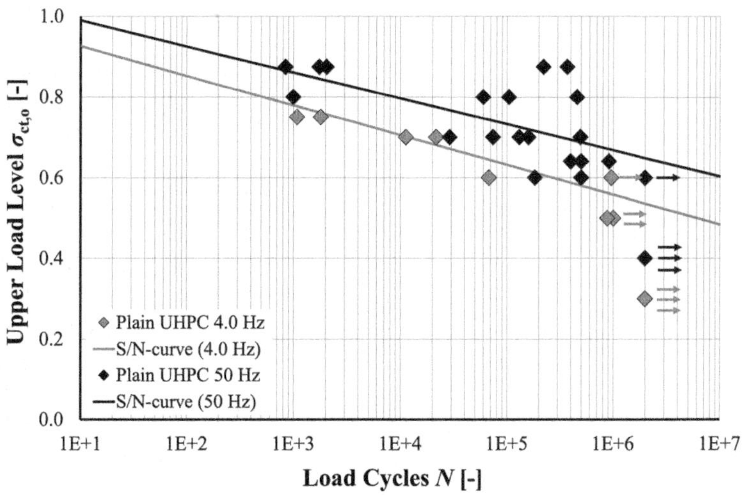

Fig. 2 Influence of loading frequency for plain UHPC under tensile loading

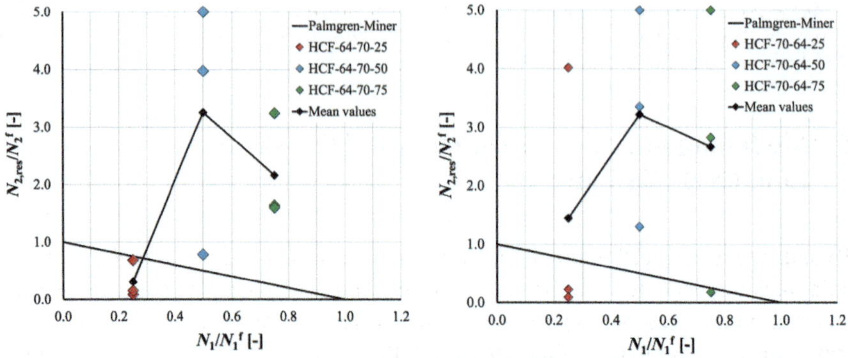

a b

Fig. 3 Influence of dual-stage loading on plain UHPC: **a** Increasing load amplitude and **b** decreasing load amplitude

The S/N-diagram shows that a higher loading frequency causes higher numbers of load cycles to failure for plain UHPC. However, a significant scatter can be observed for the test results.

Figure 3 shows the influence of a dual-stage loading on the fatigue behaviour of plain UHPC regarding an increasing and a decreasing load amplitude. The goal was to provoke a failure after both loading stages in high-cycle-fatigue (HCF), i.e. after at least 1,000,000 load cycles. That is why the upper load levels were calculated to 0.64 for stage one and 0.70 for stage two, respectively vice versa.

The upper load levels were taken from the S/N-curve in Fig. 2. The pre damages were selected to 25% (0.25), 50% (0.50) and 75% (0.75) in the first stage of the cyclic loading (N_1). It is assumed that the damages could be added according to the linear damage accumulation hypotheses of Palmgren and Miner (black line) [9, 10]. N_1/N_1^{f} stands for the damage of the specimen in the first stage of cyclic loading (see above mentioned pre damages 0.25, 0.50 and 0.75) according to the linear damage accumulation hypotheses of Palmgren and Miner. N_2/N_2^{f} stands for the additional damage in the second stage of cyclic loading.

The results from dual-stage testing under cyclic loading (N_2) show high scattering in the number of cycles to failure. However, the mean values of the results show the same behaviour for increasing and decreasing load amplitude. With a pre damage of 25%, the total damage is mostly under 100%, i.e. dual-stage loading could result in an adverse influence on the fatigue behaviour of plain UHPC. Regarding the pre damages of 50% and 75% the total damage is above 100% resulting in a favourable fatigue behaviour.

Besides the plain UHPC, the steel fibre was tested under dual-stage cyclic tensile loading. The testing machine did not allow high testing frequencies (approx. 1.0 Hz, see [6]), so the failure should occur before 1,000,000 load cycles, i.e. in low-cycle-fatigue (LCF). The pre damage was also defined at 25%, 50% and 75%. The load

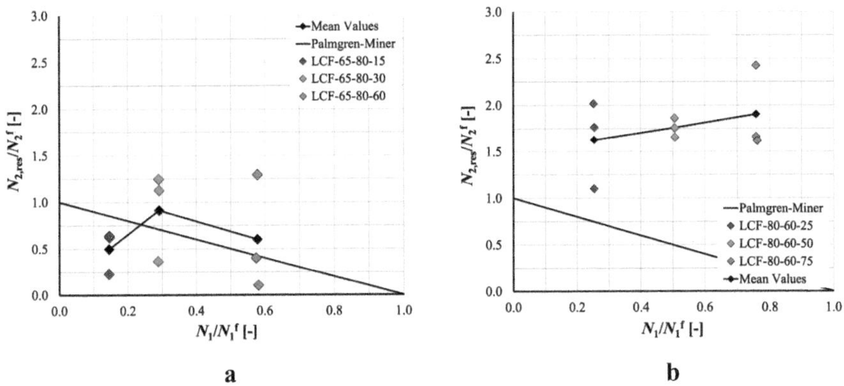

Fig. 4 Influence of dual-stage loading on bare steel fibres: **a** Increasing load amplitude and **b** decreasing load amplitude

amplitudes were 0.60 and 0.80, resp. vice versa. Figure 4 shows the results from dual-stage cyclic tensile testing.

The fatigue behaviour of the bare steel fibre (mean values) with an increasing load amplitude shows that a small pre damage results in a total damage smaller than 100%. For a pre damage of 50% and 75% in the first loading stage, the total damage scatters around 100%. Despite high scattering, it could be assumed that a decreasing load amplitude has a positive influence on the fatigue behaviour of the bare steel fibre.

2.3 *Pullout Tests*

The experimental programme was thoughtfully designed to encompass various fibre orientations. Five angles and two embedment depths were investigated. The angles were spaced by 15° intervals, excluding 15° due to production challenges (90° equals straight pullout).

Initially, three load amplitudes were planned, but testing limitations necessitated adjustments. Inertia-related issues arose at higher frequencies, affecting load accuracy. The load amplitude was thus determined from measurement data.

For run-outs ($N > 100,000$) at a load amplitude level, the next higher level was not explored. If failure occurred before 100,000 cycles, the load amplitude was lowered. Table 1 lists the programme of the pullout tests.

The manufacturing process of the specimens and the test set up is described in detail in [6] and should not be repeated in this contribution anymore. The load cycles and the related load amplitude are inserted into an S/N-diagram. Figure 5 shows the S/N-diagram with regression line for every fibre orientation.

Table 1 Test programme of pullout tests

Series	Number [–]	Orientation [°]	Planned embedded length [mm]	Related load amplitude [–]
VM30	19	30	3.25/6.50	0.12 to 0.76
VM45	17	45	3.25/6.50	
VM60	20	60	3.25/6.50	
VM75	14	75	3.25/6.50	
VM90	14	90	3.25/6.50	

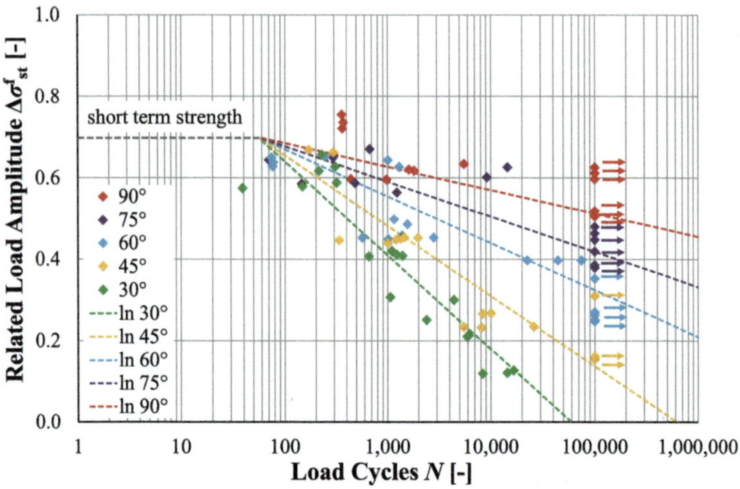

Fig. 5 S/N-diagram with regression lines for every fibre orientation

Figure 5 shows that the fatigue behaviour of embedded fibres depends on the orientation of the fibres. The lower the orientation angle is the lower are the endured number of load cycles. The regression lines in Fig. 5 could be expressed mathematically by the general Eq. 1:

$$\Delta\sigma^{\mathrm{f}}_{\mathrm{st}} = -A \cdot \ln N + B \quad (1)$$

A and *B* are parameters that should be related to the orientation of the fibres. The correlation between the orientation angle α and the parameters A and B are shown in Eq. 2.

$$\begin{aligned} A &= -0.07 \cdot \ln\alpha + 0.34 \\ B &= -0.28 \cdot \ln\alpha + 2.05 \end{aligned} \quad (2)$$

Inserting the Equations for parameter A and B into Eq. 1 results in a parametric S/N-curve for embedded high-strength micro steel fibres in UHPC given in Eq. 3.

$$\Delta\sigma_{st}^{f} = (0.07 \cdot \ln\alpha - 0.34) \cdot \ln N_{ult} + (-0.28 \cdot \ln\alpha + 2.05) \tag{3}$$

After the test execution, it was observed that fibre rupture occurred. The rupturing occurred at pullout tests where the fibres were inclined compared to the loading direction. Samples, where the fibres were orientated parallel to the loading direction (90°) showed no fibre rupture. Therefore, the fibre rupture must have come from the inclination of the fibre. When an inclined fibre is pulled out of the concrete, the fibre is redirected. The redirection of the fibre causes an additional bending stress σ_B on the one side of the fibre and a reducing tensile stress on the other side of the fibre. The background of the additional strength is shown in Fig. 6.

Therefore, the local tensile stress in the fibre is higher at the redirection point than in the other parts of the fibre. Since the additional bending stress is present at the upper and bottom load level, only the level of the load amplitude and not the load amplitude itself would be increased. Therefore, another assumption is necessary. It is assumed that there is no bending stress present at the bottom load level, i.e. the mandrel diameter in Fig. 7 (red circle) is infinite. This assumption allows adding the whole additional bending stress to the load amplitude of the fibre.

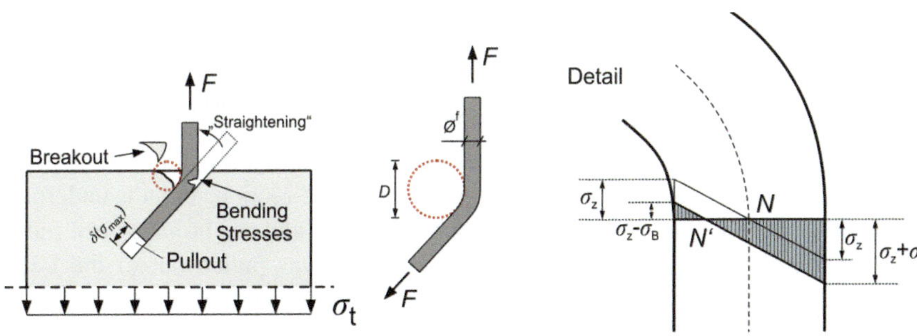

Fig. 6 Theoretical background for fibre rupture at inclined fibre pullout

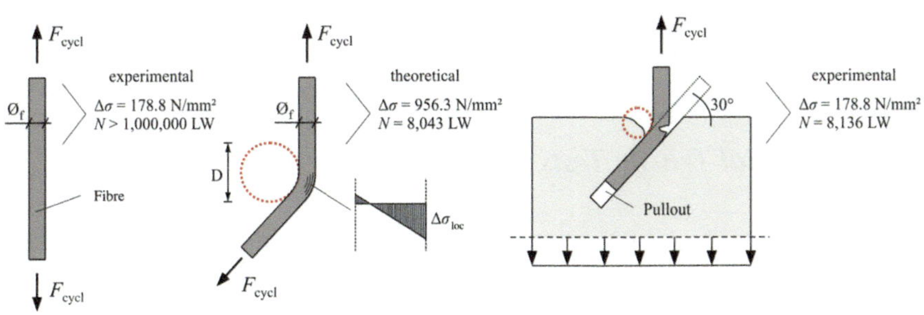

Fig. 7 Comparison of experimental and theoretical load cycles

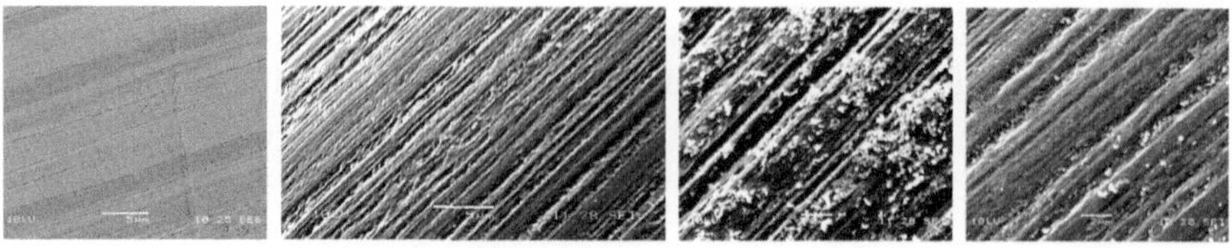

a b c d

Fig. 8 SEM images of the steel fibre (from [6], licensed under CC-BY 4.0): **a** as provided, **b** and **c** after monotonic pullout and **d** after cyclic pullout

The background of the calculation of the additional bending stress is in [11]. According to [7], a bar steel fibre showed no failure under 1,000,000 load cycles, when the load amplitude did not exceed 178.8 N/mm^2. An embedded fibre with an orientation of 30° reached 8136 load cycles at a load amplitude of 178.8 N/mm^2. The local maximum tensile stress when considering the additional bending stresses was 956.3 N/mm^2. Implementing this load amplitude into the S/N-curve for the bar steel fibre, the number of load cycles was 8042, which was similar to the load cycles of the embedded fibre. Figure 7 shows the comparison between the experimental and theoretical determination of the load cycles, while considering the bending stresses.

The fibre had been investigated with modern measuring techniques like a scanning electron microscope (SEM). The images serve to understand what happens with the fibre if it is pulled out of the concrete under monotonic and cyclic loading. Figure 8 shows SEM images of fibres provided by the manufacturer and pulled out under monotonic and cyclic loading.

Figure 8a shows the surface of the fibre in a condition as it was provided by the manufacturer. Although, the surface has a brass coating, it is not smooth. Small grooves are noticeable in longitudinal and transversal direction. What is underneath shows Fig. 8b. The grooves are deeper than expected and might have been hidden by the brass coating. Figure 8c shows a fibre that was pulled out of the UHPC under monotonic loading. Concrete components are located on the surface of the fibre mostly next to the grooves. A micro interlocking is possible. In Fig. 8d, the UHPC particles are mainly in the grooves. The cyclic loading may have caused a movement of the particles into the grooves. It is assumed that the movement also caused an abrasion of the particles so that an effective micro interlocking is not possible anymore.

2.4 *Uniaxial Tensile Tests on UHPFRC*

The test programme includes direct (uniaxial) tensile tests on so-called bones, also referred to as convex-concave specimens in the literature. Two test series have been produced and tested, one under monotonically increasing loading, and the other under cyclic loading. The test series differ in fibre content. On the one hand a moderate steel fibre content of 100 kg/m^3 (1.25% by volume) and on the other hand, a high steel fibre content of 200 kg/m^3 (2.50% by volume) were tested. Table 2 shows the test programme for the UHPFRC.

The upper and lower loads were determined based on three previous reference tensile tests (see Fig. 9). The upper loads, or amplitudes, were selected in 20% increments to provide a general indication of the load bearing behaviour under cyclic loading and to provide an S/N-line if necessary. The bottom load level was set at a constant 10% of the ultimate load of the reference tests.

The stress–strain curves initially increase linearly and the load bearing behaviour is determined by the UHPC matrix. The specimen is uncracked. When the uni-axial tensile strength of the concrete matrix is exceeded, the UHPC matrix cracks and the micro steel fibres are activated.

At a steel fibre content of 100 kg/m^3 (Test series VZS1), a plateau appears initially in the curve. The tensile stress remains almost constant within this plateau as deformation increases (0.20 mm to 0.75 mm). Multiple cracks form due to fibre activation, leading to secondary cracks if fibre stress surpasses the UHPC matrix strength. Similar behaviour is seen with a steel fibre content of 200 kg/m^3. The load can slightly increase after reaching the concrete matrix strength. Afterwards the curves drop nearly nonlinearly. Despite twice the fibre content, VZS1 and VZS2 series show similar maximum tensile stress due to fibre interactions. The UHPC matrix's tensile strength matches almost the post-crack strength. No significant stress change occurs during the transition to cracked state.

Figure 10 shows the results of the cyclic loaded tensile tests on UHPFRC with a 3D-fibre distribution.

Table 2 Test programme of uniaxial tensile tests

Series	Number [–]	Fibre content [Vol.%]	Upper load level [–]	Bottom load level [–]
VZ1	9	1.25	0.7/0.5/0.3	0.1
VZ2	12	2.50		

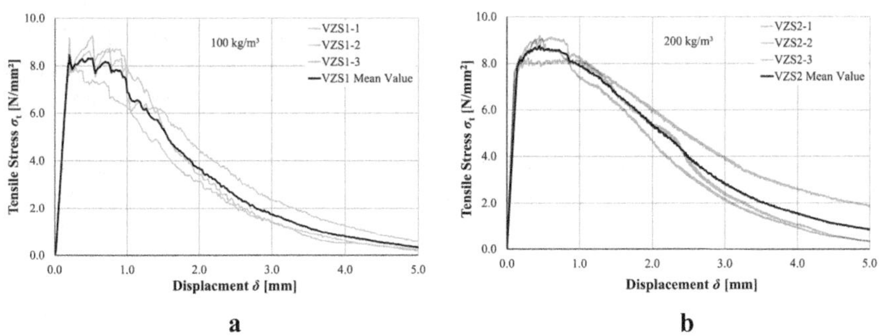

Fig. 9 Tensile stress-displacement-curves from uniaxial tensile tests: **a** 1.25 Vol.% Fibre content (100 kg/m^3) and **b** 2.50 Vol.% fibre content (200 kg/m^3)

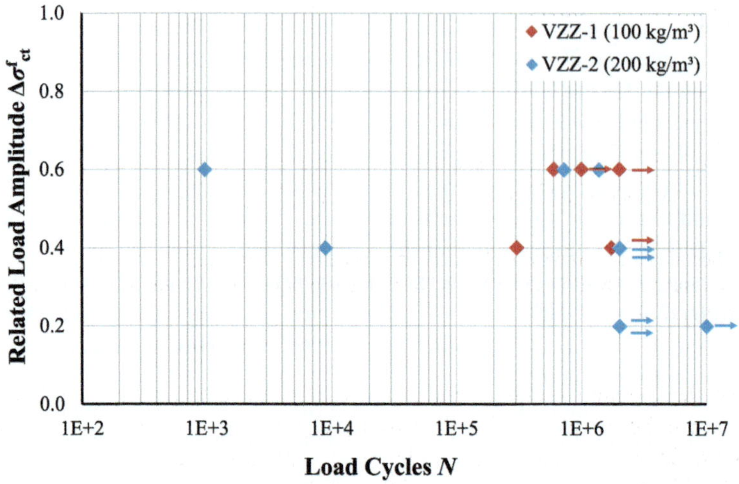

Fig. 10 S/N-curve of the uniaxial tensile tests

Figure 10 illustrates the notable variability in the number of cycles to failure. In the VZZ1 test series, failure transpired after 1 million and 2 million load cycles at a related load amplitude of $\Delta\sigma^{f}_{ct} = 0.6$. Meanwhile, at an amplitude of $\Delta\sigma^{f}_{ct} = 0.4$, one test led to failure after 300,000 load cycles. Despite the reduced load, the cycles until failure were notably fewer. A similar pattern emerged in the VZZ2 test series. Even though resistance to cyclic tensile loading was adequate at an amplitude of 0.2, results beyond this threshold exhibited significant dispersion. Consequently, these findings do not allow for a dependable interpretation of the load-bearing behaviour during cyclic loading of fibre-reinforced UHPC.

Besides the mechanical behaviour of the UHPFRC, the fibre orientation has been analysed. Therefore, the specimens were sawed into small pieces for CT-scanning. Since it was assumed that the fibres were aligned in the dumbbell specimen, reference samples were prepared where no alignment was assumed. Figure 11 shows the location of samples sawed from different specimens.

The fibre orientation was determined with a novel software called VGSTUDIO MAX by Hexagon [12]. Table 3 lists the fibre orientation factors (vectors) of the analysed samples.

It should be noted that the values in Table 3 are mean values of three individual analyses. The fibre orientation factors are named as vectors in Table 3. The values show that the dumbbell specimens show an alignment of the fibres (z-axis equals specimen axis). Furthermore, no alignment is noticeable in the samples sawed from the beam specimens.

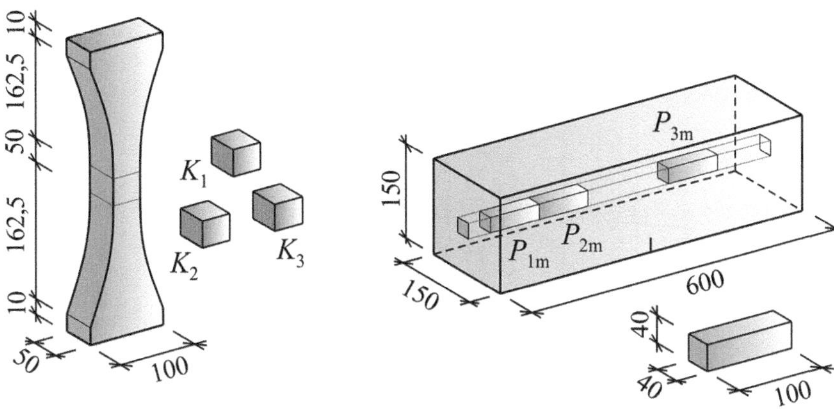

Fig. 11 Location of samples in specimens for fibre orientation analysis

Table 3 Results of fibre orientation analysis, i.e. fibre orientation factors

Series	Number [–]	Fibre content [Vol.%]	Vector x [–]	Vector y [–]	Vector z [–]
VZ1-K	Dumbbell	1.25	0.32	0.30	0.79
VZ1-P	Beam	1.25	0.50	0.45	0.56
VZ2-K	Dumbbell	2.50	0.27	0.25	0.84
VZ2-P	Beam	2.50	0.44	0.41	0.63

2.5 *Large-Scale Bending Tests*

The testing protocol encompasses six large-scale 3-point bending tensile tests, consisting of four beams subjected to quasi-static loading (used as a reference) and two beams subjected to cyclic single-stage loading. The details of the test programme can be found in Table 4.

The specimen had dimensions measuring 3000 mm in length, 300 mm in height, and 150 mm in width. Additionally, there is a notch situated directly beneath the point of loading, featuring a width of 10 mm and a height of 50 mm. This notch

Table 4 Test programme of large scale bending tests

Series	Fibre content [Vol.%]	Loading [–]	Maximum load [kN]	Load amplitude [–]
DEMO-1	1.25	Monotonic	23.59	–
DEMO-2	1.25	Monotonic	28.71	–
DEMO-3	2.50	Monotonic	58.74	–
DEMO-4	2.50	Monotonic	48.61	–
DEMO-5	1.25	Single stage	–	0.45
DEMO-6	2.50	Single stage	–	0.35

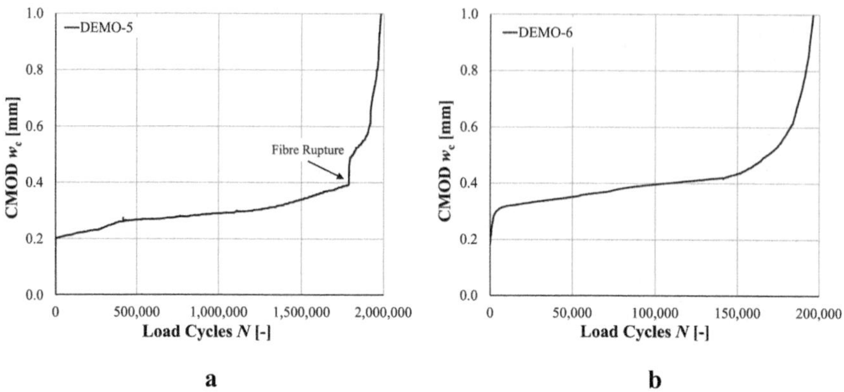

Fig. 12 Development of CMOD depending on load cycles: **a** DEMO-5 and **b** DEMO-6

serves as a deliberate crack to facilitate the measurement of the crack mouth opening displacement (CMOD). Detailed information of the test configuration (execution, measuring devices etc.) are in [6].

Figure 12 shows the development of the CMOD over the load cycles of the single stage cyclic loading tests. Before the cyclic loading was applied, the beams were "cracked" so that the fibres were activated.

DEMO-6 exhibits the characteristic response of concrete subjected to cyclic loading, particularly in cases of compressive loading. The fatigue behaviour, which is outlined in detail in reference [13], can be observed through the evolution of the CMOD as it relates to the number of load cycles.

In contrast, DEMO-5 demonstrates a more pronounced trilinear behaviour. Notably, the transitions between these phases occur more abruptly compared to DEMO-6. This abrupt change in behaviour can be attributed to the rupture of fibres, becoming evident with a rapid increase in displacement after approximately 1,750,000 load cycles. To support the hypothesis of fibre rupture, the beam of DEMO-6 was cracked open after the test execution. The fibres on the crack areas were examined with a digital microscope. Figure 13 shows a section of the cracked area where two fibres ruptured.

In both halves of the cracked areas, corrosion products are visible. Furthermore, the steel of the fibres is still recognizable in the channels of the fibre, which gives a hint of fibre rupture.

DEMO-5 failed after approximately 1,900,000 load cycles at a related load amplitude of 0.45 and DEMO-6 failed after 190,000 load cycles at a related load amplitude of 0.35. Although the load was higher, DEMO-5 endured ten times more load cycles. This relation shows the high scattering of UHPFRC under cyclic tensile loading.

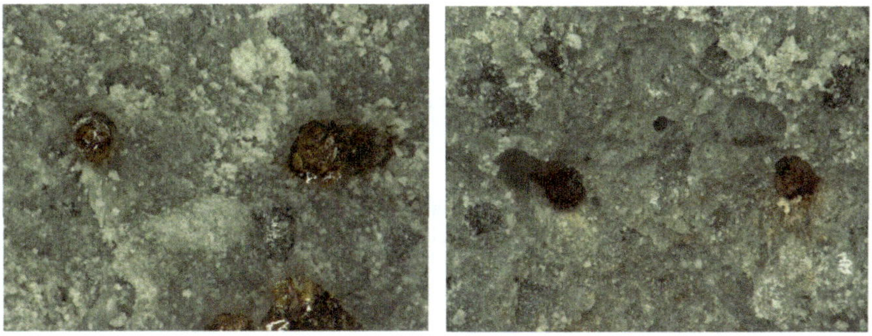

a b

Fig. 13 Cracked areas from DEMO-6 with two ruptured fibres (from [6], licensed under CC-BY 4.0): **a** one half and **b** other half

3 Conclusion

This contribution presents some of the key research results achieved during the project. The experimental investigations provide information on the fundamental load-bearing behaviour of micro steel fibres embedded in UHPC under monotonic and cyclic tensile loading. The main findings of the experimental investigations can be summarised as follows:

1. Dual stage cyclic tensile tests on plain UHPC showed a favourable fatigue behaviour at pre damages from 50%. Dual stage loading with decreasing amplitudes lead to a positive influence of the fatigue behaviour on bare steel fibres.
2. Fibre pull-out tests under cyclic tensile loading revealed fibre ruptures of inclined fibres due to locally increased bending stresses.
3. Despite adequate resistance at lower amplitudes, the results of uniaxial tensile tests showed considerable scatter at higher amplitudes. This variability makes it difficult to reliably interpret the load-bearing behaviour of fibre-reinforced UHPC under cyclic tensile loading.

The investigations presented in this chapter are part of the cooperative project with the Institute of Structural Analysis, TU Braunschweig (see chapter by L. Gietz et al., this volume).

References

1. Reichel, M. M.: Dünnwandige Segmentfertigteilbauweisen im Brückenbau aus gefasertem Ultrahochleistungsbeton (UHFB)—Tragverhalten, Bemessung und Konstruktion. Ph.D Thesis, TU Graz, Graz (2010)

2. Lappa, E. S.: High strength fibre reinforced concrete—static and fatigue behaviour in bending. Ph.D Thesis, TU Delft, Delft (2007)
3. Lohaus, L., Elsmeier, K.: Fatigue behaviour of plain and fibre reinforced ultra-high performance concrete. In: Proceedings of HiPerMat: Ultra-High Performance Concrete and Nanotechnology in Construction, Kassel (2012)
4. Makita, T.: Fatigue behaviour of UHPFRC and R-UHPFRC—RC composite members. Ph.D Thesis, École Polytechnique Fédérale de Lausanne, Lausanne (2014)
5. Fitik, B.: Ermüdungsverhalten von ultrahochfestem Beton (UHPC) bei zyklischen Beanspruchungen im Druck-Zug-Wechselbereich. Ph.D Thesis, TU München, München (2012)
6. Lanwer, J.-P.: Zum Tragverhalten von ultrahochfestem Stahlfaserbeton unter zyklischer Zugbelastung. Ph.D. Thesis, Technische Universität Braunschweig, Braunschweig (2023)
7. Lanwer, J.-P., Empelmann, M.: Investigations on the performance of micro steel fibres in ultra-high-performance concrete under monotonic and cyclic tensile loading. Appl. Sci. **11** (2021)
8. Oettel, V., Lanwer, J.-P., Empelmann, M.: Auszugverhalten von Mikrostahlfasern aus UHPC unter monoton steigender und zyklischer Belastung. Bauingenieur **96**(01–02), 01–10 (2021)
9. Miner, M. A.: Cumulative damage in fatigue. J. Appl. Mech. **12**(3), 159–164. (1945)
10. Palmgren, A.: Die Lebensdauer von Kugellagern. Z. Ver. Dtsch. Ing. **68**(14), 339–341 (1924)
11. Lanwer, J.-P., Empelmann, M.: Performance-based fibre design for ultra-high performance concrete (UHPC). Appl. Sci. **12**, 8559 (2022)
12. Volume Graphics GmbH.: VGSTUDIO MAX reference manual (2020)
13. König, G., Tue, N. V., Zink, M.: Hochleistungsbeton. Bemessung, Herstellung und Anwendung. Ernst & Sohn, Berlin (2001)

Material Modelling of Ultra-High-Performance Concrete Under Tensile Loading

Lena Gietz, Svenja Matz, Ursula Kowalsky, and Dieter Dinkler

Abstract The contribution deals with the modelling of the mechanical degradation processes of ultra-high-performance fibre-reinforced concrete (UHPFRC) with respect to tensile loading. Micro steel fibres embedded in UHPC significantly improve the post cracking behaviour due to the crack-bridging load-bearing effect. If UHPFRC is subjected to cyclic tensile loading, the irreversible degradation of the bond limits the potential of the composite material. The investigations aim to describe the stress-deformation behaviour of UHPFRC under tensile loading by developing microphysically based and mechanism-oriented models at the meso- and macroscale. Thereby the description of the debonding of the micro steel fibres from the UHPC-matrix on the mesoscale and the development of concrete cracks and fibre pull-out on the macroscopic scale are a particular challenge. The models presented here are validated by comparing numerical simulations with respective experiments.

Keywords Ultra-high-performance fibre reinforced concrete · Concrete cracking · Fibre pull-out · Material modelling · Fracture mechanics · Hybrid-mixed finite element method

1 Introduction

Sustainability and efficiency of civil engineering constructions as well as research with respect to climate-friendly concrete construction methods are becoming increasingly important. Researchers in construction engineering have focussed on ultra-high-performance fibre reinforced concrete (UHPFRC) in recent years. This building material combines excellent mechanical properties with high resistance to environmental influences. Due to the mechanical properties, members can be designed to be slender, more resource efficient and robust. The improved durability increases the service life of structures, resulting in lower building maintenance and repair costs.

L. Gietz · S. Matz · U. Kowalsky (✉) · D. Dinkler
Institute of Structural Analysis, TU Braunschweig, Braunschweig, Germany
e-mail: u.kowalsky@tu-braunschweig.de

N. Oneschkow and L. Lohaus (eds.), *Cyclic Deterioration of High-Performance Concrete*, https://doi.org/10.1007/978-3-032-19394-0_14

Furthermore, studies show that the use of UHPFRC has a positive impact on the environmental aspects during the service life of bridges [1, 2].

Plain ultra-high-performance concrete (UHPC) is characterised by its high compressive strength but brittle deformation behaviour with respect to tension, while the additional embedding of micro steel fibres leads to high tensile strength of the composite material and a significantly improved ductility in the post-cracking region. At the initial crack formation of the UHPC-matrix the steel fibres are activated and take over the tensile load transfer. If the bond limit between the UHPC-matrix and the micro steel fibres is reached, fibre pull-out occurs, which effects a controlled material rupture and a ductile macroscopic deformation behaviour. The crack-bridging load-bearing effect of the micro steel fibres can be taken into account in the design rules, if fundamental knowledge about the failure mechanisms leads to a quantification of the post-cracking behaviour.

A major disadvantage of the design of slender structures made of UHPFRC is their susceptibility to fatigue due to the unbalanced ratio between dead weight and cyclic loading compared to normal reinforced concrete constructions. Examples are slender bridges i.e. pedestrian bridges, wind energy installations or crane girders. Experimental and numerical results indicate that the fibres in UHPC exhibit remarkable resistance to cyclic loading [3–5], although the fibre orientation has a non-negligible effect [6, 7]. The exact load-bearing behaviour of fibres in UHPC components under cyclic loading is hardly quantifiable since only limited data are available which is insufficient for uniform standardisation.

Research on the structural behaviour of UHPFRC has so far concentrated primarily on the macro scale, whereby the material is described by means of simplifying homogenised isotropic parameters [3, 8]. Studies show a significant influence of the bond quality between the micro steel fibres and the UHPC-matrix [9] as well as the anisotropic fibre distribution on the tensile load-bearing behaviour of UHPFRC [10, 11], which has to be taken into account in suitable material models.

Thus, the development of models for the description and prediction of the material behaviour on the meso- and macro scale are useful, since numerical simulations can supplement experimental tests on fatigue behaviour of material, which are often costly and difficult to realise. As part of Priority Programme SPP 2020, several models have already been expanded or newly developed [7, 12, 13]. The investigations focus on the characterisation of microstructural damage in UHPC and gaining knowledge on the fibres' load-bearing effect under monotonic and cyclic tensile loading. In addition to experiments, see the chapter by J.-P. Lanwer et al., this volume, numerical methods are developed to investigate the tensile fatigue behaviour of UHPFRC in the context of an experimental virtual lab. The material is considered to be a composite comprising the UHPC-matrix, the fibres and the processes taking place in the bond zone between the two material components.

The models and numerical investigations presented here aim to contribute to the understanding of the load-bearing behaviour of micro steel fibres embedded in UHPC with respect to the meso- and macro-scale.

2 Mesoscale Modelling

The numerical simulation of the material behaviour of UHPFRC on the mesoscale requires a very careful modelling of the bond behaviour. A geometrically and physically nonlinear model is presented, which takes into account different bond properties describing the slip between micro steel fibres and the UHPC-matrix. Detailed information about the model can be found in [14, 15].

2.1 Material Model

The local behaviour of the bond zone is described assuming an additive split into rates of elastic and inelastic relative displacements between concrete and fibre

$$\dot{\delta} = \dot{\delta}^{\mathrm{el}} + \dot{\delta}^{\mathrm{in}}. \tag{1}$$

The rigid bond is modelled by assuming a linear relation between elastic relative displacements and the bond stresses: $\boldsymbol{\tau} = \boldsymbol{E}_{\mathrm{r}} \cdot \boldsymbol{\delta}^{\mathrm{el}}$. The sliding bond is described by irreversible relative displacement rates $\dot{\delta}^{\mathrm{in}}$, and takes into account the damage-induced degradation of the bond capacity τ_{lim}. The initiation of sliding is described by means of a failure surface considering lateral compression and tension, which depends on the shear stress between concrete and fibre $\tau_{||}$, and on the normal stress $\tau_{\perp}$ perpendicular to the fibre surface, see Fig. 1a. The failure criteria for transverse compression and tension are given as a function of the coefficient of friction μ and the anisotropy coefficient λ to take into account the two-dimensional anisotropy of the bond zone, whereat d describes the bond zone damage

$$F_{\mathrm{c}} = \left|\tau_{||}\right| + \mu(d) \cdot \tau_{\perp} - \tau_{\mathrm{lim}}(d) \leq 0, \tag{2}$$

$$F_{\mathrm{t}} = \tau_{||}^2 - \mu(d) \cdot (\tau_{\mathrm{lim}}(d) - \tau_{\perp})^2 + (1 - \mu(d)) \cdot \left(\left(\frac{\tau_{\perp}}{\lambda}\right)^2 - \tau_{\mathrm{lim}}^2(d)\right) \leq 0. \tag{3}$$

The rates of the inelastic relative displacements are defined by the non-associated slip rules using the potential functions Q_c and Q_{t}, which are formulated similarly to the respective failure criteria, but include the coefficient of lateral pressure evolution $\eta(d)$ instead of $\mu(d)$ [15]

$$\dot{\delta}_{\mathrm{c}}^{\mathrm{in}} = \dot{\lambda}_{\mathrm{c}} \cdot \frac{\partial Q_{\mathrm{c}}}{\partial \tau}, \; \dot{\delta}_{\mathrm{t}}^{\mathrm{in}} = \dot{\lambda}_{\mathrm{t}} \cdot \frac{\partial Q_{\mathrm{t}}}{\partial \tau}. \tag{4}$$

The degradation of the maximum bond capacity τ_{max} is caused by the deterioration of adhesion and micro aggregate interlocking, whereas small particles between concrete and fibre are responsible for a post failure resistance against sliding. The

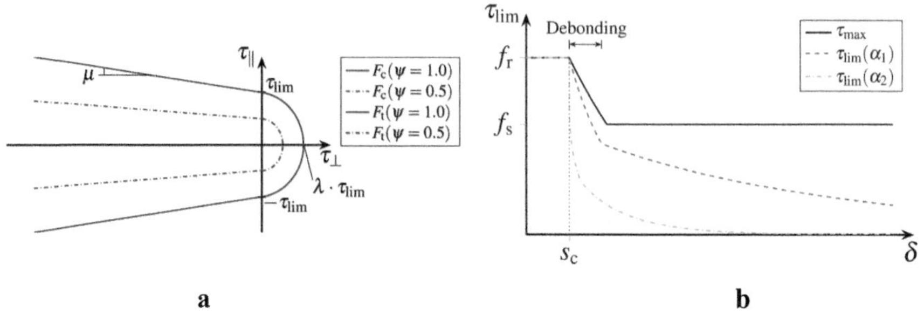

Fig. 1 **a** Failure surface and **b** evolution of bond capacity as function of constant damage evolution scaling factor α (from [15], licensed under CC-BY 4.0)

damage mechanisms of the bond zone are considered by a linear degradation function $\psi(d)$, where d describes the bond zone damage: $\psi(d) = 1 - d$. Figure 1b represents the evolution of the bond capacity τ_{lim} with damage to the bond zone as a function of the constant scaling factor α of the degradation function. The bond strengths f_{s} and f_{r} characterise the rigid and the sliding bond respectively.

The proposed bond zone model does not include any enforcement of the damage evolution due to lateral compression in the bond zone. However, in the model exists an interaction between contact pressure and bond zone resistance, which in turn regulates the magnitude of the inelastic relative displacement rates.

2.2 *Fibre-Matrix Bond with Respect to Monotonic Tensile Loading*

The numerical analysis of the fibre-matrix bond at the mesoscale requires a discrete description of the geometries of fibre and concrete matrix shifting against each other, see Fig. 2.

The geometrical and material parameters for the simulation of the fibre pull-out test are given in Table 1. Concrete and fibre are assumed to be elastic and are

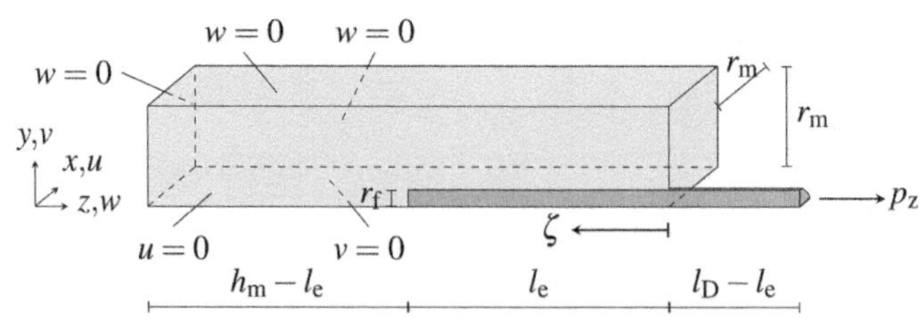

Fig. 2 Test sample for the simulation of pull-out experiments (from [15], licensed under CC-BY 4.0)

Table 1 Model parameters of fibre pull-out (Fig. 2)

Parameter	Symbol	Unit	Value
Fibre diameter	d_f	mm	0.19
Wire length	l_D	mm	16.5
Embedded fibre length	l_e	mm	6.5
Matrix radius	r_m	mm	2.0
Matrix height	h_m	mm	40.0
Bond strength (rigid bond)	f_r	N/mm^2	13.0
Bond strength (sliding bond)	f_s	N/mm^2	10.3
Initial Young's modulus of bond zone[2]	E_r	N/mm^2	13,000
Begin of sliding bond	s_c	mm	0.001
Coefficient of anisotropy	Λ	–	1.0
Coefficient of friction	M	–	0.15
Coefficient of lateral pressure evolution	H	–	0.001
Damage evolution coefficient	A	–	0.025

discretised with isoparametric 3D continuum elements. The bond zone is modelled by means of isoparametric zero-thickness interface elements, which discretise the model equations of debonding and sliding. A node-mapping algorithm generates a connectivity matrix assigning the opposite contact surfaces of concrete and fibre, and the active displacement situation [14, 16].

To verify the material model and to validate its material parameters, results from numerical simulations are compared to experimental data [14]. Figure 3a represents the 3D-discretisation of the sample with the finite element method. The simulation of the experiment with respect to monotonic loading leads to a linear increase of the displacement until achieving the failure surface and the debonding condition. After debonding the pull-out behaviour offers a linear load-slip behaviour, because the resistance per unit length remains constant whereby the bond length is reduced. The numerical simulation maps the experimental results quite well, although the experiment offers a rough response due to the stochastic bond properties at the micro scale, see Fig. 3b.

2.3 *Fatigue Behaviour of Fibre-Matrix Bond*

The fatigue behaviour of the bond zone with respect to cyclic loading is characterised by a progressive detachment of the micro steel fibres from the UHPC-matrix. The numerical simulation of the degradation is possible, if the model parameters are adjusted to the conditions of loading and unloading. Since experimental results show a hysteresis due to unloading and reloading and an increase of the irreversible

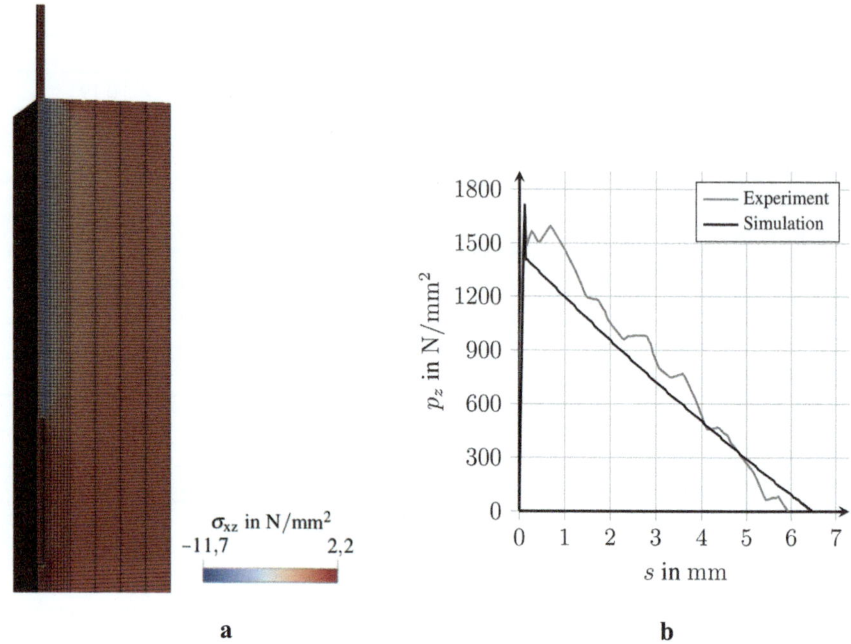

Fig. 3 Simulation of fibre pull-out at monotonic loading: **a** FE-mesh and plot of bond stresses during the fibre pull-out phase (from [14], licensed under CC-BY 4.0) and **b** comparison of load-slip curves of experiment and simulation

displacements, the model parameters τ, Q_c and Q_t are directly connected to the number of cycles, which governs the degradation function.

The results of the numerical simulation of the first cycles are plotted in Fig. 4a. The stress amplitude is chosen by $p_z = 1\,500$ N/mm^2 which is slightly below the load-bearing capacity of the fibre-matrix bond to partially detach the fibre from the UHPC-matrix. The bond stresses along the related bond length ζ are plotted in Fig. 4b due to loading and unloading conditions and due to an accumulation of the cycles. It turns out, that the stress distribution indicates residual stresses after unloading, which change slightly if the number of cycles n increases.

3 Macroscale Modelling

The numerical simulation of the load-deformation behaviour of realistic structures as multi-story frameworks needs a material model which is able to describe the failure mechanisms on the macroscale instead of taking into account single fibres and cracks between concrete aggregates on the micro- and mesoscale. Continuum damage mechanics and phase field models offer approaches to describe the stiffness

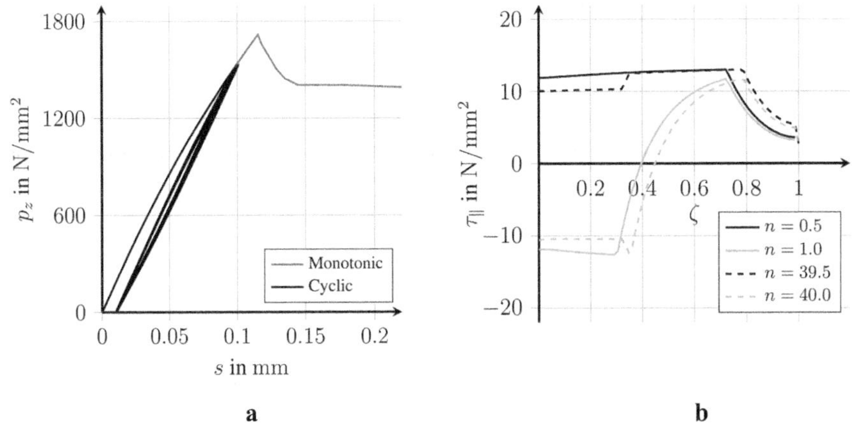

Fig. 4 Simulation of fibre pull-out at cyclic loading (from [15], licensed under CC-BY 4.0) **a** Fibre pull-out at cyclic loading **b** Shear stresses along the fibre length for different load cycles

reduction and the localisation of strains with respect to micro cracking in a very advanced theoretical framework [13, 17, 18]. Nonetheless, the internal length seems to be a nonphysical parameter, which governs the localisation zone and therefore the numerical effort with respect to the discretisation method. Own applications to steel structures [19, 20] and concrete under cyclic loading [21] demonstrate the possibilities of the methods but the limitations with respect to realistic structures as well. Thus, an alternative modelling is developed which combines the physical approaches of fracture mechanics [22, 23] and modern finite element methods.

The macroscopic material behaviour of UHPFRC is captured with a model for an efficient numerical analysis at the structural level, taking into account a smeared description of crack formation and fibre pull-out. Findings from the mesoscale modelling and the experiments are incorporated into the development of the model. Basis of the macroscopic description of the material behaviour of UHPFRC under tensile loading is a mechanism orientated modelling, see Fig. 5.

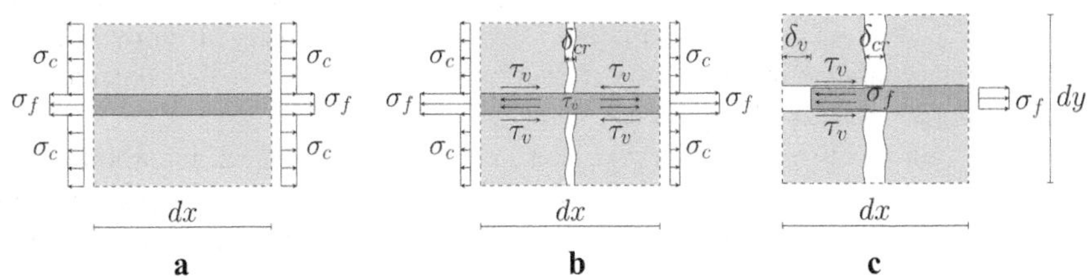

Fig. 5 Deformation mechanisms due to tensile loading **a** at linear-elastic material behaviour, **b** at initial crack formation and fibre activation and **c** at fibre pull-out in the post-cracking region

3.1 Material Model

The macroscopic description smears the material behaviour of the UHPC-matrix and of the micro steel fibres with respect to their volume contents n_c and n_f. The behaviour of both components and the bond between them is described by means of internal variables, which take into account the essential phenomena as cracking of concrete and the fibre pull-out. As known from Castigliano's theorem [24] the material equations may be developed, if the stored energy of the volume element is derived with respect to the strains and the stresses, respectively. Therefore, the stored energy Π of the UHPFRC volume element has to be established, which comprises the fractions of the UHPC-matrix Π_c and of the micro steel fibres Π_f, which are weighted by means of their volume contents n_c and n_f:

$$\prod = \int_V \pi dV = \int_{V_c} \pi_c dV_c + \int_{V_f} \pi_f dV_f = \int_V n_c \cdot \pi_c + n_f \cdot \pi_f dV. \tag{5}$$

The stored energy density π is given by a mixed stress–strain formulation, which has been published first by Hellinger [25] and further developed by Reissner [26]. The formulation offers the possibility to directly address strains and stresses of the components as internal variables. The energy density formulation of the elastic behaviour of both components is extended to the description of concrete cracking and fibre pull-out applying the method of Lagrangian multipliers and is given in total with

$$\begin{aligned} \pi = {} & n_c \left\{ \boldsymbol{\sigma}_c \boldsymbol{\varepsilon} - \frac{1}{2} \boldsymbol{\sigma}_c \boldsymbol{E}_c^{-1} \boldsymbol{\sigma}_c \right\} + n_f \left\{ \boldsymbol{\sigma}_f \boldsymbol{\varepsilon} - \frac{1}{2} \boldsymbol{\sigma}_f \boldsymbol{E}_f^{-1} \boldsymbol{\sigma}_f \right\} \\ & - n_c \boldsymbol{\varepsilon}_{cr} \left(\boldsymbol{\sigma}_c - \frac{1}{2} \boldsymbol{E}_{cr} \boldsymbol{\varepsilon}_{cr} \right) - n_f \boldsymbol{\varepsilon}_v \left(\boldsymbol{\sigma}_f - \frac{1}{2} \boldsymbol{K}_v \boldsymbol{\varepsilon}_v \right). \end{aligned} \tag{6}$$

Dealing with ε_{cr} and ε_v as internal variables, which are defined as being smeared over the volume dV, the Lagrangian multipliers have the meaning of inelastic crack strains ($\boldsymbol{\lambda}_{cr} \rightarrow \varepsilon_{cr}$) in the crack process zone of the UHPC-matrix and of the inelastic deformations due to the fibre pull-out of the micro steel fibres ($\boldsymbol{\lambda}_v \rightarrow \varepsilon_v$). The first constraint describes the energy release, since

$$\boldsymbol{\varepsilon}_{cr} \left(\boldsymbol{\sigma}_c - \frac{1}{2} \boldsymbol{E}_{cr} \boldsymbol{\varepsilon}_{cr} \right) = \frac{1}{2} \boldsymbol{\sigma}_c \boldsymbol{E}_{cr}^{-1} \boldsymbol{\sigma}_c. \tag{7}$$

holds, and the integration over the volume yields

$$\int_V \frac{1}{2} \boldsymbol{\sigma}_c \boldsymbol{E}_{cr}^{-1} \boldsymbol{\sigma}_c dV \rightarrow \left(\frac{\boldsymbol{\sigma}_c^2}{2 \boldsymbol{E}_{cr}} A \right) l_x. \tag{8}$$

Here, the term inside the brackets is directly comparable to the energy release of the Griffith-model [22], where a circular crack region A is assumed. The modulus $\boldsymbol{E}_{\mathrm{cr}}$ takes into account the softening behaviour instead of a purely brittle behaviour. The extreme condition of the mixed formulation yields the model equations, if the variation of π vanishes with respect to all macroscopic variables $\mathbf{z}$

$$\frac{\partial \pi}{\partial z} \cdot \delta z = 0 \mathrm{with} z = [\boldsymbol{\varepsilon} \boldsymbol{\sigma}_{\mathrm{c}} \boldsymbol{\sigma}_{\mathrm{f}} \boldsymbol{\varepsilon}_{\mathrm{cr}} \boldsymbol{\varepsilon}_{\mathrm{v}}]. \tag{9}$$

The conditions can be clearly structured in a system of equations

$$\begin{matrix} \delta\boldsymbol{\varepsilon}: \\ \delta\boldsymbol{\sigma}_{\mathrm{c}}: \\ \delta\boldsymbol{\sigma}_{\mathrm{f}}: \\ \delta\boldsymbol{\varepsilon}_{\mathrm{cr}}: \\ \delta\boldsymbol{\varepsilon}_{\mathrm{v}}: \end{matrix} \begin{bmatrix} & n_{\mathrm{c}} & n_{\mathrm{f}} & & \\ n_{\mathrm{c}} & -n_{\mathrm{c}} \cdot \boldsymbol{E}_{\mathrm{c}}^{-1} & & -n_{\mathrm{c}} & \\ n_{\mathrm{f}} & & -n_{\mathrm{f}} \cdot \boldsymbol{E}_{\mathrm{f}}^{-1} & & -n_{\mathrm{f}} \\ & -n_{\mathrm{c}} & & n_{\mathrm{c}} \cdot \boldsymbol{E}_{\mathrm{cr}} & \\ & & -n_{\mathrm{f}} & & n_{\mathrm{f}} \cdot \boldsymbol{K}_{\mathrm{v}} \end{bmatrix} \cdot \begin{bmatrix} \varepsilon \\ \boldsymbol{\sigma}_{\mathrm{c}} \\ \boldsymbol{\sigma}_{\mathrm{f}} \\ \boldsymbol{\varepsilon}_{\mathrm{cr}} \\ \boldsymbol{\varepsilon}_{\mathrm{v}} \end{bmatrix} = 0. \tag{10}$$

The differentiation with respect to the strains provides the stresses required for equilibrium. The differentiations according to the stresses supply the deformation conditions and the differentiations according to the crack strains and the fibre pull-out strains supply the equations for phenomenological description of the post-cracking behaviour.

3.2 *Characterisation of the Post-cracking Behaviour*

The constraints in Eq. 6 impose additional conditions on the stresses $\boldsymbol{\sigma}_{\mathrm{c}}$ and $\boldsymbol{\sigma}_{\mathrm{f}}$. These constraints lead to an additive split of the kinematics into the elastic strains $\varepsilon_{\mathrm{c}}{}^{\mathrm{el}}$ and $\varepsilon_{\mathrm{f}}{}^{\mathrm{el}}$ and the inelastic crack strain $\varepsilon_{\mathrm{cr}}$ as well as the fibre pull-out strainε_{v}. The characteristic of the material softening is described by the additionally defined moduli $\boldsymbol{E}_{\mathrm{cr}}$ and $\boldsymbol{K}_{\mathrm{v}}$. Simplifying the crack formation is described in mode I by the crack modulus $\boldsymbol{E}_{\mathrm{cr}}$ when the tensile strength f_{ct} of the UHPC-matrix has been reached [27]. $\boldsymbol{E}_{\mathrm{cr}}$ is defined by a decreasing exponential function, see Fig. 6a.

Experiments [28] show that the material failure of UHPC under tension is very brittle, which means that softening behaviour cannot be identified in the stress–strain diagrams once the tensile strength f_{ct} has been reached. By calibrating the parameters a, b and c, the abrupt failure and the associated stress drop in the UHPC-matrix can be modelled. The concrete stresses converge towards zero due to the used exponential function. For the simulation of the UHPC used in the experiments the following values supply good results: $a = 1.6$, $b = f_{\mathrm{ct}} \cdot 1\mathrm{e}^{-5}$ and $c = 1.0$. The model for the bond behaviour presented in Sect. 2 describes the bond mechanisms very detailed in three phases. In order to be able to consider the bond behaviour at the macroscopic level a simplified model based on the equilibrium between the fibre normal force

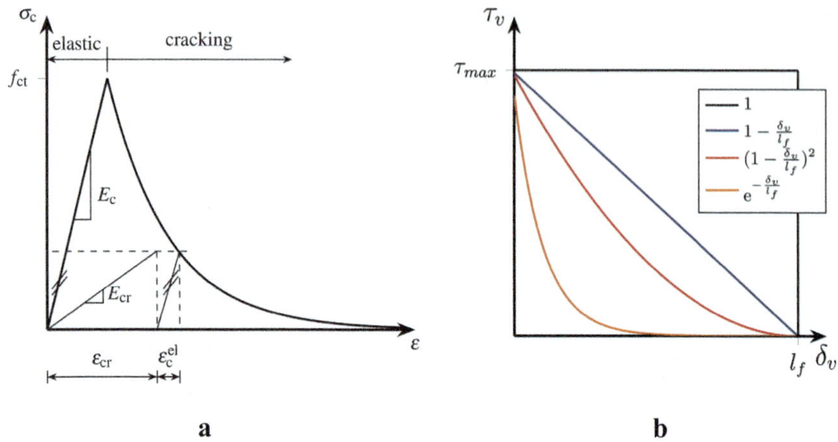

Fig. 6 **a** Stress–strain-diagram for the description of the concrete softening and **b** Development of the bond resistance τ_v with respect to the fibre pull-out length

and the bond force is taken into account. The equilibrium of the forces allows to express the fibre tensile stress as follows, taking into account the fibre geometry and the embedded fibre length l_v

$$\sigma_f = \frac{\tau_v \cdot 2l_v}{r_f} \text{with} l_v = l_f - \delta_v, \tag{11}$$

where δ_v describes the fibre pull-out length and l_f describes the effective fibre length. Based on experimental pull-out tests, half of the total fibre length l_f^{tot} is assumed to be the effective length [28]. Replacing the fibre stress in the constraint $-\sigma_f + K_v \cdot \varepsilon_v = 0$ see Eq. 10, the bond modulus K_v follows as a function of δ_v

$$K_v(\delta_v) = \frac{\tau_v}{\varepsilon_v} \cdot \frac{2l_v}{r_f} \geq 0, \tag{12}$$

The own experimental and numerical investigations show that the bond zone is increasingly damaged by abrasion in the pull-out channel as fibre pull-out progresses, especially for cyclic tensile loading, see Sects. 2.2 and 2.3. The frictional resistance to fibre pull-out is reduced. To account for this phenomenon, it is useful to describe the bond resistance τ_v as a function of the fibre pull-out length

$$\tau_v = \tau_{max} \cdot f(\delta_v) \text{with} \tau_{max} = \frac{\sigma_{max} \cdot r_f}{2l_f}, \tag{13}$$

where the maximum bond resistance τ_{max} is determined with the maximum force in the fibre before the fibre pull-out starts ($l_v = l_f$, $\delta_v = 0$), expressed by the bond limit σ_{max} according to Eq. 11. The function $f(\delta_v)$, describes the decrease in bond stresses with increasing fibre pull-out length, allowing the degradation of the bond resistance with respect to the fibre pull-out, see Figs. 6b and 9a.

3.3 Hybrid-Mixed Finite Element Concept

The finite element method (FEM) in a hybrid-mixed formulation is basis for the numerical investigation of structures. Since the energy equation and the first variation equation give the material equations with strains and stresses the kinematics have to be taken into account as well, see Eqs. 6 and 10. Applying linear kinematics using the operator matrix **D** for the differentiation rule

$$\boldsymbol{\varepsilon} = \boldsymbol{D} \cdot \boldsymbol{u} \tag{14}$$

yields the fundamental equations for the discretisation with the finite element method. Applying a hybrid-mixed element concept, based on [29, 30], the element domain describes the crack process zone. For the numerical implementation of the material equations, the strains $\boldsymbol{\varepsilon}_{\mathrm{v}}$ due to fibre pull-out is connected to the fibre pull-out lengths $\boldsymbol{\delta}_{\mathrm{v}}$ via the element length l_{el}. The macroscopic variables **z** are split into system and element variables

$$z = \left[z_{\mathrm{sys}}(\boldsymbol{u}) \;\; z_{\mathrm{elem}}(\boldsymbol{u}, \boldsymbol{\sigma}_{\mathrm{c}}, \boldsymbol{\sigma}_{\mathrm{f}}, \boldsymbol{\varepsilon}_{\mathrm{cr}}, \boldsymbol{\delta}_{\mathrm{v}})\right]. \tag{15}$$

The displacements **u** are defined at the element intersections, element edges and in the element domain, while the element variables $\mathbf{z}_{\mathrm{elem}}$ are only defined in the element domain, see Fig. 7. A 3D hexahedron continuum element is used for the discretisation of the model equations. Two different types of shape functions are employed. The displacement standard shape functions $\boldsymbol{\Omega}_{\mathrm{j}}$ are normalised to the displacements of the element nodes j and are used within the element and at the intersections, see Eq. 16. The element variables are approximated with shape functions ψusing generalised free values Eq. 17. The elimination of the element variables $\mathbf{z}_{\mathrm{elem}}$ follows a static condensation scheme and yields the element stiffness matrix. After assembling the system stiffness matrix the solution follows with the displacement degrees of freedom at the element nodes [27]. The solution for the element variables is reproduced in a subsequent analysis.

$$\Omega_{\mathrm{i}} = \frac{1}{8}\left(1 + \xi_{\mathrm{j}}\xi\right)\left(1 + \eta_{\mathrm{j}}\eta\right)\left(1 + \zeta_{\mathrm{j}}\zeta\right) \tag{16}$$

$$\begin{aligned} \boldsymbol{\psi}_{\xi} &= [1 \;\; \eta \;\; \zeta \;\; \eta\zeta] \quad \boldsymbol{\psi}_{\xi\eta} = [1 \;\; \zeta] \\ \boldsymbol{\psi}_{\eta} &= [1 \;\; \xi \;\; \zeta \;\; \xi\zeta] \quad \boldsymbol{\psi}_{\xi\zeta} = [1 \;\; \eta] \\ \boldsymbol{\psi}_{\zeta} &= [1 \;\; \xi \;\; \eta \;\; \xi\eta] \quad \boldsymbol{\psi}_{\eta\zeta} = [1 \;\; \xi] \end{aligned} \tag{17}$$

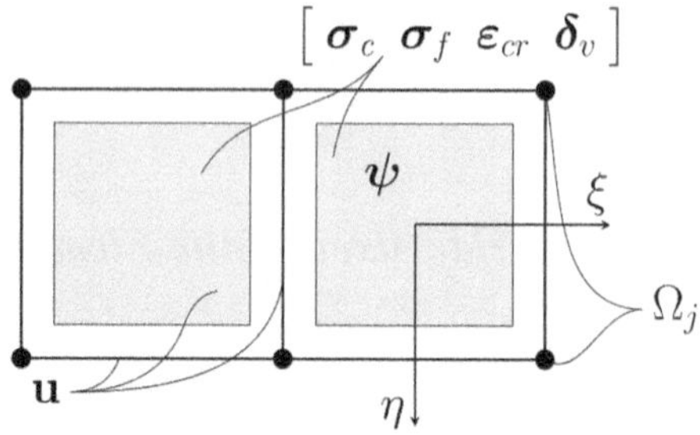

Fig. 7 Scheme of hybrid-mixed finite element concept

The hybrid-mixed element concept is advantageous, since the element domain is decoupled from the element intersection, which gives a J-integral of zero [23] and the possibility to describe the energy release due to Fig. 7 inside the element domain.

3.4 Simulation of a Rod Under Tension

Since the material behaviour offers physical instabilities due to the descending stress–strain behaviour, the regularisation of the discretised equations has to be established. Thus, the mesh independence of the numerical solution is investigated loading a rod with displacement-controlled uni-axial tension. In order to initiate a localisation of the strains an imperfection is placed at one element, see Fig. 8. The dependence of the solution is analysed for the number of elements and the size of the imperfect element, respectively. The boundary value problem is represented in Fig. 8 and the model parameters are given in Table 2.

Figure 9a represents the stress-displacement solutions for the different discretisations, which converge with decreasing element size. The reason for this unique property of the chosen mixed formulation is the fact that the energy release rate is described directly with the extreme condition of the energy density. A regularisation is not necessary. By reaching the tensile strength of the UHPC-matrix inside the imperfect element, concrete cracking is smeared over the element size. The sudden failure of the UHPC-matrix causes an abrupt drop of stiffness in the affected element, what leads to a rapid drop of the total stress, see Fig. 9b. In the subsequent iteration step, a stable equilibrium is established by activating the micro steel fibres, which increase

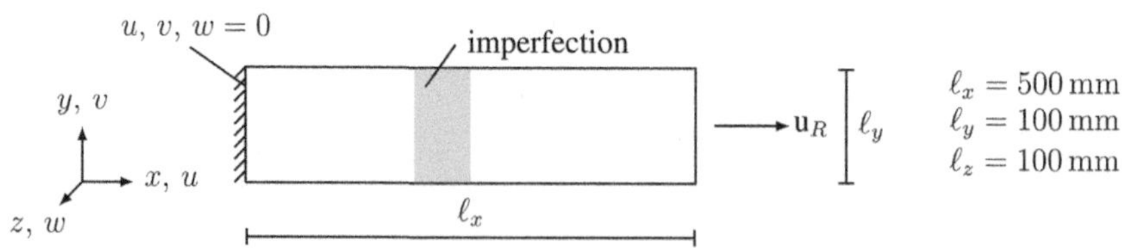

Fig. 8 Boundary value problem of a rod under tension

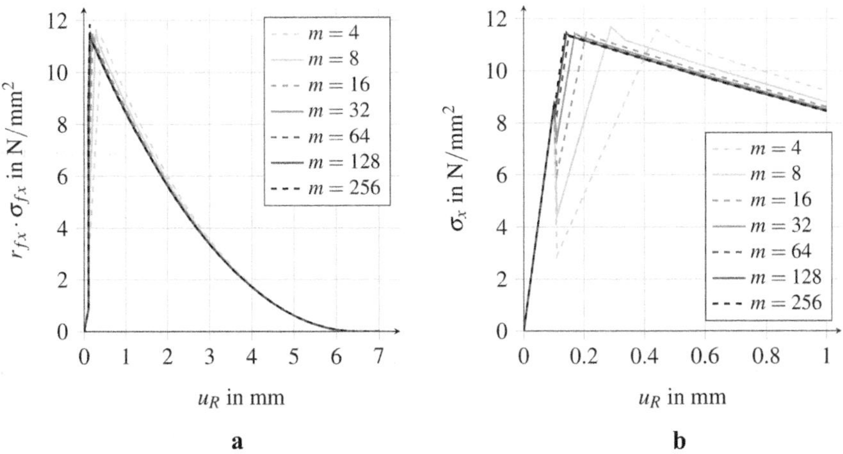

Fig. 9 **a** Stress-displacement diagram of the imperfect element under tension and **b** Zoom of the diagram in the range up to 1 mm displacement u_R

Table 2 Model parameters of the rod (Fig. 8) and of the beam bending tests (Fig. 10)

	Parameter	Symbol	Unit	Rod	DEMO
UHPC-matrix	Young's modulus	E_c	N/mm^2	40,000	43,300
	Poisson ratio	ν_c	–	0.0	0.221
	Tensile strength	f_{ct}	N/mm^2	8.0*	10.7
Micro steel fibres	Young's modulus	E_f	N/mm^2	170,000	169,000
	Fibre volume content	n_f	Vol.-%	2.5	1.25[1]/2.5[2]
	Fibre length	l_f	mm	6.5 (13)	6.5 (13)
	Fibre radius	r_f	mm	0.095	0.095
	Maximum bond resistance	τ_{max}	N/mm^2	3.325	8.0
	Fibre orientation factor	r_x	–	1.0	0.33

*Imperfect Element, [1] DEMO-1, [2]DEMO-2

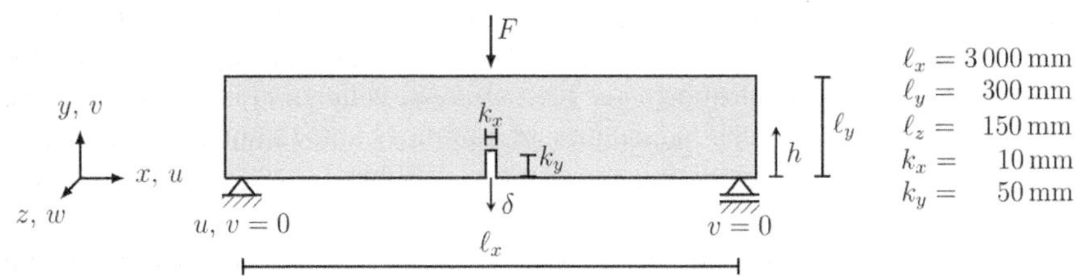

Fig. 10 Boundary value problem and dimensions of the large scale three-point-bending test

the overall stiffness and cause a further increase of macroscopic stresses at increasing displacements. In the case of larger elements, the stress drops more strongly due to the constant distribution of the crack strain over the element, which leads to larger displacements in the fibre activation phase, see Fig. 9b. By reaching the bond limit τ_{max} the fibre pull-out starts with a decreasing of the fibre stress σ_f and the bond length l_v. Thus, the deformation conditions are fulfilled in a weak sense, compare Eq. 10. Nonetheless, the overall solution is unique for different discretisations.

3.5 Simulation of Large Scale Three-Point-Bending Tests

Several large scale three-point-bending tests under monotonically increasing loading are presented in [5]. Two of them are simulated applying the macroscopic material models. DEMO-1 has a fibre volume content of 1.25 Vol.-% while DEMO-2 has a fibre volume content of 2.5 Vol.-%. The boundary value problem with the dimensions and boundary conditions is shown in Fig. 10. The supports are realised as lines along the depth of the beam. In order to weaken the singularity of the concentrated unit load $F = 1$ kN at the centre of the beam, the load is distributed over the surrounding nodes. The boundary value problem is solved using the arc-length method with respect to the overall displacement field. The beam is discretised using a structured 3D finite element mesh, and in the area of the notch, the element lengths are selected to 10 mm according to the width of the notch.

In addition to the huge number of experiments to determine strengths and Young's Moduli of the UHPC reference mix and the steel fibres of the manufacturer *Weidacon,* fibre pull-out tests with varying alignments are conducted to investigate the bond behaviour between the material components [28]. The specific test setup for the multiple fibre pull-out tests minimises the scattering of the results. The results of the fibre pull-out with an angle of 90° lead to a constant bond stress-slip relation, which is assumed in the material model with a constant function $f(\delta_v)$, see Fig. 6b. This assumption leads to a linear relationship between the fibre tensile stress and the fibre pull-out length. According to the experimental investigations [5], the maximum bond resistance τ_{max} is chosen to 8.0 N/mm^2 correlating to a fibre normal stress of $f_v = 1\,094$ N/mm^2, see Eq. 13. In order to make a reasonable assumption regarding the spatial distribution of the fibres, the fibre orientation factor r_x is set to 33% of the total fibre volume in the longitudinal direction of the beam. The material and model parameters used for the simulation are specified in Table 2.

Figure 11 compares the load-deflection behaviour between the experiments and the simulation. In the case of DEMO-1 a very good agreement is achieved. The initial stiffness and the load bearing capacity around 23 kN are well approximated, see Fig. 11a.

The simulation of DEMO-2 also reproduces the load-deflection curve qualitatively well and, as expected, shows twice the load-bearing capacity of 46 kN due to the doubled fibre volume content, but underestimates the load-bearing behaviour of the beam compared to the experiment. The specified flow direction and compaction

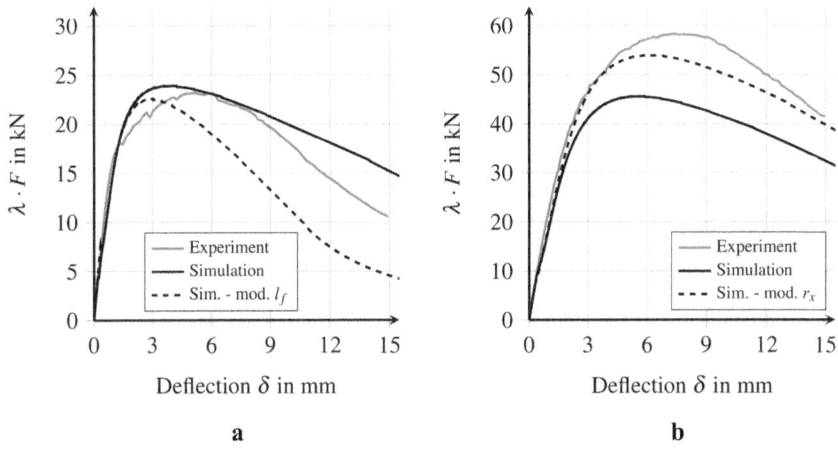

Fig. 11 Comparison of load-deflection-behaviour of three-point-bending tests between experiment and simulation: **a** DEMO-1 and **b** DEMO-2

during the manufacturing process [31] of the beam may cause the fibres to align preferentially in the longitudinal direction, which increases the load-bearing capacity. Therefore, another simulation is presented with a modified fibre orientation factor $r_x = 0.40$ in order to better capture the load-deformation behaviour of the experiment, see dashed curve in Fig. 11b.

In comparison to the experiments both simulations show a slightly stiffer behaviour after crack initiation. Since a fibre pull-out length of half of the fibre length is assumed for each fibre, a maximum resistance of the bond zone is given. In experiments the fibre pull-out length is arbitrarily distributed but generally smaller than half of the fibre length, what leads to a reduction of the resistance and therefore to a lesser stiffness with respect to increasing deformations. To assess this effect, a comparative simulation was performed with one-quarter of the fibre length (3.25 mm) taken as effective, see dashed curve in Fig. 11a.

Furthermore, at a deflection of 9.5 mm (DEMO-1) and 9.0 mm (DEMO-2), the compressive strength of about 150 N/mm^2 is reached at the upper edge in the compression zone. Since the model currently only takes into account linear elastic material behaviour of concrete under compressive loading, the damage of the compression zone is not taken into account. Both simplifications explain the deviation of the simulation from the experiment in the range between 9 and 15 mm. Overall, the numerical results lead to the following insight:

- The fibre orientation with respect to the principal direction of the tensile stresses has a very strong influence on the ultimate load and the ductility.
- Longer fibres activate a longer fibre pull-out length, which increases the ductility.

Figure 12a–f refer to the simulation of DEMO-1 exemplarily. Fig. 12a–c show the stress distributions over the cross-section above the notch for three different load levels λ. For each load level the total stress distribution (black-dashed), the concrete- (blue) and the fibre stress (red) distribution are shown separately. The concrete- and the fibre stresses are represented with respect to their volume fractions and to the fibre orientation factor in the longitudinal direction of the beam ($r_{cx} = (1\text{-}n_f) \cdot r_x$, $r_{fx} = n_f \cdot r_x$), which give in sum the total stresses. As soon as the concrete stresses reach the tensile strength at the lower edge, they drop to zero and the cross-section begins to crack over the height like a zipper.

At load level $\lambda = 9$ the cross-section is already cracked at the lower edge up to the half. The fibres completely take over the load transfer in this region ($r_{fx} \cdot \sigma_f = \sigma$), while in the compression zone, the load is mainly transferred by the concrete due to the much higher volume fraction, see Fig. 12a. During the fibre activation phase, the fibres in the full bond elongate and bridge the crack. This allows the composite material to increase the load bearing capacity of the cross section resulting in a further increasing of the total stresses, see Fig. 11a.

Figure 12b shows the stress distribution shortly before reaching the ultimate load, where the cross-section is almost completely cracked. In the upper part of the cross-section the compression zone is still intact, and slightly below, in the area between $h =$ 0.20–0.25 m, the bond between fibres and UHPC-matrix is still intact, which leads to crack bridging due to the fibres. At the lower cross-section between $h = 0.05-0.20$ m the fibre stresses decrease, which indicates the beginning of the fibre pull-out after the bond capacity is reached, correlated to the fibre normal stress of $f_v = 1\ 094$ N/mm^2, see Eq. 13. The combination of the progressive fibre pull-out and the fibre activation phase continues to enable stable material behaviour in the post-cracking region, compare Fig. 11a.

If the bond capacity of all fibres at the cracking zone is reached the unstable failure of the composite material occurs. Figure 12c represents the distributions of the stresses for load level $\lambda = 20$ shortly after exceeding the load bearing capacity, where the cross section is almost completely cracked and the fibre pull-out is very progressed. Nonetheless, the complete crack up to the upper edge of the beam cannot be modelled, as this would no longer result in equilibrium.

Figure 12d summarises the development of the total stresses, where the shifting of the neutral axes to the upper cross section is clearly recognisable, and the effect of a small fibre volume content with respect to the stiffness becomes obvious.

In Fig. 12e the effective fibre stress distribution over the cross-sectional height is represented for the three load levels, which is directly connected to the bond stresses between the UHPC-matrix and the micro steel fibres, cf. Equations 11–13. The chosen constant relationship between the bond stresses and the fibre pull-out length δ_v leads to linearly decreasing effective fibre stresses, which is recognisable from the stress distribution over the cross-sectional height, see Fig. 12e.

The constraints for the crack formation and the fibre pull-out are connected to the kinematic conditions, see Eq. 10. The development of the fibre pull-out length follows the crack opening, which is represented related to the element length *lel* of the finite element discretisation, see Fig. 12f. At the very end the fibre pull-out length converges to the related crack strain, whereby the load capacity of the fibre tends

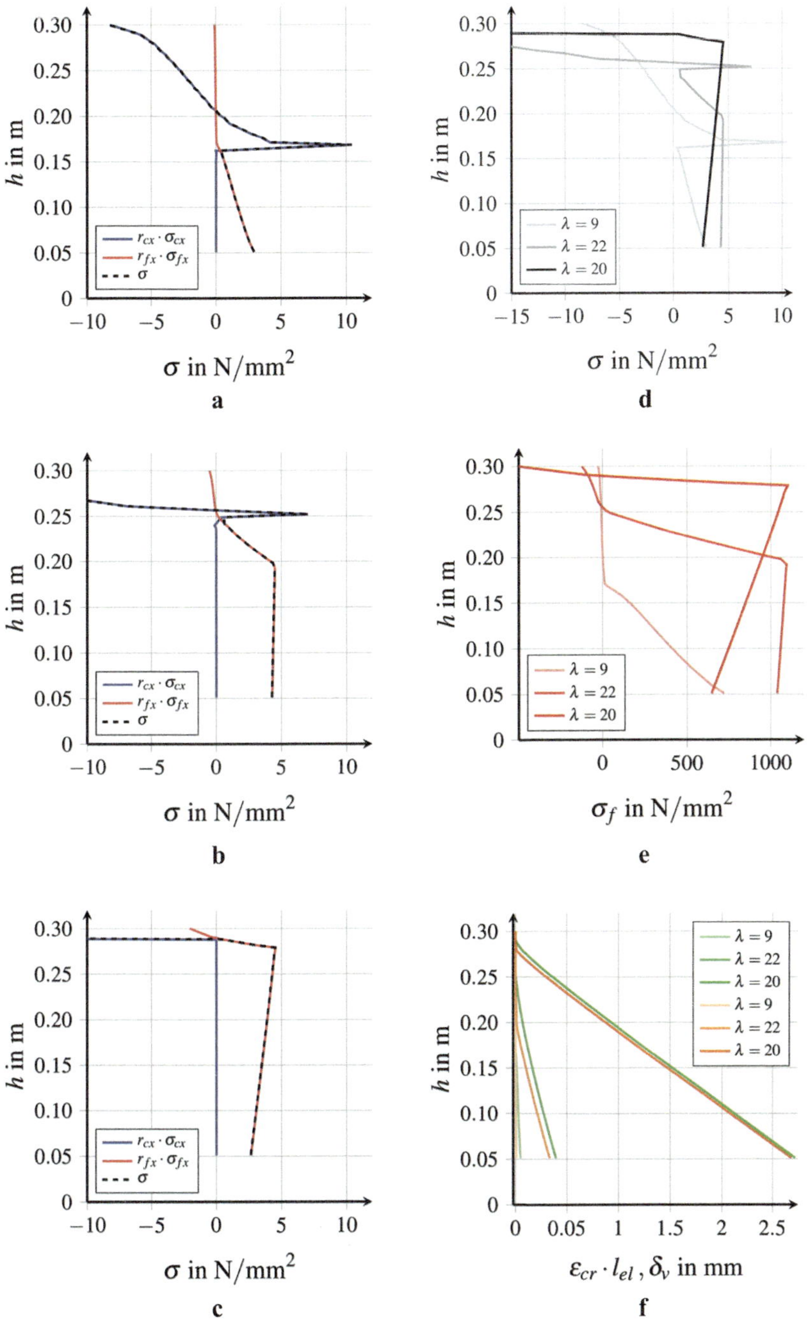

Fig. 12 Distributions of concrete, fibre, and total stresses over the cross-section above the notch for the load levels **a** $\lambda = 9$, **b** $\lambda = 22$, **c** $\lambda = 20$ and development of the element variables over the cross-section above the notch **d** total stress, **e** fibre stress **f** crack-strain related to element length (green) and fibre pull-out length (orange)

to zero. The developments of the crack strain and the fibre pull-out length over the cross-sectional height are correlated to each other at progressive fibre pull-out.

3.6 *Modelling the Tensile Fatigue Behaviour*

The extension of the material model for the description of material behaviour under cyclic tensile loading is current focus of research. The concept provides the formulations of the maximum bond resistance τ_{max} and the fibre pull-out length δ_v as a function of the number of load cycles *n*, the average bond stress τ_m and the bond stress amplitude $\Delta\tau_v$

$$\tau_{max} = \tau_{max}(n, \tau_m, \Delta\tau_v, \delta_v), \tag{17}$$

$$\delta_v = \delta_v\left(n, \tau_m, \Delta\tau_v, \delta_{v,max}\right) \tag{18}$$

Based on the mesoscale model for the fibre-matrix bond, the damage of the fibre channel at cyclic loading is to be taken into account. The approach is intended to determine the structural state of a component after a certain number of cycles. The concept comprises for elastic loading and unloading in the first load cycle to establish all parameters for the start of the cyclic loading. The approach aims for simulating high-cycle fatigue loads with little numerical effort and without analysing each individual load cycle, respectively.

4 Conclusion

The contribution at hand presents the modelling and numerical simulations of the load-bearing behaviour of micro steel fibres embedded in UHPC under monotonic and cyclic tensile loading. The numerical analyses of the bond behaviour between the UHPC-matrix and micro steel fibres on the mesoscale and of the post-cracking behaviour on the macroscale lead to the following results:

1. The bond zone model describes the bond behaviour between the micro steel fibres and the UHPC-matrix and provides information about the processes at the mesoscale as well as the bond stress distribution along the fibre and the residual stresses in the bond zone at unloading.
2. The maximum pullout resistance is present at the beginning of the fibre pull-out, which leads to a decreasing load-bearing capacity of the fibre during the micro crack opening in the UHPC-matrix.
3. Applying a smeared description of the crack development and the fibre pull-out behaviour at the macro scale enables numerical investigations of real structures. In comparison to classical continuum damage models, the post-cracking behaviour

is described with model equations, which are equivalent to fracture mechanics, and which allow for the damage mechanisms of the two material components to be specified separately.
4. A hybrid-mixed element formulation is presented, which yields a mesh-independent numerical solution of the material failure.
5. Simulations of three-point bending beam tests applying physically meaningful model parameters achieve good agreement with the experimentally load-bearing behaviour, capturing the scatter of the fibre distribution and the effective fibre pull-out length. A variation of the model parameters may fit the experiments much better if required. The focus of the investigations is more on the general modelling of the load-bearing behaviour.
6. The analysis of the macroscopic variables at the cross-section provides detailed information on how the crack bridging is effected by the fibres.
7. The approaches to model the damage behaviour can be applied for the development of new fibre designs, if the bond properties are known from experiments.

The investigations presented in this chapter are part of the cooperative project with the Institute of Building Materials, Concrete Construction and Fire Savety, Division of Concrete Construction, TU Braunschweig (see chapter by J.-P. Lanwer et al., this volume).

References

1. Wilkening, M., Joachim, L., Oettel, V.: Investigations on the fatigue loading of thin-walled and resource-efficient UHPFRC segmental bridges. Eng. Struct. (2024). https://doi.org/10.1016/j.engstruct.2024.117858
2. Bertola, N., Küpfer, C., Kälin, E., Brühwiler, E.: Assessment of the environmental impacts of bridge designs involving UHPFRC. Sustainability (2021). https://doi.org/10.3390/su132212399
3. Makita, T., Brühwiler, E.: Tensile fatigue behaviour of ultra-high performance fibre reinforced concrete (UHPFRC). Mater. Struct. (2014). https://doi.org/10.1617/s11527-013-0073-x
4. Lanwer, J.-P., Empelmann, M.: Fundamental investigations on the performance of micro steel fibres in ultra-high-performance concrete under monotonic and cyclic tensile loading. Appl. Sci. (2021). https://doi.org/10.3390/app11209377
5. Lanwer, J.: Zum Tragverhalten von ultrahochfestem Stahlfaserbeton unter zyklischer Zugbeanspruchung, Doctoral Thesis, Technische Universität Braunschweig, Braunschweig, Germany (2024)
6. Zhan, J., Nussbaumer, A., Brühwiler, E.: Influence of fiber orientation on the high cycle tensile fatigue resistance of ultra-high performance fiber reinforced cementitious composites (UHPFRC). Int. J. Fatigue (2024). https://doi.org/10.1016/j.ijfatigue.2023.108103
7. Gudžulić, V.: Virtual lab for mechanical testing of steel and steel-fiber reinforced concrete, Ruhr-Universität Bochum (2024)
8. Krahl, P.A., Carrazedo, R., El Debs, M.K.: Mechanical damage evolution in UHPFRC: experimental and numerical investigation. Eng. Struct. (2018). https://doi.org/10.1016/j.engstruct.2018.05.064
9. Lanwer, J.-P., Oettel, V., Empelmann, M., Höper, S., Kowalsky, U., Dinkler, D.: Bond behavior of micro steel fibers embedded in ultra-high performance concrete subjected to mono-tonic and cyclic loading. Struct. Concr. (2019). https://doi.org/10.1002/suco.201900030

10. Leutbecher, T., Heck, L., Schleiting, M.: Prüfung von UHFB nach DIN EN 14651—Zum Einfluss der Probekörperherstellung auf die Prüfergebnisse. Beton und Stahlbetonbau (2022). https://doi.org/10.1002/best.202200027
11. Wang, Y., Qiao, P., Sun, J.: An Chen: Influence of fibers on tensile behavior of ultra-high performance concrete: a review. Constr. Build. Mater. (2024). https://doi.org/10.1016/j.conbuildmat.2024.136432
12. Storm, J., Ranjbarian, M., Mechtcherine, V., Scheffler, C., Kaliske, M.: Modelling of fibre-reinforced composites via fibre super-elements. Theoret. Appl. Fract. Mech. (2019). https://doi.org/10.1016/j.tafmec.2019.102294
13. Schröder, J., Pise, M., Brands, D., Gebuhr, G., Anders, S.: Phase-field modeling of fracture in high performance concrete during low-cycle fatigue: numerical calibration and experimental validation. Comput. Methods Appl. Mech. Eng. (2022). https://doi.org/10.1016/j.cma.2022.11518
14. Höper, S.: Modellierung und numerische analyse des faser-matrix-verbun-des in UHPFRC unter Zugbeanspruchung, Doctoral Thesis, Technische Universität Braunschweig, Braunschweig, Germany (2021).
15. Lanwer, J.-P., Höper, S., Gietz, L., Kowalsky, U., Empelmann, M., Dinkler, D.: Fundamental investigations of bond behaviour of high-strength micro steel fibres in ultra-high performance concrete under cyclic tensile loading. Materials (2021). https://doi.org/10.3390/ma15010120
16. Höper, S., Kowalsky, U., Dinkler, D.: Micro-structure related modelling of ultra-high-performance fibre reinforced concrete (UHPFRC) subjected to cyclic tensile loading. Proc. Appl. Math. Mech. (2018). https://doi.org/10.1002/pamm.201800375
17. de Borst, R., Verhoosel, C.V.: Gradient damage vs phase-field approaches for fracture: similarities and differences. Comput. Methods Appl. Mech. Eng. (2016). https://doi.org/10.1016/j.cma.2016.05.015
18. Donnini, J., Lancioni, G., Chiappini, G., Corinaldesi, V.: Uni-axial tensile behavior of ultra-high performance fiber-reinforced concrete (uhpfrc): Experiments and modeling. Compos. Struct. (2021). https://doi.org/10.1016/j.compstruct.2020.113433
19. Velde, J., Kowalsky, U., Zümendorf, T., Dinkler, D.: 3D-FE-Analysis of CT-specimens including viscoplastic material behavior and nonlocal damage. Comput. Mater. Sci. (2009). https://doi.org/10.1016/j.commatsci.2009.03.019
20. Kowalsky, U., Meyer, J., Heinrich, S., Dinkler, D.: A nonlocal damage model for mild steel under inelastic cyclic straining. Comput. Mater. Sci. (2012). https://doi.org/10.1016/j.commatsci.2012.05.062
21. Monnamitheen Abdul Gafoor, A.H., Dinkler, D.: Modeling damage behavior of concrete subjected to cyclic and multiaxial loading conditions. Struct. Concrete (2022). https://doi.org/10.1002/suco.202100109
22. Griffith, A.A.: VI. The phenomena of rupture and flow in sol-ids. Phil. Trans. R. Soc. Lond. A (1921). https://doi.org/10.1098/rsta.1921.0006
23. Rice, J.R.: A path independent integral and the approximate analysis of strain concentration by notches and cracks. J. Appl. Mech. (1968). https://doi.org/10.1115/1.3601206
24. Castigliano, A.: Théorie de l'équilibre des systèmes élastiques et ses applications. Negro, Turin (1879)
25. Hellinger, E.: Die allgemeinen Ansätze der Mechanik der Kontinua. In: Klein, F., Müller, C. (eds.) Mechanik, pp. 601–694. Vieweg+Teubner Verlag, Wiesbaden (1907)
26. Reissner, E.: On a variational theorem for finite elastic deformations. J. Math. Phys. (1953). https://doi.org/10.1002/sapm1953321129
27. Gietz, L., Kowalsky, U., Dinkler, D.: A phenomenological model to describe cracking and fibre pull-out in ultra high-performance fibre reinforced concrete. Proc. Appl. Math. Mech. (2023). https://doi.org/10.1002/pamm.202200321
28. Oettel, V., Lanwer, J.-P., Empelmann, M.: Auszugverhalten von Mikrostahlfasern aus UHPC unter monoton steigender und zyklischer Belastung/Pullout behaviour of micro steel fibres of UHPC under monotonic and cyclic loading. Bauingenieur (2021). https://doi.org/10.37544/0005-6650-2021-01-02-27

29. Pian, T.H.H., Tong, P.: Basis of finite element methods for solid continua. Numer. Meth Eng. (1969). https://doi.org/10.1002/nme.1620010103
30. Pian, T.H.H., Sumihara, K.: Rational approach for assumed stress finite elements. Numer. Meth Eng. (1984). https://doi.org/10.1002/nme.1620200911
31. Maya Duque, L.F., Graybeal, B.: Fiber orientation distribution and tensile mechanical response in UHPFRC. Mater. Struct. (2017). https://doi.org/10.1617/s11527-016-0914-5

Multiscale Investigations of the Degradation Progress in the Localised Fracture Zone of Carbon Short-Fibre Reinforced High-Performance Concrete Subjected to High-Cycle Tensile and Flexural Fatigue Load Regimes

Oliver Fischer, Dirk Volkmer, Christian U. Grosse, Malte A. Peter, Philipp Lauff, Ante Džolan, Roland Niedermeier, Matthias Rutzen, Michael Engelhard, Polina Pugacheva, Manuel Raith, Ursula Weiss, and Tanja Lochner

Abstract The complex mechanisms leading to degradation of a 3D-printed carbon short-fibre reinforced high-performance concrete which—due to the orientation of the carbon fibres—achieves very high tensile and flexural strengths, are investigated. For this purpose, multiple measurement methods acting on different scales were combined. At macro level, acoustic emission analysis and coda wave interferometry were used in cyclic, uniaxial tensile tests alongside conventional measurement methods such as digital image correlation to visualise internal damage. In addition, conductivity of the carbon fibres enabled conclusions to be drawn about the damage by measuring electric resistances. At micro and meso scale, the microstructure of unloaded specimens and development of microcracks were recorded in bending and

O. Fischer (✉) · P. Lauff · A. Džolan · R. Niedermeier
Chair of Concrete and Masonry Structures, Technical University of Munich (TUM), Munich, Germany
e-mail: oliver.fischer@tum.de

D. Volkmer · M. Rutzen · M. Engelhard
Chair of Solid State and Materials Chemistry, Institute of Physics, University of Augsburg (UNA), Augsburg, Germany

C. U. Grosse · P. Pugacheva · M. Raith
Chair of Non-Destructive Testing (NDT), Technical University of Munich (TUM), Munich, Germany

M. A. Peter · U. Weiss
Centre for Advanced Analytics and Predictive Sciences (CAAPS), University of Augsburg (UNA), Augsburg, Germany

M. A. Peter · T. Lochner
Institute of Mathematics, University of Augsburg (UNA), Augsburg, Germany

N. Oneschkow and L. Lohaus (eds.), *Cyclic Deterioration of High-Performance Concrete*, https://doi.org/10.1007/978-3-032-19394-0_15

in tension tests utilising computer tomography and microscopic examinations. Based on measurement data, microstructure-orientated unit cells allowing for the development of microcracks were derived and integrated into a computational homogenisation scheme capable of reproducing the material behaviour and relevant failure mechanisms.

Keywords Carbon short-fibre reinforced concrete · Oriented fibres · Multiscale fatigue testing and modelling · Degradation · Localisation of failure · NDT methods

1 Introduction

The extrusion of a carbon short-fibre reinforced mortar paste through a small nozzle leads to the fibres being aligned in the direction of the material flow (i.e. the direction of movement of the printing nozzle). This fact can be utilised to adjust the fibre orientation with the main tensile trajectories and thus to an anticipated loading scenario. In previous research, this concept has been proven to be very effective and that cementitious composite specimens can be produced with outstanding tensile and flexural strength, e.g. [1, 2]. Having also shown that up-scaling—starting from small cementitious samples (micro-scale)—to a carbon short-fibre reinforced concrete (CSFRC) is possible [3, 4], a very versatile high-performance material is available which will be investigated experimentally and theoretically with regard to high-cycle fatigue under different types/levels of loading as well as in terms of damage development, failure mechanisms and localisation.

1.1 General Research Procedure

One main objective was to gain new in-depth insights into the damage behaviour and the main failure mechanisms of the CSFRC material under different loading regimes (static, cyclic, variable) by means of experimental investigations using centric tensile as well as 3-point bending tests on different scales (micro, meso and macro specimens) and to be able to interpret and compare the scale levels. While the micro specimens offer advantages for analytics and are easier to access for mathematical simulations, meso-size samples have particularly proven suitable for AE methods and correlations with LDV. Large-scale experiments that are closer to real-life applications, are essential for identifying potential scale effects and also offer advantages for various measurement techniques such as electrical and fibre-optical methods or DIC. On the basis of the extensive test and measurement results gained from the project, suitable mathematical models were derived representing the damage and failure mechanisms as realistically as possible while remaining computationally feasible with standard commercial software packages. Modelling and simulation initially starts at micro level and with static loads and consequently aims to predict experimental results derived with macro samples as well as under dynamic fatigue loading and to calibrate the models accordingly.

While the focus of the experimental investigations was firstly laid on simple static and monotonic tests (performed on micro 3-point bending and macro tensile specimens) as well as on identifying, generally assessing and evaluating the complex damage and failure mechanisms of CSFRC and the establishment of suitable measuring technologies, later the focus was laid on scale transition (micro tensile, macro/meso 3-point bending specimens) and the extension to more complex fatigue loads, more refined measurements and the combination of different sensor techniques.

1.2 Material, Testing Specimens

Following previous investigations, carbon short-fibres (CSF) with 7 μm in diameter and a length of 3 mm were utilised as "reinforcement" of the cementitious material after desizing and oxidation of the fibre surface, please refer to e.g. [1, 2, 5]. Hereby, two types of fibres were used, Tenax-J HT C261 carbon fibres (for micro and most of the meso samples) and Zoltek (Toray) PX35 fibres (macro samples). Regarding the concrete mix, for the micro samples a suitable concrete paste was deployed being already investigated for several years to achieve optimum extrudability and fibre dispersion in the 3D-printing process of small-scale samples [1]: 61.0 weight percent (wt.%) Schwenk CEM I 52.5 R, 21.0 wt.% EFACO silica fume, 3.0 wt.% superplasticiser BASF ACE 430 and 15.0 wt.% of water (w/o aggregates). The concrete mix design for the meso/macro specimens was composed of 34.7 wt.% Holcim Sulfo 52.5 R cement, 21.7 wt.% Sika Silicol P fume, 21.7 wt.% quartz powder SF 500, 7.7 wt.% quartz sand, 2.5 wt.% superplasticiser ACE 460 and 11.7 wt.% of water. While the fibre content of the micro samples varied between 0 (unreinforced) and 1–3 vol.%, all meso/micro specimens had a fibre content of 1 vol.%.

Regarding the geometry of the different specimens, the macro tensile specimens (tested in uniaxial tension) had a dog-bone shape with dimensions shown in Fig. 1a, while the smaller micro and meso specimens tested in uniaxial tension had reduced dimensions (scale factor 10 and 20%). All samples tested in 3-point bending (flexural tensile strength) were prismatic with square cross-sections and the load being applied at midspan (symmetry axis). While the macro samples (bending) had a cross-section of 40 × 40 and 180 mm free span, the corresponding dimensions of the micro samples were 5 × 5 mm (span 15 mm) and those of the meso samples 15 × 15 mm (span 90 mm). For manufacturing, a special 3D printer was developed at TUM for the *macro samples* [6, 7]. Through the printing process, the desired fibre orientation was achieved both in the dog-bone shaped tension specimens (Fig. 1b, c) and the prismatic flexural samples (fibre orientation parallel to lower edges, i.e. in the direction of the principal tensile trajectories in the midspan tensile zone) (Fig. 2).

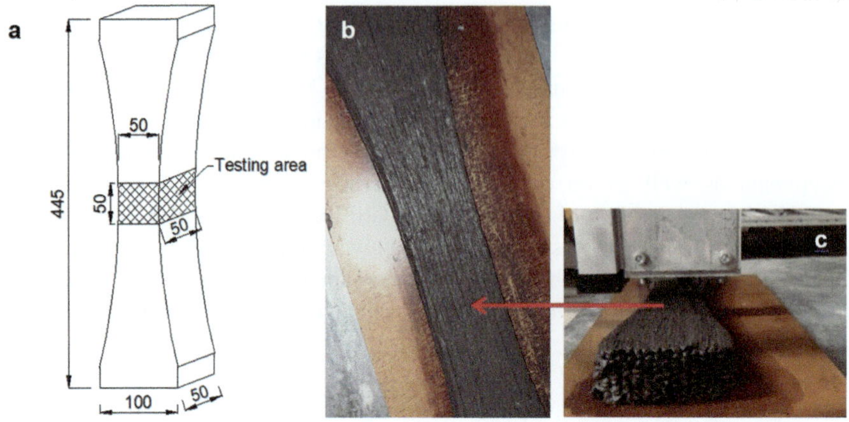

Fig. 1 Shape/dimensions (mm) of *macro specimens* tested in centric tension (**a**), 3D-printed dog-bone shaped sample (**b**) and printing process (**c**)

Fig. 2 3D printing of meso size flexural beam specimens with CSFRC material

2 Experimental Investigations on CSFRC Specimens

The present section firstly outlines which experimental tests were carried out with the CSFRC material, refer to Table 1. Please note that only the highly instrumented principal experiments are included therein, while numerous additional tests aiming on optimising the testing and measurement setup are not listed. It then addresses static loading tests, cyclic monotonic tests and fatigue tests with varied stress levels

Table 1 Overview: static/cyclic experimental investigations on the CSFRC material (number of highly instrumented tests performed)

Sample size	Static load	Cyclic loading (monotonous)	Cyclic loading (varied load, sequence effects)
Micro	> 60	25	–
Meso	48	10	1
Macro	30	36	26

and the analysis of sequence effects. In each case, it is specified both whether micro or macro/meso test specimens were used and if uniaxial tensile tests or 3-point bending tests were performed. Hereby, key findings from the various tests are explained and discussed directly. In addition, Sect. 3 provides a detailed summary of all the measurement techniques utilised (including their combination) and the findings that can be derived from the measuring results with regard to load-bearing behaviour, damage development, fatigue behaviour and the different failure mechanisms of the CSFRC material. Finally, Sect. 4 addresses mathematical modelling aspects and the numerical simulation of the load-bearing behaviour of the material.

2.1 Static Loading Tests, General Behaviour of the Material

To illustrate the general load-bearing and deformation behaviour and the significant difference to fibre-free samples, Fig. 3 illustrates typical stress–strain curves in both static tensile tests (Fig. 4) and in 3-point bending tests for different fibre content using the CSFRC micro test specimens as an example.

Micro beam specimens without reinforcement show purely linear-elastic behaviour and a flexural strength of a maximum of about 22 MPa (Fig. 3a). Addition

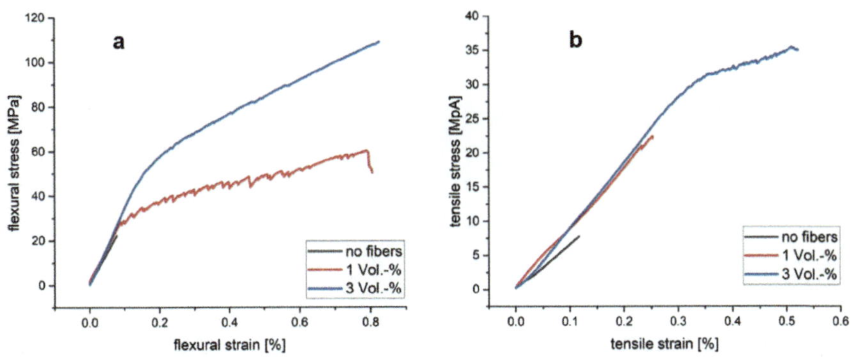

Fig. 3 Stress–strain curves with different fibre content; **a** extruded micro beam specimens (3-point bending test [8]) and **b** micro tensile test samples

of 1 volume percent (vol.%) of CSF leads to a sharp increase in flexural strength up to 60 MPa, if 3 vol.% are included in the mortar a flexural strength of the specimens of up to 110 MPa can be achieved, also see e.g. [9]. In comparison, tensile micro samples with 1 vol.% CSF achieve a uniaxial tensile strength of up to 23 MPa, while with 3 vol.% some 35 MPa of tension can be reached before failure (Fig. 3b).

Specimens reinforced with CSF show a pronounced strain-hardening behaviour. For tensile micro specimens with 1 vol.% it can be stated that the critical fibre content is not yet reached. Nevertheless, a distinct strain hardening can be observed and was therefore investigated in detail (see e.g. Sects. 3.1–3.4), because a higher fibre content would have made segmentation impossible (see Sect. 3.4, Fig. 23b) and thus, no reliable basis would have been available for the multiscale modelling (see Sect. 4). The specific deformation behaviour is linked to the CSF being able to bridge the cracks. Leaving the linear-elastic regime instantly leads to unstable crack growth (and thus to failure) in the unreinforced material, crack growth is inhibited in the fibre-reinforced specimens. As the fibres bridge the cracks, there is only a minor change in the overall tensile load transfer along the specimens which in turn causes the formation of multiple microcracks. Materials showing such a behaviour have been described and discussed in the literature since the early 1990s under the terms 'Engineered Cementitious Composites' (ECCs) or 'Strain Hardening Cementitious Composites' (SHCCs), see e.g. [10–13].

In addition to 3-point bending tests, also dog-bone shaped *micro specimens* were analysed in *static tensile tests*. The main reason for this was to ensure the feasibility of the in-situ tensile testing in the computer tomograph (refer to Sect. 3.4), which

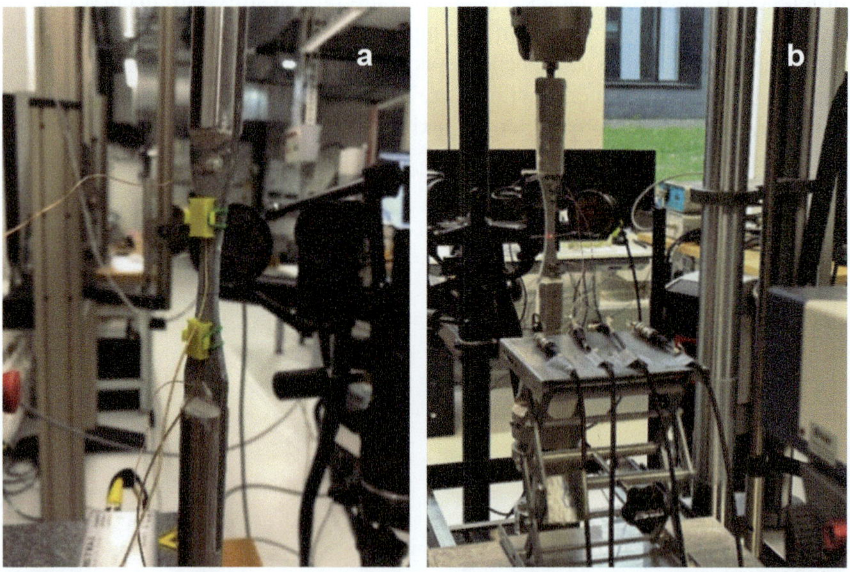

Fig. 4 Static tensile test setup on micro sized specimens

gained more insights and also facilitated a multiscale modelling (see Sect. 4). It was possible to integrate various measurement methods in-situ. Tensile tests on micro specimens were carried out using a universal testing machine with a 5 kN load cell produced by Zwick&Roell (Ulm, Germany) whereas the meso size samples were tested using an Instron (Norwood, USA) ElectroPuls E10000 testing machine.

The experiments were partly monitored by AE detectors (also see Sect. 3.3), 2D or 3D camera systems for DIC and by means of LDV. Additional measurements were performed to determine the material properties, while testing methods were improved. In this context, it has to be stated that the meso size specimens were specifically designed for investigations concerning the correlation of AE and LDV (see Sect. 3.3).

2.2 Cyclic Tests (Monotonous)

In the beginning, a variety of uniaxial *tensile fatigue tests* (monotonic load cycles) have been carried out with *macro specimens*, please e.g. refer to [14]. These tests were systematically accompanied utilising several measuring techniques in order to gain as much insight as possible to internal damaging mechanisms and the general fatigue process of the novel material, see e.g. [15]. In the following, test series 'E6' is discussed, which was tested at a frequency of 5 Hz, a cyclic load varying between 40 and 80% of the static tensile strength ($f_{ct} = 13.48$ MPa) and with 1 vol.% CSF. Specimen age was between 29 and 35 days at beginning of testing; the total number of load cycles (until failure) varied from 0.4 to 5.1 million.

Compared to standard concrete under fatigue loading, which exhibits three clearly distinguishable phases of damage evolution, i.e. initiation of microcracks (phase 1), stable crack growth (phase 2) and instable crack development (phase 3), CSFRC behaves differently—generally and scale-dependent (sample size). Put simply: the larger the scale, the less distinct the phases can be recognised (especially phase 3).

As can be seen in Fig. 5a (strain gauge measurement), phase 1 is very short and barely visible in tests E6.1 and E6.6. The subsequent phase 2 accounts for most of the testing time, with crack growth tending to stagnate or even 'decrease', as in test E6.6. Phase 3 is only visible in test E6.4, while samples in the other tests break more or less abruptly without any noticeable increased cracking. A closer look at test E6.4 reveals the reason for the high strain rate shortly before failure (see Fig. 6). During the load cycles, the diagram shows the evolution of strain at the respective points of maximum load with temperature compensation (index T). In this test, the crack that led to failure ran across the strain gauge DMS1U.

The strain increases rapidly as the crack reached the respective gauge at about 270,000 cycles, whereas the other strain gauges remain in their slowly increasing tendency. DMS3U is located very close to the crack, however, only a slight effect on the strain development can be seen at the end of the fatigue test. This evaluation shows that the development of microcracks is highly localised and has little effect

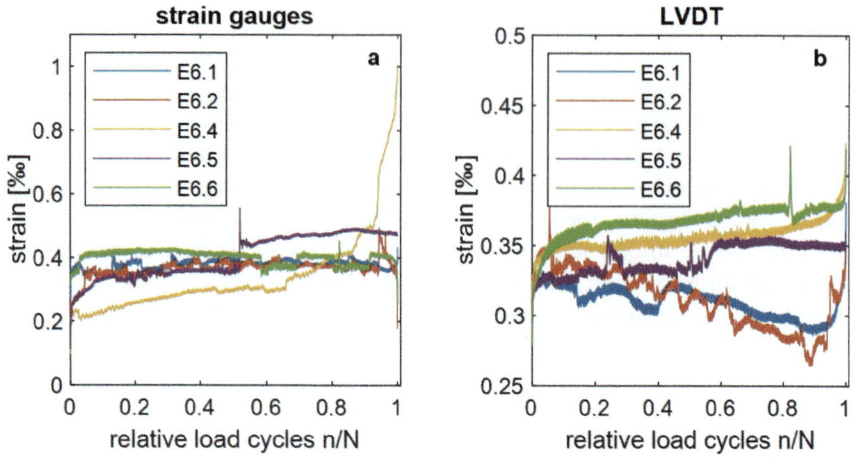

Fig. 5 Strain development during fatigue test (macro tensile samples, series E6): **a** average value from four strain gauges with virtually abrupt failure; **b** average value from three inductive displacement sensors (LVDT) measuring the strain integrally over the specimen's length

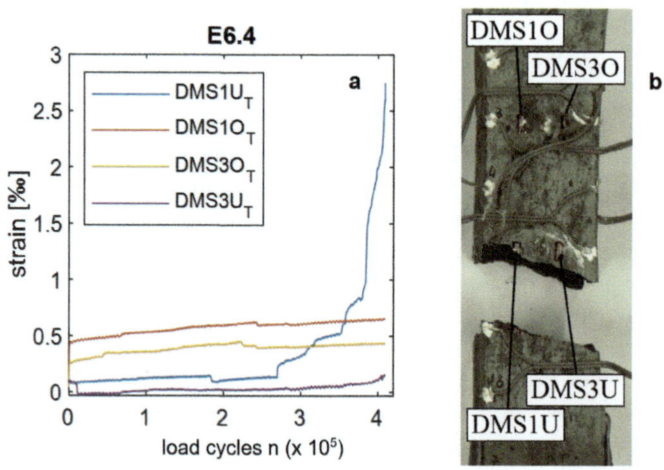

Fig. 6 Examination of individual strain gauge measurements for test E6.4 (**a**), where gauge DMS1U being located directly above the final failure crack; position of individual strain gauges with 3 mm measuring length and final crack (**b**)

on neighbouring areas. If no strain gauge crosses the microcrack leading to failure, the proximity to fracture cannot be detected.

Figure 15 shows the strain measurement with DIC for the same specimen E6.4. The gauges are located on the side opposite the speckle pattern or DIC camera, respectively (also see Fig. 15). Two areas with major damage and microcrack widening are clearly visible. The one further down in the DIC image leads to final failure. On closer inspection of this area, multiple parallel microcracks can be clearly identified, which define a small damage zone but are locally limited in the horizontal direction. Further DIC evaluations for this test (not shown here) reveal that the critical failure area formed at the beginning of the fatigue test due to the merging of several microcracks.

Looking at the results obtained with inductive displacement transducers (LVDT), measuring integrally over the entire height of the specimen, a more general trend in strain development can be observed (Fig. 5b). Almost all tests show a flattening curve in phase 1, with four tests also showing a clear transition to phase 3. The tests E6.5 and E6.6, however, fail and hence show a similar behaviour as the local strain gauge measurement without indicating phase 3. A considerable factor is the decreasing strain development in tests E6.1 and E6.2, the cause of which is still unknown.

The crack surface analysis by means of electron microscopy reveals the fact that final failure of the specimen mainly occurs due to fibre breakage, not due to fibre pull-out. This finding also fits to the size of the microcracks. Although the final failure crack is widening significantly compared to other regions of the tested sample, this crack is smaller than 100 μm and therefore is barely visible by human eyes.

In summary, it can be said that, compared to normal concrete, a basically similar fatigue behaviour is observed, but that there are significant deviations in phase 3.

In addition to monotonous tests on macro specimens, also a huge number of dynamic experiments was performed on *micro samples* subjected to *cyclic loading* in *3-point bending* tests, particularly in the first phase of the research, see e.g. [8, 9]. In the following, a typical experiment is shown with selected representative results, which are subsequently discussed.

The positive influence of the fibres again leads to an increased fatigue resistance under cyclic loads. Figure 7b shows a Woehler S–N-diagram for unreinforced micro beam specimens and those with 1 vol.% carbon fibres. For better comparability, the stress values in the diagram are normalised to the flexural strength of the respective test specimen. Specified numbers correspond to the upper load limit (i.e. 60% of the flexural strength); the lower load limits can be obtained by subtracting the selected load amplitude of 40%. The tests were carried out using a DMA testing machine with a load frequency of 100 Hz. In general, the number of cycles to failure increases by a factor of approximately 100 when comparing the micro beam specimen without fibres (which, depending on load level, typically withstands between 10^3 and 10^5 cycles until failure) with the sample containing 1 vol.% CSF (which fail after approximately 10^6–10^7 load cycles).

A more micro-structural view on the fatigue behaviour can be obtained by observing various parameters that correlate with the damage accumulated in the specimen during the load cycles. It has been shown that observation of viscoelasticity represented by Storage Modulus (SM; representing the elastic properties, similar to the Young's Modulus) and Loss Modulus (LM; representing the viscous properties and thus pointing to irreversible changes) allows for meaningful predictions. A more in-depth explanation of the significance of SM and LM during cyclic experiments is given in [16]. Figure 8 shows the development of these moduli over the duration of a fatigue test (here: upper load limit 60% of the beam's flexural strength).

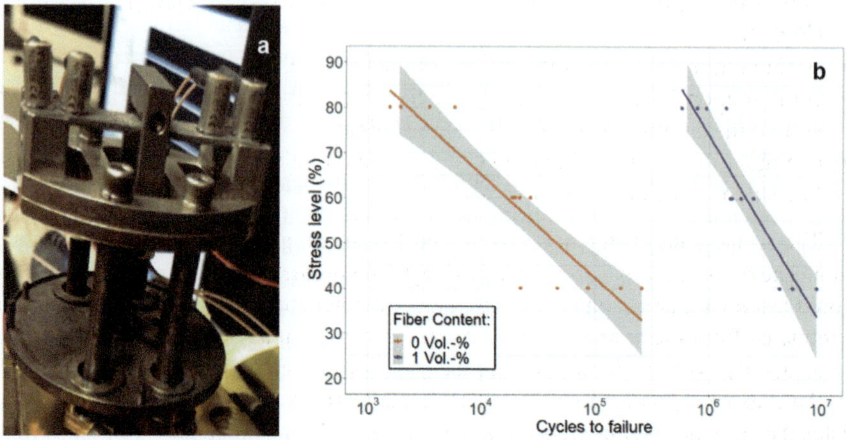

Fig. 7 Cyclic 3-point micro bending test setup within an DMA Q800, TA Instruments (**a**), Woehler diagram comparing the fatigue life of extruded micro beam specimens without fibres to those with 1 vol.% of carbon short-fibres (**b**). Adapted from [16], with modifications

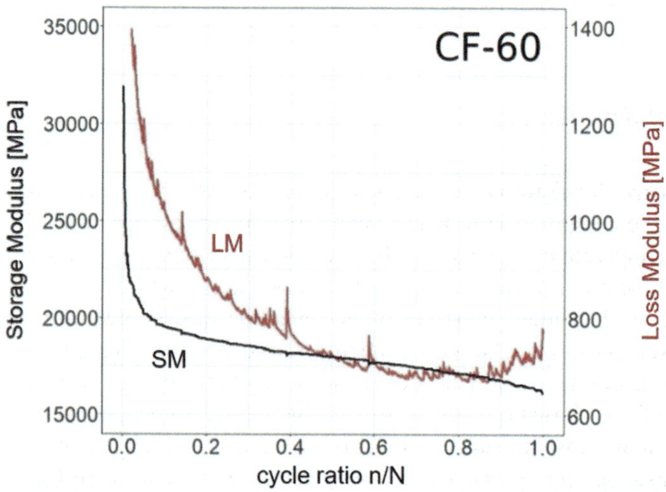

Fig. 8 Development of SM and LM during fatigue test of a CSFRC micro beam specimen (1 vol.% CSF, upper load limit 60% (i.e. load type CF-60), lower load limit 20% of flexural strength). Adapted from [16], with modifications

The moduli show a specific and repeatable pattern during the runtime of the fatigue test. SM and LM both start to fall rapidly very early, corresponding to a considerable loss in stiffness due to crack formation. After about 10% of the experiment's runtime, both moduli decrease at a steady and slow speed. Some sharp peaking in LM can be seen, indicating non-reversible inelastic events, which likely show the formation of further cracks. After some 80% of the runtime, the specimen enters a third phase of behaviour, where SM continues to decrease slightly faster while LM starts to increase again. This behaviour proved to be the indication of the end of the micro beam's fatigue life. As the increase in modulus appears quite early, it might prove a useful indicator during real-world application.

An identical experiment was carried out on another micro beam sample. Hereby, the test was interrupted at the cycle ratios indicated by dashed lines in Fig. 8 and an X-ray CT scan was performed to assess the damage caused by the cyclic loading. The results can be seen in Sect. 3.4 (refer to e.g. Figure 24). The scanned area shows a length of some 7 mm, taken at the midspan area of the beam thus representing the location subjected to the highest bending moment. Since several AE sensors were attached to the surface of the sample, it was also possible to correlate the AE activity (see Sect. 3.3) with the specimen's mechanical behaviour.

2.3 *Advanced Cyclic Tests (Varied Loading, Sequence Effects)*

Tensile and flexural tests were mainly performed with macro samples (please refer to Table 1) using four different schemes of cyclic loading in order to investigate sequence effects by varied loading. These schemes are described in Table 2 and schematically shown in Fig. 9 for load type 1. Hereby, the background colours for the individual load levels are also used in Fig. 10. The concrete stresses σ_u (i.e. upper) and σ_l (lower load/stress level) are related to the tensile strength f_{ct} under flexural or centric loading; n is the intended number of load cycles at each load level.

The average *flexural strength* based on the results gained from eight static tests was $f_{ct} = 37.90$ MPa. During the flexural fatigue tests, the frequency of cyclic harmonic load was 25 Hz. The average *tensile strength* based on results of static uniaxial tensile tests, was $f_{ct} = 14.90$ MPa [17]. During the axial tensile fatigue tests, the frequency of the applied cyclic harmonic load was 10 Hz.

Table 2 Cyclic testing schemes applied in tensile and flexural tests were mainly performed with macro samples (varied loading, sequence effects)

Load type	Load level 1			Load level 2			Load level 3		
	$\frac{\sigma_u}{f_{ct}}$	$\frac{\sigma_l}{f_{ct}}$	$\frac{n}{10^6}$	$\frac{\sigma_u}{f_{ct}}$	$\frac{\sigma_l}{f_{ct}}$	$\frac{n}{10^6}$	$\frac{\sigma_u}{f_{ct}}$	$\frac{\sigma_l}{f_{ct}}$	$\frac{n}{10^6}$
1	0.8	0.2	3	0.6	0.2	2	0.4	0.2	1
2	0.4	0.2	1	0.6	0.2	2	0.8	0.2	3
3	0.8	0.4	3	0.6	0.2	2	0.4	0.1	1
4	0.4	0.1	1	0.6	0.2	2	0.8	0.4	3

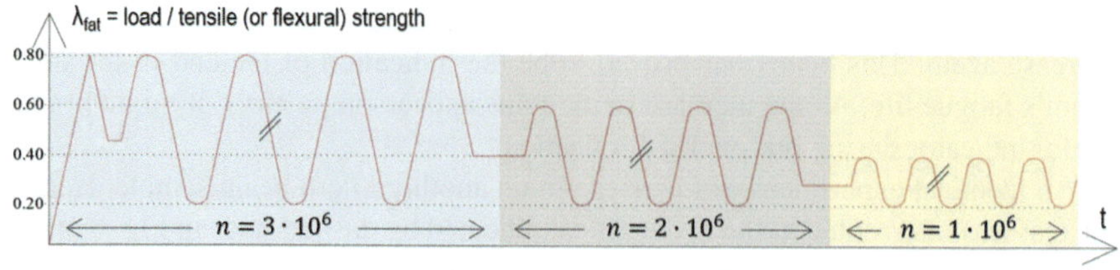

Fig. 9 Cyclic loading scheme, exemplarily shown for load type 1 (for details see Table 2)

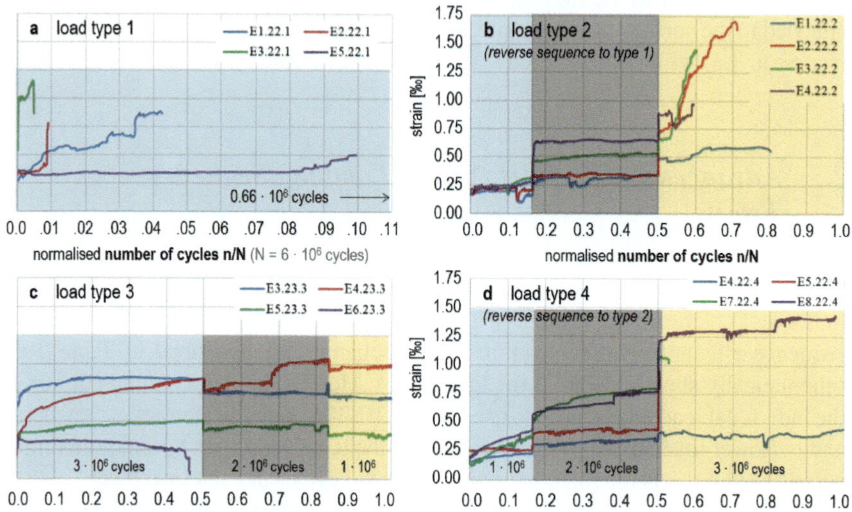

Fig. 10 Average values of strain: **a** load type 1, **b** load type 2, **c** load type 3 and **d** load type 4; the underlying maximum number of cycles N is 6 million

2.3.1 Tensile Fatigue Tests

In the (axial) tensile fatigue tests, strains were measured with four strain gauges, two of which were attached at the upper and two at the lower edge of the testing area, refer to Fig. 1a. They were orientated in the direction of the tensile stresses. Four samples were tested for every load type and the average strain for the individual specimens was determined from the measuring data of all four strain gauges. The experimental results are displayed in Fig. 10a–d.

Significant non-matching of results for different specimens can be noticed from the very beginning of the fatigue test. Results given in Fig. 10a for load type 1 show that all specimens failed during load level 1 (high initial $\sigma_u/f_{ct} = 0.8$ and high initial $\Delta\sigma = (\sigma_u - \sigma_l)/f_{ct} = 0.6$) before reaching the intended number of 3 million cycles. This indicates that the specimens were already cracked during static load increase before they were subjected to the harmonic load. Hence, the tested samples were in the range of non-linear behaviour during the entire harmonic load (fatigue test).

Results for load type 3 (high initial $\sigma_u/f_{ct} = 0.8$ but lower initial $\Delta\sigma = 0.4$) compared to type 1, Fig. 10b show a disagreement of results for different specimens from the very beginning of the fatigue test, too. Thus, cracks also appeared there during the static load increase having preceded to harmonic load. In contrast to the specimens of load type 1, the samples subjected to load type 3 withstood a significantly higher number of cycles. Three of them survived 6 million cycles, while one sample broke shortly before reaching 3 million. Figure 10b also shows a slight relaxation of strain, which may indicate closing cracks.

Results in Fig. 10c (load type 2; maximum $\Delta\sigma = 0.6$ in load level 3), and Fig. 10d (load type 4; maximum $\Delta\sigma = 0.4$ in level 3) show a good agreement for different specimens in the first million cycles (load level 1; both with moderate $\sigma_u/f_{ct} = 0.4$) indicating a global linearity of the specimens. However, a slight deviation of the results may be identified, though, which may be a consequence of the local micro-nonlinearity of the specimens. This minor deviation is a little more emphasised with specimens subjected to load type 4 which indicates a more significant local micro-nonlinearity compared to specimens subjected to load type 2. After 1 million cycles and the increase of the upper load level (to $\sigma_u/f_{ct} = 0.6$ in both cases), deviation of the individual results for both load types 2 and 4 is considerably more pronounced than in the first million cycles, although not as significant as in Fig. 10a, b. This indicates global non-linearity of the specimens caused by the microcrack pattern. However, when comparing type 2 and 4, the results are generally similar up to and including load level 2, which is to be expected due to the basically identical loading (only exception: in load level 1, lower $\Delta\sigma = 0.2$ for type 2 compared to 0.3 for type 4). After 3 million cycles and an increase in the upper load level to $\sigma_u/f_{ct} = 0.8$, no correlation can be established between the results of the different specimens (due to fully developed cracks as is also the case with the specimens subjected to load type 1 and 3) and no individual sample reaches the intended maximum number of cycles of 6 million with load type 2. But this is also to be expected because $\Delta\sigma$ for load type 2 ($\Delta\sigma = 0.6$) is significantly higher than for type 4 ($\Delta\sigma = 0.4$). Finally, a comparison of load type 1 and 2 (reversed load sequence compared to type 1) shows more favourable results when starting with initially lower upper load, i.e. type 2. This effect is significantly less pronounced when comparing load types 3 and 4 (reversed sequence compared to type 3), which may also be attributed to the stress amplitude ($\Delta\sigma = 0.4$) with type 3, which is considerably lower compared with type 1 ($\Delta\sigma = 0.60$; early fatigue failure in load stage 1). In summary, it can generally be concluded that the tensile fatigue investigations with varied loading schemes certainly provide some indications of existing sequence effects, although these cannot be quantified on the basis of the available experimental data.

2.3.2 Flexural Tensile Fatigue Tests

In flexural tests, strains were captured by one strain gauge positioned on the bottom surface of the specimens (the surface opposite to the external load) at midspan and again orientated in direction of the expected principle tensile stresses. Since specimens in load types 2 and 4 broke before reaching the maximum load level or immediately upon reaching it, only two samples could be tested with the planned number of cycles. Further, specimens tested with load type 1 and 3 respectively, a significant disagreement of the results was achieved, indicating that fully formed cracks appeared during the static load increase prior to the fatigue testing. Thus, as also observed in the tensile fatigue tests, the specimens were in the non-linear range throughout the entire harmonic load. Since there were already fully formed cracks, the energy introduced by the applied external loading was dissipated on these cracks without opening microcracks and caused failure at a relatively low number of cycles. Unlike tensile tests, in flexural fatigue tests, specimens subjected to load type 3 also broke before reducing the initial upper load level to lower value.

Overall, the flexural fatigue tests showed similar trends to the tensile tests discussed in Sect. 2.3.1, and there were also slight indications of sequence effects. However, the general scatter and deviation of results were much greater and it may be concluded that the carbon short-fibre reinforced concrete (CSFRC) has a *more favourable fatigue behaviour in tension* than under flexural tensile loading.

3 Non-destructive Measuring Techniques and Combinations

This section focuses on special NDT methods and their combinations. These techniques were used in addition to load cells, deformation/strain sensors (e.g. surface-mounted transducers, strain gauges or fibre-optic quasi-continuous sensors) in order to derive new findings on the damage behaviour and local/global damage development inside the test specimens. These results form an important additional basis for the mathematical-numerical modelling and simulation, addressed in Sect. 4.

3.1 Digital Image Correlation (DIC)

The camera-based DIC measuring technique is a very versatile method to acquire detailed information on the deformation and resulting strains on the surface of structural members. In the present project, DIC was utilised in static and dynamic tests as well as with all types of specimens, particularly with micro and macro samples. In the following, some results gained from micro/macro samples and different loading schemes are exemplarily presented; for more detailed information and a broader range of measurement results please refer to publications resulting from the present research project, e.g. [4, 8, 9, 12, 14–16, 18–20].

Regarding *micro tensile specimens* the ARAMIS 4M DIC system (GOM GmbH) was used in-situ to continuously record surface deformation/strain of the specimen with high resolution. In doing so, two CCD cameras were employed to capture the strain on the specimen surface, to which a stochastic pattern was applied. This so-called speckle pattern is customised to the specific requirements and also allows for 3D measurements that can reduce errors caused by uneven surfaces. An initial image is taken as reference before starting a test, with no force being applied [21, 22].

A sampling rate of 2 Hz was considered appropriate for capturing sufficient images during the test, while avoiding overwhelming the measurement system with storable data. Furthermore, the ARAMIS system was synchronised with the universal testing machine (Zwick Roell) to initiate recording at an initial force of 20 N, thereby establishing a direct correlation of DIC and the force and displacement signals. DIC data was analysed by using GOM Correlate Professional 2020, enabling deformation/strain evaluation in all directions through the described methodology.

Figure 11a illustrates the force and deformation in the tensile direction relative to the cumulative number of captured images. As the data was continuously recorded at 2 Hz, the x-axis can be interpreted as time. The blue lines marked with asterisks in diagram correspond to the deformation states in Fig. 11b, where Fig. 11c represents the image of the right CCD camera after final failure.

Initially, it becomes evident that a linear-elastic region is departed around image number 535. Subsequently, the strain hardening behaviour of the CSFRC specimen becomes apparent through an increasing number of developing microcracks. These cracks mostly occur abruptly, as indicated by both an immediate drop in force and a sudden increase in deformation (also see multiscale modelling, Sect. 4). Hereby, the material's capacity for energy absorption increases by a multitude, compared to unreinforced references (see e.g. Figure 3). Microcrack formation occurs until the stress exceeds the strength of the composite material. The resulting macro crack leading to final failure was usually found in a previously undamaged region, suggesting a strong fibre-matrix interface.

In addition, DIC measurement was also used both with statically and dynamically loaded *macro tensile (and bending) specimens* to confirm and complement the results obtained from conventional strain gauges. Exemplarily, Fig. 12 shows DIC results for a dog-bone shaped specimen under cyclic varying tensile forces (loading type 3; please refer to Sect. 2.3); all DIC measurements were performed using a Stemmer camera (type Dalsa Genie Nano-M4040).

In Fig. 12a, fully formed cracks can be seen in the specimen before applying the harmonic load. The state of cracks following 3 million cycles is given in Fig. 12b. By comparing Fig. 12b with a, both new cracks opening can be noticed, but also relaxation of existing cracks, i.e. the new cracks are opened, while the width of existing cracks is reduced. After reducing the upper load level in accordance with loading type 3 (maximum cyclic load at the beginning) a further relaxation or even closure of existing cracks can be noticed, see Fig. 12c, d. These DIC results correspond very well with those obtained from conventional strain gauges.

In addition to uniaxial tensile loading, the results of cyclic 3-point *bending tests* on *macro samples* are presented below. To this end, Fig. 13 exemplarily depicts the temporal development of strain and cracking for the one bending specimen subjected to the varying cyclic loading type 4 (refer to Sect. 2.3). A strain state being able to

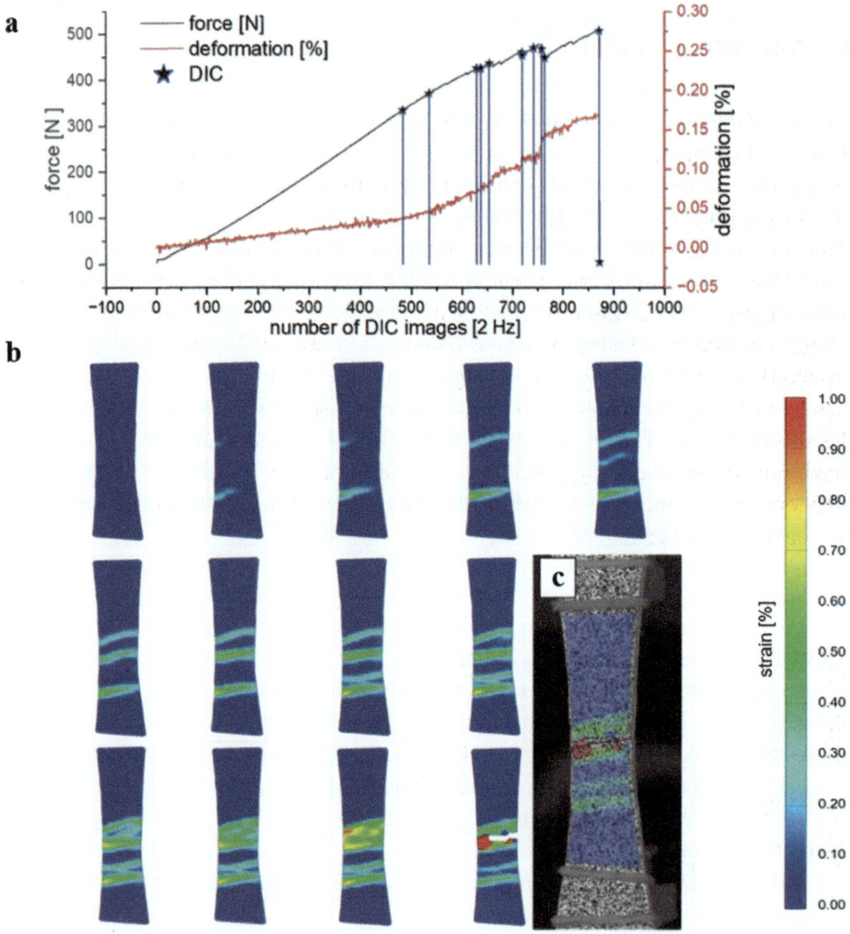

Fig. 11 Force–time–deformation diagram (**a**) recorded with ARAMIS 4M DIC, asterisks represent pictures (**b**) (no. of acquired images: 483–535–629–637–654–719–720–742–758–764–872), **c** final failure

cause appearance of microcracks can be seen after some 400,000 cycles (Fig. 13a), whereas no fully developed crack may be identified at this stage of the loading regime. Upon increasing the upper load level to 60% of the flexural tensile strength, see Fig. 13b, c, a merging of microcracks into fully formed cracks becomes apparent. When the upper load level reaches the maximal value (i.e. 80% of the flexural tensile strength), further widening of existing cracks and opening of new cracks occurs, see Fig. 13d. After utilising energy dissipation capacity of the cracking pattern developed (micro and fully formed cracks), the specimen breaks. Obviously, these results confirm those obtained from the strain gauges.

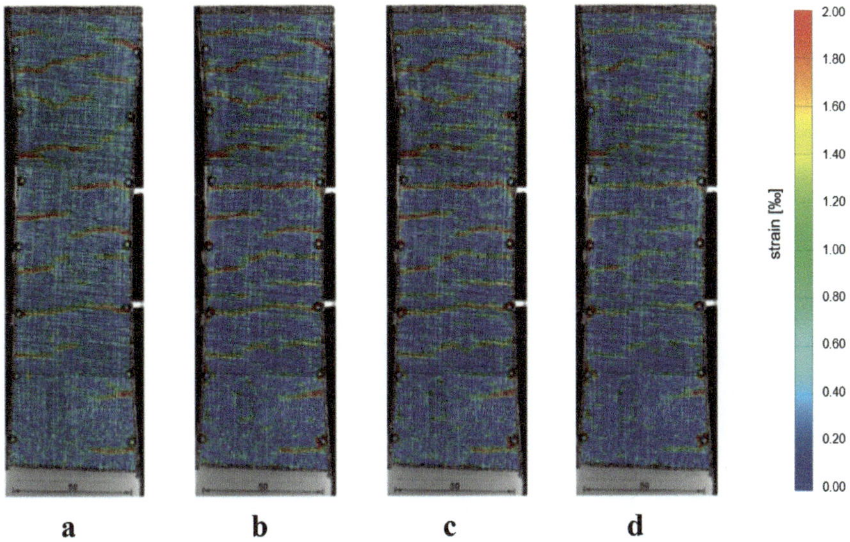

Fig. 12 DIC images for the load type 3 (refer to Fig. 9): **a** before start of harmonic load, **b** after 3 million cycles, **c** after 5 million cycles, **d** after 6 million cycles (i.e. immediately before failure)

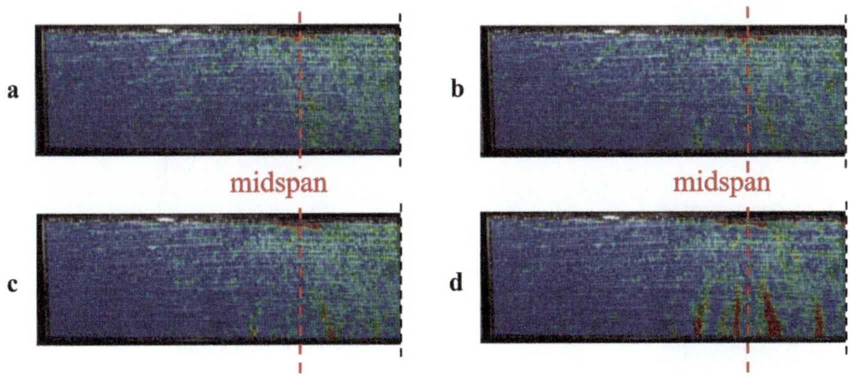

Fig. 13 DIC results for load type 4 (also refer to Table 2): **a** after 400,000 cycles, **b** after 410,000 cycles, **c** after 1 million cycles, **d** just before failure; strain scale, see Fig. 12

3.2 *Electrical Measurements*

In order to detect damage during load cycles, up to 17 electrodes were attached to the *macro specimens* (Fig. 14) to detect changes in voltage at several positions on the specimen's surface. The main objective was to investigate whether changes in voltage (induced e.g. by fibre pull-out or fibre breakage) can be used to detect damage propagation sooner as compared to other measurement methods. In the described uniaxial tension test with monotonic load cycles the individual electrodes are arranged on three edges of the sample with a vertical distance of 25 mm to each other. A voltage of 10 V (DC current) is supplied via a reference resistor of 10 kΩ to the electrode U_0. In [19] it is described that this measuring technique has a high resolution and fits very well to fatigue tests as single load cycles can be detected. Further information to the setup can be found within the cited paper.

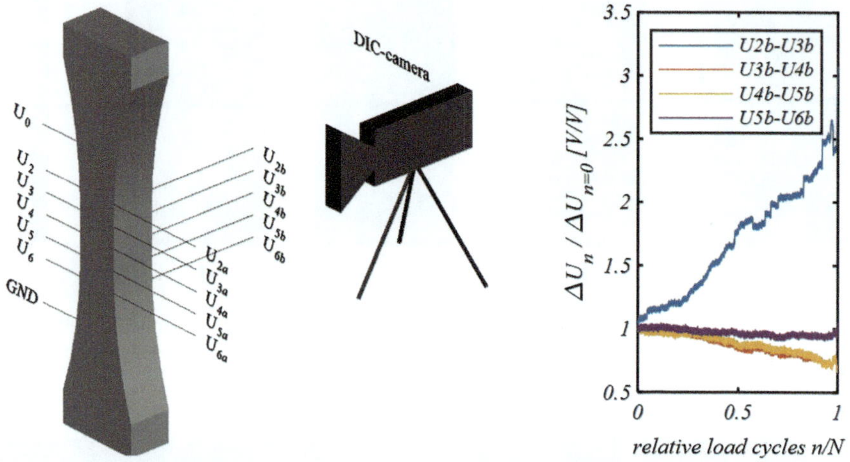

Fig. 14 Schematic sketch of measurement layout during fatigue test showing the location of the electrodes and the position of the DIC camera—current driving from U_0 to GND; relative voltage evolution with clearly revealing areas of damage progression (N = 409,063 cycles)

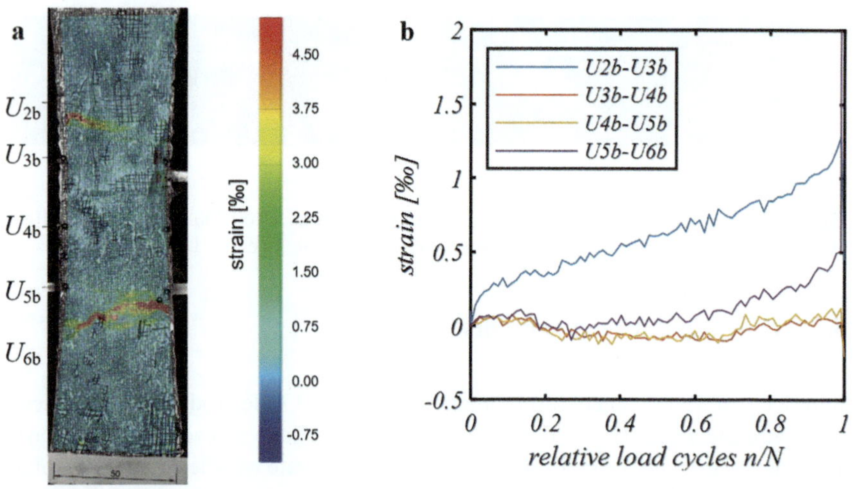

Fig. 15 DIC strain measurement (**a**) (sample surface; electrodes U_{2b}-U_{6b} attached on left side); the diagram **b** displays strain between the electrodes, which shows very similar results as in Fig. 14

As current is applied only to one individual electrode, all voltage measurements at the other electrodes depend on changes in voltage of this specific electrode. This is because changes in overall resistance of the specimen lead to a different relationship between the specimen's resistance and the constant reference resistor and therefore the applied voltage divides differently. To cancel out this effect, the change of two adjoined electrodes is examined. In order to improve further, the change in voltage between two adjacent electrodes is related to their primary voltage drop in the first load cycle as shown in the diagram in Fig. 15. This can be described with Eq. (1), where i denotes the electrode number and n the load cycle.

$$\frac{\Delta U_{\mathrm{n}}}{\Delta U_{\mathrm{n}=0}} = \frac{U_{\mathrm{i,n}} - U_{\mathrm{i+1,n}}}{U_{\mathrm{i,n}=0} - U_{\mathrm{i+1,n}=0}} \tag{1}$$

Comparing this result with digital image correlation (DIC) proves very well the reliability of the electrical conductivity measurement. The electrodes were attached to the left side as shown in Fig. 15. The image was taken at a relative load cycle of 0.99 (i.e. shortly before collapse). In this stage there are many microcracks all over the specimen's surface visible in a light blue colour. Two major cracks (though with a very small crack opening of some 40 μm) arose during a fatigue test with a high tensile strain marked in red. The crack in the lower area between electrodes U_{5b} and U_{6b} is not totally proceeding to the left edge, while the one in the upper area (at electrodes U_{2b} and U_{3b}) seems to originate from there. This perception matches the curve progression in Fig. 14. Evaluation U_{3b}-U_{4b} and U_{4b}-U_{5b} behave simultaneously while the other two detect some change in electrical conductivity. As the upper crack is very close to the electrodes, the current is forced to find a way around the crack and the measured voltage difference increases vastly. As the lower crack barely reaches the electrodes on the specimen's left edge, the measurement of sensor U_{5b}-U_{6b} is hardly affected by the crack formation. Compared to the diagram given in Fig. 15, the strain evolution matches the resistivity revolution very well.

To improve both testing and localisation, additional investigations are necessary. Connecting electrodes on all four sides of the specimen or with a fine-meshed sensor grid distributed over the surface can help to detect fatigue damage more accurately. Also applying the voltage during fatigue tests to different electrodes for only a short time while measuring the voltage at all other electrodes in a cyclic/repeating manner might reveal internal damage. This technique is called electrical resistivity tomography (ERT) and has already been used in geophysical surveys for decades.

3.3 Acoustic Emission (AE) and Laser Doppler Vibrometry (LDV)

The AE measuring method is based on elastic wave propagation rediated from internal cracking and described in the literature, e.g. [23–25]. Since AE can be applied under operating conditions without stopping machines and equipment, it is a highly efficient tool for remote control. Understanding the damage process can be significantly improved by signal-based processing techniques that allow the extraction of the maximum amount of information recorded in the entire AE signal waveforms, refer to e.g. [26]. AE is beneficial in fatigue testing of high-performance materials, e.g. [14, 27, 28]. On the one hand, these materials have excellent durability and can resist destruction for long periods of time, thereby providing enough time to detect damage and to monitor its dynamics. On the other hand, the damage accumulation in heterogeneous materials is accompanied by a more active and complicated AE response, which is influenced by the properties of the composite constituents, their volume fractions, and spatial distribution [9, 29–31]. Dynamic load tests show that the fatigue process in fibre-reinforced material generally occurs in several stages, differing in AE signal characteristics. There, the first elastic deformation is followed by the so-called strain-hardening stage [32, 33]. It is characterised by the occurrence and growth of multiple cracks within the matrix, which cannot spread far due to the presence of fibres. The carbon short-fibres, which are stronger than the matrix, take on further stress and maintain the material's resistance in tension (fibres bridge the cracks) until it exceeds the maximum bonding capacity (anchorage of the fibres on both sides of the cracks) or the ultimate tensile strength of the fibres (fibre failure). The cracks' distribution over quite a large area is a typical damage pattern developing in high-performance concrete—in contrast to standard concrete, being characterised by localised damage, e.g. [15]. Many studies accompanied by AE analysis show that low-energy and -frequency AE events are associated with matrix micro-cracking, while high-amplitude and -frequency signals are caused by fibre breakage and may be predictors of fatigue failure, see e.g. [34, 35]. The most promising strategy for analysing internal damage mechanisms is data clustering, where AE signals are combined into groups with identical characteristics (frequency, amplitude, duration, etc.) corresponding to a specific damage mechanism, see e.g. [28, 36–38].

3.3.1 Static Tensile Tests Equipped with AE Sensors

In this study, both AE and DIC were applied jointly to better characterise the damage progress in CSFRC. In the first phase, the material's damage behaviour was studied in tensile tests subjected to monotonous fatigue loading. To ensure stability of the data acquisition, four small passive piezoelectric AE sensors were glued to the specimen (with 1 vol.% CSF) and then connected to the acquisition system.

The root-mean-square value of a raw acoustic signal RMS_{AE} was used to represent the intensity of the AE because it is directly proportional to the strength of the acoustic signal and thus, its energy. Analysis of the AE activity is based on the interpretation of one of the most relevant parameters of the AE signal, the cumulative number of acoustic emission events. One-dimensional localisation was performed using MATLAB©. AE signals were localised in linear geometry based on differences in signal arrival time between the different sensors. Onset times of AE signals were

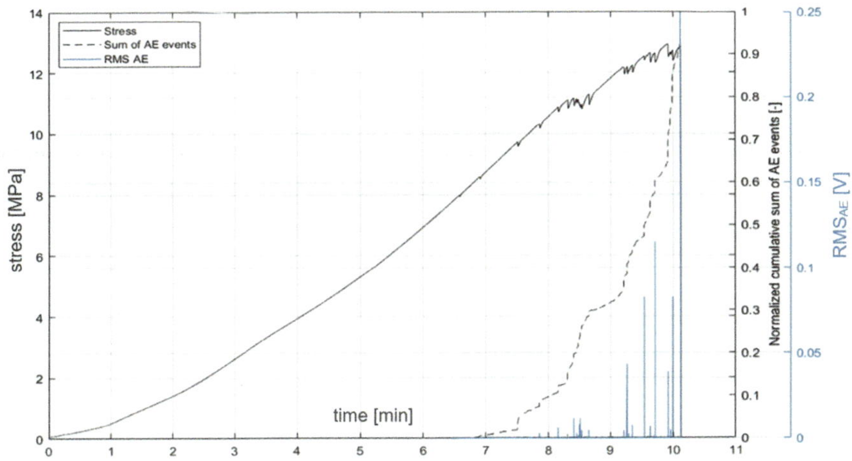

Fig. 16 Characteristics of AE: normalized cumulative number of AE events (in black, dashed curve), RMS_{AE} (in blue), and a mechanical parameter, stress (in black), as a function of time

picked automatically using an autoregressive algorithm as time points where Akaike Information Criterion (AIC) values were minimal.

Figure 16 shows the representative response of acoustic emission signals, normalised cumulative number of AE events, RMS_{AE} and a mechanical parameter, stress (in black), plotted as a function of time for a meso tensile specimen with 1 vol.% of carbon fibres. As the graph indicated, there is no acoustic activity in the first half of the test regime. After about 6.5 min, first AE events appear, but AE activity remains low for some time (up to 80% of the test duration), suggesting that the sample reached the strain-hardening phase. The appearance of low-intensity acoustic signals is associated with the formation of microcracks in the concrete matrix, which stabilise the specimen material and allow to resist load more effectively.

Active acoustic emission mainly accompanies the last 20% of the loading time. During this period, most of the acoustic events occurred, causing a dramatic increase in AE energy with the highest peaks shortly before specimen failure (after 90% of the test time). Thus, the response of the specimen to loading is characterised by the appearance of acoustic phenomena in the second half of the test. A sharp increase in AE intensity at the last stage, manifested by high RMS_{AE} amplitudes, indicates further rapid formation of microcracks and an accelerated crack growth.

Figure 17 shows the sample after failure and distribution of localised AE events. The cause of the failure was a perpendicular fracture. For this specimen, sources of AE events are concentrated within some 3 mm, with the maximum number of events generated close to the primary fracture. The spatial distribution of the AE events approximately corresponds to the width of the damage zone. The scatter plot, see Fig. 17a, shows the temporal localisation of the acoustic emission events. It is noticeable that the first events occur after 6.5 min and correspond to the position of the main fracture. As the test time progresses, the number of events increases at the

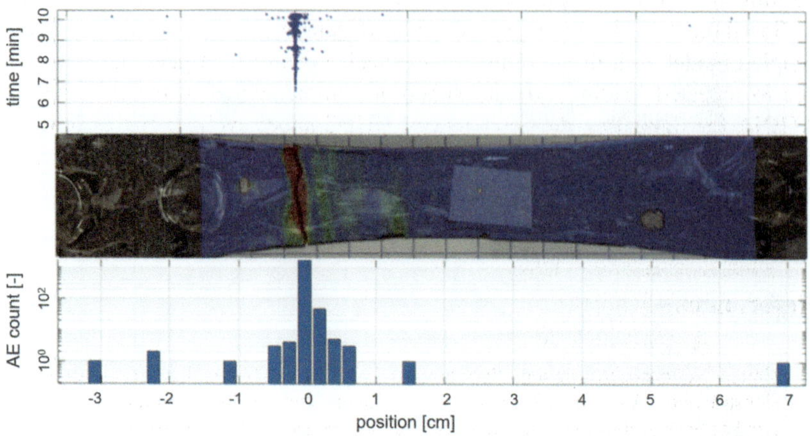

Fig. 17 Scatter plot **a** illustrates temporal distribution of AE events; image **b** displays crack pattern visible by overlaying the picture of the specimen with the recorded DIC data; histogram **c** showing the spatial distribution of the AE events along the specimen

location of the primary fracture and within a radius of some 5 mm right of it, where additional cracks begin to form. When reaching the final time (approximately after 10.2 min), the main fracture leads to the failure of the whole sample.

3.3.2 Static Flexural Test with AE and Laser Doppler Vibrometer

During this 3-point bending test, the same measurement setup for the AE recording was used (see Fig. 18), as mentioned above in Sect. 3.3.1. The detailed analysis of the AE parameters for the flexural static test can be found [15].

In order to detect high-frequency signals resulting from the main characteristic damage mechanisms of the fibre-reinforced cementitious material (in particular fibre pull-out and/or fibre breakage), it was decided to supplement AE measurements with LDV, which makes it possible to obtain data in frequency rages of more than 1 MHz. Hereby, a LDV system from POLYTEC with OFV-505 single-point measuring head and OFV-5000 controller, was employed. Further, a DD-300 (30 kHz–24 MHz frequency range) with measurement range of approximately ± 50 nm/V and an additional output in the auxiliary field was used as a displacement decoder.

Figure 19 shows a graph of the cumulative sum of AE events combined with a stress–time curve, where it is possible to distinguish two different stages that differ in AE activity and mechanical response of the specimen. At the first stage (0–1.5 min), the linear shape of the stress curve and the absence of acoustic events indicate that the material retains the ability to deform elastically under an applied load. In the second stage, upon appearance of the first AE signals, moderate activity was recorded until approximately 3.5 min. Then, the AE activity begins to increase and reaches a maximum shortly before destruction of the specimen. The signal selected for wavelet

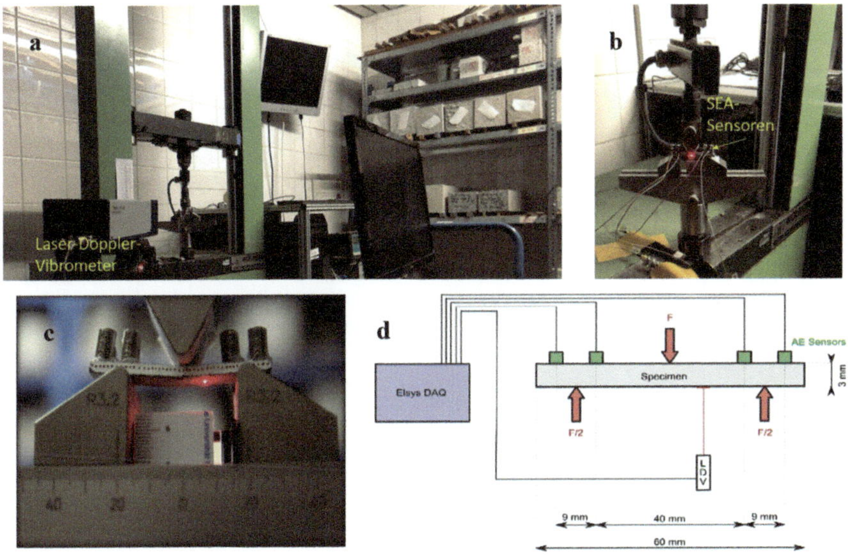

Fig. 18 Experimental setup for static 3-point test with AE and LDV (**a**, **b**); **c** loaded specimen with four sensors glued onto it; **d** specimen with dimensions and AE sensor arrangement (sketch)

analysis was recorded at a point of time of about 20% before the specimen's failure. At this time, the curve of the cumulative sum of AE events becomes steeper, reflecting the rapid progression of damage with potential fibre involvement.

In order to determine the underlying damage mechanisms, the signal's frequency components were analysed using the analytic Morlet wavelet transform, which decomposes the signal into a series of small waves localised in time. This offers advantages over other time–frequency analysis methods, such as the Fourier transform, which implies that the frequencies are maintained throughout the entire signal. The same signal recorded in the last quarter of the measurement time was selected for analysis (see Fig. 19). A wavelet transform was performed on the AE and LDV signals, see Fig. 20a, b. The results of an analytic Morlet wavelet transform of these signals are shown in Fig. 20c, d. Thus, it is possible to identify the contribution of the frequencies 100, 280 and 800 kHz in the frequency spectrum. Hereby, the AE signal showed lower frequencies (280 and 100 kHz), while the vibrometer recorded the signal with higher frequency of 800 kHz. Signals of different frequencies can be attributed to different destructive and damaging processes, which occur simultaneously and suggest a complex failure mechanism involving damage to both matrix and fibre phases. The lower frequencies of 100 kHz can be associated with matrix cracking, while the higher frequencies can be attributed to pull-out of fibres and fibre breakage. More information on this topic can be found e.g. in [8].

The present investigations confirm that laser Doppler vibrometry can be used in addition to acoustic emission to draw conclusions on complex fracture mechanisms.

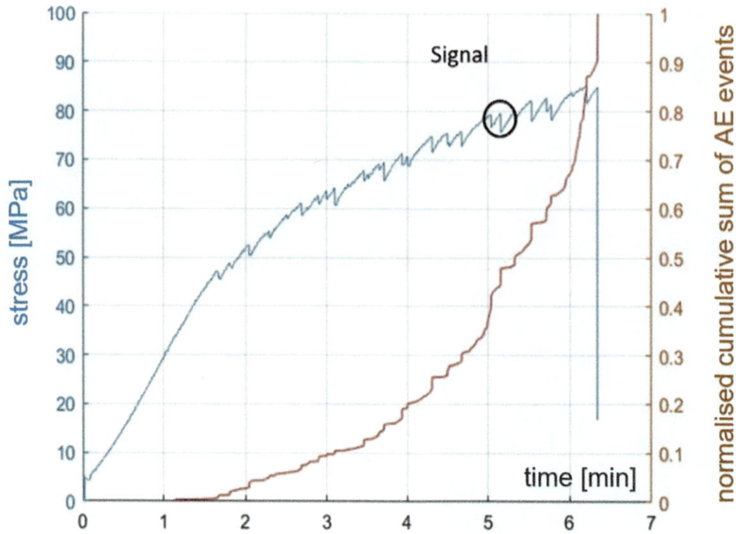

Fig. 19 Recorded parameters for one sample during 3-point bending test (normalised cumulative no. of AE events (in red) and stress curve (in blue) as a function of time) with the approximate position of the selected signal for further wavelet transform

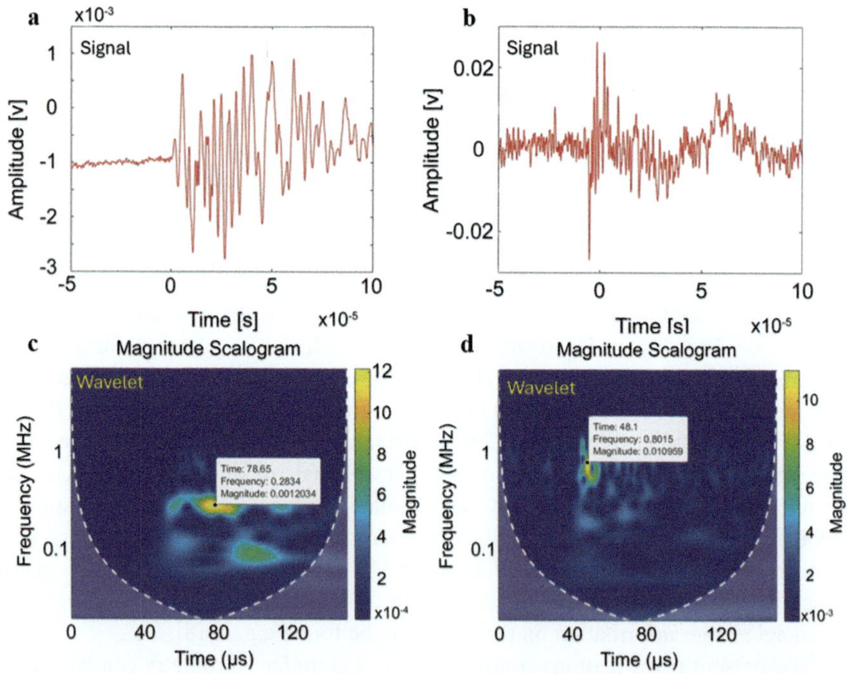

Fig. 20 **a** AE signal used in wavelet transform; **b** LDV signal used in wavelet transform; **c** analytic Morlet wavelet transform of AE signal **a**; **d** analytic Morlet wavelet transform of LDV signal **b**

3.3.3 Dynamic Flexural Test Equipped with AE Sensors

Since most engineering structures experience loads varying over time, an experiment with dynamic loading makes it possible to identify the characteristics of the fatigue behaviour of a material under conditions close to real ones.

AE measurements were carried out using four small-sized passive piezoelectric sensors AE VS700-D (150–800 kHz), placed at the edges of a micro bending beam specimen and coupled with hot glue. The tested sample (with 1 vol.% of fibres) had dimensions of 50 × 12 × 1.5 mm. Measurements were carried out with a DMA machine with a lower force of 4.3 N, an upper force of 13.2 N, and a frequency of 10 Hz. The data acquisition system "Elsys" recorded AE signals at a sampling rate of 20 MHz and converted them to digital form. The experimental setup with the DMA can also be seen in Fig. 7 (left), the arrangement of the AE sensors is in analogy to Fig. 18d. The cumulative sum of AE events/activity was investigated through time domain analysis of the AE signal characteristics. Further, the emission intensity was analysed by estimating the root mean square value of the original RMS_{AE} signal which is proportional to the strength and the energy of the AE signal.

Figure 21 shows the characteristics of the AE signal, the RMS_{AE}, the normalised cumulative sum of AE events and a mechanical parameter, the strain ratio, plotted against the cycle ratio. Based on the AE analysis, four periods of different AE activity which correlate with the specimen's mechanical behaviour can be observed. In the initial period of the test, i.e. 0–0.07 of the total number of cycles, the acoustic emission activity is low. Only sporadic events with low energy levels are recorded during this time. In the second period (0.07–0.3 of the cycles), the number of AE events increases sharply. The slope of the cumulative curve of AE events becomes almost vertical, suggesting that the AE event rate accelerates rapidly. This coincides with a noticeable increase in the AE energy. The intensity of the AE signal began to fluctuate more, and the first signals with higher energy appeared. During the further loading cycles (0.3–0.8 of cycles), the acoustic activity decreases and remains very low for a long time, as being indicated by the nearly horizontal cumulative curve. During this period, the strain ratio did not noticeably increase, constantly remaining within 0.95–0.96 of its maximum values. At approximately 0.78 of the load time, high-energy signals appear on the RMS_{AE} graph. This clearly correlates with the onset of a rapid increase in strain. During the last 20% of the loading time, AE signals again become more frequent and intense, with the highest energy signal occurring at the moment of specimen destruction.

Thus, after the peak of acoustic emission in the early cycles, there is a relatively quiet period with minimal AE activity, which continues throughout most time of the test. The increase in the number of AE events and the appearance of high AE energy peaks in the last 20% of the test time can be considered a sign of impending material failure. Hence, analysis of the characteristics of the AE signal makes it possible to distinguish the stages of damage accumulation in the material, namely the formation of microcracks in early loading cycles, a further state of equilibrium, indicating the ability of the material to resist deformation, and, ultimately, in the final part of the test, accelerated damage with further crack formation and their (unstable) growth.

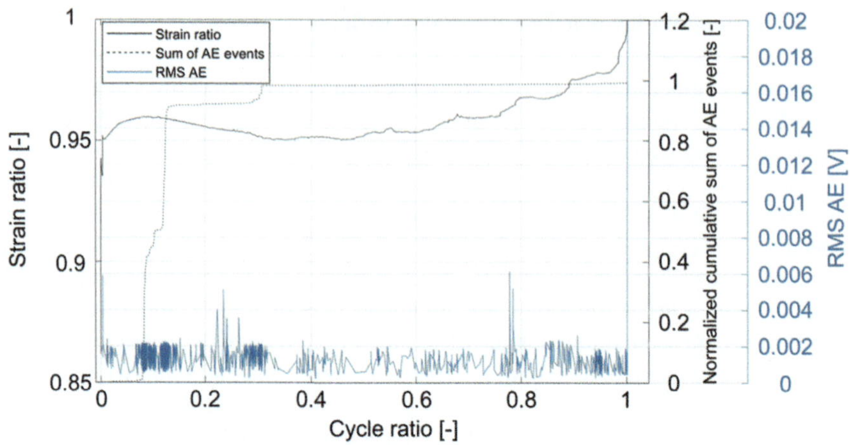

Fig. 21 AE signal characteristics, RMS_{AE} (in blue), cumulative sum of AE events (black dashed curve) and strain curve (in black) for micro beam specimen, plotted against the cycle ratio

3.4 X-Ray Computed Tomography (CT)

All X-ray images were taken with a Phoenix nanotom 180 m scanner (GE Sensing & Inspection Technologies GmbH). To correlate the mechanical behaviour with the fibre alignment present within the CSFRC samples, the distribution of the alignment angles was investigated using X-ray CT [32]. Figure 22a shows the distribution of in-plane angles with standard deviation shaded in red. Figure 22b presents a scan excerpt, showing the segmented fibres with the distribution being highlighted in blue. For the purpose of this publication, fibres being aligned with the principle tensile stresses are designated as having an alignment angle of 0°.

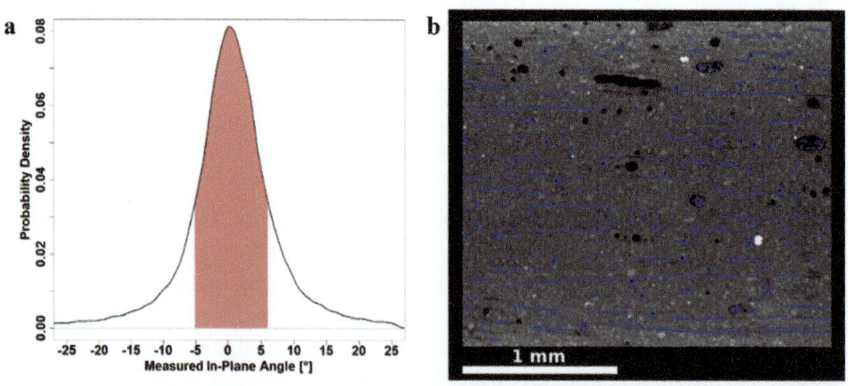

Fig. 22 Fibre alignment in a sample measured with X-ray CT: **a** distribution of in-plane fibre angles (red area: standard deviation); **b** tomographic scan (segmented fibre material in blue)

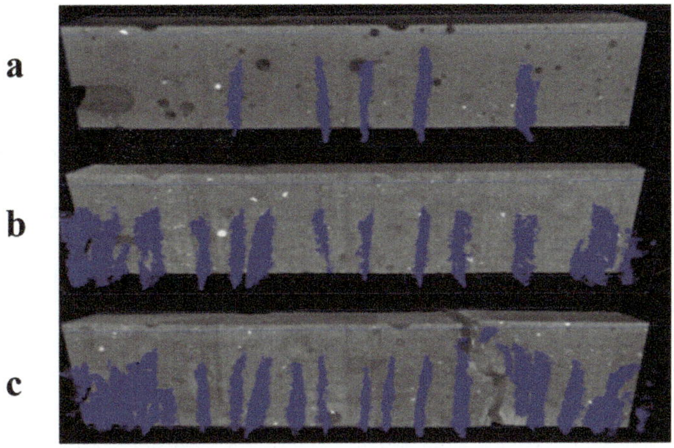

Fig. 23 Segmented cracking patterns of an extruded micro beam specimen during fatigue loading. Numbered intervals correspond to points annotated in Fig. 8. (1) Crack pattern after 0.2% of runtime of the fatigue experiment; (2) cracking after 24% of runtime, and (3) crack pattern after failure. Adapted from [16], with modifications

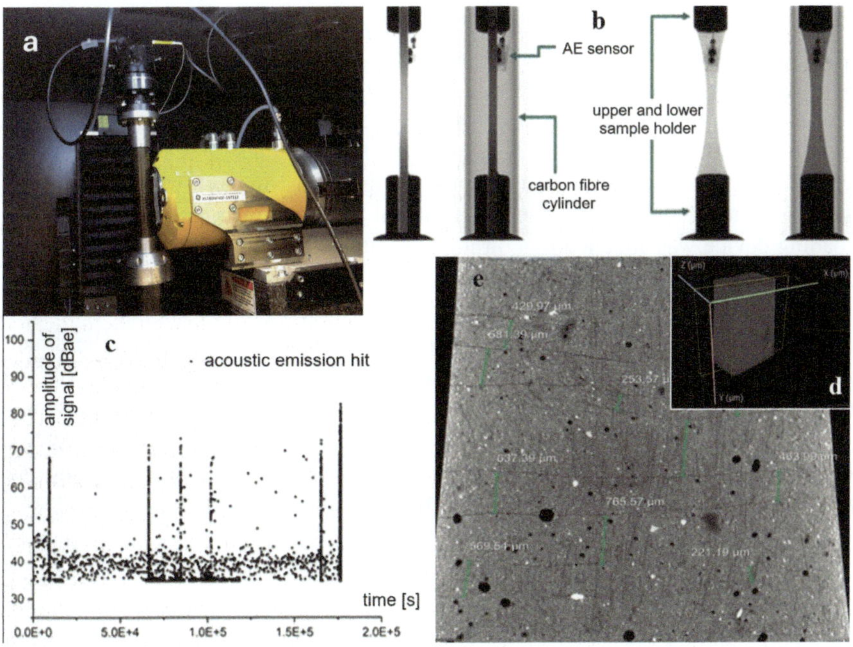

Fig. 24 **a** In-situ CT loading setup (micro tensile samples); **b** X-ray CT scan of specimen with 90° and 0° in different greyscales, right ones showing sample chamber; **c** AE hits during tensile test with peaks (formation of microcracks); **d** reconstruction of 3D image where plane **e** is extracted

As can be seen from Fig. 22a, the in-plane alignment scatters around the desired angle of 0°. Further, the standard deviation shows, that 68% of the fibres are aligned within ± 5° of the direction of the expected tensile force. Previous research suggests, that as long as the alignment reinforcement does not deviate by more than 10° from the direction of the acting stresses, a very strong composite material can be created [14]. The resulting potency of the aligned reinforcement can be seen in Fig. 3a, showing stress–strain diagrams of micro flexural beams reinforced with up to 3 vol.% of fibres, in comparison with an unreinforced specimen for reference.

The cracking patterns visualised in Fig. 24 (corresponding measurement data gained from segmenting the cracks can also be seen in Table 3) support the conclusion drawn from the assessment of SM and LM in Fig. 8. The scan corresponds to a runtime of 0.2% or just a few hundred load cycles. Cross-referencing with Fig. 8 reveals that this is during the first loading period where the micro beam exhibits a drastic decrease in both moduli and a lot of the initial damage seems to occur. At this point, the test beam shows five segmentable cracks with a relatively wide spacing of 933 μm in average. Point (2) was taken during the steady-state period at a cycle ratio of 24.4%, after the initial damage has formed. The crack number triples to 15 and consequently, crack spacing becomes a lot tighter (550 μm). As expected, the formation of new cracks is mostly confined to the earlier stages of fatigue. After failure (and thus at a cycle ratio of 100%) at point (3), a number of 23 segmentable cracks can be identified, as well as a large critical crack causing final failure. Crack spacing also only decreased relatively little to around 400 μm.

As the former experiments describe ex-situ X-ray investigations, it was also possible to do in-situ tests. An X-ray CT in-situ tensile test was carried out using a load stage that was developed in-house by the MRM Institute at UNA and integrated in the nanotom [39]. The experimental setup is illustrated in Fig. 24a. The loading device was attached to a turntable positioned in the CT, such that the specimen inside a carbon fibre reinforced polymer tube was directly in front of the X-ray source. Figure 24b shows sample holders, where the right pictures have a higher greyscale to show the experimental setup within the sample chamber. The dimensions of the chamber also allowed for attaching an AE sensor to the sample, enabling continuous monitoring. An HD2WD sensor (Mistras) was connected to the sample surface with Korasilon paste aided by 3D-printed sample holders, and connected to a differential preamplifier (Mistras) [40]. Before onset of loading, a reference CT scan was performed. After starting the experiment, the force increase was interrupted after many AE hits were recorded and another CT scan was taken. The very last AE

Table 3 No. of cracks and crack spacing extracted from CT scan given in Fig. 23

Cycle ratio (%)	Mean crack distance (μm)	No. of segmented cracks
0.2	933 (± 359)	5
24.4	549 (± 193)	15
100.0	401 (± 161)	23

peak in Fig. 24c is the final failure. Further, Fig. 24e shows the plane taken from the 3D image (marked in yellow in Fig. 24d), which was the last scan before failure.

The average crack spacing is about 497 μm while average opening is some 7 μm but can vary from 10 to 2 μm, although latter value is close to the resolution limit, but the important difference is that the force was not reduced during the scan, which gives us strong evidence that the carbon fibres bridge microcracks and absorb stress although the matrix is already widely cracked. For more details, please refer to [18].

4 Multiscale Modelling, Simulation

The CSFRC material involves different length scales, most importantly fibres with 7 μm diameter preferably oriented in direction of the principle tensile trajectories, the fibre length of about 3 mm and the sample size of several cm. The aim is to characterise the failure behaviour of the composite material exposed to increasing tension numerically based on the experimental observations, performed with different specimens and on different scales (micro, meso, macro). By modelling and simulation, it is possible to derive a better understanding of the local load-bearing and failure mechanisms on the microscale and how these mechanisms are related to the macroscopic material properties and stages of damage. Due to the novelty of the material, the damage mechanisms involved in failure of the bulk material have not yet been determined sufficiently in detail. Accordingly, it is of fundamental interest to explore further which of the mechanisms typically occurring in fibre composites, such as debonding and anchorage failure (i.e. fibre pull-out), fibre breakage or matrix failure, are relevant for the overall behaviour of the material. Thus, different possible damage scenarios are derived and validated from the experimental results aiming to develop a general failure model. The experimental data, particularly from CT and DIC analyses of tension tests (see Sect. 3.4), deepen the understanding of relevant damage mechanisms on the microscale in the material and provide the information for developing the failure model. Here, the different involved length scales require the application of mathematical multiscale methods since direct simulation of the whole specimen including the resolution of the microstructure is not feasible. Therefore, a structure-oriented numerical multiscale model of the material is developed.

In contrast to typical FE^2 methods (see e.g. [41] or, for a recent overview, [42]), adapted methods based on periodic homogenisation on the continuum scale in conjunction with discretisation by finite elements are used to simulate the behaviour of a specimen in the experiment, where the local material law in each macroscopic point is determined by solving so-called cell problems on representative volume elements (RVE) on the fibre scale (see Fig. 25).

This derived local effective macroscopic material law is used for the simulation at the macro scale, i.e. of the entire specimen. Following this general idea, it is not only possible to simulate the undamaged material (see Sect. 4.2) but also to include different failure stages in the specimen according to the damage behaviour observed in the experiments by allowing damage to evolve in the RVEs (compare Sects. 4.2,

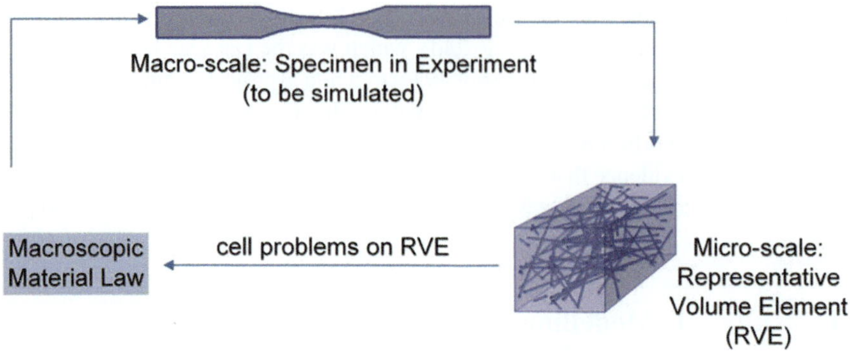

Fig. 25 Adapted homogenisation method

4.3 and Fig. 28). An essential criterion for the numerical model is its simplicity and efficiency with a sufficiently accurate representation of the relevant mechanisms. Following this guiding principle, the implementation is conducted in the sense of an experimental virtual lab, not requiring too much computational capacity and being based on the standard engineering software COMSOL Multiphysics® [43] as a basis to enable a potentially broad use. The investigations are limited to material with 1 vol.% of carbon fibres. Hereby, the small fibre diameter leads to a relatively high number of fibres, causing various problems in mesh generation the more fibres the material contains. Further, the occurring mechanisms are structurally the same but the individual effects of damage overlap with higher fibre content.

4.1 Homogenisation Method

The concept of homogenisation treats numerical issues involving composite materials where the macroscopic scale is far larger than the characteristic length of the microstructure. Resolving the whole specimen including the microstructure for simulation is numerically not feasible. Therefore, analytical techniques are used to derive an effective continuum model describing an artificial (homogenous) material with approximately the same macroscopic properties as the original problem by a rigorous limit procedure. As a result, the linear elasticity equation for the unknown displacement u is obtained as

$$-\nabla \cdot \sigma = f, \quad \left(\text{with: } \sigma = A^{hom} e(u); \quad e(u) = \frac{1}{2}\left(\nabla u + (\nabla u)^T\right)\right) \tag{2}$$

with the stress tensor σ and A^{hom} representing the local effective material tensor of the artificial homogenous material while $e(u)$ is the linearized strain tensor and f is the force density (with respect to volume). The macroscopic relationship of the stress and the strain tensors described via the homogenised tensor A^{hom} can be computed by mean values of quantities determined by so called cell-problems in local representative volume elements Y (see RVE in Fig. 26 and the RVE description in Sect. 4.3.1) as follows

$$A^{hom} = \left(a_{ijkl}^{hom}\right)_{1 \le i,j,k,l \le 3} \tag{3}$$

with

$$a_{ijkl}^{hom}(x) = \frac{1}{|Y|} \int_Y A(x, y) e_{ij} \left(e_{kl} - e_y\left(w^{kl}\right)(x, y)\right) dy,$$

where x denotes the macroscopic spatial variable in the specimen and y is the microscopic spatial variable in the RVE Y. The corrector functions w^{kl}, $k, l \in \{1, 2, 3\}$ are unique solutions of the cell problem

$$-\nabla_y \cdot \left(A(x, \cdot)\left(e_y\left(w^{kl}\right)(x, \cdot) - e_{kl}\right)\right) = 0, \tag{4}$$

subject to periodic boundary conditions and having zero mean value, where A describes the material properties of the different phases in the RVE. For more details, please refer to e.g. [44, 45].

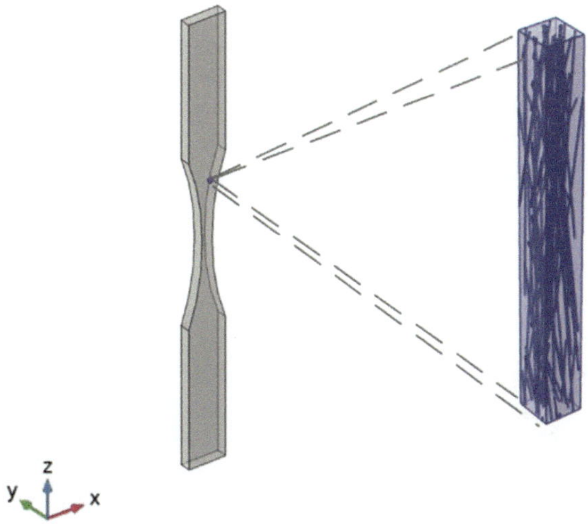

Fig. 26 Principle of homogenisation with local macroscopic undamaged RVE

4.2 Undamaged Material

A crucial aspect of models based on mathematical homogenisation approaches is the determination of RVEs which are representative for the entire material, here consisting of concrete and 1 vol.% of oriented carbon fibres. Further, the efficiency of the numerical model depends on the choice of the mesh, which is just fine enough to resolve all relevant mechanisms. Careful specification of an appropriate RVE and the optimum mesh size is therefore the starting point not only for simulation of undamaged material but also for further modelling of the damage mechanisms. Based on preliminary studies, aggregates on fibre scale and pores on fibre as well as specimen scale are neglected, meaning that the RVEs only consist of the concrete matrix and the fibres. In addition, it is assumed that the bonding between fibre and concrete matrix is perfect. The case where weaker bonding is assumed and modelled via slip-displacement conditions is investigated in [46]. The derived homogenised problem shows only marginal impact of the weaker bond on the macroscopic behaviour in the case of undamaged concrete. In [15], the authors studied the sufficient size of the RVE, which is as small as possible and still representative for the material. The fact that the fibres are small in diameter (7 μm) but long (about 3 mm) leads to long RVEs of small cross-section. With these results and from the statistical data concerning angular orientation of the fibres, a final RVE of the size 317 × 317 × 3000 μm with oriented carbon fibres was constructed (see also Fig. 27 for the RVE of the undamaged material). Further detailed convergence studies on the size of the mesh elements were conducted. In the simulations presented below, the mesh consists of 195,105 tetrahedral mesh elements resulting in 834,717 degrees of freedom (quadratic Lagrange elements). By using this RVE, the macroscopic material law for the undamaged material is derived according to Sect. 4.1, which is relevant for the first phase of the tensile tests (see Sect. 3.3.1). Based on these results and the RVE for the undamaged material, further adaptations are made for considering damage mechanisms relevant in stage II of the degradation process.

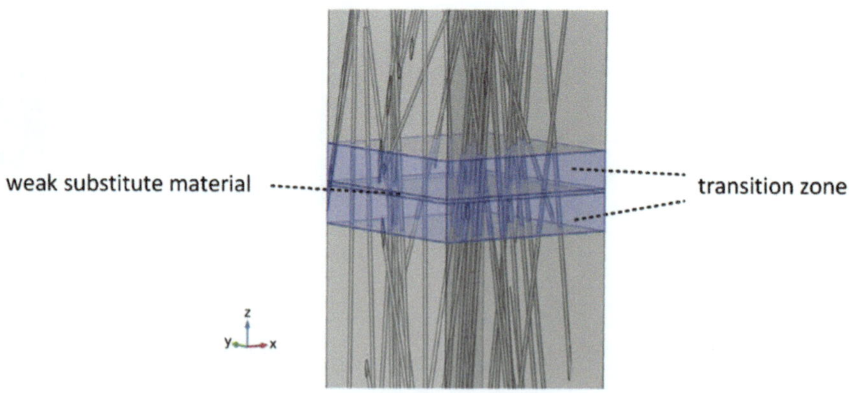

Fig. 27 Damage model via weak substitute material and transition zone

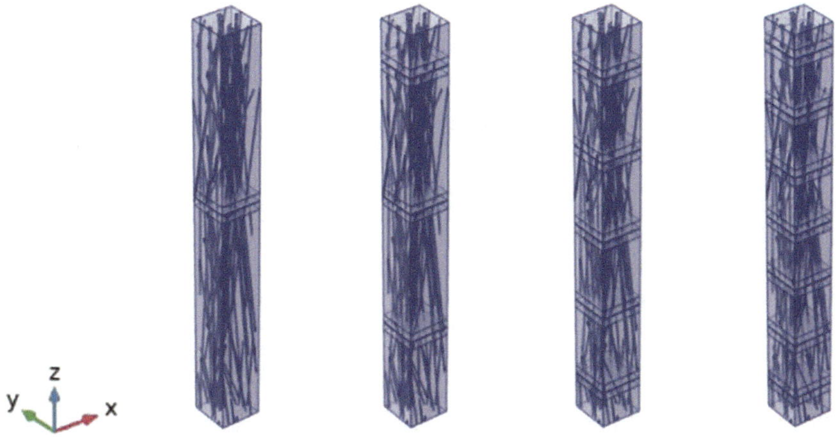

Fig. 28 Different stages of RVE damage with 1, 3, 5 and 7 cracks (left to right)

4.3 Modelling of Damage Behaviour and Relevant Mechanisms

The main goal is to understand the principles leading to fatigue and failure of the material and to implement a model which reproduces the relevant mechanisms. The experimental results serve as input for the numerical model. The comparison of the stress–strain diagram (see Fig. 11a) and DIC images evaluated at the corresponding points in time (e.g. Figure 11b) show a direct correspondence of sudden damage in the material and the drops in measured stress. Furthermore, it is seen from these experimental results of samples with fibre content of 1 vol.% that, between these events, the material behaviour can be considered as piecewise linear. Similarly, in experiments with samples with higher fibre content, only higher crack density is observed until final failure, where each cracking event is still associated with a drop of stress, leading to a smoother appearance of the measured stress–strain diagram. The DIC results further show regions with different strain, which is associated with different amounts of material damage, i.e. crack density. Comparing the DIC and the CT images (see Figs. 11 and 24), it can be seen that, during the tensile tests with the small specimen, complete crack propagation through the matrix of the component takes place without inducing further crack opening in fibre direction or complete failure of the component which allows for further damage in form of cracks at other positions. Also, the fact that, as stated in [40], the initial crack does not necessarily develop as fatal crack leads to the conclusion that the oriented fibres firstly bridge the progressing cracks in the concrete matrix. This substantiates the hypothesis that following onset of cracking little to no fibre-related damage, e.g. fibre pull-out or breakage, occurs in the cracked area until final failure and the fibres bridge the material. It is deduced that the material-typical strain-hardening-like behaviour [15], which is typically observed in samples with a significant volume content of oriented fibres, results from superposition of multiple events causing stress drops.

4.3.1 Modelling

From these results/observations, the following assumptions are extracted: the relevant 'failure' mechanism on microscale level is the sudden cracking of the concrete matrix without fibre-related damage. Crack propagation on RVE scale is instantaneous during the considered time scales. Thus, on macro scale, locally only either un-damaged RVEs or RVEs with cracks in the matrix perpendicular to the fibre orientation through the whole RVE are considered. The number of cracks per RVE varies in the experiments according to the length of the respective RVE, from 1 to 7 cracks.

The macroscopic damage and successive development of damaged zones in the specimen can thus be modelled locally by associated RVEs. In more detail, cracks of small opening in the cement phase are modelled by weak substitute material with Young's modulus of order 10^{-9} less than the cement paste (also compare experimental observations shown in Fig. 24). This procedure is of no practical consequence but preserves the mathematical foundation of the applied homogenisation method. Hereby, a designated transition zone, where the fibres and the matrix are disconnected, ensures correct material behaviour in the vicinity of the cracks (see e.g. Figure 27). The local macroscopic amount of damage is modelled depending on a criterion based on local von Mises stresses via different stages of RVE damage, characterised by zero to seven cracks of the cement matrix within the RVE (see Fig. 28).

The proof of principle is restricted to simulating a failure in the relevant part of the test specimen in a displacement-controlled tensile test as shown in Fig. 29.

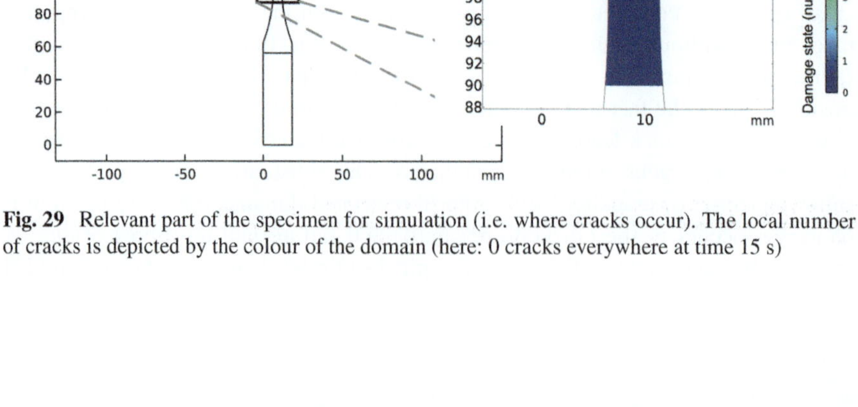

Fig. 29 Relevant part of the specimen for simulation (i.e. where cracks occur). The local number of cracks is depicted by the colour of the domain (here: 0 cracks everywhere at time 15 s)

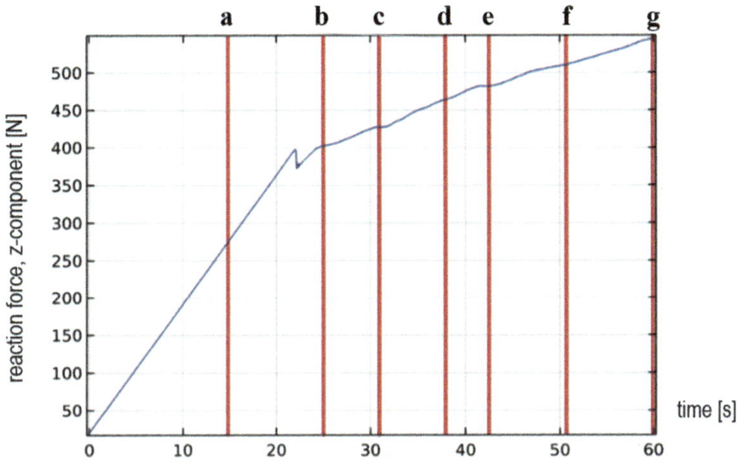

Fig. 30 Reaction force versus time. The red lines indicate, where the number of cracks in the relevant part of the sample are plotted in Figs. 29 (no crack) and 31b–g

4.3.2 Results

In the simulation of path-controlled tensile tests, the specimen is fixed at the lower part and the displacement is prescribed at the upper part following the experimental setup of the static tensile test. This results in the reaction force–time diagram depicted in Fig. 30. The graph shows at first a linear-elastic behaviour followed by several drops at different time steps, each connected by a linear material response.

Figure 31 shows the number of cracks in the relevant part of the sample (refer to Fig. 29) at different simulation times as indicated in Fig. 30. The combination of both figures implies that the drops happen as soon as the spread of the damaged zone associated with a certain no. of cracks extends across the entire specimen. Further, it can be seen that successive damage reduces the gradient of the curve, which corresponds to the strain-hardening like material response seen in the experiments.

5 General Conclusion, Outlook

Through extensive static and dynamic/cyclic (fatigue) testing series both on bending and tensile specimens of different scales, a wide range of valuable and new findings on the complex load-bearing, damage and failure behaviour of carbon short-fibre reinforced concrete (CSFRC) were gained. Strain-hardening of the material (and the transition to a non-linear stress–strain relationship) is achieved by the carbon fibres and their extremely good bond properties, leading to tiny microcracks and a very narrow crack spacing—providing a significant increase both in static/dynamic performance and flexural and tensile strength, respectively. Further, compared to

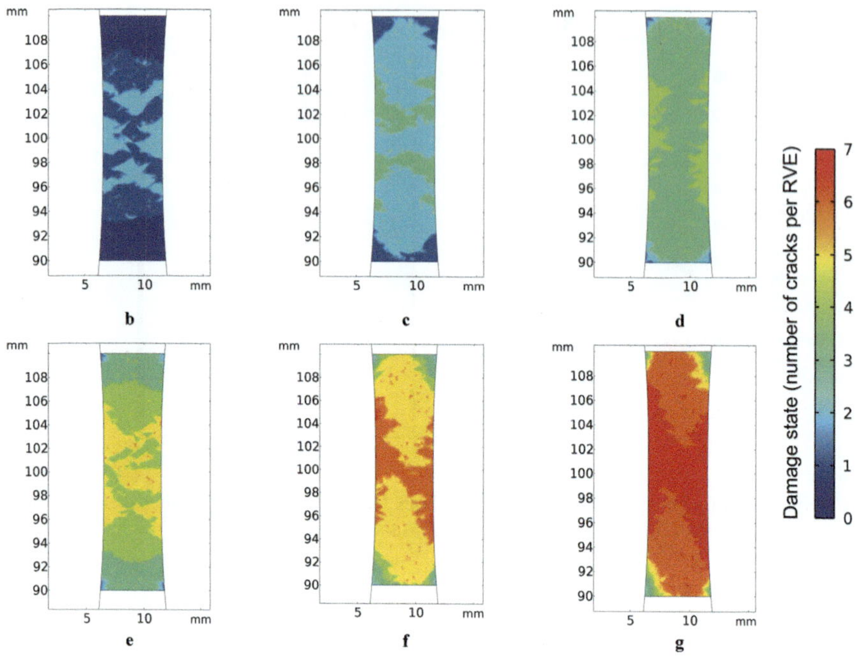

Fig. 31 No. of cracks at different simulation times: **b** 25 s, **c** 31 s, **d** 38 s, **e** 42.5 s, **f** 50.5 s, **g** 60 s

conventional steel fibres, the carbon fibres are already significantly activated in the un-cracked cross-section (i.e. before crack initiation at some 0.3–0.4‰).

Extensive NDT measurements (e.g. DIC, AE, electric sensors, X-ray CT, LDV, CWI, fibre-optics), including combined interpretations, significantly contributed to a deeper understanding of damage and fracture mechanisms. While DIC was primarily utilised to provide a sound overall survey of the (near-surface) damage and crack development, AE and electric measurements were particularly helpful to discriminate individual source types and to localise events, respectively—with different advantages/disadvantages dependent of the sample size and experimental setup. Further, ex-situ CT images were used to gain additional valuable information on the type of failure (fibre breakage or pull-out). Other techniques such as Coda Wave Interferometry (CWI) can help to indicate the progress of deterioration but has drawbacks in localisation and classification of defects. Lessons learned comprise information on best sensor positions, coupling, selection of sensors, and, in particular, frequency transfer functions of sensors.

The principle mechanical behaviour and structural damage as well as the damage development in cross-sectional areas under tension could be confirmed for all sample sizes/types and loading regimes (tension, flexure) with only minor differences. With regard to the specific type of failure, the static tests showed that the final specimen failure was initiated by progressive fibre breakage combined with fibre pull-out (i.e. bond/anchorage failure, especially of fibres with shorter bond length on one crack edge). In the cyclic fatigue tests, both damage evolution phase 2 (stable crack development) and phase 3 (unstable crack growth) could be observed across all loading types and scales, where the individual phases tended to be more pronounced and better identifiable with decreasing scale, particularly the generally very short phase 3. Thereby, the individual NDT measurements and their combinations suggest that phase 2 is mainly characterised by gradual fibre pull-out, while the final failure initiation in phase 3 seems to be primarily associated with fibre breakage. Within the additional fatigue tests with non-monotonous cyclic (varying) loading schemes on macro samples the results indicate tendencies towards sequence effects. Furthermore, it can be concluded that the specimens exhibit a more favourable fatigue behaviour under axial tension than with harmonic bending loading.

With regard to the mathematical multiscale modelling the results show that the developed model is capable of reproducing both the material behaviour and the relevant failure mechanisms. RVEs were derived from μ-CT and convergence tests determined the required RVE sizes and discretisation parameters, which turned out to give a numerically efficient computational model of the mechanical behaviour of the material. In combination with the experimental results, the model was then used to characterise the failure mechanisms. In particular, it was shown that the correlation between drops in reaction force and successive damage observed in the experiments can be reproduced by locally differently damaged RVEs related to different macroscopic crack densities associated with cracks in the concrete matrix and weakening of the surrounding fibre-matrix interface. Failure of the entire sample is finally caused by subsequent failure of the carbon fibres and/or fibre anchorage. Further refinement of the model like inclusion of pores or other macroscopic inhomogeneities or different sizes of the transition zones can be made straightforwardly if this level of detail is required. The cyclic tests generally show similar damage mechanisms as in the static domain. Therefore, by incorporating a specific fatigue parameter (taking into account the load/damage history) into the localised damage model developed, it is also possible to describe the failure process under cyclic loading in principle. However, an in-depth alignment beyond the investigations possible within the scope of the project with specific experimental fatigue data can contribute to develop further and calibrate generally the numerical model accordingly.

References

1. Hambach, M., Möller, H., Neumann, T., Volkmer, D.: Portland cement paste with aligned carbon fibers exhibiting exceptionally high flexural strength (> 100 MPa). Cem. Concr. Res. (2016). https://doi.org/10.1016/j.cemconres.2016.08.011
2. Hambach, M.: Hochfeste multifunktionale Verbundwerkstoffe auf Basis von Portlandzement und Kohlenstoffkurzfasern. Dissertation (2016)
3. Fischer, O., Volkmer, D., Lauff, P., Hambach, M., Rutzen, M.: Zementgebundener kohlenstofffaser-verstärkter Hochleistungswerkstoff (Carbonkurzfaserbeton). Bericht F 3178, BBSR, Fraunhofer (2019). ISBN 978-3-7388-0422-5

4. Lauff, P., Fischer, O.: Effizienter Ultrahochleistungsbeton mit innovativer trajektorienorientierter "Bewehrung." Ce/papers **3**, 82–88 (2019)
5. Hambach, M., Volkmer, D.: Properties of 3D-printed fiber-reinforced Portland cement paste. Cem. Concr. Compos. (2017). https://doi.org/10.1016/j.cemconcomp.2017.02.001
6. Lauff, P.: Zur Herstellung und zum Trag- und Ermüdungsverhalten von Car-bonkurzfaserbeton. Dissertation, Technical University of Munich, in process
7. TUM: 3D Printing in Concrete—Additive Manufacturing at TUM. Online: https://www.youtube.com/watch?v=H3aD0VZ3ESU. Accessed 17 Mar 2021
8. Rutzen, M., Lauff, P., Niedermeier, R., Fischer, O., Raith, M., Grosse, C.U., Weiss, U., Peter, M.A., Volkmer, D.: Influence of fiber alignment on pseudoductility and microcracking in a cementitious carbon fiber composite material. Mater. Struct. **54**, 58 (2021). https://doi.org/10.1617/s11527-021-01649-2
9. Rutzen, M., Schulz, M., Mossburger-Will, J., Lauff, P., Fischer, O., Volkmer, D.: 3D printing as an automated manufacturing method for a carbon fiber reinforced cementitious composite with outstanding flexural strength (105 N/mm^2). Mater. Struct. **54**, 234 (2021). https://doi.org/10.1617/s11527-021-01827-2
10. Curosu, I., Liebscher, M., Mechtcherine, V., Bellmann, C., Michel, S.: Tensile behavior of high-strength strain-hardening cement-based composites made with high-performance polyethylene, aramid and PBO fibers. Cem. Concr. Res. (2017). https://doi.org/10.1016/j.cemconres.2017.04.004
11. Li, V.C., Leung, C.K.Y.: Steady-state and multiple cracking of short random fiber composites. J. Eng. Mech. (1992). https://doi.org/10.1061/(ASCE)0733-9399(1992)118:11(2246)
12. Li, V.C.: On engineered cementitious composites (ECC). ACT (2003). https://doi.org/10.3151/jact.1.215
13. Van Zijl, G.P.A.G., Wittmann, F.H.: On durability of SHCC. ACT (2010). https://doi.org/10.3151/jact.8.261
14. Lauff, P., Raith, M., Große, C., Rutzen, M., Volkmer, D., Reischmann, L., Weiß, U., Peter, M., Fischer, O.: Investigation of localised damage indicators of a carbon short-fibre reinforced high performance concrete under dynamic and flexural load. In: Proceedings of fib Symposium, Krakow/Poland, pp. 1952–1959 (2019)
15. Lauff, P., Pugacheva, P., Rutzen, M., Weiß, U., Fischer, O., Volkmer, D., Peter, M.A., Grosse, C.U.: Evaluation of the behavior of carbon short fiber reinforced concrete (CSFRC) based on a multi-sensory experimental investigation and a numerical multiscale approach. Materials **14**(22), 7005 (2021)
16. Rutzen, M., Volkmer, D.: Viscoelasticity and energy dissipation as indicators of flexural fatigue behavior in a ductile carbon fiber-reinforced cementitious composite. Int. J. Fatigue (2022). https://doi.org/10.1016/j.ijfatigue.2022.106839
17. Džolan, A., Lauff, P., Fischer, O., Šunjić, G.: Influence of concrete shrinkage on the behavior of carbon short-fiber-reinforced concrete (CSFRC) under tension. Appl. Sci. **13**(12), 7081 (2023). https://doi.org/10.3390/app13127081
18. Engelhard, M., Kalytta-Mewes, A., Volkmer, D.: Testing methods for deformation characterization on 3D printed carbon fiber reinforced cementitious composites. In: Proceedings of 14th fib PhD Symposium, Rome/Italy (2022)
19. Lauff, P., Fischer, O.: Damage indication by electrical resistivity measurement of carbon short fiber reinforced concrete in fatigue testing. In: Proceedings of the 13th fib International PhD Symposium in Civil Engineering (2021). ISBN 978-2-940643-11-0
20. Rutzen, M., Volkmer, D., Weiß, U., Reischmann, L., Peter, M.A., Lauff, P., Fischer, O., Raith, M., Große, C.: Microstructural analysis of crack growth caused by static and dynamic loads in a carbon fiber reinforced cement paste. In: Zingoni, A. (ed.) Proceedings of Advances in Engineering Materials, Structures and Systems: Innovations, Mechanics and Applications. CRC Press (2019)
21. Sause, M.G.: In-Situ Monitoring of Fiber-Reinforced Composites, vol. 242. Springer International Publishing, Cham (2016)

22. Schukraft, J., Lohr, C., Weidenmann, K.A.: 2D and 3D in-situ mechanical testing of an interpenetrating metal ceramic composite consisting of a slurry-based ceramic foam and AlSi10Mg. Compos. Struct. (2021). https://doi.org/10.1016/j.compstruct.2021.113742
23. Aggelis, D.G., Dassios, K.G., Kordatos, E.Z., Matikas, T.E.: Damage accumulation in cyclically-loaded glass-ceramic matrix composites monitored by acoustic emission. Sci. World J. **12**, 869467 (2013)
24. Gautschi, G.: AE sensors. In: Piezoelectric Sensorics: Force Strain Pressure Acceleration and Acoustic Emission Sensors Materials and Amplifiers, pp. 199–207 (2002)
25. Schumacher, T., Linzer, L., Grosse, C.: Signal-based AE analysis. In: Acoustic Emission Testing: Basics for Research—Applications in Engineering, pp. 73–116 (2022)
26. Grosse, C.U., Ohtsu, M., Aggelis, D.G., Shiotani, T. (eds.): Acoustic Emission Testing: Basics for Research—Applications in Engineering. Springer Nature (2021)
27. Sause, M.: Identification of Failure Mechanisms in Hybrid Materials Utilizing Pattern Recognition Techniques Applied to Acoustic Emission Signals. mbv Verlag (2019)
28. Sause, M.G., Gribov, A., Unwin, A.R., Horn, S.: Pattern recognition approach to identify natural clusters of acoustic emission signals. Pattern Recogn. Lett. **33**(1), 17–23 (2012)
29. Ospitia, N., Korda, E., Kalteremidou, K.A., Lefever, G., Tsangouri, E., Aggelis, D.G.: Recent developments in acoustic emission for better performance of structural materials. Dev. Built Environ. **13**, 100106 (2023)
30. Bian, C., Wang, J.Y., Guo, J.Y.: Damage mechanism of ultra-high performance fibre reinforced concrete at different stages of direct tensile test based on acoustic emission analysis. Constr. Build. Mater. **267**, 120927 (2021). https://doi.org/10.1016/j.conbuildmat.2020.120927
31. Wang, J.Y., Guo, J.Y.: Damage investigation of ultra high performance concrete under direct tensile test using acoustic emission techniques. Cem. Concr. Compos. **88**, 17–88 (2018). https://doi.org/10.1016/j.cemconcomp.2018.01.007
32. Ohtsu, M., Watanabe, H.: Quantitative damage estimation of concrete by acoustic emission. Constr. Build. Mater. **15**(5–6), 217–224 (2001). https://doi.org/10.1016/S09050-0618(00)00071-4
33. Van Steen, C., Verstrynge, E.: Signal-based acoustic emission clustering for differentiation of damage sources in corroding reinforced concrete beams. Appl. Sci. **12**(4), 2154 (2022). https://doi.org/10.3390/app12042154
34. Larisa, U., Solbon, L., Sergei, B.: Fiber-reinforced concrete with mineral fibers and nanosilica. Procedia Eng. **195**, 147–154 (2017)
35. Li, V.C., Wang, S., Wu, C.: Tensile strain-hardening behavior of polyvinyl alcohol engineered cementitious composite. ACI Mater. J. **98**(6), 483–492 (2001)
36. Gutkin, R., Green, C.J., Vangrattanachai, S., Pinho, S.T., Robinson, P., Curtis, P.: On acoustic emission for failure investigation in CFRP: pattern recognition, peak frequency analyses. Mech. Syst. Signal Process. **25**(4), 1393–1407 (2011)
37. Kravchuk, R., Landis, E.N.: Acoustic emission-based classification of energy dissipation mechanisms during fracture of fiber-reinforced ultra-high-performance concrete. Constr. Build. Mater. **176**, 531–538 (2018). https://doi.org/10.1016/j.conbuildmat.2018.05.039
38. Pomponi, E., Vinogradov, A.: A real-time approach to acoustic emission clustering. Mech. Syst. Signal Process. **40**(2), 791–804 (2013). https://doi.org/10.1016/j.ymssp.2013.03.017
39. Thum, F., Potstada, P., Sause, M.G.: Development of 25 kN in-situ load stage combining X-ray computed tomography & acoustic emission measurement. KEM (2019). https://doi.org/10.4028/www.scientific.net/KEM.809.563
40. Potstada, P.-E.: Konzeption und Durchführung von in-situ Computertomographie in der Anwendung auf Aluminiumlegierungen und Faserverbundwerkstoffe. Dissertation, Universität Augsburg (2021)
41. Schröder, J.: A numerical two-scale homogenization scheme: the FE2-method. In: Schröder, J., Hackl, K. (eds.) CISM Course 550, Plasticity and Beyond, pp. 1–64. Springer (2014). https://doi.org/10.1007/978-3-7091-1625-8_1
42. Raju, K., Tay, T.-E., Tan, V.B.C.: A review of the FE2 method for composites. Multiscale Multidiscip. Model. Exp. Des. **4**, 1–24 (2021). https://doi.org/10.1007/s41939-020-00087-x

43. COMSOL Multiphysics® 5.6. COMSOL AB, Stockholm. www.comsol.com
44. Cioranescu, D., Donato, P.: An Introduction to Homogenization. In: Oxford Lecture Series in Mathematics and Applications. Oxford University Press (1999)
45. Lochner, T., Peter, M.A.: Identification of microstructural information from macroscopic boundary measurements in steady-state linear elasticity. Math. Methods Appl. Sci. **46**, 1295–1316 (2023). https://doi.org/10.1002/mma.8581
46. Lochner, T., Peter, M.A.: Homogenization of linearized elasticity in a two-component medium with slip displacement conditions. J. Math. Anal. Appl. **483**, 123648 (2020). https://doi.org/10.1016/j.jmaa.2019.123648

Enhancement of Fatigue Resistance of Strain-Hardening Cement-Based Composites by Means of Experimental-Virtual Multiscale Material Design

Dominik Junger, Johannes Storm, Rodolfo Piña, Rydell Riko, Michael Kaliske, and Viktor Mechtcherine

Abstract This chapter explores the enhancement of fatigue resistance in strain-hardening cement-based composites (SHCC) through a combined experimental and numerical approach. The studies at hand focussed on investigating fibre-matrix interactions, crack propagation, and damage mechanisms. Experimental tests, such as uniaxial tension, pull-out, and relaxation tests, were conducted to understand SHCC's mechanical behaviour subjected to different environmental and loading conditions. These findings were compared to numerical simulations using the micro-layer approach, fibre super-element, cohesive interface models and phase-field fracture, which simulate complex behaviours like fibre pull-out and crack formation. To address the computational challenges of simulating high-cycle fatigue, time homogenisation techniques were used to reduce the number of calculations required. The results show strong agreement between experimental data and numerical simulations, highlighting the effectiveness of the multiscale modelling approach. The findings of the studies presented in this chapter contribute to advancing the understanding of the fatigue behaviour of these composites and provide practical insights for developing more durable, fatigue-resistant materials for infrastructure exposed to cyclic stresses, setting a precedent for improving multiscale material design approaches.

Keywords Fatigue · SHCC · Degradation mechanism · Numerical modelling · Super-element · Microlayer approach

D. Junger · V. Mechtcherine (✉)
Institute of Construction Materials, TUD Dresden University of Technology, Dresden, Germany
e-mail: viktor.mechtcherine@tu-dresden.de

J. Storm · R. Piña · R. Riko · M. Kaliske
Institute of Structural Analysis, TUD Dresden University of Technology, Dresden, Germany

N. Oneschkow and L. Lohaus (eds.), *Cyclic Deterioration of High-Performance Concrete*, https://doi.org/10.1007/978-3-032-19394-0_16

1 Introduction

The increasing demand for more durable and resilient construction materials has led to extensive research into SHCC, a special class of highly ductile, microfibre reinforced cementitious composites exhibiting superior tensile strength and strain capacity as well as exceptional crack-bridging behaviour [1]. These characteristics make them favourable for applications under dynamic loading scenarios, such as cyclic loading. However, fatigue resistance is critical for infrastructure exposed to repeated stress, where long-term durability and resistance to crack propagation are essential. Previous experimental studies already described the fatigue performance and the degradation mechanisms of SHCC on different observation levels [2–4].

To deepen the observations made in those previous studies, this chapter focuses on the enhancement of SHCC's fatigue resistance through a combined experimental-virtual multiscale material design. By leveraging both experimental data and advanced numerical simulations, the research seeks to improve the understanding of how it behaves under cyclic loading, particularly at the micro and mesoscale levels. The core of this investigation includes fibre-matrix interaction, fibre inclination, fatigue damage mechanisms, and the overall mechanical response of these materials under both monotonic and cyclic loading conditions.

The research integrates a variety of tests, including uniaxial quasi-static, cyclic tension and alternating cyclic loading tests as well as single fibre pull-out, and relaxation tests to characterise the mechanical properties of the fibres and the composite matrix depending on various influencing factors. Numerical modelling approaches, such as the microlayer approach, super-element and cohesive interface models, are used to simulate the behaviour at different scales, with special attention given to crack formation, fibre pull-out, and damage progression. Additionally, advanced computational techniques, including time homogenisation, are explored to reduce the computational burden of simulating high-cycle fatigue, making the analysis more efficient while maintaining accuracy.

Through these combined efforts, the investigations aim to provide a deeper understanding of SHCC's performance under cyclic loading, ultimately contributing to the development of more durable, fatigue-resistant materials.

2 Materials

Two distinct material compositions were used for the investigations, namely a normal-strength (M1) and a high-strength matrix (M2), that were previously developed at the Institute of Construction Materials, TU Dresden [5]; see also Table 1. Matrix M1 was mainly used for the micromechanical tests with polyvinyl alcohol (PVA) fibres, while the mesoscopic investigations were performed on the high-strength matrix (M2) reinforced with high-performance polyethylene fibres (PE).

Table 1 Composition of the normal-strength (M1) and high-strength (M2) SHCC matrices (in kg/m^3)

Component	M1	M2
CEM I 42.5R-HS	505	–
CEM I 52.5R-SR3/NA	–	1460
Fly ash	621	–
Micro silica ELKEM 971	–	292
Quartz sand 0.06–0.2 mm	536	145
SP BASF MasterGlenium ACE30	10	–
SP BASF MasterGlenium ACE460	–	35
VMA Sika UW-Compound-100	4.8	–
Water	338	315
PVA Kuraray Kuralon K-II REC 15	26	–
PE DSM Dyneema SK78 (2 vol.%)	–	20

Table 2 Properties of the fibres under investigation (according to the manufacturer)

Fibre type	Diameter d_f (μm)	Length l_f (mm)	Tensile strength f_t (MPa)	Young's modulus E_f (GPa)
PVA	40	12	1600	40
PE	20	12	3300–3900	109–132

Table 2 shows the dimensions and mechanical parameters of the fibres used in the various experiments described. PE fibres show significantly higher strengths with an increased elastic modulus compared to the PVA fibres. Due to the smaller diameter, PE fibres cause a crack pattern with increased number of cracks showing smaller crack widths then PVA fibres.

3 Theoretical Background

3.1 Constitutive Model of the Concrete Matrix

The microplane model distributes material deformations onto a set of oriented planes, where simplified constitutive relations are evaluated. This approach is designed to capture key mechanisms, such as damage and plasticity-induced anisotropy, by relating macroscopic material behaviour to the microstructure. Over time, this model has been extended to incorporate various advanced deformation mechanisms in concrete, as discussed in [6–9].

However, recent microplane models present several inconsistencies, such as issues with the volumetric-deviatoric (V-D) deformation split, static constraints, and stress homogenisation relations. Despite its empirical success, the microplane model lacks physical meaning for its micromechanical variables. To address these limitations, the microlayer model is introduced in [10], using Blanco's variational homogenisation framework [11] to apply micromechanical assumptions, building on the original ideas proposed in [6].

The representative volume element (RVE) in the microlayer model imposes fewer restrictions. For example, this model allows for arbitrary convex polyhedrons for rigid aggregates, finite thicknesses for deformable layers, and the application of standard constitutive models (see Fig. 1). The domain of the aggregate Ω^K along with the domains of the deformable layers $\Omega^{(i)}$ collectively define the microscale domain Ω. In the contribution at hand, the distance between the geometric centre of the microscale and the outermost layer surface $r^{(i)}$ is characterised by the aggregate radius $r^{K(i)}$ and the layer height $h^{(i)}$. Each layer is further described by a normal distance vector $\boldsymbol{r}^{(i)} = r^{(i)}\boldsymbol{N}^{(i)}$, where $\boldsymbol{N}^{(i)}$ represents the unit normal vector associated with the layer. This flexibility preserves the computational efficiency of the model while allowing more realistic representations of concrete microstructures. However, the main advantage of the microlayer model lies in its ability to derive macroscopic material behaviour through a homogenisation framework. This enables physically meaningful micromechanical variables, which can be directly compared to experimental data.

As illustrated in Fig. 2, the aggregate radius r and the layer thickness h in the RVE significantly influence the macroscopic material behaviour properties that the microplane model is unable to capture. Additionally, as discussed in [12], the microlayer model has been extended to fibre reinforced concrete to capture the combined effects of compression and tension loads.

The model provides a more accurate representation of damage initiation and failure loads under multiaxial loading conditions, since it presents a stronger decoupling of spatial directions, as demonstrated in Fig. 3. A finite deformation formulation of the microlayer model is provided in [12].

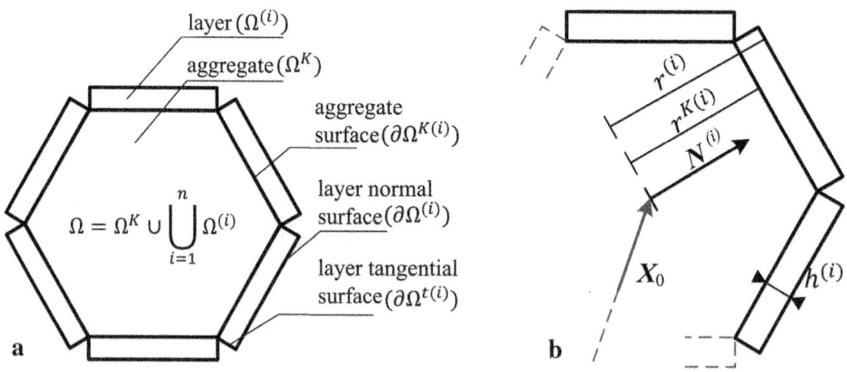

Fig. 1 Microlayer model: **a** domain definition, **b** geometry definition [10]

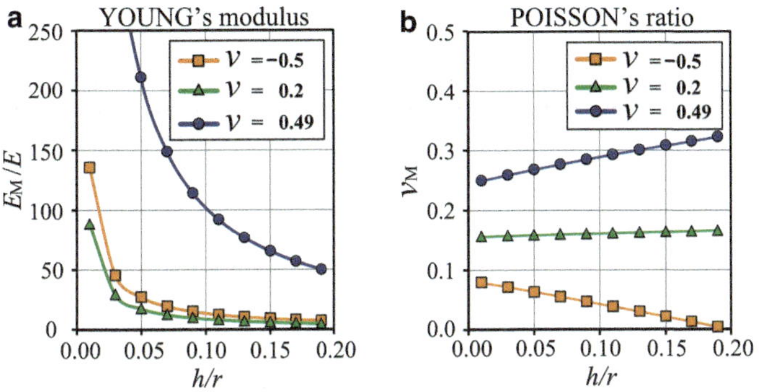

Fig. 2 **a** Relative macroscopic stiffness E_M/E and **b** macroscopic Poisson's ratio υ_M depending on the ratio of the layer thickness h to the aggregate radius r [10]

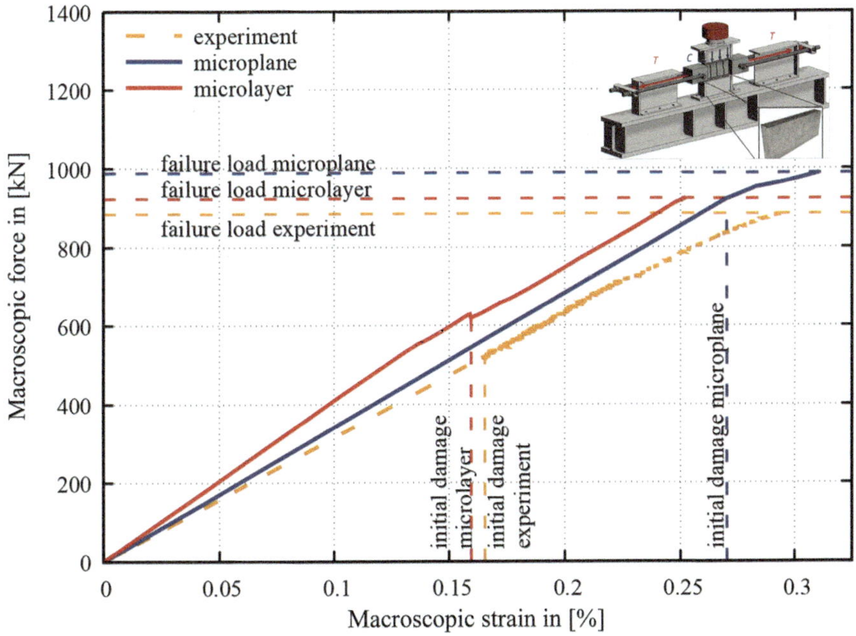

Fig. 3 Physical validation of the microplane and the microlayer model (from [12] licensed under CC-BY 4.0)

3.2 *Phase-field Formulation for Brittle Fracture*

The phase-field approach for elastic brittle fracture models the system's total energy functional

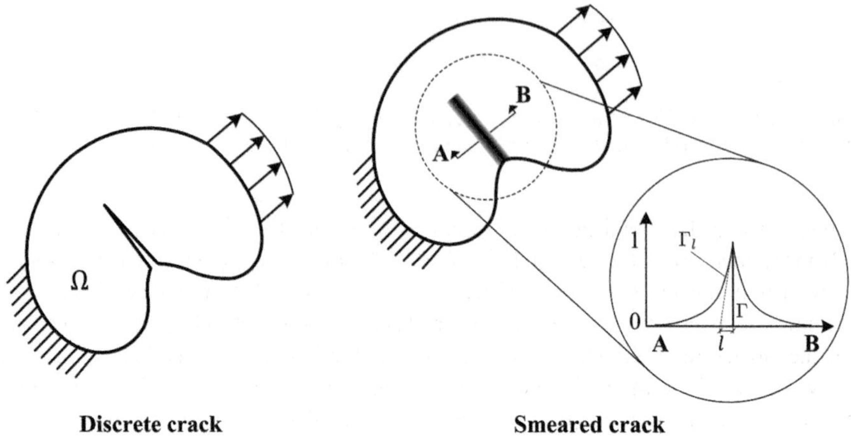

Fig. 4 Profile of the phase-field through a crack

$$E_{\text{total}} = \int\limits_{\Omega\backslash\Gamma} \psi \mathrm{d}\Omega + G_c \int\limits_{\Gamma} \mathrm{d}A - E_{\text{ext}} \tag{1}$$

As shown in Fig. 4, Ω denotes the entire material domain, Γ represents the crack surface within this domain, and $\Omega\backslash\Gamma$ corresponds to the remaining intact material excluding the crack surface. The material's fracture toughness is given by G_c. The energy functional also includes the external contributions E_{ext}, as the contributions from external forces and body forces.

The fracture energy can be regularised into a volume integral, incorporating a crack density function γ_l. The representative crack element (RCE), as presented in [13, 14], is used as a physics based approach for the energy split method, leading to

$$\begin{aligned} E_{\text{total}} = & \int\limits_{\Omega} \left[\psi^{\text{crack}} + g(d)\left(\psi^{\text{solid}} - \psi^{\text{crack}}\right)\right] \mathrm{d}\Omega \\ & + G_c \int\limits_{\Omega} \gamma_l(d) \mathrm{d}\Omega - E_{\text{ext}} \rightarrow \min, \end{aligned} \tag{2}$$

where $g(d)$ is the degradation function and d is the phase-field variable representing damage (ranging from 0 for intact to 1 for fully damaged material).

To account for fatigue effects, an additional degradation function $f(\overline{\alpha})$ is defined, and it modifies the fracture energy, reducing $G_{c,0}$ to $G_c = f(\overline{\alpha}) \cdot G_{c,0}$ according to the material's loading history. The damage driving energy α is

$$\alpha = g'(d)\left(\psi^{\text{solid}} - \psi^{\text{crack}}\right) \tag{3}$$

and the cumulative fatigue damage $\overline{\alpha}(x, t)$ is integrated over time, considering only the loading phases where $\dot{\alpha} \geq 0$. The fatigue degradation function $f(\overline{\alpha})$ adapts based on the accumulated damage, as described in [15].

3.3 Constitutive Description of the Fibres

Polymer fibres, such as PVA fibres, exhibit transversely isotropic material behaviour due to their microstructure, which is composed of filaments and manufacturing related surface treatments; see also Fig. 5.

The phenomenological constitutive law for these fibres is derived from specific deformation modes observed in experiments, which can be related to the invariants of the Cauchy-Green strain tensor and various damage mechanisms in the fibres [17]. The hyperelastic and viscoelastic behaviour of PVA fibres can be described by an empirical potential function

$$\psi(\boldsymbol{E}) = \sum_{i=2}^{n} a_i (I_4)^i, \quad I_4 = \boldsymbol{n} \cdot \boldsymbol{E} \cdot \boldsymbol{n}, \tag{4}$$

where ψ is the strain energy function, $\boldsymbol{E}$ is the strain tensor, a_i are parameters, and $\boldsymbol{n}$ represents the fibre direction. The viscoelastic behaviour can be modelled using a generalised Maxwell approach

$$\sigma(t) = \sigma_0(t) + \sum_{j=1}^{n} \int_0^t \gamma_j \exp\left(-\frac{t-s}{\tau_j}\right) \frac{\partial \sigma_0(s)}{\partial s} \mathrm{d}s, \tag{5}$$

where $\sigma(t)$ is the stress at time t, $\sigma_0(t)$ is the elastic stress, γ_j are normalised moduli, and τ_j are relaxation times.

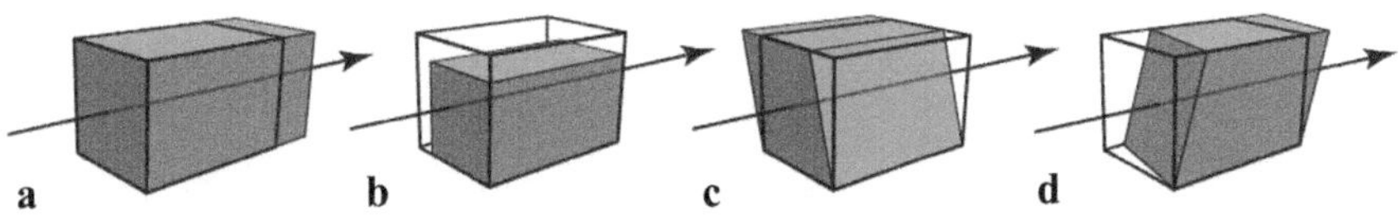

Fig. 5 Deformation modes: **a** longitudinal tension, **b** transversal compaction, **c** transversal shear, **d** longitudinal shear [16]

3.4 Cohesive Interface Model

Interface models are well established in the literature [18–21]. However, simulating interfaces, such as those between millions of fibres in fibre-reinforced concrete, is computationally expensive. For this reason, the cohesive interface model from [22] is adopted for small-scale investigations.

As shown in Fig. 6, the model considers a solid body $\mathcal{B}$ consisting of various sub-bodies, $\mathcal{B}_{t_i}$ and $\mathcal{B}_{t_k}$, having different mechanical properties such as the concrete matrix and fibres, with an interface region Ω, that is separated by a separation vector $\boldsymbol{\Delta}$ into the surfaces Ω^+ and Ω^-. Following [22], the interface mechanisms are incorporated into the linear momentum balance principle for the interface region as

$$\boldsymbol{G}_{\text{kinetic}} = \mathbf{A}_{E=1}^{N} \left[\int_{\mathcal{B}} \boldsymbol{B}^T \boldsymbol{\sigma} \mathrm{d}v + \int_{\mathcal{B}} \boldsymbol{N}^T \rho \boldsymbol{N} \mathrm{d}v \ddot{\boldsymbol{u}} + \int_{\mathcal{B}} \boldsymbol{N}^T \rho \boldsymbol{b} \mathrm{d}v - \int_{\partial\mathcal{B}} \boldsymbol{N}^T \boldsymbol{t} \mathrm{d}a - \int_{\Omega^+} \boldsymbol{N}_{\Omega^+}^T \boldsymbol{t}^+ \mathrm{d}\Omega^+ - \int_{\Omega^-} \boldsymbol{N}_{\Omega^-}^T \boldsymbol{t}^- \mathrm{d}\Omega^- \right] = 0, \tag{6}$$

where the damage evolution parameter d and the total elastic traction separation energy $\boldsymbol{\Phi}_0$ are defined, according to [23], as

$$d = 1 - \exp\left[-\frac{\max_{\tau \in [0,t]} \boldsymbol{\Phi}_0(\boldsymbol{\tau})}{\eta_d} \right], \quad \boldsymbol{\Phi}_0 = \frac{1}{2} k_s |\boldsymbol{\Delta}|^2. \tag{7}$$

In this formulation, $\boldsymbol{N}$ and $\boldsymbol{B}$ are the shape function and its derivative. The terms $\boldsymbol{\sigma}$, $\boldsymbol{b}$, and $\boldsymbol{t}$ represent stress, body force, and surface traction, respectively, with ρ denoting the material density. The interface stiffness k_s and damage intensity parameter η_d are also considered.

This model uses penalty terms to prevent fibre penetration but does not account for fibre-channel interruptions caused by cracks or channel curves in the concrete

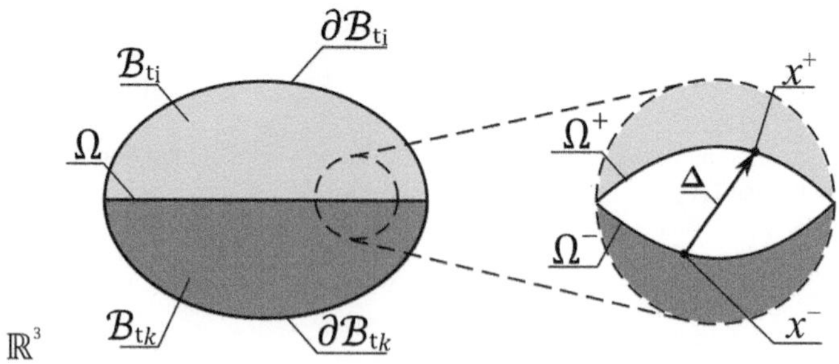

Fig. 6 Schematic graph of the cohesive interface model by [22]

matrix. The cohesive elements are supplemented with a global node-to-surface contact formulation using Lagrange multipliers.

3.5 Fibre Super-element

Modelling fibre-reinforced structures traditionally results in a high computational burden due to numerous degrees of freedom. However, this can be reduced by decoupling the discretisation of the concrete matrix from the fibres, as demonstrated in [24]. In this approach, two networks are superimposed using kinematic coupling (conditions as proposed in [17]) between the fibre super-element and the concrete's mesh. Fibre debonding and pull-out are treated as embedded problems, solved by means of the FE^2 method (Fig. 7).

Each fibre is represented by one super-element, subdivided into isoparametric truss sub-elements with linear shape functions as defined in [24]. The displacements of the fibre and concrete coincide where the bond remains intact. However, as debonding progresses, virtual nodes are introduced, with the deformable fibre regions kinematically coupled to the bonded areas.

The frictional force $\hat{f}$ is governed by the relative displacement between the concrete and fibre. The friction in the fibre channel is modelled using Coulomb friction. As explained in [25], a bi-linear friction model is defined for the kinetic friction f^{kin} as

$$f^{\mathrm{kin}}\left(\tilde{u}^{\mathrm{irr}}\right) = \begin{cases} f^{\mathrm{kin},1} - \frac{\tilde{u}^{\mathrm{irr}}}{\Delta\tilde{u}^{\mathrm{trans}}}\left(f^{\mathrm{kin},1} - f^{\mathrm{kin},2}\right), & \text{if } \tilde{u}^{\mathrm{irr}} \leq \Delta\tilde{u}^{\mathrm{trans}} \\ f^{\mathrm{kin},1}, & \text{else} \end{cases}, \tag{8}$$

where $f^{\mathrm{kin},1}$ and $f^{\mathrm{kin},2}$ refer to the initial and final kinetic friction force and the transition slip displacement is $\tilde{u}^{\mathrm{trans}}$.

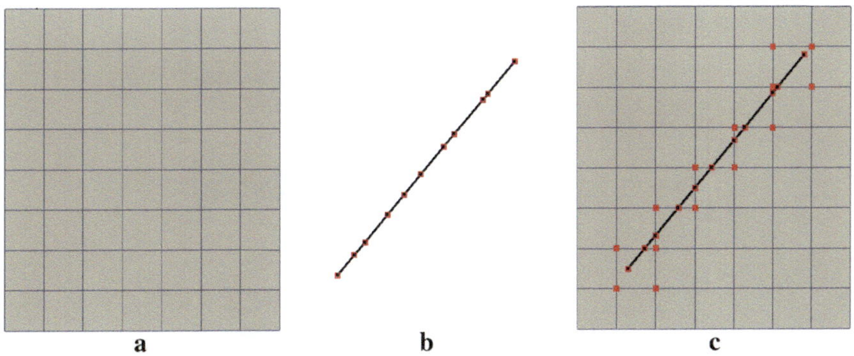

Fig. 7 Kinematic coupling of **a** the concrete matrix's mesh and **b** the superimposed fibre's mesh into **c** a composite mesh using kinematic constraints at the nodes (in red) [24]

Fibre debonding is controlled by the interfacial bond strength f^c. When the fibre force exceeds f^c, the bond front propagates, and a new equilibrium state is determined based on the bond front location.

3.6 Time Homogenisation (TH)

To accelerate the analysis of long-term structural and material responses under periodic loading, where accumulative effects such as fatigue occur, it is crucial to integrate the TH technique with the phase-field framework. Since phase-field simulations for fracture are inherently computationally demanding, particularly when involving multiple cycles, this integration is even more significant.

TH, as discussed in [26, 27], addresses systems with distinct temporal multiscale characteristics by separating them into micro and macro time scales, as illustrated in Fig. 8.

As explained in [28], the procedure begins with full resolution simulations that analyse the behaviour of internal variables y over a representative period τ_{micro}. From these results, a homogenised rate of change of internal variables $\langle \dot{y}_0 \rangle$, which reflects the system's characteristics, is computed as

$$\langle \dot{y}_0 \rangle = \frac{1}{\tau_{\text{micro}}} \int_0^{\tau_{\text{micro}}} \dot{y}(\tau) \mathrm{d}\tau \tag{9}$$

The macro time computation then proceeds with a sequence of micro and macro steps, where internal variables are extrapolated during the micro step and adjusted to maintain equilibrium during the macro step. The size of each time step in the macro level simulation $\Delta T_{\max}$ is adaptive and it is determined based on the sensitivity of the internal variables. The overall objective is to capture the main characteristics

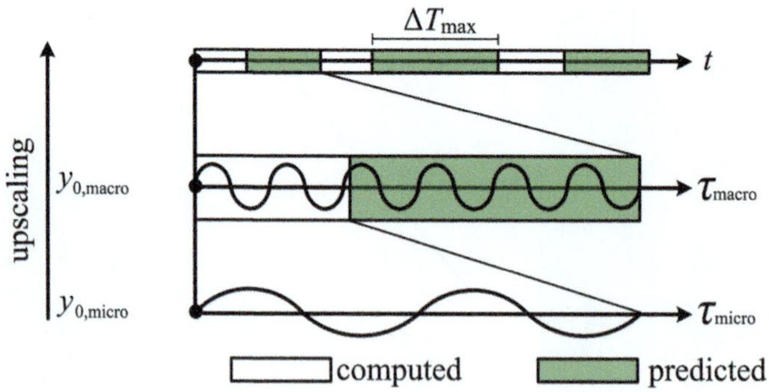

Fig. 8 Illustration of the different time scales

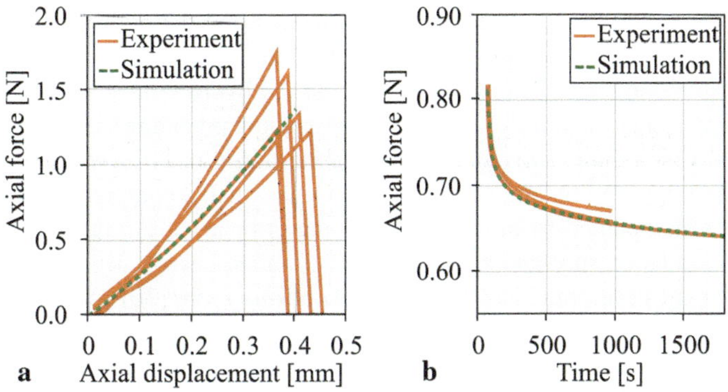

Fig. 9 Comparison of the fibre behaviour obtained by experiment and simulation: **a** uniaxial tension, **b** relaxation [24]

and behaviour of the system's internal variables in a simplified manner, making the analysis computationally efficient while maintaining essential features of the system.

4 Experimental and Numerical Findings

The following section provides a brief overview on the experimental findings, such as crack-bridging performance and cyclic behaviour of the presented materials on the micro- and mesoscale.

4.1 Microscopic Material Behaviour Under Cyclic Loading

4.1.1 Uniaxial Single Fibre Tension Tests

To determine the mechanical properties of the PVA fibres and validate the numerical model, preliminary uniaxial tension and relaxation tests were conducted on single fibres using a specialised test setup [29]. As outlined in Sect. 3.3 and [24], the PVA fibre model incorporates both viscoelastic and hyperelastic behaviour. The numerical results are compared to experimental data in Fig. 9, showing good agreement.

4.1.2 Single-Sided, Single Fibre Pull-Out (SPO)

In a next step, single-sided, single fibre pull-out tests were performed to investigate the pull-out behaviour and the bond/interface properties between the PVA fibres and the surrounding normal-strength matrix (M1), as well as to identify degradation mechanisms of the polymer fibres during pull-out [30]. A comprehensive description of the test setup and procedures is provided in [29].

The results indicate a strong bond between the hydrophilic PVA fibres and the cementitious matrix, leading to increased fibre rupture potential even at short embedding lengths of 2 mm, as evidenced by an abrupt drop in force without significant fibre debonding; see Fig. 10a (green curve). Occasionally, complete debonding was observed, accompanied by an increase in force prior to fibre rupture, suggesting a distinct degradation mechanism; cf. Fig. 10a (orange curve). This phenomenon is attributed to the development of the so-called *Locking Front* [30]—an interlocking of the microfibres due to superficial abrasion of fibre material, leading to an excavation and a subsequent accumulation of this material behind the affected area with a length of 100–400 μm (Fig. 10b). This bump impedes further pull-out of the fibre, thereby increasing pull-out resistance. Due to the reduction in cross-section, the fibre potentially ruptures in this area, leading to the second drop in force and the subsequent constant pull-out of the unruptured fibre section, as seen in Fig. 10a. The observations indicate an uneven interaction between the fibres and the surrounding matrix, resulting in local fibre damage as described above. For a detailed explanation of the mechanism, please refer to Ranjbarian et al. [30].

A microstructural numerical model was used to simulate the fibre-locking mechanism. The fibre-matrix bond is described as a cohesive layer (as introduced in Sect. 3.4) based on a traction–separation model, as presented in [23], and by a contact model including surface friction. The contact model follows a node-to-surface approach, based on [31], and it is in charge of capturing the fibre slip and fibre-matrix friction. The fibre-locking hypothesis allows to describe the entire increase in the pull-out force after complete debonding of the fibre in the simulation. The model comprises an isotropic fibre, concrete matrix and an interface (see Fig. 11a), which captures the debonding process.

This micromechanical model can capture the total pull-out behaviour of the fibre, the debonding process (between the start and Point I), the relaxation of the elastic deformation (between Points I–II), the increased force due to locking effect (between

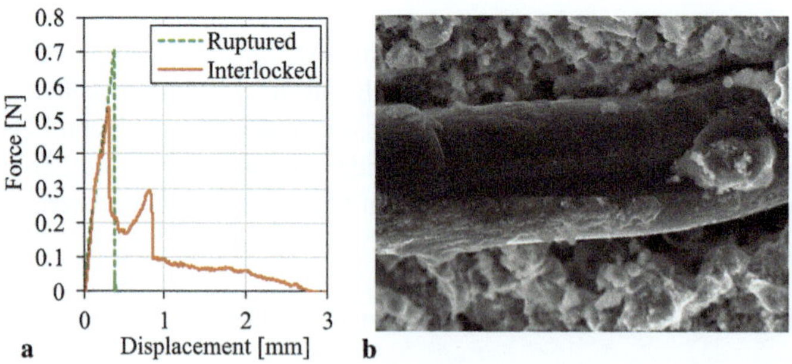

Fig. 10 Results from the single-sided single fibre pull-out tests: **a** force–displacement-characteristics showing the development of the *Locking Front*, **b** micro-excavation and bump (originally published in [30], edited by the authors)

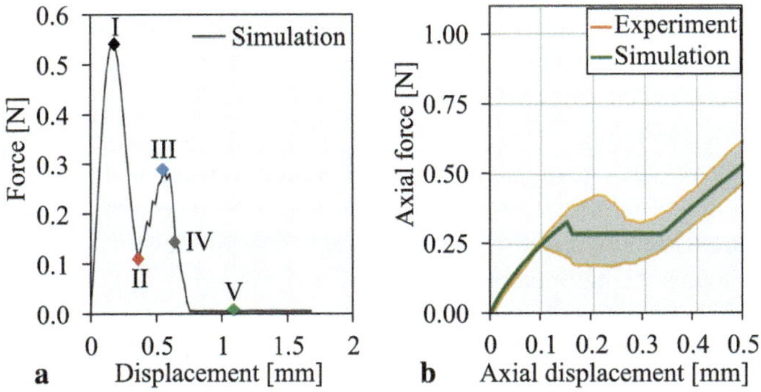

Fig. 11 Simulation of the *Locking Front* effect using: **a** the cohesive interface approach and **b** the fibre super-element model (originally published in [24, 30], edited by the authors)

Points II–III), the reduction of the aforementioned force (Point IV) and the full extraction (Point V). The same effect can be observed when using the super-element approach, as shown in Fig. 11b. Further explanation regarding the simulation is also presented in [30].

Due to the random orientation of microfibres in fibre-reinforced composites, it is crucial to investigate the impact of different fibre inclination angles on the material degradation. For this, single-sided single fibre pull-out tests were performed at fibre inclination angles of 0, 30, and 60°—relative to the load direction—under both monotonic and cyclic loading conditions [32].

The results of the quasi-static pull-out tests indicate a significant increase in force at fibre rupture (Fig. 12a) and displacement at fibre rupture (Fig. 12b) alongside a considerable increase in scattering of the test results with increasing inclination.

These observations can be traced back to stress concentrations in the fibre at the transition from embedded to free fibre length due to its deflection at this point.

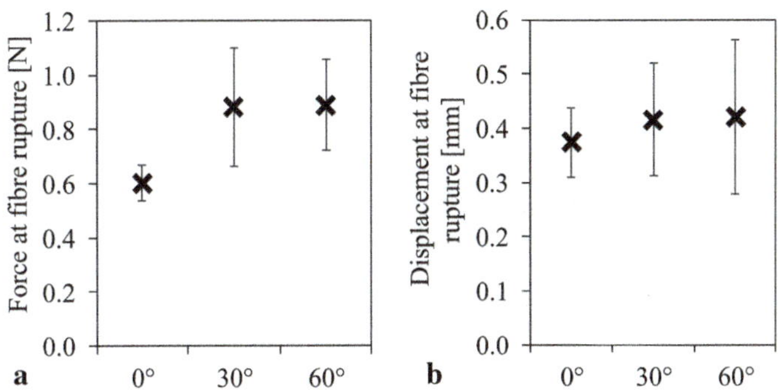

Fig. 12 Pull-out parameters under different inclination angles: **a** force at rupture, **b** displacement at rupture [32]

Higher deflections, i.e., fibre inclination, correlate with increased stress concentrations, leading to an increased force at rupture. This stress concentration can result in substantial superficial abrasion of the fibre, reducing its cross-section. Hence, failure potentially occurs in regions of high stress concentrations.

Under pure cyclic tension, only minor differences could be noticed when altering the fibre inclination. Similar to the monotonic tests, the maximum pull-out displacement increased with increasing inclination angle. Despite severe deterioration under monotonic loads, the inclined fibres were able to resist up to 230,000 loading cycles, indicating a slow degradation process under tension-swelling loads.

On the basis of those single fibre pull-out tests, quasi-static and cyclic tests were conducted using the high-strength matrix with PE fibres on the mesoscale; see Sect. 4.2.2.

The aforementioned test was simulated using the cohesive interface model (described in Sect. 3.4) and the super-element model (discussed in Sect. 3.5). The results of these simulations are presented in Fig. 13a, b corresponding to fibre lengths of $l_f = 20$ mm and $l_f = 30$ mm, respectively. The material model considered is described in Sect. 3.3. Refer to [25] for more details about the material properties and parameters of the simulation.

The super-element model is notably more computationally efficient than the cohesive interface model, achieving similar results in less than a second, compared to the 5 min required for a single-core run of the cohesive interface model. In the framework of this research, the elasto-plastic phase-field model for fracture was applied to simulate pull-out tests of straight steel fibres embedded in HPC, as introduced in [33]. This approach delivers highly accurate results, but it also comes with a significant computational cost. The phase-field method requires the inclusion of extra degrees of freedom, which substantially increases the computational complexity. Consequently, modelling fibre-reinforced composites structures within this framework will be extremely expensive.

Another approach presented in this chapter for analysing fibre pull-out involves representing fibres as Timoshenko beams linked to the concrete matrix using a

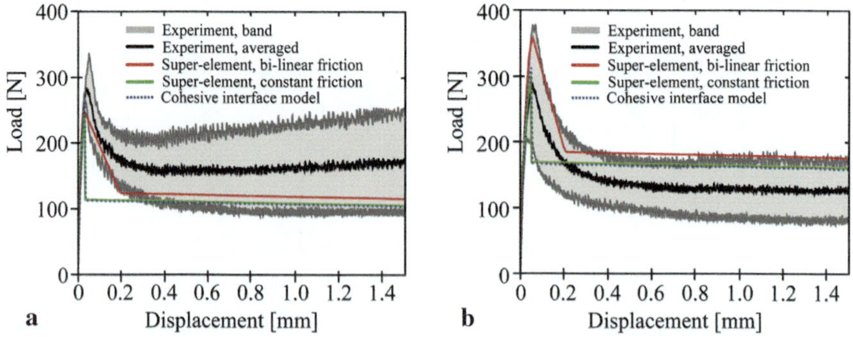

Fig. 13 Force–displacement characteristics comparing experimental and numerical simulation results for fibres: **a** $l_f = 20$ mm, **b** $l_f = 30$ mm

penalty-based frictional tying algorithm, as described in [34]. This formulation reduces computational costs compared to traditional fully 3D simulations. However, this simplification comes at the expense of certain details, such as the precise distribution of traction along the fibre-matrix interface.

4.1.3 Double-Sided, Single Fibre Pull-Out (DPO)

To address limitations of single-sided pull-out (SPO) tests, particularly regarding crack-bridging behaviour and material performance under alternating cyclic loading conditions, Ranjbarian and Mechtcherine [29] developed a novel double-sided, single fibre pull-out (DPO) test setup using a notched specimen with a centred single fibre being embedded 2 mm on the pull-out side and 6 mm on the other side; see also Fig. 14a.

The DPO test was simulated using the super-element model, as described in Sect. 3.5. The results of these simulations are presented in Fig. 14b, where it can be seen that the simulation and the experiment are in good agreement and, therefore, the model is able to capture the main behaviour of the pull-out at very low computational cost. Analogously to the simulations done in Sect. 4.1.2, the mechanical properties of the materials can be seen in more detail in [25].

The results from the DPO tests reveal that at a displacement of 0.02 mm the PVA fibres start to effectively bridge the cracks, leading to slip-hardening behaviour in the post-cracking stage (Fig. 15a). Comparative analysis of single-sided and double-sided pull-out test results indicates an increase in rupture force coupled to reduced displacement at rupture due to potential different boundary conditions in the distinct sample types caused by the distinct free lengths of the fibres. Furthermore, analysis of the force-crack opening-characteristics allows for a more accurate determination

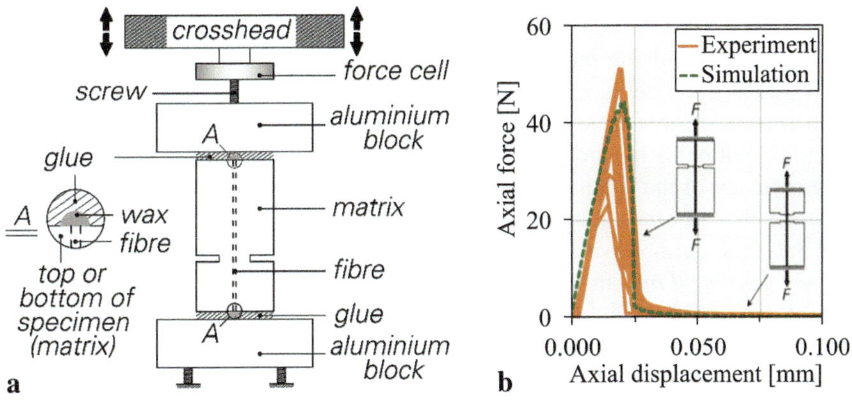

Fig. 14 Double-sided, single fibre pull-out test: **a** experimental set-up [29], **b** experimental and numerical results (originally published in [24], edited by the authors)

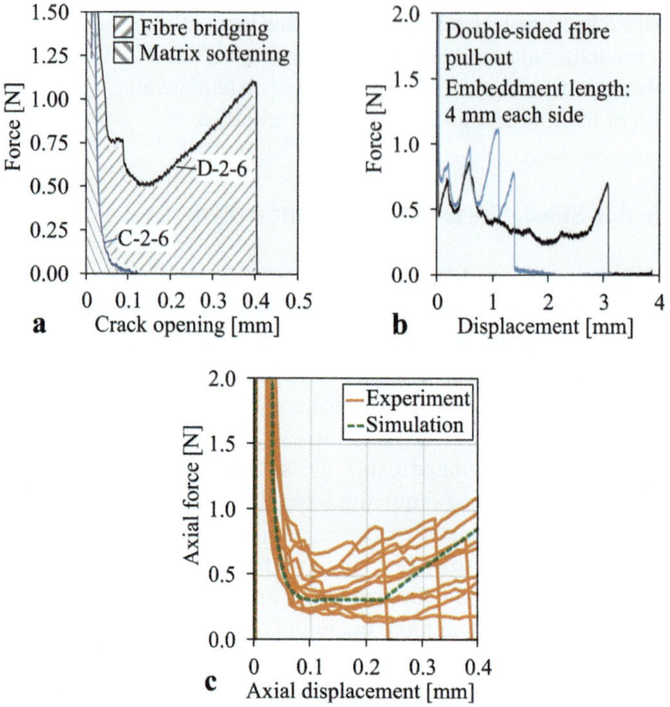

Fig. 15 Test results for double-sided single fibre pull-out tests: **a** comparison of force-crack opening-characteristics of samples with (D-2-6) and without (C-2-6) PVA fibre, **b** two-way fibre pull-out in sample D-2-6 and **c** numerical results (originally published in [29], edited by the authors)

of the frictional bond strength between the fibre and matrix. Additionally, the investigation of varying embedded lengths revealed a two-way pull-out of the micro fibres from both embedded sites, see Fig. 15b.

This crack-bridging mechanism was also observed in the simulation using the super-element model (see Fig. 15c), where slip-hardening behaviour similar to the experimental results could be identified. The model accurately captures the transition from the softening to hardening stage in the post-cracking phase, further validating the accuracy of the simulation approach. Refer to [25], for more details about the characteristics of the simulation.

Based on the observation of the degradation phenomena, e.g. the development of the *Locking Front* in the monotonic SPO tests and the implementation of the novel double-sided single fibre pull-out test setup, the mechanical response of the micro fibres subjected to alternating cyclic loading regimes was investigated [35, 36]. The cyclic tests were performed with constant upper force levels and varying lower reversal points, i.e. no (NC), moderate (MC) and high (HC) compression levels with an initial crack width of 100 μm according to [36].

The results from these cyclic tests after 2000 load cycles (Fig. 16a) revealed a reduced force- and displacement-at-rupture as the repeated closure of the cracks results in significant fibre buckling due to the lack of lateral support that could induce severe deterioration, e.g., defibrillation (Fig. 16b). Higher compression levels intensify this degradation increasing the potential fibre failure due to its complete crushing in-between the crack edges. However, the tension-swelling loads are capable of diminishing the *Locking Front* mechanism as the continuous pull-out and push-in of the fibre triggers the removal of the accumulated fibre material that causes this interlocking effect, allowing fibre pull-out without rupture.

Given that the absence of lateral support is a critical factor for fibre degradation, understanding the influence of initial crack width on the deterioration process under cyclic loading is essential. Hence, a study was carried out using the same experimental procedure (as described above) but with a smaller initial crack opening of 40 μm to assess the effect of the crack width on the degradation [37]. The results are displayed in Fig. 17.

As can be clearly seen from the diagrams, unlike large crack widths (> 100 μm), small crack widths do not significantly affect the mechanical characteristics of

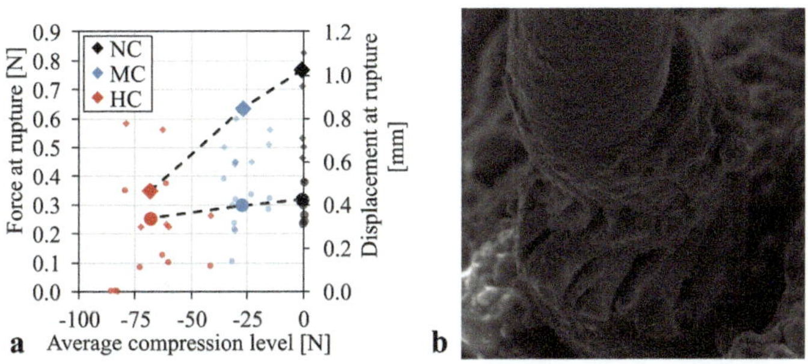

Fig. 16 Cyclic DPO tests: **a** force and displacement-at-rupture induced by different compression levels after 2000 loading cycles (force—diamond, displacement—circle), **b** severe buckling and defibrillation caused by repeated compression (**b** published in [36])

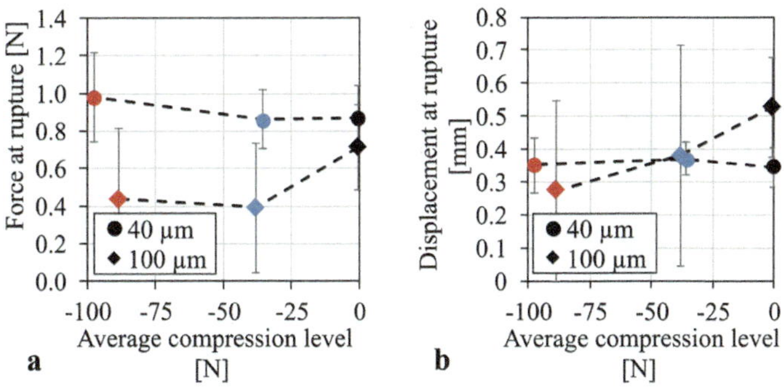

Fig. 17 Results for DPO tests with varied crack widths (40 μm—circle, 100 μm—diamond): **a** force-at-rupture versus average compression level, **b** displacement-at-rupture versus average compression level (see also [37], licensed under CC-BY 4.0)

the fibres, irrespective of the compression level as the increased lateral support in narrower cracks reduces fibre buckling, therefore, preventing fibre deterioration. Hence, it can be concluded that the degradation increases with increasing crack width, highlighting the importance of minimising initial crack widths in SHCC potentially through the use of alternative fibres, such as the suggested polyethylene fibres.

4.2 Mesoscopic Material Behaviour Under Cyclic Loading

4.2.1 Influence of Compression Level and Loading Frequency

Based on the findings of the microscale studies, cyclic tension-swelling and tension–compression tests were performed on the composite level with 150,000 load cycles on the high-strength matrix (M2) with PE fibres to assess the transferability of findings from micromechanical studies to mesoscopic scales with a varied material.

First, the effect of the loading frequency (1 and 20 Hz) at a constant load level was examined, followed by the evaluation of the impact of different compression levels on deterioration mechanisms, as previously discussed in Sect. 4.1.3. For experimental details, please refer to [38].

The variation of loading frequency in the tension-swelling tests causes only minor changes in the material behaviour, closely resembling quasi-static behaviour even at this number of load cycles (Fig. 18a). An increase in loading frequency results in a reduction in maximum displacement (Fig. 18b) due to an increased bond strength at higher loading rates [39] resulting in less deterioration of the post-cyclic material behaviour, characterised by lower maximum displacements and higher mean residual load-bearing capacity. This suggests that material deterioration is more pronounced at lower loading rates but remains limited at the tested number of cycles.

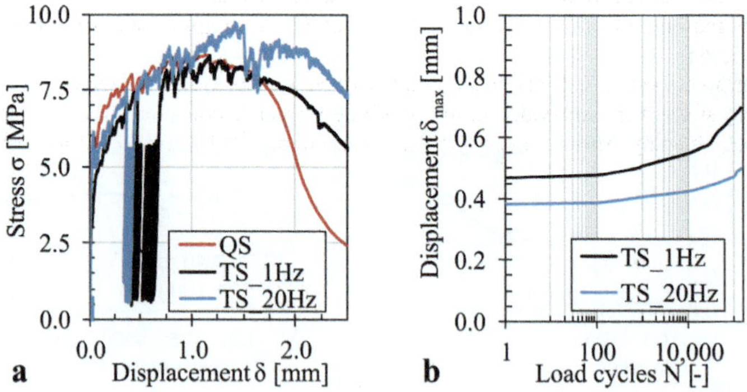

Fig. 18 Results of tension-swelling tests: **a** stress versus displacement, **b** maximum displacement during load cycles. QS, TS and Alt are the short terms for the quasi-static, tension-swelling and alternating cyclic test series

In contrast, a shift towards alternating cyclic loading conditions provokes a considerable decrease in material performance; see Fig. 19. The repeated forceful closing of the cracks causes a progressing loss of the structural integrity, i.e. increasing deformation and decreasing stress transfer that finally results in premature material failure after a few hundred or thousand cycles for compression levels of 50% and 25% of the mean compressive strength, respectively.

The microscopic analysis of the crack surfaces revealed significant differences in the degradation phenomena induced by the distinct loading regimes; cf. Fig. 20. While under repeated tensile loads, the fibres are typically pulled out from the matrix with minimal damages, a few ruptured fibres with a scale-like surface texture could be detected in areas that are possibly exposed to higher local stresses (Fig. 20a). The repeated pull-out and push-in of the fibre causes the development of multiple small bumps due to abrasion effects that reduce the cross-section and, therefore, result in fibre rupture.

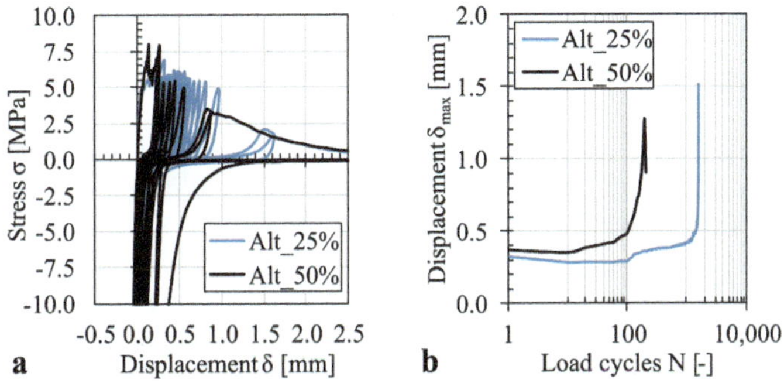

Fig. 19 Results of tension–compression tests: **a** stress versus displacement, **b** maximum displacement during load cycles

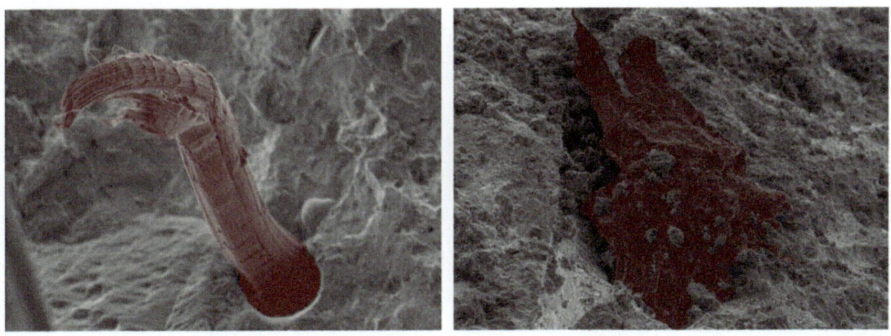

Fig. 20 Microscopic images of PE fibres after test: **a** tension-swelling loads, **b** alternating tension–compression loads (see also [38])

Under alternating tension–compression loads, the repeated closing of the cracks triggers fibre buckling followed by complete crushing of the fibres between the crack faces leading to substantial mechanical deterioration; see also Fig. 20b.

4.2.2 Influence of Fibre Orientation

As was already seen on the microscopic level (cf. Section 4.1.2), the fibre orientation considerably affects both the mechanical response and the deterioration of crack-bridging fibres. To further investigate this phenomenon, quasi-static and cyclic tension-swelling tests with 20,000 loading cycles were performed on notched, prismatic specimens with defined fibre inclination angles of 0, 30 and 60° with respect to the load direction [40]. The results from these tests are summarised in Table 3.

The quasi-static tests indicated that increasing inclination, i.e., deflection of the PE fibres at the crack edges, markedly reduces the tensile strength and deformation capacity of the composite due to an increased potential of severe abrasion and fibre rupture; see Fig. 21a. Notably, in contrast to microscale studies, the specimens exhibit the weakest performance at an inclination of 30°, rather than 60°.

Table 3 Overview on the test results of different fibre orientation under monotonic and cyclic loads

Angle (°)	Testing regime	Tensile strength f_t (MPa)	Max. deformation u_{max} (mm)
0	Quasi-static	9.51 (± 1.11)	0.61 (± 0.23)
	Cyclic	9.06 (± 0.83)	0.75 (± 0.34)
30	Quasi-static	6.79 (± 1.13)	0.33 (± 0.13)
	Cyclic	9.09 (± 1.25)	0.48 (± 0.18)
60	Quasi-static	6.84 (± 0.69)	0.42 (± 0.24)
	Cyclic	7.03 (± 1.00)	0.51 (± 0.12)

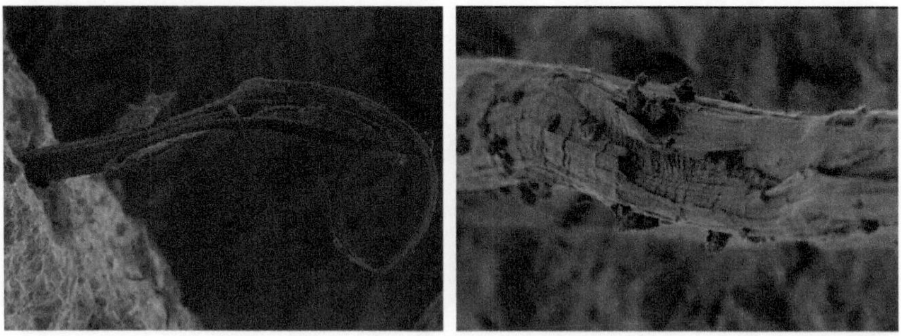

a b

Fig. 21 Microscopic images of PE fibres after testing: **a** quasi-static load, **b** cyclic tension-swelling load [40]

However, when shifting to a tension-swelling loading regime, the 30°-series demonstrates the highest tensile strength, showing a significant increase compared to the monotonic tests while simultaneously exhibiting the lowest deformation. The microscopic analysis revealed substantial superficial abrasion on the fibre surface (Fig. 21b) similar to the pattern observed in Fig. 10 that could probably cause an interlocking of the fibre in the tunnel due to the accumulation of the removed material leading to an increase in pull-out resistance accompanied by reduced deformation. In contrast, at 60°, stress concentrations at the deflection point are high enough to cause matrix spalling, therefore, leading to higher deformations compared to 30° fibre inclination.

These findings suggest that variations in material combinations and fibre orientation lead to distinct material responses under different loading regimes compared to the microscale tests; cf. Section 4.1.2.

4.2.3 Impact of Temperature

To evaluate the influence of varying environmental temperature on the mechanical performance under cyclic loading conditions, quasi-static and cyclic tension-swelling tests with 100,000 load cycles were conducted on the M2-PE-samples at constant temperatures of − 20, 20, 40 and 80 °C referring to potential temperature scenarios during a structure's service life [41]. Representative stress-displacement-curves are given in Fig. 22.

The temperature-dependent properties of PE fibres—specifically, a decrease in strength and an increase in strain capacity with rising temperature—resulted in a marked increase in ductility from − 20 to 80 °C, accompanied by reduction in tensile strength under monotonic loads, as shown in Fig. 22a. This temperature effect also led to greater deformations during the cyclic loading phase of the tension-swelling tests, with higher temperatures accelerating fibre degradation (Fig. 22b) due to increased crack widths caused by the increased fibre deformation. Consequently, the ultimate

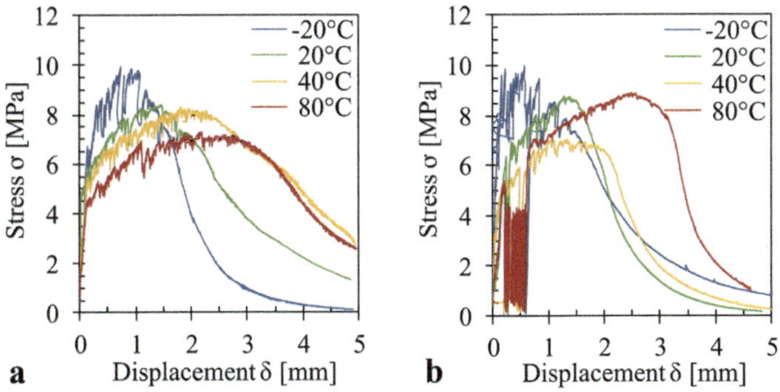

Fig. 22 Temperature-dependent stress-displacement-characteristics of the M2-PE samples under **a** quasi-static load, **b** cyclic tension-swelling load [41]

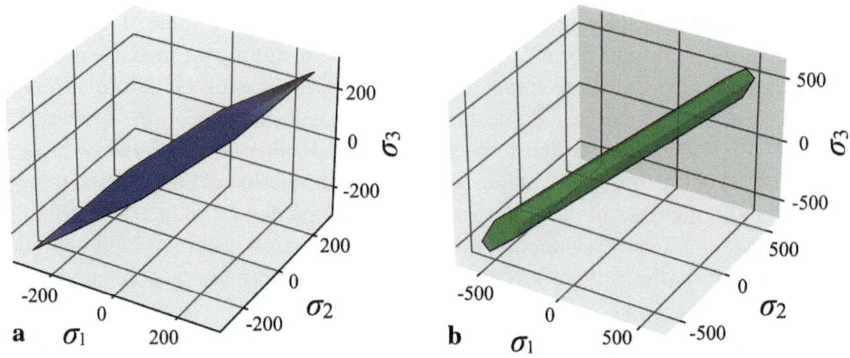

Fig. 23 Isometric perspective plot of the macroscopic yield surface: **a** microlayer model, **b** microplane model (stresses are given in MPa) (from [12], licensed under CC-BY 4.0)

tensile strength in the post-cyclic stage is significantly diminished, particularly at 40 °C. At 80 °C, although the diagram may not fully capture this aspect, the severe fibre deterioration due to elevated deformations substantially increases the likelihood of premature material failure. Overall, the results clearly demonstrate the adverse impact of elevated temperatures, driven by temperature-induced alterations in the mechanical properties of polymer fibres.

4.3 Macroscopic Concrete Damage Modelling

Using the Von-Mises yield function as described in [12], the macroscopic yield surface and its cross-sections along the hydrostatic axis are compared to the microplane model to highlight key differences (see Figs. 23 and 24). The yield surfaces of the two models exhibit significant differences in shape, including their deviatoric cross-sections. A key distinction is the microplane model's non-physical prediction, where the yield surface remains independent of the hydrostatic stress state. In contrast, the microlayer model, based on a consistent homogenisation framework, more realistically predicts the closed, elongated form observed in experiments. The findings based on the microlayer result have been validated based on FE^2 simulations in [12].

4.4 Concrete Specimen Subjected to High-cycle Fatigue

A concrete beam subjected to fatigue, see Fig. 25a, was simulated using the phase-field model for fracture extended to incorporate fatigue effects, as discussed in Sect. 3.2. This approach requires the simulation of multiple loading cycles,

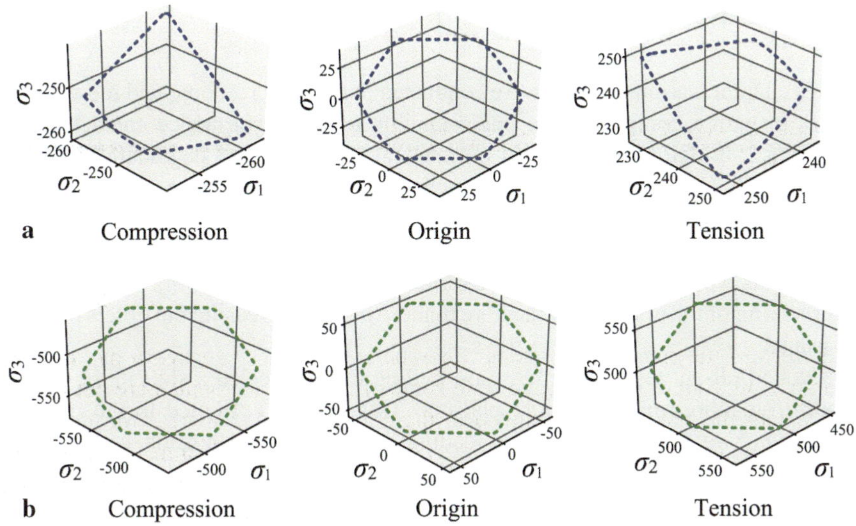

Fig. 24 Macroscopic yield surface in the deviatoric stress planes at different hydrostatic stress levels: **a** microlayer model, **b** microplane model [12] (deviatoric and hydrostatic stresses are given in MPa) (from [12], licensed under CC-BY 4.0)

which is computationally expensive, thereby making the implementation of the time homogenisation scheme (introduced in Sect. 3.6) essential as an acceleration technique. Refer to [28] for more details about the simulation and the analysis.

The time homogenisation algorithm is still under development. Nevertheless, preliminary results for high-cycle fatigue are promising. In this particular case, the homogenised solution achieved a computational cost reduction of 94.5% while maintaining a relative error of only 0.66% compared to the full simulation. Figure 25b illustrates the outcomes of both the full simulation and the homogenised simulation, demonstrating similar crack propagation patterns.

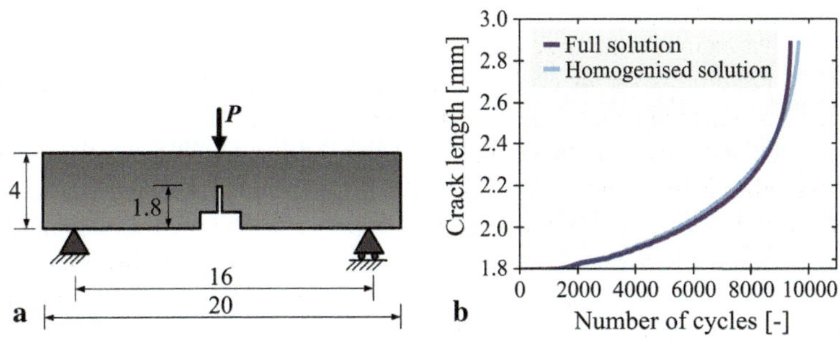

Fig. 25 **a** Dimensions of the bending specimen (all dimensions in mm), **b** crack propagation along cycles

5 Conclusions and Outlook

In this chapter, various studies were conducted at both microscopic and mesoscopic scales to gain a comprehensive understanding of the performance characteristics (e.g., crack-bridging capacity and mechanical response) and degradation mechanisms in SHCC under monotonic and cyclic loading conditions across a broad range of influential factors.

In quasi-static pull-out tests, fibre rupture was identified as the predominant failure mode. This was accompanied by the formation of a *Locking Front*, which induces interlocking of the fibre within the fibre tunnel, thereby increasing the fibre's pull-out resistance.

Cyclic pull-out tests under tension-swelling loads revealed a shift in the failure mode from fibre rupture to fibre pull-out. This shift is attributed to the removal of the *Locking Front* under cyclic loading with alternating tension–compression loads. Significant fibre degradation was observed, primarily caused by fibre buckling and crushing between the crack faces. The severity of this degradation is strongly influenced by both the initial crack width and the level of compression, with larger values resulting in more severe degradation.

Furthermore, increased fibre inclination relative to the load direction leads to stress concentrations within the fibre, affecting both its mechanical response and the extent of fibre deterioration. However, this only induces a slow progression of degradation.

The temperature-dependent properties of polymer fibres play a critical role in their crack-bridging performance and mechanical behaviour. Higher temperatures enhance the deformation capacity of the fibres but simultaneously reduce their tensile strength due to the increased degradation potential.

From an experimental perspective, future work will focus on hybrid fibre reinforcement to mitigate fibre deterioration, particularly under alternating cyclic loads. Additionally, polymer fibres with higher stiffness, such as Aramid or PBO fibres, will be investigated to improve mechanical performance under varying temperature conditions. Another key focus will be on repairing pre-damaged SHCC samples by filling cracks with mineral- and polymer-based suspensions, followed by subsequent cyclic loading.

The numerical models, which have been developed in this project, such as the microlayer model for damage in the concrete matrix and the super-element for fibre-matrix interaction, have demonstrated their effectiveness in simulating SHCC under various loading conditions. These models successfully capture the mechanics of fibre pull-out and the impact of the *Locking Front* on fibre performance, aligning well with experimental findings. The super-element model exhibits high accuracy while maintaining computational efficiency compared to traditional models. The incorporation of time homogenisation techniques, coupled with phase-field models for fracture, has shown potential in enhancing computational efficiency for simulating the effects of long-term cyclic loading on SHCC fatigue and will be further developed until the end of the project.

Future work will focus on applying an implicit time homogenisation framework to enhance the efficiency of simulating SHCC. Within the time homogenisation algorithm, the extrapolation equations should be refined to better capture the non-linear behaviour of the crack propagation phenomenon. Additionally, the adaptive step size determination in the numerical simulations should be optimised to enhance the accuracy of predictions while maintaining a low computational cost. This improved framework will be applied to the RVE of fibre reinforced concrete to accelerate predictions of material behaviour subjected to fatigue.

Moreover, the microlayer model, which incorporates damage induced by plasticity in concrete, will also be employed to simulate SHCC. This enables future research focusing on the cyclic behaviour of SHCC. Furthermore, this approach has the potential to be extended to multiphysical formulations. Coupling implicit time homogenisation with the microlayer model will provide a deeper understanding of the effects of the microstructural interactions.

References

1. Li, V.C.: From micromechanics to structural engineering—the design of cementitious composites for civil engineering applications. J. Struct. Mech. Earthq. Eng. **10**, 37–48 (1993)
2. Zhang, J., Li, V.C.: Monotonic and fatigue performance in bending of fiber-reinforced engineered cementitious composite in overlay system. Cem. Concr. Res. **32**, 415–423 (2002). https://doi.org/10.1016/S0008-8846(01)00695-0
3. Jun, P., Mechtcherine, V.: Behaviour of strain-hardening cement-based composites (SHCC) under monotonic and cyclic tensile loading. Cem. Concr. Compos. **32**, 810–818 (2010). https://doi.org/10.1016/j.cemconcomp.2010.08.004
4. Müller, S., Mechtcherine, V.: Fatigue behaviour of strain-hardening cement-based composites (SHCC). Cem. Concr. Res. **92**, 75–83 (2017). https://doi.org/10.1016/j.cemconres.2016.11.003
5. Curosu, I., Mechtcherine, V., Millon, O.: Effect of fiber properties and matrix composition on the tensile behavior of strain-hardening cement-based composites (SHCCs) subject to impact loading. Cem. Concr. Res. **82**, 23–35 (2016). https://doi.org/10.1016/j.cemconres.2015.12.008
6. Bažant, Z.P., Oh, B.H.: Microplane model for progressive fracture of concrete and rock. J. Eng. Mech. **111**, 559–582 (1985). https://doi.org/10.1061/(ASCE)0733-9399(1985)111:4(559)
7. Leukart, M.: Kombinierte anisotrope Schädigung und Plastizität bei kohäsiven Reibungsmaterialien. Doctoral Thesis, Institut für Baustatik, Universität Stuttgart, Stuttgart (2005)
8. Platen, J., Zreid, I., Kaliske, M.: A nonlocal microplane approach to model textile reinforced concrete at finite deformations. Int. J. Solids Struct. **267**, 1–12 (2023). https://doi.org/10.1016/j.ijsolstr.2023.112151
9. Zreid, I., Kaliske, M.: A gradient enhanced plasticity–damage microplane model for concrete. Comput. Mech. **62**, 1239–1257 (2018). https://doi.org/10.1007/s00466-018-1561-1
10. Storm, J., Kaliske, M.: A consistent multi-scale derivation of a micro-plane model within the framework of RVE homogenisation. Proc. Appl. Math. Mech. **18**, 1–2 (2018)
11. Blanco, P.J., Sánchez, P.J., de Souza Neto, E.A., Feijóo, R.A.: Variational foundations and generalized unified theory of RVE-based multiscale models. Arch. Comput. Methods Eng. **23**, 191–253 (2016). https://doi.org/10.1007/s11831-014-9137-5
12. Platen, J., Storm, J., Bosbach, S., Claßen, M., Kaliske, M.: The microlayer model: a novel analytical homogenisation scheme for materials with rigid particles and deformable matrix—applied to simulate concrete. Comput. Struct. **293**, 1–22 (2024). https://doi.org/10.1016/j.compstruc.2023.107258

13. Storm, J., Yin, B., Kaliske, M.: The concept of representative crack elements (RCE) for phase-field fracture: transient thermo-mechanics. Comput. Mech. **69**, 1–12 (2022). https://doi.org/10.1007/s00466-021-02135-w
14. Storm, J., Supriatna, D., Kaliske, M.: The concept of representative crack elements for phase-field fracture: anisotropic elasticity and thermo-elasticity. Int. J. Numer. Methods Eng. **121**, 779–805 (2020). https://doi.org/10.1002/nme.6244
15. Carrara, P., Ambati, M., Alessi, R., De Lorenzis, L.: A framework to model the fatigue behavior of brittle materials based on a variational phase-field approach. Comput. Methods Appl. Mech. Eng. **361**, 112731 (2020). https://doi.org/10.1016/j.cma.2019.112731
16. Charmetant, A., Vidal-Sallé, E., Boisse, P.: Hyperelastic modelling for mesoscopic analyses of composite reinforcements. Compos. Sci. Technol. **71**, 1623–1631 (2011). https://doi.org/10.1016/j.compscitech.2011.07.004
17. Storm, J., Götze, T., Hickmann, R., Cherif, C., Wießner, S., Kaliske, M.: Homogenisation by cylindrical RVEs with twisted-periodic boundary conditions for hybrid-yarn reinforced elastomers. Int. J. Solids Struct. **139–140**, 283–301 (2018). https://doi.org/10.1016/j.ijsolstr.2018.02.006
18. Barenblatt, G.I.: The mathematical theory of equilibrium cracks in brittle fracture. Adv. Appl. Mech. **7**, 55–129 (1962). https://doi.org/10.1016/S0065-2156(08)70121-2
19. Fleischhauer, R., Behnke, R., Kaliske, M.: A thermomechanical interface element formulation for finite deformations. Comput. Mech. **52**, 1039–1058 (2013). https://doi.org/10.1007/s00466-013-0862-7
20. Geißler, G., Kaliske, M.: Time-dependent cohesive zone modelling for discrete fracture simulation. Eng. Fract. Mech. **77**, 153–169 (2010). https://doi.org/10.1016/j.engfracmech.2009.09.013
21. Hillerborg, A., Modéer, M., Petersson, P.-E.: Analysis of crack formation and crack growth in concrete by means of fracture mechanics and finite elements. Cem. Concr. Res. **6**, 773–781 (1976). https://doi.org/10.1016/0008-8846(76)90007-7
22. Fleischhauer, R., Qinami, A., Hickmann, R., Diestel, O., Götze, T., Cherif, C., Heinrich, G., Kaliske, M.: A thermomechanical interface description and its application to yarn pullout tests. Int. J. Solids Struct. **69–70**, 531–543 (2015). https://doi.org/10.1016/j.ijsolstr.2015.04.030
23. Miehe, C.: Discontinuous and continuous damage evolution in Ogden-type large-strain elastic materials. Eur. J. Mech.-A/Solids **14**, 697–720 (1995)
24. Storm, J., Ranjbarian, M., Mechtcherine, V., Scheffler, C., Kaliske, M.: Modelling of fibre-reinforced composites via fibre super-elements. Theoret. Appl. Fract. Mech. **103**, 102294 (2019). https://doi.org/10.1016/j.tafmec.2019.102294
25. Storm, J., Pise, M., Brands, D., Schröder, J., Kaliske, M.: A comparative study of micro-mechanical models for fiber pullout behavior of reinforced high performance concrete. Eng. Fract. Mech. **243**, 107506 (2021). https://doi.org/10.1016/j.engfracmech.2020.107506
26. Behnke, R., Wollny, I., Hartung, F., Kaliske, M.: Thermo-mechanical finite element prediction of the structural long-term response of asphalt pavements subjected to periodic traffic load: tire-pavement interaction and rutting. Comput. Struct. **218**, 9–31 (2019). https://doi.org/10.1016/j.compstruc.2019.04.003
27. Behnke, R., Kaliske, M.: Square block foundation resting on an unbounded soil layer: long-term prediction of vertical displacement using a time homogenization technique for dynamic loading. Soil Dyn. Earthq. Eng. **115**, 448–471 (2018). https://doi.org/10.1016/j.soildyn.2018.07.045
28. Pina-Torres, R., Zhao, D., Storm, J., Kaliske, M.: Time homogenization: an acceleration scheme for phase-field modeling of fatigue. Comput. Struct. **315**, 107824 (2025). https://doi.org/10.1016/j.compstruc.2025.107824
29. Ranjbarian, M., Mechtcherine, V.: A novel test setup for the characterization of bridging behaviour of single microfibres embedded in a mineral-based matrix. Cem. Concr. Compos. **92**, 92–101 (2018). https://doi.org/10.1016/j.cemconcomp.2018.05.017
30. Ranjbarian, M., Mechtcherine, V., Zhang, Z., Curosu, I., Storm, J., Kaliske, M.: Locking front model for pull-out behaviour of PVA microfibre embedded in cementitious matrix. Cem. Concr. Compos. **103**, 318–330 (2019). https://doi.org/10.1016/j.cemconcomp.2019.04.031

31. Zienkiewicz, O.C., Tylor, R.L.: The Finite Element Method: Solid Mechanics, vol. 2. Butterworth-Heinemann, Oxford, England (2000)
32. Ranjbarian, M., Mechtcherine, V., Storm, J., Kaliske, M.: Influence of fibre orientation on pull-out behaviour of PVA microfibre embedded in cement-based matrix under monotonic and cyclic loading. In: Zingoni, A. (ed.) Advances in Engineering Materials, Structures and Systems: Innovations, Mechanics and Applications, pp. 1433–1438. CRC Press, London (2019)
33. Gebuhr, G., Pise, M., Sarhil, M., Anders, S., Brands, D., Schröder, J.: Analysis and evaluation of the pull-out behavior of hooked steel fibers embedded in high and ultra-high performance concrete for calibration of numerical models. Struct. Concr. **20**, 1254–1264 (2019). https://doi.org/10.1002/suco.201900034
34. Gudžulić, V., Daadouch, K., Meschke, G.: Modeling of damage processes in concrete under monotonic and cyclic loading. In: Proceedings of the 11th International Conference on Fracture Mechanics of Concrete and Concrete Structures. Bangalore, India (2023)
35. Ranjbarian, M., Mechtcherine, V.: Cyclic damage to PVA microfibre embedded in cementitious matrix in alternating tension-compression regime. In: Fibre Reinforced Concrete: From Design to Structural Applications, pp. 321–329. Desenzano, Lake Garda, Italy (2018)
36. Ranjbarian, M., Mechtcherine, V.: Influence of loading parameters in cyclic tension-compression regime on crack-bridging behaviour of PVA microfibres embedded in cement-based matrix. Constr. Build. Mater. **228**, 116760 (2019). https://doi.org/10.1016/j.conbuildmat.2019.116760
37. Ranjbarian, M., Ma, X., Mechtcherine, V.: Influence of crack width in alternating tension-compression regimes on crack-bridging behaviour and degradation of PVA microfibres embedded in cement-based matrix. Materials **13**, 4189 (2020). https://doi.org/10.3390/ma13184189
38 Junger, D., Mechtcherine, V.: Cyclic loading effects on high-strength strain-hardening cement-based composites with PE fibres (PE-SHCC): crack development and material degradation analysis. Mater. Struct. **58**, Article number 330 (2025). https://doi.org/10.1617/s11527-025-02857-w
39. Mechtcherine, V., Silva, F., Butler, M., Zhu, D., Mobasher, B., Gao, S.-L., Mäder, E.: Behaviour of strain-hardening cement-based composites under high strain rates. J. Adv. Concr. Technol. **9**, 51–62 (2011). https://doi.org/10.3151/jact.9.51
40. Junger, D., Mechtcherine, V.: Influence of fiber orientation on the cyclic behavior of strain-hardening cement-based composites (SHCC). In: Proceedings of the 11th International Conference on Fracture Mechanics of Concrete and Concrete Structures. Bangalore, India (2023)
41. Junger, D., Mechtcherine, V.: Impact of temperature on the cyclic behavior of strain-hardening cement-based composites (SHCC). In: Mechtcherine, V., Signorini, C., Junger, D. (eds.) Transforming Construction: Advances in Fiber Reinforced Concrete, pp. 423–430. Springer Nature Switzerland, Cham (2024)
42. Junger, D., Mechtcherine, V.: Influence of loading frequency and force level on the cyclic performance of strain-hardening cement-based composites (SHCC). In: Kunieda, M., Kanakubo, T., Kanda, T., Kobayashi, K. (eds.) Strain Hardening Cementitious Composites, pp. 178–187. Springer International Publishing, Cham (2023)

Numerical Benchmark: Simulation of the Flexural Tensile Test for Fibre Reinforced Concrete

Johannes Storm, Vladislav Gudžulić, Lena Gietz, Gregor Gebuhr, Jan-Paul Lanwer, Michael Kaliske, Günther Meschke, Ursula Kowalsky, Dieter Dinkler, Steffen Anders, and Martin Empelmann

Abstract The development of numerical models to predict the deterioration behaviour of fibre-reinforced high-performance concrete presents significant challenges. Existing approaches balance predictive accuracy and computational efficiency in varying ways. This chapter compares different models developed within the DFG Priority Programme SPP 2020. The performance of these models is evaluated against experimental results of flexural tensile tests on steel-fibre-reinforced concrete and under low-cycle fatigue loading.

Keywords Steel-fibre-reinforced concrete · Hybrid-mixed FEM · Numerical benchmark · Cohesive zone model · Discrete fibre model

1 Motivation and Aims

Fibre-reinforced high-performance concrete exhibits pseudo-ductile material behaviour beyond fracture of the concrete matrix. Characterizing and modelling this composite material is particularly challenging due to the interaction of various deformation, degradation, and failure mechanisms occurring within the matrix, the

J. Storm (✉) · M. Kaliske
Institute for Structural Analysis, Technische Universität Dresden, Dresden, Germany
e-mail: johannes.storm@tu-dresden.de

V. Gudžulić · G. Meschke
Institute for Structural Mechanics, Ruhr University Bochum, Bochum, Germany

L. Gietz · U. Kowalsky · D. Dinkler
Institute of Structural Analysis, Technische Universität Braunschweig, Braunschweig, Germany

G. Gebuhr · S. Anders
Institute for Structural Engineering, Bergische Universität Wuppertal, Wuppertal, Germany

J.-P. Lanwer · M. Empelmann
Institute of Building Materials, Concrete Construction and Fire Safety, Technische Universität Braunschweig, Braunschweig, Germany

N. Oneschkow and L. Lohaus (eds.), *Cyclic Deterioration of High-Performance Concrete*, https://doi.org/10.1007/978-3-032-19394-0_17

reinforcements, and their interfaces, especially under fatigue loading. During cyclic loading, the material structure undergoes significant changes, such as crack nucleation and propagation, delamination at the fibre-concrete interface, damage and failure of both fibres and concrete, as well as fibre pull-out.

Experimental studies reveal that the degradation mechanisms of fibre-reinforced concrete differ significantly comparing low-cycle and high-cycle fatigue. The present study focuses on numerical models for fibre-reinforced high-performance concrete under low-cycle fatigue, specifically in the context of flexural tensile testing. These models are based on experimental investigations targeting:

- the concrete matrix,
- the fibres,
- the fibre-concrete interface via fibre pull-out, and
- the fibre distribution and orientation.

The characterization of the composite's components and their interactions is carried out using various experimental setups, including the Dumbbell specimen, cylindrical compression tests, fibre tension tests, fibre pull-out tests, and computer tomography imaging.

The collected data enables the development of numerical models for the composite material. These models are applied to simulate flexural tension tests on notched beams, with different modelling strategies compared against experimental findings. Additionally, the influence of fibre orientation is numerically investigated by comparing the post-fracture behaviour observed for the measured distribution with two idealized fibre orientations. To account for the non-linear and dissipative effects observed during testing, the monotonic loading path is interrupted by multiple unloading cycles.

The benchmark experiments provided for this study were conducted with three different fibre contents, and several individual tests were performed for each variant. However, the dataset does not allow for the derivation of a mean curve or scatter band that could serve as a reliable reference for comparing the numerical models. Instead, the original experimental measurements are directly presented alongside the numerical results.

2 Materials, Test Setup and Experimental Methods

2.1 Flexural Tensile Test

The experimental investigation of the 3-point flexural behaviour of the fibre-reinforced concrete was carried out in accordance with EN 14651 at the Chair of Construction Materials at the University of Wuppertal [1, 2]. The dimensions of the experimental body and the experimental setup can be taken from Fig. 1a. The notch is 4.75–5.0 mm wide and has a depth of 25.0 mm. The support rollers each have

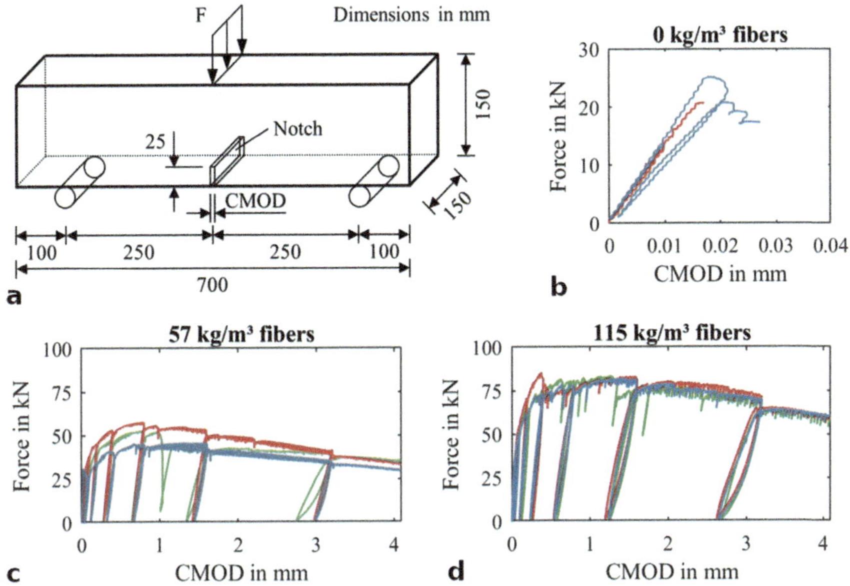

Fig. 1 **a** Experimental setup for 3-point bending tensile test, dimensions in mm, and **b–d** force-CMOD-characteristics for 3-point bending tensile test for UHPC with fibre content of 0, 57, and 115 kg/m^3

a diameter of 30 mm, and the load is applied via a steel roller with a diameter of 50 mm.

The test control is based on direct measurement of crack mouth opening displacement (CMOD) using two LVDTs mounted laterally at the level of the lower edge of the beam parallel to the beam axis. The loading protocol is summarized in Table 1. The experimental results consist of the force-CMOD characteristics, see Fig. 1.

2.2 Materials

An ultra-high-performance concrete (UHPC) was used for the tests. The composition corresponds to the UHPC reference concrete of SPP 2020 according to Table 1 in the chapter by M. Basaldella et al., this volume. The mechanical material properties for the concrete matrix were determined in Lanwer et al. [3]. Tensile strength on dumbbell specimens and compressive strength and modulus of elasticity on compression cylinders (d/l = 150 mm/300 mm) were determined. Until the test, the beams were stored in water. The mechanical properties of the material are summarized in Table 3.

The steel fibres used, type Weidacon FM 0.19/13, are straight with a round cross-section (d/l = 0.19 mm/13 mm) and without end anchoring. The mechanical properties of the fibres were determined in Lanwer et al. [3] using tensile tests on fibres

Table 1 Load protocol of the 3-point bending beam tests

Control type	Target	Speed	Explanation
Displacement (loading)	Force; 1 kN	1 mm/min	Start of experiment
CMOD (loading)	CMOD; 0.01, 0.02, 0.03, 0.04, 0.05 mm	0.005 mm/min	5 fine loading loops for accurate detection of stiffness changes during cracking
Force (unloading)	Force 1 kN	7.5 kN/min	
CMOD (loading)	CMOD; 0.1, 0.2, 0.4, 0.8, 1.6, 3.2 mm	0.2 mm/min	5 loading loops to capture to investigate the stiffness changes for large crack openings
Force (unloading)	Force 1 kN	7.5 kN/min	
CMOD (loading)	CMOD; 4 mm	0.3 mm/min	End of experiment

Table 2 Material parameters of the steel fibres

Parameter		Mean
Tensile strength	MPa	3575.8
Young's modulus	GPa	168.5

Table 3 Material parameters of the UHPC

Parameter		Mean
Tensile strength	MPa	8.2
Compressive strength	MPa	158
Young's modulus	GPa	43.3
Poisson's ratio		0.221

bonded into cannulas. The test results are presented in Oettel et al. [4]. The results are summarized in Table 2. According to previous knowledge, the fibres undergo mainly elastic deformations in the fibre-reinforced concrete under consideration.

2.3 *Fibre Pull-Out Behaviour*

The interface between concrete matrix and fibre is characterized experimentally by means of fibre pull-out tests carried out at the University of Wuppertal. The test setup and procedure were described in the SPP 2020 Benchmark—Fibre pull-out test [5]. The setup is shown in Fig. 2. The pull-out test was performed with embedment depths between 0.5 and 7 mm at a pull-out rate of 0.3 mm/min. In addition to straight pull-out (90°), tests were also performed with a fibre pull-out angle of 45°.

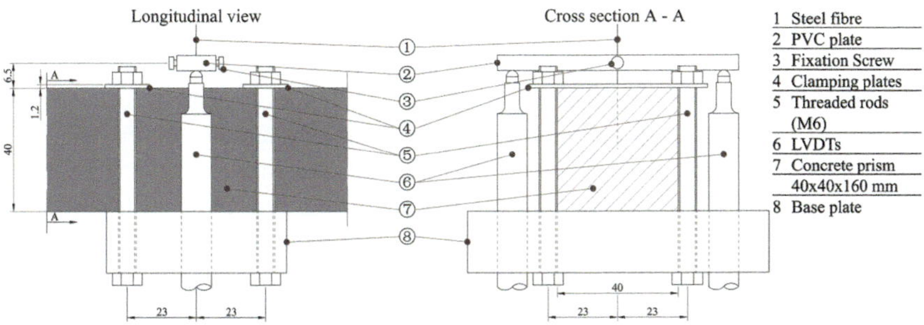

Fig. 2 Test setup single fibre pull-out tests, dimensions in mm

As shown in Fig. 3, the pull-out characteristics of the individual fibres tend to show strong scatter. In addition to the single fibre tests, experiments with several parallel fibres were presented in Lanwer et al. [3] and Oettel et al. [4]. There, considerably lower scatter in pull-out behaviour was observed, see Fig. 4.

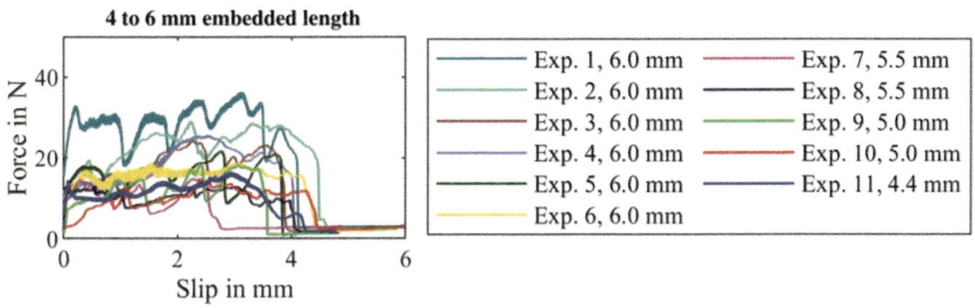

Fig. 3 Exemplary pull-out results with embedment depths between 4 and 6 mm and the pull-out angle 90° in N and mm

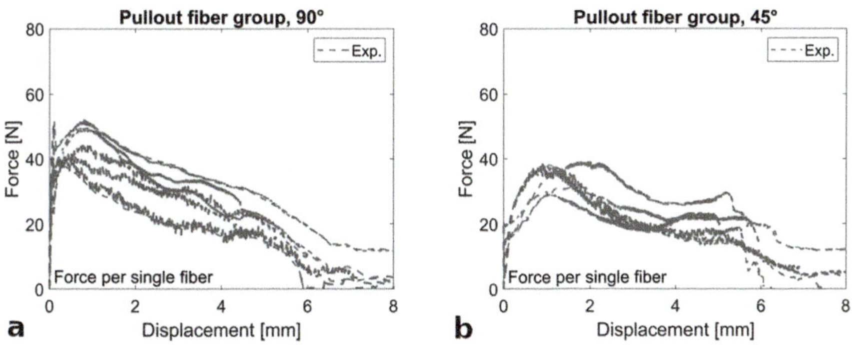

Fig. 4 Fibre group pull-out characteristics for **a** 90°, and **b** 45° inclination of the fibres to the free surface

2.4 Fibre Distribution

Specimens with fibre contents of 0, 57, and 115 kg/m^3 were prepared, corresponding to approximate volume fractions of 0.0, 0.75, and 1.5% steel fibres. CT scans, performed at the Institute of Building Materials, Concrete Structures, and Fire Protection at TU Braunschweig, were used to analyse fibre and pore distribution. The samples were extracted and scanned with a resolution of 0.07 mm/voxel using the Phoenix VTOMEXS 240 (General Electric). The sampling positions are shown in Fig. 5, chosen to capture edge effects in fibre distribution caused by production processes [6] (Fig. 6).

Two methods were employed to analyse the orientation of micro steel fibres: statistical fibre orientation using an integration mesh and surface mapping with coordinate extraction. While surface mapping enables discrete fibre simulation in FEM models, the integration mesh approach provides averaged fibre orientations relative to a specified axis.

2.4.1 Statistical Fibre Orientation via Integration Mesh

Fibre orientation was analysed in VG STUDIO MAX by overlaying the sample volume with a cubic integration mesh. The software recommends a cell edge length matching the fibre diameter (approximately 0.2 mm). However, such fine meshing across the entire sample would yield 16,875,000 cells, exceeding computational capacity. Instead, a coarser mesh with properties outlined in Table 4 was used.

Preliminary analyses using finer cell sizes (0.1, 0.2, and 0.5 mm) showed negligible deviations in fibre orientation (eigenvectors: $x = 0.39^{\pm 0.001}$, $y = 0.65^{\pm 0.0014}$, $z = 0.56^{\pm 0.0026}$). For full-sample analysis, the mesh was centrally positioned in the bore core (Fig. 7), with flexibility for alternate placements.

To examine edge effects in fibre orientation, additional integration meshes were analysed at three bore core locations: near the filling side, the midsection, and the bottom formwork side. Analyses were conducted for specimens with 57 kg/m^3 fibre content.

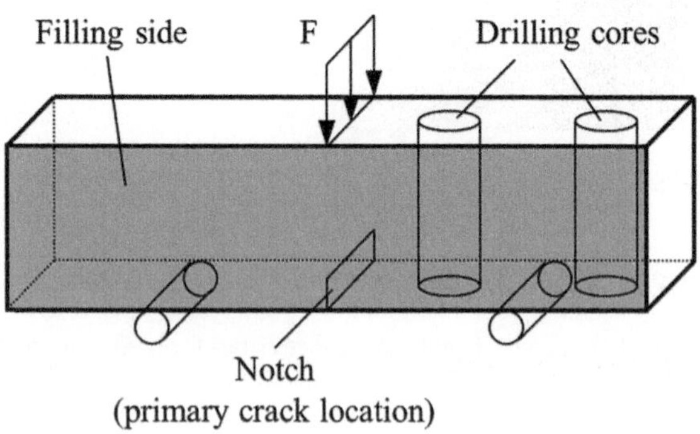

Fig. 5 Position of CT samples in 3-point bending specimen

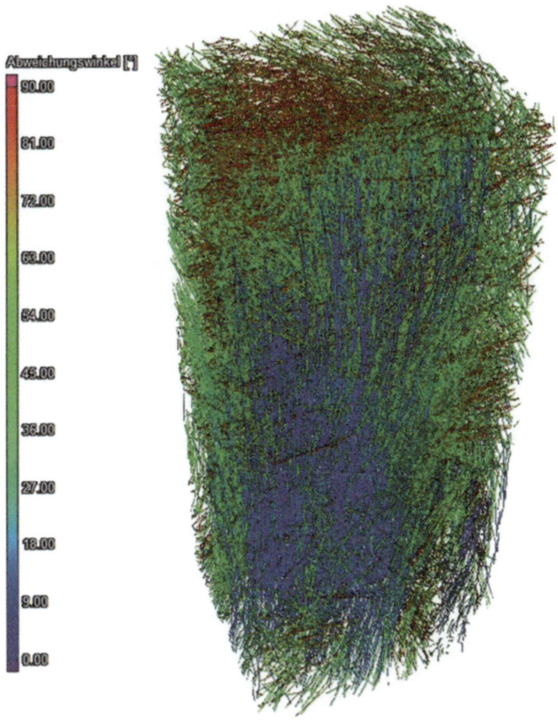

Fig. 6 Visualization of the fibre distribution in the CT sample

Table 4 Properties of the integration mesh

Number of cells		60	300	60
Cell size	mm	0.5	0.5	0.5
Origin	mm	– 15	– 75	– 15

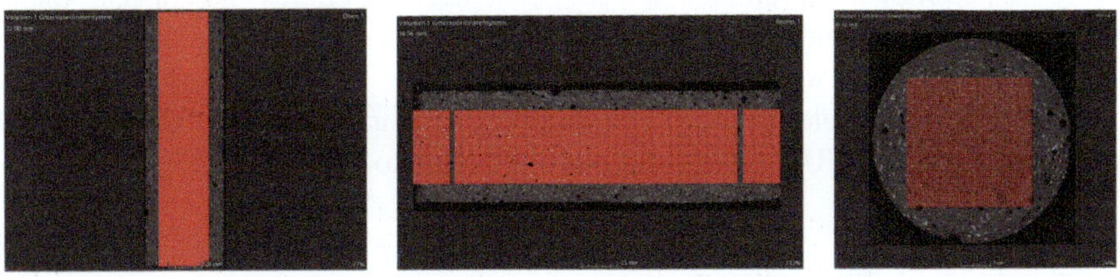

Fig. 7 Location of the integration mesh in the bore core

2.4.2 Fibre Surface Mapping

Fibre surfaces were extracted through segmentation of CT images. The process involved isolating the bright steel fibre regions in grey-scale images and applying

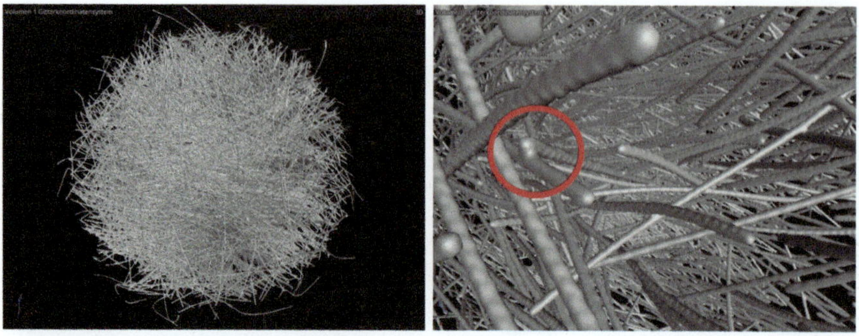

Fig. 8 Rendering of the segmented micro steel fibres in the bore core with a fibre content of 57 kg/m^3

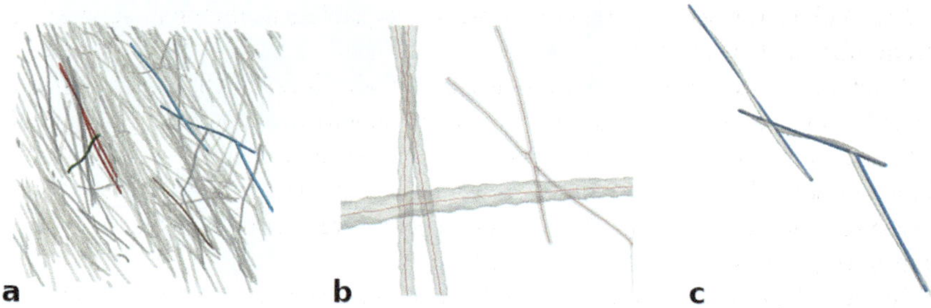

Fig. 9 Identification of individual fibres **a** via surface connectivity, **b** centre line determination, and **c** axis projection for defragmentation

isosurface extraction. Figure 8 illustrates the segmented fibres rendered as a triangular mesh for the bore core with a fibre content of 57 kg/m^3.

The fibre positions were described by endpoint coordinates, neglecting curvature. Fibre connectivity was evaluated to distinguish individual fibres and clusters. Using the Vascular Modelling Toolkit (VMTK), fibre centre lines were determined and fragmented segments were merged (Fig. 9).

The algorithm, implemented in Python with the Visualization Toolkit (VTK) and parallelized using GNU Parallel, required 3 h and up to 230 GB RAM per sample.

2.4.3 Synthetic Fibre Distribution

For numerical simulations of fibre orientation effects in 3-point bending tests, synthetic fibre distributions were generated. Two configurations were considered: a homogeneous isotropic distribution and an idealized alignment along the beam axis (Fig. 10).

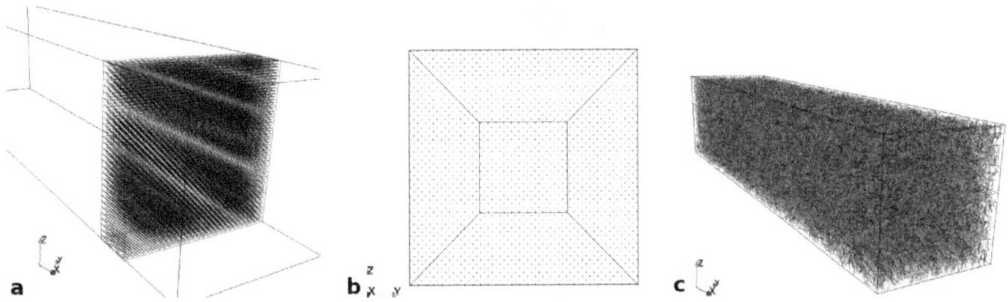

Fig. 10 Synthetic fibre distributions with **a**, **b** preferred orientation along the beam axis and **c** isotropic orientation distribution with a fibre content of 57 kg/m^3

3 Numerical Models

The height and width of the test specimen can be found in Fig. 11. The specimen depth is 150 mm. The introduced notch has a length of 25 mm and a width of 4.75–5 mm.

The load application to the specimen is modelled by three univalent supports extending along the entire depth of the specimen. Two supports are positioned symmetrically at the bottom side, and one support is centrally located at the top side. The cylindrical supports of the specimen are represented by prisms. The apex of the prisms aligns with the cylinder axes to allow rotation around the support points. This results in a height of 15 mm for the prisms. By bonding the prisms to the specimen over an area approximated by Hertzian pressure, stress concentrations are avoided. The base width of the prisms is set to 20 mm for the central support and 10 mm for the two bottom supports. The central support consists of two parts to avoid obstructing the crack opening below the support. The prisms are assumed to be rigid or elastic with a stiffness of 210 GPa.

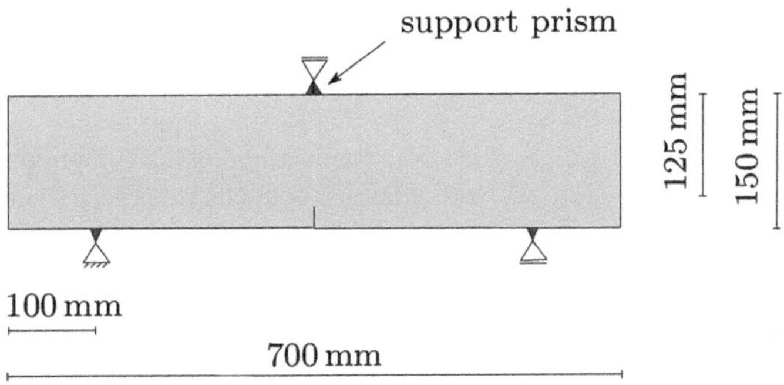

Fig. 11 Sketch of the boundary value problem

3.1 Discrete Approach for Simulating Fracture in UHPFRC

This section presents the numerical investigations of notched ultra-high-performance fibre-reinforced (UHPFRC) beams utilizing the discrete fracture and discrete fibre models developed in [7]. The finite element model used to perform virtual experiments can discretely represent cracks and fibres, utilizing zero-thickness cohesive-frictional interface elements and elastoplastic Timoshenko beam elements, respectively. Beam elements were chosen for fibre modelling, since, in addition to axial forces, they can also transfer shear forces, which play a significant role in describing various local damage mechanisms, such as spalling of the cementitious matrix, plastic hinge formation, and frictional fibre pull-out. The complete description of the model can be found in [7], and its brief outline is provided in the chapter by V. Gudžulić et al., this volume.

3.1.1 Material Properties and Model Calibration

Tensile strength f_t, compressive strength f_c, Young's modulus E, and Poisson's ratio v were experimentally determined and considered fixed parameters in simulations (see Table 3). The mean experimentally determined tensile strength was used as material strength in numerical analyses. The fracture energy was not experimentally determined. Ideally, the fracture energy should be determined using the standard RILEM procedure [8]. As a first approximation to estimate the fracture energy, the formula proposed in [9] was used $G_F = G_{F0}(1 - 0.77\, f_{c0}/f_{cm})$, with $G_{F0} = 0.18$ N/mm, $f_{c0} = 10$ MPa, and f_{cm} is the concrete mean compressive strength, which was experimentally determined as 158 MPa. This equation yields the value of 0.17 N/mm.

However, looking closely at the data used to derive this "fitting" formula in [9], one can observe no data points for concretes with strength higher than 130 MPa. Therefore, the formula for fracture energy for concrete of this strength is an extrapolation, and these recommendations should not be taken for granted. Preliminary numerical tests have revealed that the value given by this formula leads to an overestimation of the peak load (rupture strength) in both cases. Therefore, a compromise value of 0.15 N/mm, which provides satisfactory results, was selected. This parameter choice is no more uncertain than using an empirical formula based on data extrapolation.

In three-point bending analyses presented in this section, the Mode II fracture parameters β, κ, and μ are specified for completeness but are less significant due to the dominance of Mode I cracking, ensured by the presence of the notch. These parameters are chosen to be within the typical range for concrete materials (see the discussion on the choice of Mode II fracture parameters in [7]). For the progressive crack closure submodel parameters, the reference closure stress ($t_{\pi,\mathrm{ref}}$) was selected to have the same value as the tensile strength (8.2 MPa) as proposed by [10], and the reference crack opening ($[\![u]\!]_{\pi,\mathrm{ref}}$) is 1 μm. Table 5 lists the material properties used in numerical simulations. The material properties of fibres are adopted according to

Table 5 Adopted values of material properties for UHPC

Material property		Value
Young's modulus (E)	MPa	43,224.4[a]
Penalty parameter (K_c)	kN/mm^3	43,224.4
Tensile strength (f_t)	MPa	8.2[a]
Poisson's ratio (ν)		0.221[a]
Shear strength factor ($\beta = f_s/f_t$)		3.0
Fracture energy (G_F)	N/mm	0.15
Fracture energy (Mode II) factor ($\kappa = G_{F,II}/G_F$)		5.0
Friction coefficient (μ)		0.7
Reference closure stress ($t_{\pi,ref}$)	MPa	– 8.2
Reference crack opening ($[\![u]\!]_{\pi,ref}$)	μm	1.0

[a] Experimentally determined in [11]

Table 2. The mortar-steel fibre bond properties used in the simulations were calibrated by means of numerical simulations of fibre group pull-outs from Sect. 4.1.2 "Pullout of Group of Fibres" in the chapter by V. Gudžulić et al., this volume, and the identified bond properties are listed in Table 3 of the same chapter.

3.1.2 Fibre Sampling Based on CT Images and Predefined Distributions

This subsection presents the characterization and analysis of fibre orientations and contents in CT images of cylindrical cores extracted from actual tested beams. Further, the fibre samples used in the numerical analyses are presented. To that end, three fibre orientation states were considered. These are named *aligned*, *CT-image-based*, and isotropic in this work.

Figure 12 shows the fibres segmented from CT images of the drilled cylindrical cores near the notch region. Analysis of the CT images of cores has revealed that the actual fibre volume fractions were 0.97 and 1.53% ($\approx$ 74 and $\approx$ 117 kg/m^3) instead of the nominal values of 0.75 and 1.5% (57 and 115 kg/m^3). This deviation indicates the possibility of strong non-uniformity of fibre distribution during the casting for the case with the nominal fibre volume fraction of 0.75%. Whether this fibre concentration increase is only local, i.e., just in the region near the extracted core or more spread out, remains unresolved. Also, it is possible that the deviation occurred in only one specimen.

Figure 12b, d show the distribution of orientations of fibres segmented from CT images of two cores with nominal fibre contents of 0.75 and 1.5% (57 and 115 kg/m^3). Polar plots show the stereographic projections of the fibre orientations plotted on a unit hemisphere. These plots show points where the fibre would intersect the unit hemisphere if the fibre endpoint were located in the centre of the corresponding unit sphere. CT images show a general fibre alignment in the beam span direction,

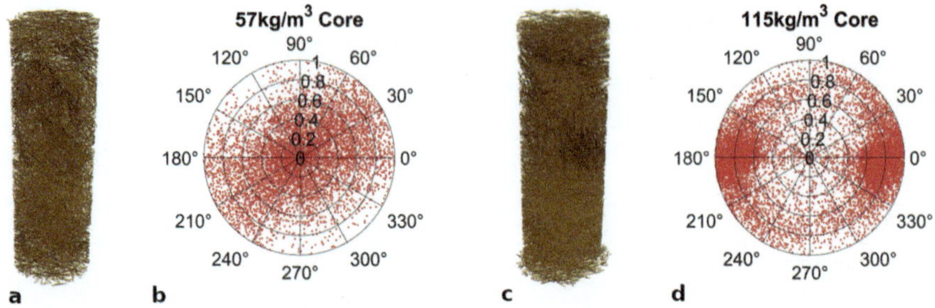

Fig. 12 Visualization of fibres segmented from CT images and fibre orientation analysis for nominal volume fractions of **a**, **b** 0.75% (57 kg/m^3), and **c**, **d** 1.5% (115 kg/m^3). In figures **b** and **d** stereographic projections of the fibre orientations for two considered cases plotted on a unit hemisphere (from [7], licensed under CC-BY 4.0)

a consequence of the casting procedure, which was designed so that most of the fibres align in the direction of tensile stresses that would ensue from bending. For the 57 kg/m^3, Fig. 12a shows an orientation bias in the direction of around 30° to the horizontal axis. The fibres are predominantly oriented in the beam span direction (0 and 180°) for the 115 kg/m^3 case (Fig. 12b).

All fibre realizations used in numerical simulations were sampled according to the nominal volume fractions, and fibres were generated in the small region around the notch. For the 0.75% volume fraction, approximately 28,000 fibres were generated, which resulted in about 205,000 beam elements (226,000 nodes) after discretization. For the 1.5% volume fraction, around 53,000 fibres were generated, yielding approximately 438,000 beam elements (485,000 nodes) after discretization. More details on fibre orientation analysis, generation of fibre samples, and discretization can be found in [7].

3.1.3 Results and Discussion

Figure 13 shows a comparison between experimental and calculated load–displacement characteristics of studied UHPFRC beams for two different fibre contents and three different fibre orientation cases.

For both volume fractions, the load–displacement results in Fig. 13 show that the *aligned* and *CT image* fibre orientations capture the overall response well, while the *isotropic* case significantly underestimates the residual strength of the beam. In the case of the fibre content of 57 kg/m^3, the experimental response shows a non-negligible scatter, and the numerically calculated response of the beam with *CT-image-based* fibre orientations is on the bottom bound of the experiments. As mentioned earlier, the analysis of the CT image of a core from one of the beams reveals that the volume fraction of fibres in that core is 0.97% instead of 0.75%, which means that there exists at least one beam that has substantially more fibres in the cracking region than expected, leading to a higher post-cracking residual strength.

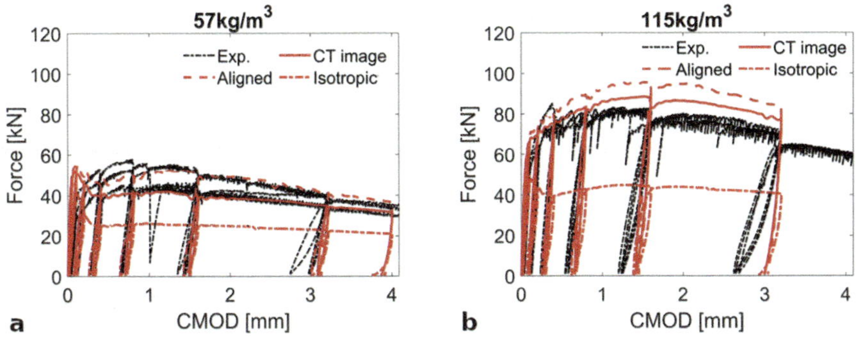

Fig. 13 Comparison of experimentally and numerically obtained load-CMOD characteristics for UHPFRC beams subjected to cyclic three-point bending for various fibre contents and orientations: **a** 57 kg/m^3, **b** 115 kg/m^3

In the numerical simulations, fibre samples were generated based on the volume fraction of approximately 0.75%; hence, the lower calculated residual strength is not surprising.

One feature observable in calculated load–displacement characteristics that does not appear in experimental data is a pronounced peak followed by a drop in force, after which the residual strength stabilizes. This behaviour, characterized by post-peak softening, is characteristic for sub-critical fibre contents. The sub-critical response can be typically expected for fibre contents of 20–60 kg/m^3 [12], and this test case falls within this range. Nevertheless, the post-peak softening or hardening can be also influenced by fibre type and orientations, such as observable in Fig. 13b, where the *aligned* and *CT-image-based* fibre orientation cases corresponding to the 115 kg/m^3 case show super-critical post-cracking behaviour. In contrast, the *isotropic* case shows sub-critical post-cracking behaviour.

Fibre orientation significantly influences the size and inclination of the hysteresis during unloading/reloading. In beam bending, an interplay between crack surface friction, which occurs for cracks with small CMODs, and the sliding and plastic bending of fibres, which is relevant in zones with large CMODs near the notch, influences the size of the observed hysteresis. For the fibre content of 57 kg/m^3, the unloading/reloading stiffness is slightly overestimated for the aligned case, while *the CT-image-based* and *isotropic* cases replicate the unloading/reloading stiffness well. However, the *isotropic* case somewhat overestimates the hysteresis size, possibly due to the substantial bending of fibres misaligned to the cracks. In the 115 kg/m^3 case, the *aligned* and *CT-image-based* cases show somewhat overly stiff unloading behaviour, while the *isotropic* case captures the change of stiffness well. As mentioned earlier, the hysteresis size and its inclination are not directly controlled by the model parameters but by an interaction between different modelled mechanisms. The interplay of mechanisms makes the exact replication of the experimentally observed hysteresis behaviour challenging. The model can be extended by introducing damage accumulation in the bond, as well as a submodel for fibre fatigue, which was shown by

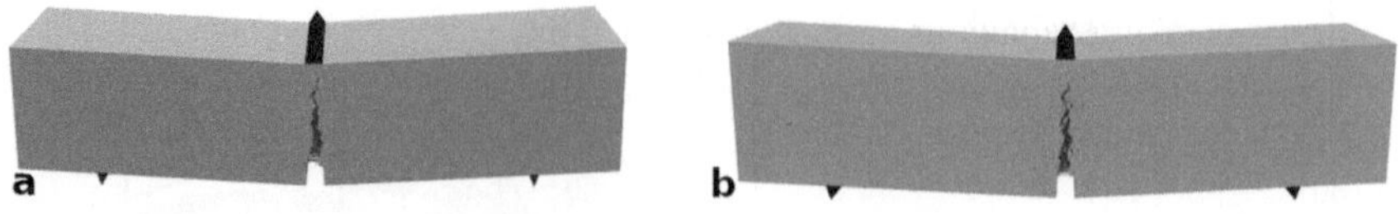

Fig. 14 3D view of the crack pattern at failure of UHPFRC beams (CT image fibre orientations) for different fibre contents: **a** 57 kg/m^3 and **b** 115 kg/m^3. Deformations are scaled fivefold (from [7], licensed under CC-BY 4.0)

Lanwer and Empelmann [13] to play a significant role in cyclic pull-out tests, leading often to premature fibre failure. In this context, accounting for fibre failure and bond damage could improve the prediction of stiffness degradation over successive loading cycles.

Figure 14 shows the crack patterns at the end of numerical simulations for the beam with *the CT-image-based* fibre sample for both fibre contents. In all cases, the numerical failure is characterized by a single localized crack bridged by fibres and multiple arrested microcracks in the region near the notch. Crack growth is arrested in compressive zone, resulting in local branching.

3.2 *Modelling the Stress–Strain Behaviour of UHPFRC Using a Hybrid-Mixed FEM*

The contribution presents numerical results of notched three-point bending beam tests, which are analysed using the macroscopic material model of UHPFRC described in the chapter by L. Gietz et al., this volume. In addition, the calibration of the individual components of the composite material is discussed, which is conducted at the material point level under predefined strain conditions.

3.2.1 Model and Discretization

The material model of UHPFRC is based on a mixed stress–strain formulation of the strain energy density

$$\pi = n_c \left\{ \boldsymbol{\sigma}_c\, \boldsymbol{\varepsilon} - \frac{1}{2} \boldsymbol{\sigma}_c\, \boldsymbol{E}_c^{-1}\, \boldsymbol{\sigma}_c \right\} + n_f \left\{ \boldsymbol{\sigma}_f\, \boldsymbol{\varepsilon} - \frac{1}{2} \boldsymbol{\sigma}_f\, \boldsymbol{E}_f^{-1}\, \boldsymbol{\sigma}_f \right\} - n_c \boldsymbol{\varepsilon}_{cr} \left\{ \boldsymbol{\sigma}_c - \frac{1}{2} E_{cr} \boldsymbol{\varepsilon}_{cr} \right\} - n_f\, \boldsymbol{\varepsilon}_v \left\{ \boldsymbol{\sigma}_f - \frac{1}{2} \boldsymbol{K}_f \boldsymbol{\varepsilon}_v \right\}, \tag{1}$$

applying the Lagrange multiplier methodology for the description of the discrete crack formation and the fibre pull-out evolution. The complete models for the cracking and fibre pull-out are presented in the chapter by L. Gietz et al., this volume.

The post-cracking behaviour of the concrete and the resistance against fibre pull-out are characterized by the modules $\boldsymbol{E}_{\mathrm{cr}}$ and $\boldsymbol{K}_{\mathrm{v}}$, whereat $\varepsilon_{\mathrm{cr}}$ describes the crack-strain and ε_{v} the strain due to fibre pull-out. The volume content of the fibres n_{f} and the volume content of the UHPC n_{c} are explicitly included in the model equations and the distribution of the fibres is considered as an orthotropic material [14, 15]. The full set of model equations follows with the variation of π to the macroscopic variables

$$\frac{\delta\pi}{\delta z} \cdot \delta z = 0 \text{ with } z = \left[\,\boldsymbol{\varepsilon}\ \boldsymbol{\sigma}_c\ \boldsymbol{\sigma}_f\ \boldsymbol{\varepsilon}_{cr}\ \boldsymbol{\varepsilon}_v\,\right]. \tag{2}$$

For the numerical analysis of the benchmark with the finite element method, a hybrid-mixed finite element method is applied by splitting the macroscopic variables into system and element variables

$$z = \left[\,z_{sys}(\boldsymbol{u})\ z_{elem}\left(\mathbf{u}, \boldsymbol{\sigma}_c, \boldsymbol{\sigma}_f, \boldsymbol{\varepsilon}_{cr}, \boldsymbol{\delta}_v\right)\right]. \tag{3}$$

The approach requires shape functions for all variables, cf. the chapter by L. Gietz et al., this volume. For the displacements, standard linear shape functions are used, while the element variables are approximated with general approach functions balanced to the shape functions of the displacements. For the numerical implementation of the material equations, the strains $\boldsymbol{\varepsilon}_{\mathrm{v}}$ due to fibre pull-out is connected to the fibre pull-out lengths $\boldsymbol{\delta}_{\mathrm{v}}$ via the element length l_{el}. Thus, the material behaviour is described as smeared over the element domain.

3.2.2 Model Calibration

In the first step, the model was calibrated by analysing the various material components of UHPFRC (UHPC, micro steel fibre and fibre-matrix bond) at the material point level under predefined strain conditions. A comparison of numerical investigations was performed with uniaxial tensile tests and fibre pull-out tests from project-related experimental investigations [4]. The measured material parameters include the tensile strengths f_{ct} and f_{ft}, the transverse strain coefficient ν_{c} of the UHPC as well as the Young's moduli E_{c} and E_{f}. The parameters used for the calibration of the model are provided in Table 6.

Characterizing both material components and the fibre-matrix bond is essential for accurately mapping the macroscopic behaviour of the material. The geometry of the fibres is incorporated into the bond model to describe the fibre pull-out behaviour. It is assumed that the embedded length of the fibres is half their total length, ensuring optimal utilization of the bonding capacity. Furthermore, it is assumed that the fibres are pulled out only perpendicular to the crack, corresponding to the straight fibre pull-out experiment. According to [16], a constant bond stress of 8.0 N/mm^2 is assumed to describe the friction between concrete and fibre when it is fully debonded from the UHPC-matrix. The pull-out resistance is accounted for by means of the bond

Table 6 Model parameters for calibration

	Parameter			Value
UHPC-matrix	Young's modulus	E_c	N/mm^2	43,300
	Tensile strength	f_{ct}	N/mm^2	10.7
Steel fibers	Young's modulus	E_f	N/mm^2	169,000
	Tensile strength	f_{ft}	N/mm^2	3575[a]
	Fibre length	l_f	mm	6.5 (13)
	Fibre radius	r_f	mm	0.095
	Max. bond resistance	τ_{max}	N/mm^2	8.0

[a] Not directly used in numerical simulations

modulus $\boldsymbol{K}_v$ as a function of the fibre pull-out length $\boldsymbol{\delta}_v$ (see Eq. 1), resulting in a linear decrease in the fibre stress during the pull-out process.

Figure 15a, b represent the comparison of the uniaxial material behaviour with respect to tension of experimental and numerical investigations for both material components. The linear elastic phase and the rapid failure of the UHPC-matrix are very well approximated, as shown in Fig. 15. The model describes the concrete stress converging towards zero after the UHPC matrix is cracked, which facilitates a stable material behaviour during the subsequent fibre activation phase in the numerical simulation. The behaviour of the fibre is assumed to be elastic, ensuring that the composite material fails by fibre pull-out rather than fibre rupture.

The comparison of the fibre pull-out behaviour of a multiple fibre pull-out test from [4] and the simulation are represented in Fig. 16a. In the experiment, the fibres were pulled out perpendicularly from the UHPC matrix, resulting in a nearly linear decrease of the fibre stress. The model achieves a good correlation with the experimental results, as the assumed maximum bond resistance of 8.0 N/mm^2 aligns well with the peak load observed. The increase of the bond stress up to the maximum pull-out resistance is calculated based on a linear elastic behaviour during the fibre activation phase, governed by the modulus of elasticity of the fibre E_f. This approach

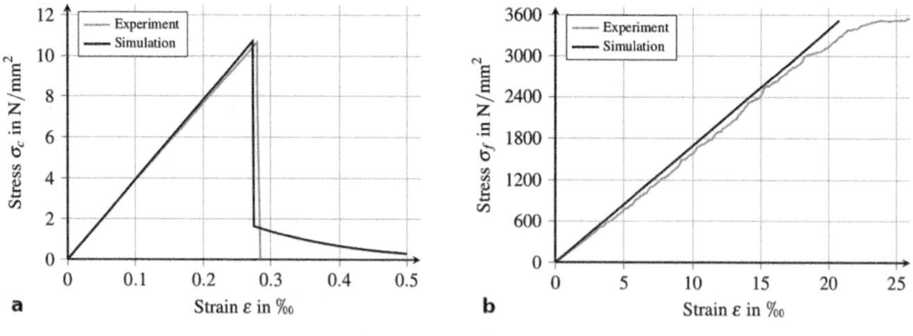

Fig. 15 Experiments [4] and simulation of material samples with respect to monotonically increasing tension: **a** UHPC, **b** micro steel fibre

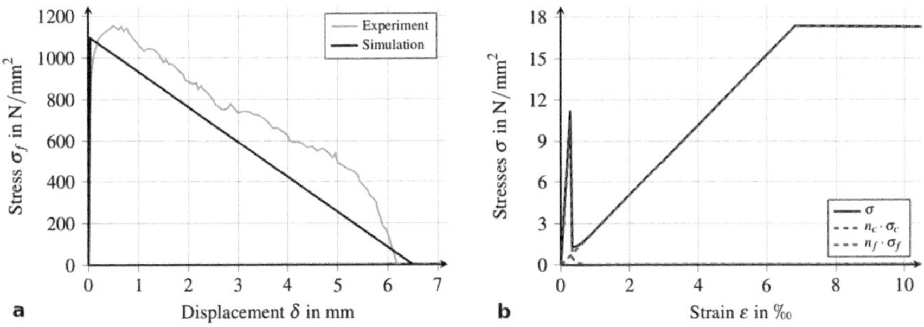

Fig. 16 **a** Experiment [4] and simulation of a fibre pull-out test, **b** total stress, concrete and fibre stresses in the case of $n_\mathrm{f} = 1.5$ Vol.-%

leads to a slight overestimation of the load-bearing behaviour during crack formation. Overall, the model effectively maps the fibre pull-out behaviour for a straight fibre pull-out scenario.

Figure 16b depicts the total stress of the composite material, taking into account both material components with a fibre volume content of $n_\mathrm{f} = 0.015$. The total stress is represented as the sum of the concrete and fibre stress in relation to their volume fractions: $\sigma = n_\mathrm{c} \cdot \sigma_\mathrm{c} + n_\mathrm{f} \cdot \sigma_\mathrm{f}$. The stress–strain diagram is limited to the range up to the beginning of the fibre pull-out. Subsequently, the fibre stress decreases over the fibre pull-out length as illustrated in Fig. 16a. The strain is directly related to the fibre pull-out length, meaning that a strain of 1.0 indicates the complete fibre pull-out. Moreover, the diagram clearly illustrates that the significant effect of the fibres with respect to the overall load-bearing behaviour of the UHPFRC only becomes obvious at the point of crack initiation.

3.2.3 Model Parameters and Setup

The model parameters used for the simulation of the benchmark are given in Table 7. In comparison to the calibration, most of the parameter are similar, as a specific reference concrete mix was employed during the project. The fibre volume content is explicitly included in the model equations, cf. Eq. 1. The benchmark is numerically investigated for fibre volume contents of 0.75 and 1.50 Vol.-%. The anisotropic fibre distribution is considered by incorporating three different orientations. Due to the sharp notch, the crack will propagate perpendicular to the notch over the cross-sectional height. Applying the orthotropic material description, it is simplistically assumed that primarily the fibres aligned in the longitudinal direction of the beam will contribute to crack bridging, which is accounted for by the fibre orientation factor r_x. Consequently, it could be assumed that 33% of the fibres participate in the case of an equal distribution in the three dimensions, whereat 100% of the fibres contribute to a uniaxial alignment. Based on data from the CT-scans, cf. Section 2.4, the realistic fibre distribution participating in the crack direction is estimated to 60%.

Table 7 Model parameters for simulation of benchmark

	Parameter			Value
UHPC-matrix	Young's modulus	E_c	N/mm^2	43,300
	Poisson ratio	ν_c		0.221
	Tensile strength	f_{ct}	N/mm^2	8.2
Steel fibres	Young's modulus	E_f	N/mm^2	169,000
	Fibre volume content	n_f	Vol.-%	0.75/1.5
	Fibre orientation factor	r_x		0.33/0.60/1.0
	Fibre length	l_f	mm	6.5 (13)
	Fibre radius	r_f	mm	0.095
	Max. bond resistance	τ_{max}	N/mm^2	8.0

The solution of the discretized model equations is obtained applying an arc-length method by increasing the unit load F of 1 kN by the load factor λ. The load is applied by distributing it to the surrounding nodes in a triangular pattern, according to the formulated boundary value problem, cf. Figure 11. The sharp notch is discretely modelled within the finite element model, which includes 1345 finite elements. The load is only applied quasi-statically here. The beams are not subjected to cyclic loading. The target is to map the envelope curve of the experimental investigations.

3.2.4 Results and Discussion

Figure 17 represents the comparison of the crack-mouth opening displacement (CMOD) of the experimental and numerical tests with respect to the applied force *F*. The CMOD describes the difference of the horizontal displacements of the opposite nodes of the notch at the lower edge of the beam. The deformation behaviour is qualitatively well approximated for all fibre distributions. As expected, twice the bearing loads are achieved for a fibre content of 1.5 Vol.-% than for a fibre content of 0.75 Vol.-%. The scattering of the experimental results of the beam with 0.75 Vol.-% can be mapped by varying the aligned fibre content of the numerical simulation between 60 and 100%. The isotropic fibre distribution tends to underestimate the material's post-cracking behaviour for both fibre volume contents, whereat the fully aligned fibre distribution overestimates the load capacity and post-cracking behaviour for the beam containing 1.5 Vol.-% fibres. The estimated realistic fibre distribution provides the best comparison between the experimental and numerical results of the beam with 1.5 Vol.-% fibres.

In all simulations, the crack propagation is initiated at the notch. The maximum load is reached during the crack propagation through the cross-section, whereat the fibres ensure the transmission of tensile forces by bridging the crack. Despite the damage, an increase of the load is possible until the bond between concrete and fibres is weakened and fibre pull-out starts. Beginning at the bottom of the beam the

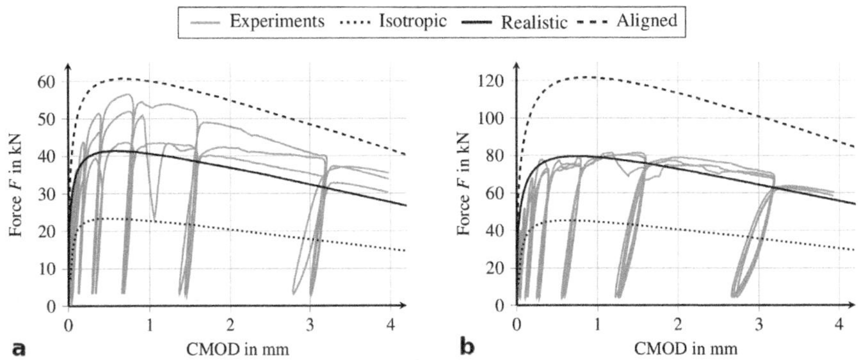

Fig. 17 Comparison of experimental and numerical results of the benchmark: **a** 0.75 Vol.-% fibres, **b** 1.5 Vol.-% fibres

fibre pull-out propagates over the cross-section until the final failure of the beam. This phenomenon ensures the very ductile post-cracking behaviour of UHPFRC samples, which exhibits a controlled failure at CMOD in the millimetre range.

The experimental and numerical results show that both the fibre volume content and especially the fibre distribution have a significant influence on the load-bearing behaviour of UHPFRC. The fibre distribution is random, although various experimental investigations show an influence of the manufacturing process [17–19]. In particular, the fibres align and orient in slender structures parallel to the main dimension. This observation supports the numerical results of the significantly underestimated load-bearing behaviour, if an isotropic distribution of fibres is assumed.

The chapter by L. Gietz et al., this volume, offers furthers results on three-point bending beams, for which a good agreement with the experimental data is mapped by setting an isotropic fibre distribution. The fibre volume contents are similar, but the beam geometry differs with an effective length size factor of approximately 6.0 and a height size factor of 2.0. This suggests that the structure dimensions may influence the fibre distribution during the manufacturing process. The results indicate that appropriate measures must be taken to ensure a consistent ductile load-bearing behaviour in the post-fracture domain.

Further research aims to enhance the model to account for material behaviour under cyclic loading. A concept for describing the residual bond resistance during the fibre pull-out process as a function of load cycles is introduced in the chapter by L. Gietz et al., this volume.

4 Comparison and Conclusions

The numerical simulations of the flexural tensile tests on fibre-reinforced concrete beams have been conducted using two distinct modelling approaches: a mesoscale model in Sect. 3.1 and a macroscale model in Sect. 3.2. Both models provide valuable insights into the structural behaviour of ultra-high-performance fibre-reinforced concrete (UHPFRC) under flexural loading, yet they differ in their methodologies, computational demands, and practical applications.

The mesoscale model employs a highly detailed discrete approach, explicitly modelling cracks and fibres using zero-thickness cohesive-frictional interface elements and elasto-plastic Timoshenko beam elements. This methodology enables an explicit representation of various damage and toughening mechanisms, including crack initiation and propagation, fibre debonding, plastic bending, concrete spalling, and fibre pull-out. Consequently, the model effectively captures the unloading and reloading behaviour and provides valuable insights into the interaction between cracks and fibres. However, this high level of detail comes at a significant computational cost.

In contrast, the macroscale model employs a hybrid-mixed finite element method that describes the phenomenological behaviour of UHPFRC at the macroscale using a stress–strain-based formulation. Instead of explicitly modelling individual fibres, it accounts for fibre reinforcement through volume fractions and statistical fibre orientation. This approach ensures computational efficiency and robust performance for large-scale structures. Furthermore, the model relies exclusively on experimentally measurable material parameters, simplifying calibration and enhancing practical applicability. While this method effectively captures the overall mechanical response of fibre-reinforced concrete, it lacks the fine-scale resolution necessary to investigate local effects. The numerical implementation allows for an element-wise distribution of material parameters such as fibre volume content, fibre orientation, and concrete tensile strength, enabling the investigation of size effects.

The results from both numerical models highlight the substantial impact of fibre orientation on the structural response of UHPFRC beams. The simulations demonstrate that the same fibre volume fraction can lead to either sub-critical (softening) or super-critical (hardening) post-cracking responses, depending on fibre alignment. This underscores the importance of fibre distribution and the casting process in determining the material's load-bearing capacity and ductility. To further assess the effectiveness of the proposed models, further systematic experimental–numerical studies investigating the influence of fibre content and casting procedures would be beneficial.

Associated data is published via the link: https://doi.org/10.25532/OPARA-851.

References

1. Gebuhr, G., Anders, S., Pise, M.T., Sahril, M., Brands, D., Schröder, J.: Deterioration development of steel fibre reinforced high performance concrete in low-cycle fatigue. In: Proceedings of the Seventh International Conference on Structural Engineering, Mechanics and Computation (SEMC), Cape Town, South Africa (2019)
2. Gudžulić, V., Neu, G.E., Gebuhr, G., Anders, S., Meschke, G.: Numerisches Mehrebenen-Modell für Stahlfaserbeton: Von der Faser- zur Bauteilebene. Beton- Stahlbetonbau **115**, 146–157 (2020). https://doi.org/10.1002/best.201900067
3. Lanwer, J.-P., Oettel, V., Empelmann, M., Höper, S., Kowalsky, U., Dinkler, D.: Bond behavior of micro steel fibers embedded in ultra-high performance concrete subjected to monotonic and cyclic loading. Struct. Concr. **20**, 1243–1253 (2019). https://doi.org/10.1002/suco.201900030
4. Oettel, V., Lanwer, J.-P., Empelmann, M.: Auszugverhalten von Mikrostahlfasern aus UHPC unter monoton steigender und zyklischer Belastung (Pullout behaviour of micro steel fibres of UHPC under monotonic and cyclic loading). Bauingenieur **96**, 1–10 (2021). https://doi.org/10.37544/0005-6650-2021-01-02-27
5. Brands, D., Pise, M., Schröder, J., Gudžulić, V., Meschke, G., Storm, J., Kaliske, M., Höper, S., Dinkler, D., Weiß, U., Peter, M., Gebuhr, G., Anders, S.: Zyklische Schädigung in Hochleistungsbetonen im Experimental-Virtual-Lab. SPP2020 Benchmark 1—Faserauszugversuch. Technischer Bericht, Essen (2019)
6. Ferrara, L.: Fiber Reinforced SCC, pp. 161–219. Springer International Publishing, Cham (2014). https://doi.org/10.1007/978-3-319-03245-0_6
7. Gudžulić, V.: Virtual lab for mechanical testing of steel- and steel-fiber reinforced concrete. Ph.D. thesis, Ruhr University Bochum, Bochum (2024)
8. RILEM: Determination of the fracture energy of mortar and concrete by means of three-point bend tests on notched beams. Mater. Struct. **18**, 285–290 (1985)
9. Müller, H., Aitcin, P.-C., Bentur, A., Chiorino, M. A., Clementand, J.-L., Curbach, M., Faust, T., Hammer, T. A., Jaccoud, J.-P., Kessler-Kramer, C., Rinder, T., Rokugo, K., Taerwe, L., Ueda, T., Walraven, J.: Constitutive modelling for high strength/high performance concrete. Technical report. fédération internationale du béton (fib), Lausanne (2008)
10. Sellier, A., Casaux-Ginestet, G., Buffo-Lacarriere, L., Bourbon, X.: Orthotropic damage coupled with localized crack reclosure processing. Part I: constitutive laws. Eng. Fract. Mech. **97**, 148–167 (2013)
11. Lanwer, J.-P., Oettel, V., Empelmann, M., Höper, S., Kowalsky, U., Dinkler, D.: Bond behavior of micro steel fibers embedded in ultra-high performance concrete subjected to monotonic and cyclic loading. Struct. Concr. **20**, 1243–1253 (2019). https://doi.org/10.1002/suco.201900030
12. Holschemacher, K., Dehn, F., Müller, T., Lobisch, F.: Betonkalender 2017. Faserbeton, Ernst und Sohn, Berlin (2017)
13. Lanwer, J.-P., Empelmann, M.: Fundamental investigations on the performance of micro steel fibres in ultra-high-performance concrete under monotonic and cyclic tensile loading. Appl. Sci. **11**, 9377 (2021). https://doi.org/10.3390/app11209377
14. Gietz, L., Kowalsky, U., Dinkler, D., Lanwer, J.-P., Empelmann, M.: Homogenisation of the material behaviour of UHPFRC under tension. In: Current Perspectives and New Directions in Mechanics, Modelling and Design of Structural Systems, pp. 1313–1319. CRC Press, London (2022). https://doi.org/10.1201/9781003348443-214
15. Gietz, L., Kowalsky, U., Dinkler, D.: A phenomenological model to describe cracking and fibre pull-out in ultra high-performance fibre reinforced concrete. Proc. Appl. Math. Mech. **22**, e202200321 (2023). https://doi.org/10.1002/pamm.202200321
16. Lanwer, J.-P.: Ein Beitrag zum Tragverhalten vonfaser bewehrtem UHFB unter zyklischer Zugbeanspruchung. Ph.D. thesis, Technische Universität Braunschweig, Braunschweig (2024)
17. Leutbecher, T., Heck, L., Schleiting, M.: Prüfung von UHFB nach DIN EN 14651—Zum Einfluss der Probekörperherstellung auf die Prüfergebnisse. Beton- Stahlbetonbau **117**, 410–421 (2022). https://doi.org/10.1002/best.202200027

Overview of Principle Investigators and Projects Within the SPP 2020

PIs	Affiliation	Project	Project number
Prof. Dr.-Ing. Ludger Lohaus Dr.-Ing. Nadja Oneschkow	Leibniz University Hannover	Coordination funds	353531623
Dr.-Ing. Nadja Oneschkow Prof. Dr.-Ing. Stefan Löhnert	Leibniz University Hannover; TUD Dresden University of Technology	Composition influenced damage development in high-strength concrete due to cyclic loading	353530889
Dr.-Ing. Thorsten Leusmann Prof. Dr. Laura De Lorenzis	TU Braunschweig	Multiscale investigation of fatigue crack growth in high-performance concrete based on computer tomography and phase-field fracture modelling	353942799
Prof. Dr. Rostislav Chudoba Prof. Dr.-Ing. Martin Classen	RWTH Aachen University	Experimental and numerical framework for characterization of high-strength concrete fatigue accounting for local dissipative mechanisms at subcritical load levels	441550460
Prof. Dr.-Ing. Frank Dehn Prof. Eduardus Koenders, Ph.D. Prof. Dr.-Ing. Matthias Pahn	Karlsruhe Institute of Technology (KIT); Technical University of Darmstadt; RPTU Kaiserslautern-Landau	Visualisation and micromechanical modelling of the microstructural changes of cyclically stressed high-performance concretes with special consideration of hygric and thermal boundary conditions	353981539

(continued)

N. Oneschkow and L. Lohaus (eds.), *Cyclic Deterioration of High-Performance Concrete*, https://doi.org/10.1007/978-3-032-19394-0

(continued)

PIs	Affiliation	Project	Project number
Prof. Dr.-Ing. Michael Haist Prof. Dr.-Ing. Ludger Lohaus Prof. Dr.-Ing. Peter Wriggers Prof. Dr.-Ing. Fadi Aldakheel	Leibniz University Hannover	Water-induced damage mechanisms of cyclically loaded high-performance concrete	353757395
Prof. Dr.-Ing. Harald Garrecht Prof. Dr.-Ing. Holger Steeb	University of Stuttgart	Temperature- and moisture-induced damage processes due to cyclic multiaxial compressive fatigue loading	353921616
Prof. Dr.-Ing. Rolf Breitenbücher Prof. Dr. Günther Meschke	Ruhr University Bochum	Effect of microfibres on degradation in high performance concrete under cyclic loading	353819637
Prof. Dr.-Ing. Steffen Anders Dr.-Ing. Dominik Brands Prof. Dr.-Ing. Jörg Schröder	University of Wuppertal; University of Duisburg-Essen	Description and numerical modelling of the effects of steel-fibres on the degradation of high performance concrete subjected to fatigue loading	353513049
Prof. Dr.-Ing. Frank Schmidt-Döhl Dr.-Ing. Martin Ritter Prof. Dr.-Ing. Maksym Dosta	Hamburg University of Technology (TUHH)	High-resolution electron microscopic investigations on the fatigue behaviour of ultra-high-performance concrete and modelling with bonded-particle model	353408149
Dr.-Ing. Silke Scheerer Dr.-Ing. Ngoc Linh Tran	TUD Dresden University of Technology; Technical University of Darmstadt	Influence of load-induced temperature fields on the fatigue behaviour of UHPC subjected to high frequency compression loading	353981739
Prof. Dr.-Ing. Dieter Dinkler Prof. Dr.-Ing. Martin Empelmann	TU Braunschweig	Degradation processes in ultra-high performance fibre-reinforced concrete (UHPFRC) under cyclic tensile loading	353961703

(continued)

(continued)

PIs	Affiliation	Project	Project number
Prof. Dr.-Ing. Oliver Fischer Prof. Dr.-Ing. Christian Grosse Prof. Dr. Malte Andreas Peter Prof. Dr. Dirk Volkmer	Technical University of Munich (TUM); University of Augsburg	Multiscale investigations of the degradation progress in the localised fracture zone of carbon short-fibre reinforced high-performance concrete subjected to high-cycle tensile and flexural fatigue load regimes	354003768
Prof. Dr.-Ing. Michael Kaliske Prof. Dr.-Ing. Viktor Mechtcherine	TUD Dresden University of Technology	Enhancement of fatigue resistance of strain-hardening cement-based composites by means of experimental-virtual multiscale material design	352324592

This research was funded by the Deutsche Forschungsgemeinschaft (DFG, German Research Foundation) within the DFG Priority Programme SPP 2020 "Cyclic Deterioration of High-Performance Concrete in an Experimental-Virtual-Lab". The funding is gratefully acknowledged.

GPSR Compliance

The European Union's (EU) General Product Safety Regulation (GPSR) is a set of rules that requires consumer products to be safe and our obligations to ensure this.

If you have any concerns about our products, you can contact us on ProductSafety@springernature.com

In case Publisher is established outside the EU, the EU authorized representative is:

Springer Nature Customer Service Center GmbH
Europaplatz 3
69115 Heidelberg, Germany

Batch number: 10593112

Printed by Printforce, the Netherlands